T0396745

# The Future of the Large Hadron Collider

## A Super-Accelerator with Multiple Possible Lives

# The Future of the Large Hadron Collider

## A Super-Accelerator with Multiple Possible Lives

Editors

Oliver Brüning
*CERN, Switzerland*

Max Klein
*University of Liverpool, UK*

Lucio Rossi
*University of Milano, Italy & INFN, Italy*

Paolo Spagnolo
*INFN Pisa, Italy*

NEW JERSEY · LONDON · SINGAPORE · BEIJING · SHANGHAI · HONG KONG · TAIPEI · CHENNAI · TOKYO

*Published by*

World Scientific Publishing Co. Pte. Ltd.
5 Toh Tuck Link, Singapore 596224
*USA office:* 27 Warren Street, Suite 401-402, Hackensack, NJ 07601
*UK office:* 57 Shelton Street, Covent Garden, London WC2H 9HE

Library of Congress Control Number: 2023939770

**British Library Cataloguing-in-Publication Data**
A catalogue record for this book is available from the British Library.

**THE FUTURE OF THE LARGE HADRON COLLIDER**
**A Super-Accelerator with Multiple Possible Lives**

Copyright © 2024 by The Editors

*All rights reserved.*

This is an open access book published by World Scientific Publishing Co Pte Ltd. The Creative Commons License of each chapter can be found on the first page of each chapter.

ISBN 978-981-128-017-7 (hardcover)
ISBN 978-0-00-099151-5 (paperback)
ISBN 978-981-128-018-4 (ebook for institutions)
ISBN 978-981-128-019-1 (ebook for individuals)

For any available supplementary material, please visit
https://www.worldscientific.com/worldscibooks/10.1142/13513#t=suppl

Desk Editor: Carmen Teo Bin Jie

© 2024 The Author(s)
https://doi.org/10.1142/9789811280184_fmatter

# Foreword

Particle physics is in a remarkable situation. It has the wonderful Standard Model which summarizes almost all experimentally known data, but is only a model put together, with 26 parameters and including 'by hand' any new experimental discoveries or many constants taken from data. It unites the electromagnetic interaction with the weak interaction, however, the strong force is just attached to it. For example, there is no theoretical relation between all the coupling constants. The last missing building block of that Standard Model, the Higgs particle, was found by the LHC experiments ATLAS and CMS ten years ago. The SM, however, does not give any hints about which direction it could be extended and where to experimentally look for new physics. Particle physics is not yet based on a united, comprehensive theory derived from fundamental assumptions.

In the first decade of operation, as described in this book, the LHC experiments have provided an enormous amount of data in support of the SM while not observing further new particles predicted, for example, by Supersymmetry. The situation reminds me of the late 1960's and early 1970's when we had nice theories, like Regge poles, dispersion relations etc., and a lot of experimental data representing a particle zoo, but no deeper understanding of its origin nor symmetry. Initially a change towards the SM came by the more or less unexpected discoveries of partons, quarks and gluons in 1968 and the $J/\psi$ particle in 1973, as well as with the application of Yang Mills theory and the prediction of asymptotic freedom of the strong interaction as described by Quantum Chromodynamics.

Where should we look today? The main particle physics instruments to investigate nature at higher energies, which is equivalent to looking for finer details of the structure of matter, are electron (e) and proton (p) based colliders of three types; $e^+$-$e^-$, e-p, p-p (or hadrons instead of protons or antiparticles), which are complemented by a few special experiments such

This is an open access article published by World Scientific Publishing Company. It is distributed under the terms of the Creative Commons Attribution 4.0 (CC BY) License.

as those looking for axions or cosmic radiation observations. However, not only are higher energies necessary, greater collision rates are required at the same time since the quantum mechanical interaction cross sections decrease with the square of the center of mass energies. This is the principal reason for the huge experimental, technical, and theoretical effort, described in all its facets in this book dedicated to the largest hadron collider built so far, the LHC at CERN, and its future exploitation. The LHC at high luminosity (HL-LHC) is now scheduled to operate from the end of the 2020's until about 2040, possibly beyond. The technical upgrade of the facility opens prospects for its utilization for a further phase of collider and fixed target experiments, depending on the perspectives of high energy physics at CERN and worldwide.

The quest for new, higher energy colliders is perhaps not surprisingly focused on colliding the same kind of particles, $e^+$-$e^-$ or p-p. Known examples are the future circular collider (FCC) project at CERN, a similar project in China or the international linear collider (ILC) in Japan. The technologies for producing large beam currents are quite different for electrons and protons since synchrotron radiation losses increase with the fourth power of the inverse of particle mass and are thus much bigger for electrons while they become noticeable only at extreme high energies for protons heaving a much greater mass. A special possibility for a next, higher energy hadron collider consists in developing and building magnets at industrial scale of about 20 T field strength, tripling the LHC achievement. The book describes an interesting further option of using the LHC infrastructure by inserting such high field magnets, should they become available by about 2040, into the LHC tunnel, a project called HE-LHC. Besides the like particle $e^+$-$e^-$ and p-p colliders, a special interest is raised by the possibility to collide electrons off protons as was realized with HERA at DESY. HERA produced unique results especially on the distribution functions of quarks and gluons inside the nucleon in a much extended phase space.

During the last decade, a large community of physicists and engineers, partly enthusiasts from HERA, came together to study the advantages and possibilities which an energy frontier e-p facility based on the intense hadron beams of the LHC would offer from the view of new physics and at the same time exploring the practical and technical challenges.

In order to make a considerable step beyond HERA, an electron energy in the range of 50 to 100 GeV is necessary. Since the physics of this new e-p collider demanded very high luminosity, the preferred electron accelerator solution would not be a storage ring but rather the application of a new technology which reuses the energy stored in the decelerated beam. Such a 'beam energy recovery technology' is under development in various laboratories in the USA, Asia, and Europe. For advancing this technology to be used directly for an e-p collider, a test facility is being built, which could be used also for other purposes, by an international collaboration, including CERN, with the main installation of the PERLE facility at IJCLab Orsay (France).

Under the assumption that this beam energy recovery can be applied, detailed configurations for an added electron racetrack have been worked out, both for the HL-LHC, possibly the HE-LHC, and the FCC p-p version. This work included not only characteristics of the electron ring such as lattice or civil engineering but also detailed designs of the asymmetric e-p interaction region. The addition of an electron-proton and electron-ion experiment to the LHC, and later possibly the FCC, opens new horizons as to the investigation of the Higgs boson properties or the reliable precision determination of the complete set of parton distributions as is required for fully exploiting the physics of the hadron colliders.

All new collider projects under discussion require not only a scientific but also political decision. It currently is very difficult to guess what the chances of a particular facility might be. The LHC has been the most successful and the largest enterprise of particle physics so far. In this book, the striking knowledge and colossal work which has been performed by many colleagues, sometimes voluntarily and often in addition to their normal job, has been collected which illustrates the LHC achievements as well as several options for a further future of the LHC based on its ongoing upgrade to enhanced luminosity. I am convinced that the material, partly scientific, partly technical, will find useful applications in one form or another while it may also be instrumental to lead particle physics beyond the Standard Model.

Herwig Schopper
*University Hamburg, former CERN Director General*

© 2024 The Editor(s)
https://doi.org/10.1142/9789811280184_fmatter

# Contents

## I. Introduction:



## II. The First Decade of the LHC:

## III. High Luminosity LHC:

A. Accelerator Challenges



B. Physics with HL-LHC

## IV. Future Prospects:

A. Electron-Hadron Scattering



B. The High-Energy LHC

© 2024 The Author(s)
https://doi.org/10.1142/9789811280184_0001

# Chapter 1

# New Theory Paradigms at the LHC

Margarete Mühlleitner and Tilman Plehn
*Institute for Theoretical Physics, Karlsruhe Institute of Technology, Germany*
*Institut für Theoretische Physik, Universität Heidelberg, Germany*

The success of particle physics rests on precision measurements combined with precision predictions, to answer burning fundamental physics questions. Modern LHC physics combines searches for physics beyond the Standard Model with a first-principle understanding of the vast LHC dataset. Building on the Higgs discovery and a detailed understanding of weak-scale physics, the upcoming LHC runs will keep incorporating new concepts, for instance from data science, to probe the properties and interactions of all known and to-be-discovered new elementary particles.

**The Puzzles of Particle Physics** The defining features of particle physics are the big and exciting fundamental physics questions, for which we try to find answers (for example, at the LHC). Some of these questions come from the mathematical structure of quantum field theory, others are posed by cosmological observations combined with a fundamental model describing elementary particles and their interactions.

The consistency of the Standard Model (SM) as a quantum theory for the electroweak interactions has lead us directly to the discovery of the Higgs boson. The renormalizability of the electroweak gauge theory, experimentally confirmed by many LEP measurements, predicts a new scalar with a mass in the electroweak range. The Higgs boson arises as quantum excitation of the Higgs field with non-zero vacuum expectation value (VEV), which generates the masses of the gauge bosons and fermions in the SM. This has to be separated from the hadron masses, which are generated by the non-abelian structure of QCD, which is also probed at colliders.

This is an open access article published by World Scientific Publishing Company. It is distributed under the terms of the Creative Commons Attribution 4.0 (CC BY) License.

The mechanism of generating mass through spontaneous symmetry breaking has been observed in other systems and other fields of physics. One big open question that remains after the Higgs discovery is how electroweak symmetry breaking is realized in our Universe, and if Nature really follows its most economic realization with one fundamental scalar particle.

Cosmology allows us to probe physics over a vast range of energy scales by combining observations with a fundamental understanding of the thermal history of the universe. Because our Universe is a somewhat complex system, we have not been able to pin down the fundamental, dynamic mechanisms behind, for example, the observed dark matter relic density and the observed matter-antimatter asymmetry from cosmological data. However, dark matter and the matter-antimatter asymmetry have to be put into the context of elementary particle physics. Provided that these mechanisms affect physics below the TeV scale, we can test them in the controlled environment of particle physics experiments. Here, multi-purpose experiments at hadron colliders are an especially promising path to search for dark matter particles and to probe the symmetry structures behind baryogenesis.

Most generally, we need to ask the question: whether we can describe all physics effects and all measurements at the LHC in terms of a fundamental quantum field theory or its effective field theory (EFT) extension. At the parton level, we can describe the hard scattering precisely in the Standard Model or possible extensions; we also know that we can use resummed QCD predictions to describe, for instance, parton showers; in both cases the challenge is to match the experimental precision with perturbative or resummed calculations. Open questions in QCD include hadron spectra, dynamic hadronization, and parton densities from first principles. QCD should be the correct fundamental theory to describe all these effects, but for effects out of reach of perturbation theory, we would need to close the gap with non-perturbative computations and lattice gauge theory. At hadron colliders these aspects of fundamental QCD can be targeted by electron-hadron or heavy-ion collisions. The symmetry structure of the QCD Lagrangian is the main motivation for new light axions, which would, in turn, provide a portal to dark matter and link quantum field theory, collider physics, and cosmology.

**Hadron Colliders and Theory** Experimental collider physics and theory are two inseparable sides of the same coin — the path to the fundamental structure of Nature. Back in the days of the LEP, the Tevatron, and HERA, the common wisdom was that hadron colliders were discovery machines, while electron-position and electron-proton colliders were needed for precision measurements and to really understand new particles or interactions. Looking back at the discoveries in the heavy, electroweak sector this judgement is sensible. The $W$ and $Z$-bosons were discovered in 1983 at the SPS, the top quark was discovered in 1994 at the Tevatron, and the Higgs boson in 2012 at the LHC. In addition, LEP has established the electroweak SM as a predictive, renormalizable quantum field theory, predicting the Higgs discovery at the LHC. This list defines a similar task for the LHC: to discover, if at all possible, new particles by understanding all LHC data and starting from the full Standard Model.

Because of the complexity of the experimental environment, the large data sets with correspondingly small statistical uncertainties, and the long list of effects which need to be described by theory, a close interaction between theory and experiment is crucial for hadron collider physics. At a time when advanced particle physics experiment and theory (each tackling their respective challenges) tend to drift apart, a unified approach is more important than ever.

Strictly speaking, an experimental measurement targets a rate or kinematic correlations in a given fiducial phase space. Already, this measurement requires theory input for calibration or to transfer knowledge from control regions to the signal region. However, from a physics perspective a QCD-dominated total rate measurement is not interesting. What we want to measure are fundamental parameters, which need to be extracted from the original rate measurement through additional kinematic handles. This means any relevant physics measurement rests on a fundamental physics interpretation framework, and any inference requires a well-defined hypothesis which relates a measured rate to an interesting physics question.

The workhorse in theoretical predictions for the LHC are fast event generators combined with detector simulations. Multi-purpose generators take fundamental Lagrangians as inputs and generate events as we expect them in the virtual world defined by a given Lagrangian. These generators are maintained by the theory community and provide the main pipeline for realising theory ideas into experiments. Simulation-based inference then compares measured and simulated events and extracts information on the underlying theory, probing the particle content, the interactions, or the fundamental symmetry structure. This kind of fundamental analyses, combined with a precisely controlled experimental environment and equally precise theoretical prediction, has turned the LHC into the first precision hadron collider, breaking the historic split between precision-lepton-collider physics and discovery-proton-collider physics.

The appeal of future hadron colliders is driven by the immense success of the LHC. As we will discuss below, the experimental and theoretical LHC program established the notion of precision-hadron collider physics. Experimentally, the proposed HE-LHC and FCChh are defined by significant increases in energy and luminosity beyond the full LHC dataset. The LHeC and FCCeh attempt to combine the advantages of hadron collider physics with electron beams that have enough energy to induce scattering of electroweak gauge bosons. Their setup benefits from the success of modern hadron colliders, but shifts the focus from QCD to electroweak scattering. The benefits from increased luminosity are key to all future colliders, a serious challenge to the entire underlying methodology, and a trigger for exciting developments already for the final phase of the LHC.

**Precision Predictions** A key ingredient to the success story of the LHC lies in theoretical developments since around 2000.[1] The prediction of hard scattering amplitudes is now dominated by automated next-to-leading order (NLO) calculations in QCD, available essentially for all relevant signal and background processes.[2] These calculations are not only available for total rates, but for the full event kinematics, thanks to advanced subtraction schemes for soft and collinear phase space regions. Automated NLO QCD calculations as part of standard event generators are the final step of decade-long developments of analytical[3] and numerical methods and tools.[4] The next challenge will be to systematically provide NLO electroweak corrections[5] and next-to-NLO (NNLO) QCD corrections to experiment, and cover the relevant channels to next-to-NNLO (NNNLO), including as many kinematic distributions as possible. Current state of the art predictions in precision theory includes: NLO predictions to $t\bar{t}b\bar{b}$ production with all off-shell effects included,[6] NNLO for exclusive jet production[7] or Higgs production and decay in weak boson fusion,[8] combined NNLO-QCD and NLO-electroweak corrections to top pair production,[9] or first NNNLO predictions for Higgs production.[10]

In addition, Monte Carlo event generators have taken precision predictions significantly beyond the simple hard process. Complex hard processes with many particles in the final state can be described by helicity amplitudes, avoiding the CPU-consuming task of squaring scattering amplitudes analytically, and instead computing them numerically and then squaring these single numbers.[12] Multi-purpose generators like PYTHIA,[13] MadGraph,[14] Sherpa,[15] and Herwig[16] now describe events with high and variable jet multiplicities, arguably the one big challenge for LHC

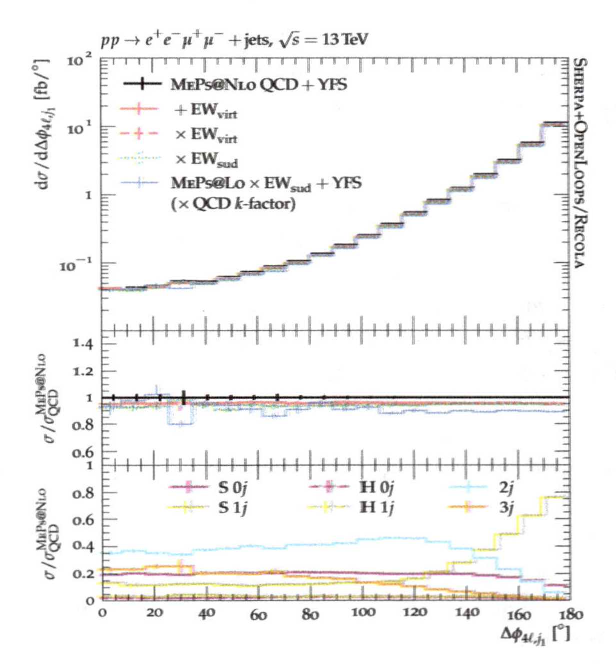


Fig. 1. Correlation for $ZZ$+jets production based on precision MC, including fixed-order QCD and electroweak corrections, as well as Sudakov logarithms. Figure from Ref. 11.

simulations. The combination of a hard matrix element with logarithmically enhanced jet radiation is solved by so-called jet merging, first solved by introducing the CKKW[17] method. In Fig. 1 we show what state of the art programs can analyse, including NLO-QCD and NLO-electroweak corrections as well as Sudakov logarithms and jet radiation in Sherpa.[11]

While this is not true for all current analyses, modern simulation-based inference approaches, as will be discussed in more detail below, require simulations of SM-backgrounds and any new physics hypotheses with the same precision. The additional challenge is that precision predictions for signal hypotheses have to cover a large model space. Here, MadGraph and Sherpa drove the development of flexible and automated event generation based on a given Lagrangian,[18] while still with the same access to automated higher-order corrections.

Finally, there still exists a part of the LHC simulation chain where our access to first-principle predictions is limited, for instance, parton densities or fragmentation. In these cases the event generators combine data and theory input using modern data science methodologies. This does not mean that we have given up on understanding these aspects through first principles forever, but that, at this stage, modelling them provides a better basis for experimental analyses.

**Higgs Discovery and EFT Properties** As mentioned above, the existence of the Higgs boson can be derived as a purely formal prediction based on the description of the massive electroweak sector in terms of a renormalizable quantum field theory.[19–21] In that sense the starting point of all Higgs physics are the precision measurements of the electroweak SM at LEP. These measurements and their interpretation went beyond the usual leading order in perturbation theory and probed quantum corrections systematically for the first time in collider-based particle physics. This legacy lives on at the LHC, where we systematically describe hadron collider data including quantum effects for the first time. Given the combined LEP results, the existence of some kind of Higgs boson was never really a question, because it is needed to ensure unitary predictions for scattering processes.

Weak boson fusion and weak boson scattering are sensitive processes at the LHC, consequently forming the core of the electroweak physics program at the LHC. The main question answered by the Higgs discovery in 2012 was about its mass. Nature's choice of 125 GeV[22,23] is perfect for the LHC program, because it is right in the middle between the light-Higgs regime with dominant Higgs decays to fermions and photons, and the heavier-Higgs regime with dominant Higgs decays to weak bosons. This means that Higgs physics, as described in this book, could move immediately from the Higgs discovery to Higgs measurements.[24,25] These measurements combine different production and decay channels to a global analysis, initially described in terms of Higgs couplings and by now upgraded to a consistent, effective quantum field theory.

Following the Higgs discovery in Run 1, the LHC Run 2 has been the first comprehensive precision program at a hadron collider. Experimentally, this refers to, for instance, the large number of measured Higgs production and decay channels and their comprehensive treatment of statistical and systematic uncertainties. On the theory side it is driven by precision predictions of rates as well as precision simulations of the entire phase space for, essentially, all LHC measurements. Additional key ingredients are precision predictions for parton densities and all other aspects of the event generation chain.

Responding to the precision of the Run 2 measurements and their sensitivity to higher orders in perturbation theory, a major shift in the theoretical interpretation has been to move from an ad-hoc and theoretically inconsistent modification of Higgs couplings to a proper description of modified interactions in terms of an effective Lagrangian. Effective field theory tracks deviations through higher-dimensional Higgs operators, induced by unspecified heavy new particles. This SM effective field theory (SMEFT) had already been established at LEP, to describe anomalous electroweak gauge couplings, and was easily extended to the gauge and Higgs sector at the LHC.[26] One of the great successes of the LHC is that the SMEFT precision measurements of electroweak Wilson coefficients outperform the corresponding LEP measurements,[27] turning the LHC into a discovery-and-precision machine. The successful SMEFT description of the electroweak and Higgs sector[28–31] has, by now, been extended to the top sector,[32–34] a combination of the two,[35,36] anomalous QCD couplings,[37] and the link between top quark and bottom quark physics.[38]

The SMEFT interpretation framework for the LHC comes with many benefits. Firstly, quantum field theory properties like renormalizability allow us to formulate the underlying hypotheses including quantum effects, or higher orders in perturbation theory. If LEP has established perturbative quantum field theory as the correct description of elementary particles, the LHC has turned this description into a systematic interpretation framework for all its data. Secondly, higher-dimensional operators do not only modify total production rates, they also affect kinematic distributions, specifically high-energy tails. This way, SMEFT allows us to describe and analyze potential deviations in a wide range of kinematic observables. Finally, SMEFT does not just modify individual sectors of physics, rather, it expands the entire SM-Lagrangian using higher-dimensional operators. This means that we can use it to answer the big global question: how well does the SM describe all LHC measurements, as well as measurements from other experiments probing similar mass scales? Figure 2 illustrates the situation in the Higgs sector after the LHC Run 2.

These features imply that a global SMEFT analysis serves as a first step towards a comprehensive analysis of the entire SM facing the full LHC dataset, as we will discuss below. On the other hand, given what we know about the shortcomings of the SM and given the success of renormalizable field theory, we should really consider global SMEFT analyses a useful limit setting tool. Once the LHC experiments observe a significant anomaly, the corresponding theory interpretations will use all our phenomenological and conceptual background knowledge to identify the fundamental structures behind such an anomaly.

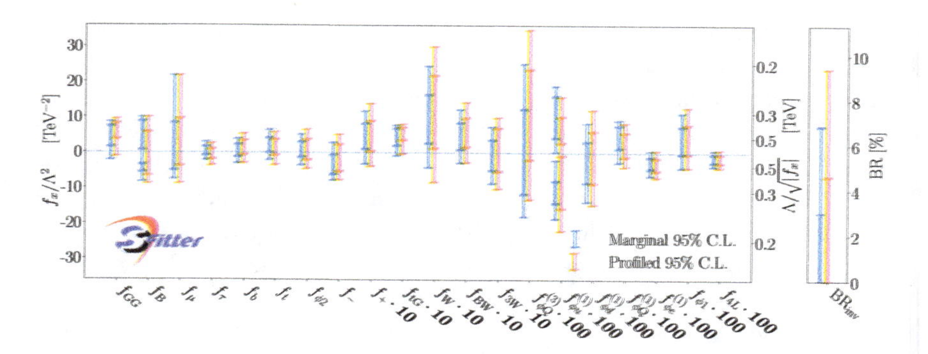


Fig. 2. Global SMEFT analysis of the Higgs and electroweak gauge sector. The two colors correspond to consistent likelihood and Bayesian marginalization frameworks for the nuisance parameters and Wilson coefficients. Figure from Ref. 39.

**Physics beyond the Standard Model** The ultimate goal of the LHC community is not to confirm the Standard Model, but to find cracks in this description and to discover new particles and interactions. Towards this goal, the discovery of a fundamental Higgs scalar does not only give us faith in the renormalizability of the SM and its structural validity to high energies, it also provides us with a framework to tackle two fundamental questions related to cosmological observations: the nature of dark matter and the origin of the matter-antimatter asymmetry of the universe. Neither of them can be answered within the SM, and both point towards renormalizable extensions of the SM. Such models predict new particles and a modified symmetry structure, and the Higgs or scalar sector is a prime candidate to accommodate these features.[40–42]

While we know that the SM does not explain all the observations we expect it to explain, the LHC has not yet found any sign of new particles. The constraints from these LHC searches either push new particle masses to larger values or their couplings to SM particles to very small values. While indirect probes of heavy new particles can often be described by an effective field theory, there will be observables, like the relic dark matter density, for which we need to work with the full models and new particles on their mass shell. Such light and weakly interacting new particles will also be produced on-shell at the LHC.[43]

Going beyond effective field theories towards well-defined renormalizable models requires a solid understanding of the underlying quantum field theory. It allows us to predict LHC signatures by studying the interplay between UV-complete models and effective field theories. The very specific questions about dark matter and baryogenesis then suggest what to look for at the LHC. For dark matter to be produced during the thermal history of the universe, it has to interact with the SM in some way. Thermal freeze-out production forms the general basis of weakly interacting massive dark matter. While the $W$ and $Z$-bosons are ruled out as mediators, the Higgs sector provides an attractive portal to dark matter.[44] A direct consequence of such a Higgs portal could be an invisible decay of the SM-like Higgs boson, more general dark matter searches target missing transverse energy in association with jets or other SM production processes.

Alternatively, extended scalar sectors with a spectrum of Higgs particles can serve as a direct link to gauge-singlet dark matter particles. Any such additional mediators couple to the Standard Model and to dark matter, which means we can search for them as missing energy or as resonances, for example, in di-jet or di-lepton production.

Extended Higgs sectors also play an important role for baryogenesis, as they can promote the electroweak phase transition to strong first order with the SM-like Higgs boson remaining at a mass of 125 GeV. This way the matter-antimatter asymmetry can be generated dynamically through electroweak baryogenesis,[46] illustrated in Fig. 3, if besides the departure from thermal equilibrium the remaining two of the three Sakharov conditions[47] are fulfilled. These are charge (C) and charge-parity (CP) violation and baryon number violation. The first condition entails additional Higgs bosons that can be lighter or heavier than the SM-like Higgs boson. They can be searched for at the LHC. CP-violation in the Higgs sector can be probed at the LHC, either through CP-sensitive or optimal observables, or by searching for heavy Higgs decays into two SM-like Higgs bosons and into a SM-like Higgs boson plus a $Z$-boson, simultaneously.[48,49]

Coming back to a global interpretation of LHC data, model-based searches for new particles and SMEFT searches are closely related in their theory interpretations. At some point we always need to match the two theory hypotheses in phase space regions where both of them are valid. This matched description covers all channels where the new particle can be produced on its mass shell, but also light new particles remaining off-shell in $t$-channel exchange. Precision matching beyond leading order can uncover potential shortcomings of the SMEFT approach when we truncate the series at operators dimension six,[50] and it introduces a matching scale uncertainty.[51,52] Model-based searches and SMEFT are also related on the analysis side. While an observed resonance can, obviously, not be interpreted in terms of SMEFT, limits from such resonance searches for example

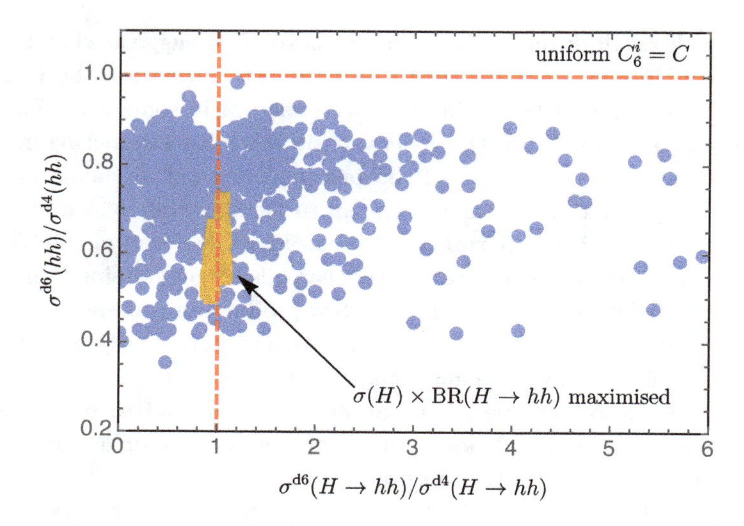


Fig. 3. Modification of Higgs pair production $gg \to hh$ and its correlation with resonance production $gg \to H \to hh$ in a 2HDM including scalar dimension-6 operators to achieve a strong first order electroweak phase transition. The Wilson coefficients are chosen uniformly, $C_6^i = C$. Highlighted are the Higgs-philic scan result points. Figure taken from Ref. 45.

in $WW$, $WZ$, or $WH$ production provide some of the most useful inputs to SMEFT analyses.[27,28] No matter if we are more interested in global analysis strategies or in finding fundamentally motivated new physics models, SMEFT and model-based searches are two sides of the same medal.

**Predictions and inference for the HL-LHC** In many ways the expected size, complexity, and precision of the HL-LHC dataset challenge our established methodology, starting from data acquisition to data processing, analysis, and theory predictions. The 10-fold increase in the integrated luminosity as compared to the combined Runs 1-3 reduces many statistical uncertainties to a level where systematic and theory uncertainties will dominate the vast majority of analyses. For theory this means that we need to avoid a situation where theory uncertainties become the limitation to experimental measurements, these measurements become purely data-driven, and this way turn to modelling rather than understanding fundamental physics. One way to tackle this challenge is to employ ideas and methods from modern data science to improve theory predictions as well as the way we make them available to analyses.[53] This task sounds technical at first, but it links two of the most exciting aspects of modern science: fundamental questions from physics and cosmology and the revolutionary tool box from data science.

The immediate motivation to use data science methods at the LHC is the combination of the size of the LHC dataset with the availability of precision simulations based on first principles. Any LHC analysis and any comparison between data and predictions already employs multi-variate methods and simple neural networks. The natural and necessary next step is to update these methods and make use of the transformative developments in data science research over the last 20 years.

The more abstract motivation for data science methods in LHC physics is that modern data science provides a common language for theory and experiment. It not only builds bridges between data science and theory or experiment individually, it also allows us to build the bridges between experiment and theory, exactly what we need to make the HL-LHC a success. Furthermore, modern machine learning provides new interdisciplinary opportunities between fundamental science and data science.

For particle theory, modern machine learning (ML) has the potential to improve all aspects of our established theory computation and simulation chain, as illustrated in Fig. 4. Standard network architectures we can employ for theory predictions include simple regression, but also classification and generative models. Critical regression tasks in LHC theory include loop integrals, libraries of standard functions, or surrogates for loop amplitudes. The NNPDF parton densities[54] have shown how machine learning does not only increase the speed of the evaluation, it should also allow for controlled precision by avoiding biases from non-perfect theory assumptions. Other modules in the forward simulation chain which we expect to benefit from machine learning are phase space sampling, parton showers, and, especially, hadronization or fragmentation models.

A second strategy to improve LHC simulations involved generative networks for event generation[53] and for detector simulations.[56] Here we

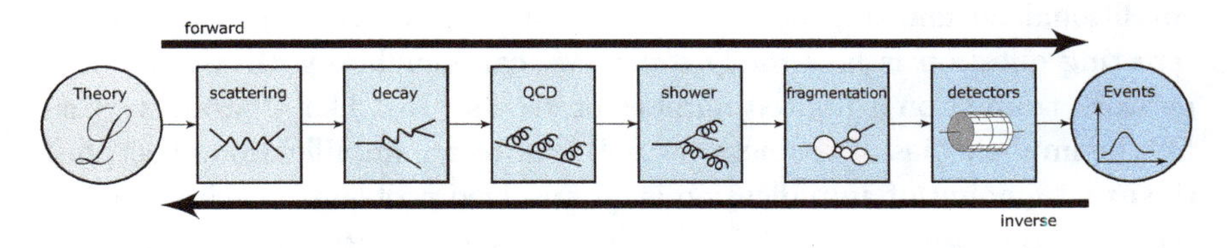


Fig. 4. Illustration of the forward simulation and the inverted simulation or unfolding-inference direction for the LHC. Figure from Ref. 53.

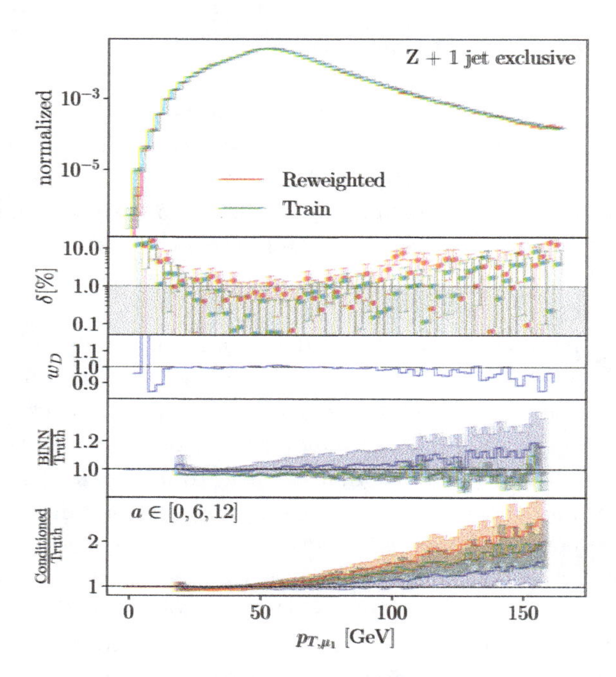


Fig. 5. Results from a INN-based NN-event generator including a comprehensive uncertainty treatment. Figure from Ref. 55.

attempt to replace the entire chain shown in Fig. 4 by a generative network. Such a ML-generator can be trained on Monte Carlo or on measured events, it can combine positive events and subtraction terms, be used to subtract entire event samples or to unweight events, transform events in control regions into events in signal regions, and it can provide an efficient way to distribute standardized event samples. The key challenge for all networks employed in LHC physics, but especially generative networks, is to ensure that they have learned all relevant phase space features and can reproduce them within a given uncertainty.[55] Results from Bayesian and conditional normalizing flows are shown in Fig. 5. Conceptually, an interesting question is how many events we can simulate with a generative network trained on a limited number of events. Just like a parameterized fit, the implicit bias of the network will lead to an amplification effect, but the exact amount of amplification is an unsolved problem,[57] where particle physics should give answers the data science community has not provided.

Obviously, we can use data science concepts for LHC inference. An orthogonal approach to model-based searches and the logical next step from global SMEFT analyses is to analyze the LHC dataset by directly comparing measured and simulated events.[58] For this strategy, detailed precision simulations are crucial, because they allow us to cover the full phase space and search for features which we would not see in rate measurements. This means that the main challenge in simulation-based inference are not the experimental setup, but suitable theory predictions.

Finally, LHC inference can be transformed by data science through inverse simulations, as illustrated in Fig. 4. One of the problems of precision LHC physics is that it is almost impossible to use the most recent theory predictions if they cannot be implemented in event generators. To identify the best simulation or data processing stage to compare theory and experiment we can add inverse simulations to the forward simulation chain. Stochastically defined inverse problems can be solved using multi-dimensional classifier reweighting[59] or conditional generative networks.[60] Inverted simulations are already used at the LHC, but as localized efforts with limited and ad-hoc techniques. They include: detector unfolding of simple kinematic distributions, jet algorithms, unfolding to the parton level process, and the matrix element method. For all of them, machine learning applied to simulations and to simulation-based inference provides us with a consistent and powerful framework to make the best use of the vast HL-LHC dataset.

**A Bright Future** There is no crisis in modern particle physics. On the contrary, the future of particle physics is bright, because we have exciting and fundamental physics questions to answer and datasets which allow us to do so. Here, we should really consider the HL-LHC as a new experiment, with a proper name, with new strengths, and with new challenges, in both theory and experiment. Modern hadron collider physics will benefit from the HL-LHC because it is all about precision measurements, precision predictions, and a theoretical interpretation in terms of fundamental Lagrangians. Exciting discoveries will be driven by these unique strengths in the particle physics and science landscape.

When interpreting LHC data we need to keep in mind that there is nothing to learn from modelling data without a fundamental interpretation framework. While it might be necessary to model effects or background kinematics to then be able to search for physics beyond the SM, a purely data-driven approach would potentially leave us with an outcome where the HL-LHC discovers no new particles, and we do not learn anything fundamentally interesting from the SM dataset either. This is illustrated by the not very well-known fact that the LHC has not only discovered the Higgs boson, but it has discovered more than 60 new particles. However, all but one of these particles are hadronic resonances without revolutionary theory implications. Only the Higgs tells us something new and structural about fundamental physics; the existence of a background field in the vacuum that is responsible for particle masses.

The fact that in this discussion of hadron collider physics and theory the invention of models for physics beyond the Standard Model hardly shows up is often interpreted as a problem for particle physics. However, what it really implies is that particle physics has entered an exciting, data-driven era. Theory is crucial, as it formulates the fundamental questions, provides precision predictions, defines consistent interpretation frameworks, and allows us to combine LHC results with a wide range of particle physics and cosmological insights. What theory cannot provide is the answers to our fundamental questions — theoretical models without data to test them make for a fun game, but physics needs relevant datasets like the one provided by the HL-LHC.

In this new, data-driven era, successful LHC physics relies on a wide range of experimental and theoretical techniques. It derives its excitement from new ideas, concepts, and tools, and their huge impact on detectors, analysis techniques, theory calculations, and simulations. Looking at the expected size of the HL-LHC dataset (and future collider designs), we have to make use of modern machine learning wherever we can. Some data science concepts might be directly applicable to LHC physics, but in most cases we will have to develop methods and tools which guarantee the control, precision and uncertainty treatment needed for the LHC. And no matter what we do, particle physics is defined by its physics questions and the close ties between experiment, analysis, and fundamental theory. These ties give us great hope that, eventually, we will discover new effects leading to new, unexpected, and exciting insights into the Nature of fundamental particles.

## References

1. G. Heinrich, Collider Physics at the Precision Frontier, *Phys. Rept.* **922**, 1–69 (2021). doi: 10.1016/j.physrep.2021.03.006.
2. C. Buttar et al., Les houches physics at TeV colliders 2005, standard model and Higgs working group: Summary report. In *4th Les Houches Workshop on Physics at TeV Colliders* (4, 2006).
3. J. A. M. Vermaseren, New features of FORM (10, 2000).

4. J. M. Campbell and R. K. Ellis, MCFM for the Tevatron and the LHC, *Nucl. Phys. B Proc. Suppl.* **205-206**, 10–15 (2010). doi: 10.1016/j.nuclphysbps.2010.08.011.
5. A. Denner and S. Dittmaier, Electroweak Radiative Corrections for Collider Physics, *Phys. Rept.* **864**, 1–163 (2020). doi: 10.1016/j.physrep.2020.04.001.
6. A. Denner, J.-N. Lang, and M. Pellen, Full NLO QCD corrections to off-shell ttbb production, *Phys. Rev. D.* **104** (5), 056018 (2021). doi: 10.1103/PhysRevD.104.056018.
7. X. Chen, T. Gehrmann, E. W. N. Glover, A. Huss, and J. Mo, NNLO QCD corrections in full colour for jet production observables at the LHC (4, 2022).
8. K. Asteriadis, F. Caola, K. Melnikov, and R. Röntsch, NNLO QCD corrections to weak boson fusion Higgs boson production in the $\mathrm{H} \to \mathrm{b}\bar{b}$ and $\mathrm{H} \to \mathrm{WW}^* \to 4\mathrm{l}$ decay channels, *JHEP.* **02**, 046 (2022). doi: 10.1007/JHEP02(2022)046.
9. M. Czakon, D. Heymes, A. Mitov, D. Pagani, I. Tsinikos, and M. Zaro, Top-pair production at the LHC through NNLO QCD and NLO EW, *JHEP.* **10**, 186 (2017). doi: 10.1007/JHEP10(2017)186.
10. X. Chen, T. Gehrmann, E. W. N. Glover, A. Huss, B. Mistlberger, and A. Pelloni, Fully Differential Higgs Boson Production to Third Order in QCD, *Phys. Rev. Lett.* **127** (7), 072002 (2021). doi: 10.1103/PhysRevLett.127.072002.
11. E. Bothmann, D. Napoletano, M. Schönherr, S. Schumann, and S. L. Villani, Higher-order EW corrections in ZZ and ZZj production at the LHC, *JHEP.* **06**, 064 (2022). doi: 10.1007/JHEP06(2022)064.
12. H. Murayama, I. Watanabe, and K. Hagiwara, HELAS: HELicity amplitude subroutines for Feynman diagram evaluations (1, 1992).
13. T. Sjöstrand, S. Ask, J. R. Christiansen, R. Corke, N. Desai, P. Ilten, S. Mrenna, S. Prestel, C. O. Rasmussen, and P. Z. Skands, An introduction to PYTHIA 8.2, *Comput. Phys. Commun.* **191**, 159–177 (2015). doi: 10.1016/j.cpc.2015.01.024.
14. J. Alwall, R. Frederix, S. Frixione, V. Hirschi, F. Maltoni, O. Mattelaer, H. S. Shao, T. Stelzer, P. Torrielli, and M. Zaro, The automated computation of tree-level and next-to-leading order differential cross sections, and their matching to parton shower simulations, *JHEP.* **07**, 079 (2014). doi: 10.1007/JHEP07(2014)079.
15. E. Bothmann et al., Event Generation with Sherpa 2.2, *SciPost Phys.* **7** (3), 034 (2019). doi: 10.21468/SciPostPhys.7.3.034.
16. J. Bellm et al., Herwig 7.0/Herwig++ 3.0 release note, *Eur. Phys. J. C.* **76** (4), 196 (2016). doi: 10.1140/epjc/s10052-016-4018-8.
17. S. Catani, F. Krauss, R. Kuhn, and B. R. Webber, QCD matrix elements + parton showers, *JHEP.* **11**, 063 (2001). doi: 10.1088/1126-6708/2001/11/063.
18. A. Alloul, N. D. Christensen, C. Degrande, C. Duhr, and B. Fuks, FeynRules 2.0 - A complete toolbox for tree-level phenomenology, *Comput. Phys. Commun.* **185**, 2250–2300 (2014). doi: 10.1016/j.cpc.2014.04.012.
19. P. W. Higgs, Broken Symmetries and the Masses of Gauge Bosons, *Phys. Rev. Lett.* **13**, 508–509 (1964). doi: 10.1103/PhysRevLett.13.508.
20. F. Englert and R. Brout, Broken Symmetry and the Mass of Gauge Vector

Mesons, *Phys. Rev. Lett.* **13**, 321–323 (1964). doi: 10.1103/PhysRevLett.13.321.

21. P. W. Higgs, Spontaneous Symmetry Breakdown without Massless Bosons, *Phys. Rev.* **145**, 1156–1163 (1966). doi: 10.1103/PhysRev.145.1156.
22. G. Aad et al., Observation of a new particle in the search for the Standard Model Higgs boson with the ATLAS detector at the LHC, *Phys. Lett. B.* **716**, 1–29 (2012). doi: 10.1016/j.physletb.2012.08.020.
23. S. Chatrchyan et al., Observation of a New Boson at a Mass of 125 GeV with the CMS Experiment at the LHC, *Phys. Lett. B.* **716**, 30–61 (2012). doi: 10.1016/j.physletb.2012.08.021.
24. S. Dawson, C. Englert, and T. Plehn, Higgs Physics: It ain't over till it's over, *Phys. Rept.* **816**, 1–85 (2019). doi: 10.1016/j.physrep.2019.05.001.
25. M. Spira, Higgs Boson Production and Decay at Hadron Colliders, *Prog. Part. Nucl. Phys.* **95**, 98–159 (2017). doi: 10.1016/j.ppnp.2017.04.001.
26. I. Brivio and M. Trott, The Standard Model as an Effective Field Theory, *Phys. Rept.* **793**, 1–98 (2019). doi: 10.1016/j.physrep.2018.11.002.
27. A. Butter, O. J. P. Éboli, J. Gonzalez-Fraile, M. C. Gonzalez-Garcia, T. Plehn, and M. Rauch, The Gauge-Higgs Legacy of the LHC Run I, *JHEP.* **07**, 152 (2016). doi: 10.1007/JHEP07(2016)152.
28. A. Biekoetter, T. Corbett, and T. Plehn, The Gauge-Higgs Legacy of the LHC Run II, *SciPost Phys.* **6** (6), 064 (2019). doi: 10.21468/SciPostPhys.6.6.064.
29. S. Kraml, T. Q. Loc, D. T. Nhung, and L. D. Ninh, Constraining new physics from Higgs measurements with Lilith: update to LHC Run 2 results, *SciPost Phys.* **7** (4), 052 (2019). doi: 10.21468/SciPostPhys.7.4.052.
30. J. de Blas, M. Ciuchini, E. Franco, A. Goncalves, S. Mishima, M. Pierini, L. Reina, and L. Silvestrini, Global analysis of electroweak data in the Standard Model (12, 2021).
31. E. d. S. Almeida, A. Alves, O. J. P. Éboli, and M. C. Gonzalez-Garcia, Electroweak legacy of the LHC run II, *Phys. Rev. D.* **105** (1), 013006 (2022). doi: 10.1103/PhysRevD.105.013006.
32. S. Brown, A. Buckley, C. Englert, J. Ferrando, P. Galler, D. J. Miller, L. Moore, M. Russell, C. White, and N. Warrack, TopFitter: Fitting top-quark Wilson Coefficients to Run II data, *PoS.* **ICHEP2018**, 293 (2019). doi: 10.22323/1.340.0293.
33. N. P. Hartland, F. Maltoni, E. R. Nocera, J. Rojo, E. Slade, E. Vryonidou, and C. Zhang, A Monte Carlo global analysis of the Standard Model Effective Field Theory: the top quark sector, *JHEP.* **04**, 100 (2019). doi: 10.1007/JHEP04(2019)100.
34. I. Brivio, S. Bruggisser, F. Maltoni, R. Moutafis, T. Plehn, E. Vryonidou, S. Westhoff, and C. Zhang, O new physics, where art thou? A global search in the top sector, *JHEP.* **02**, 131 (2020). doi: 10.1007/JHEP02(2020)131.
35. J. Ellis, M. Madigan, K. Mimasu, V. Sanz, and T. You, Top, Higgs, Diboson and Electroweak Fit to the Standard Model Effective Field Theory, *JHEP.* **04**, 279 (2021). doi: 10.1007/JHEP04(2021)279.
36. J. J. Ethier, G. Magni, F. Maltoni, L. Mantani, E. R. Nocera, J. Rojo,

E. Slade, E. Vryonidou, and C. Zhang, Combined SMEFT interpretation of Higgs, diboson, and top quark data from the LHC, *JHEP.* **11**, 089 (2021). doi: 10.1007/JHEP11(2021)089.
37. F. Krauss, S. Kuttimalai, and T. Plehn, LHC multijet events as a probe for anomalous dimension-six gluon interactions, *Phys. Rev. D.* **95** (3), 035024 (2017). doi: 10.1103/PhysRevD.95.035024.
38. S. Bißmann, J. Erdmann, C. Grunwald, G. Hiller, and K. Kröninger, Constraining top-quark couplings combining top-quark and $B$ decay observables, *Eur. Phys. J. C.* **80** (2), 136 (2020). doi: 10.1140/epjc/s10052-020-7680-9.
39. I. Brivio, S. Bruggisser, N. Elmer, E. Geoffray, M. Luchmann, and T. Plehn, To Profile or To Marginalize – A SMEFT Case Study (8, 2022).
40. R. Grober, M. Muhlleitner, M. Spira, and J. Streicher, NLO QCD Corrections to Higgs Pair Production including Dimension-6 Operators, *JHEP.* **09**, 092 (2015). doi: 10.1007/JHEP09(2015)092.
41. R. Grober, M. Muhlleitner, and M. Spira, Higgs Pair Production at NLO QCD for CP-violating Higgs Sectors, *Nucl. Phys. B.* **925**, 1–27 (2017). doi: 10.1016/j.nuclphysb.2017.10.002.
42. H. Abouabid, A. Arhrib, D. Azevedo, J. E. Falaki, P. M. Ferreira, M. Mühlleitner, and R. Santos, Benchmarking Di-Higgs Production in Various Extended Higgs Sector Models (12, 2021).
43. M. Mühlleitner, M. O. P. Sampaio, R. Santos, and J. Wittbrodt, Phenomenological Comparison of Models with Extended Higgs Sectors, *JHEP.* **08**, 132 (2017). doi: 10.1007/JHEP08(2017)132.
44. G. Arcadi, A. Djouadi, and M. Raidal, Dark Matter through the Higgs portal, *Phys. Rept.* **842**, 1–180 (2020). doi: 10.1016/j.physrep.2019.11.003.
45. Anisha, L. Biermann, C. Englert, and M. Mühlleitner, Two Higgs doublets, Effective Interactions and a Strong First-Order Electroweak Phase Transition (4, 2022).
46. W. Bernreuther, CP violation and baryogenesis, *Lect. Notes Phys.* **591**, 237–293 (2002).
47. A. D. Sakharov, Violation of CP Invariance, C asymmetry, and baryon asymmetry of the universe, *Pisma Zh. Eksp. Teor. Fiz.* **5**, 32–35 (1967). doi: 10.1070/PU1991v034n05ABEH002497.
48. D. Fontes, J. C. Romão, R. Santos, and J. a. P. Silva, Undoubtable signs of $CP$-violation in Higgs boson decays at the LHC run 2, *Phys. Rev. D.* **92** (5), 055014 (2015). doi: 10.1103/PhysRevD.92.055014.
49. S. F. King, M. Muhlleitner, R. Nevzorov, and K. Walz, Exploring the CP-violating NMSSM: EDM Constraints and Phenomenology, *Nucl. Phys. B.* **901**, 526–555 (2015). doi: 10.1016/j.nuclphysb.2015.11.003.
50. S. Dawson, S. Homiller, and M. Sullivan, Impact of dimension-eight SMEFT contributions: A case study, *Phys. Rev. D.* **104** (11), 115013 (2021). doi: 10.1103/PhysRevD.104.115013.
51. S. Dawson, S. Homiller, and S. D. Lane, Putting standard model EFT fits to work, *Phys. Rev. D.* **102** (5), 055012 (2020). doi: 10.1103/PhysRevD.102.055012.
52. I. Brivio, S. Bruggisser, E. Geoffray, W. Killian, M. Krämer, M. Luchmann,

T. Plehn, and B. Summ, From models to SMEFT and back?, *SciPost Phys.* **12** (1), 036 (2022). doi: 10.21468/SciPostPhys.12.1.036.
53. S. Badger et al., Machine Learning and LHC Event Generation (3, 2022).
54. R. D. Ball et al., The path to proton structure at 1% accuracy, *Eur. Phys. J. C.* **82** (5), 428 (2022). doi: 10.1140/epjc/s10052-022-10328-7.
55. A. Butter, T. Heimel, S. Hummerich, T. Krebs, T. Plehn, A. Rousselot, and S. Vent, Generative Networks for Precision Enthusiasts (10, 2021).
56. A. Adelmann et al., New directions for surrogate models and differentiable programming for High Energy Physics detector simulation. In *2022 Snowmass Summer Study* (3, 2022).
57. A. Butter, S. Diefenbacher, G. Kasieczka, B. Nachman, and T. Plehn, GANplifying event samples, *SciPost Phys.* **10** (6), 139 (2021). doi: 10.21468/SciPostPhys.10.6.139.
58. J. Brehmer, F. Kling, I. Espejo, and K. Cranmer, MadMiner: Machine learning-based inference for particle physics, *Comput. Softw. Big Sci.* **4** (1), 3 (2020). doi: 10.1007/s41781-020-0035-2.
59. A. Andreassen, P. T. Komiske, E. M. Metodiev, B. Nachman, and J. Thaler, OmniFold: A Method to Simultaneously Unfold All Observables, *Phys. Rev. Lett.* **124** (18), 182001 (2020). doi: 10.1103/PhysRevLett.124.182001.
60. M. Bellagente, A. Butter, G. Kasieczka, T. Plehn, A. Rousselot, R. Winterhalder, L. Ardizzone, and U. Köthe, Invertible Networks or Partons to Detector and Back Again, *SciPost Phys.* **9**, 074 (2020). doi: 10.21468/SciPostPhys.9.5.074.

© 2024 The Author(s)
https://doi.org/10.1142/9789811280184_0002

# Chapter 2

# Commissioning and the Initial Operation of the LHC

Mike Lamont

*CERN*

By the end of Run 2 in December 2018, the LHC had seen seven full years of operation and a wealth of knowledge and experience has been built up. The key operational procedures and tools were well established. The understanding of beam dynamics was profound and utilized on-line by well-honed measurement and correction techniques. Key beam-related systems have been thoroughly optimised and functionality sufficiently enhanced to deal with most of the challenges encountered. Availability had been optimised significantly across all systems. This collective experience forms the initial operational basis for Run 3 and subsequent HL-LHC operation.

A brief review of Run 1 and Run 2 is given below, firstly to outline the progress made, and secondly to highlight the issues encountered and surmounted along the way. An overview of operational features of the machine and the lessons learnt is then presented. The chapter concludes with brief look at consolidation activities in view of the need to sustain high availability and safe operation given the considerable challenges of the HL-LHC operational regime and the time-frame over which it will operate.

## 1. Overview of Run 1

Following recovery from the September 2008 incident, Run 1 saw initial commissioning at reduced energy and the steep learning curve faced in bootstrapping the operations of a 27 km superconducting collider. Nonetheless, having established the core operational and machine protection systems, healthy levels of performance were achieved. A brief overview of 2010–2013 operations follows, with the aim of highlighting the main issues addressed.

This is an open access article published by World Scientific Publishing Company. It is distributed under the terms of the Creative Commons Attribution-NonCommercial 4.0 (CC BY-NC) License.

### 1.1. *2010*

Essentially, 2010 was devoted to commissioning, sorting out the operational basics, and establishing confidence in procedures and the machine protection system.

Ramp commissioning to 3.5 TeV was relatively smooth and led to (very public) first collisions at 3.5 TeV unsqueezed on the 30$^{th}$ March 2010. Squeeze commissioning subsequently reduced the $\beta^*$ to 2.0 m in all the four main experiments. Thereafter was a period of stable operation for physics interleaved with continued system commissioning.

The decision was then made to operate with bunches of nominal intensity. Consequently, there was a halting push through the introduction of nominal bunch intensity and further operational debugging up to a total stored beam energy of around 3 MJ. Eventually, this led to a period of steady running that was used to fully verify machine protection and operational procedures.

In June, the decision has been made to go for bunches with the nominal population ($1.15 \times 10^{11}$ protons per bunch), which involved another extended commissioning period. Up to this point, in deference to machine protection concerns, only around one fifth of the nominal bunch population had been used. To further increase the number of bunches, the move to bunch trains separated by 150 ns was made and the crossing angle bumps spanning the experiments' insertion regions were deployed. Each step-up in intensity was followed by operational and machine protection validation, as well as running it for a few days to check system performance. The 2010 proton run finished with beams of 368 bunches of around $1.2 \times 10^{11}$ protons per bunch, and a peak luminosity of $2.1 \times 10^{32}$ cm$^{-2}$s$^{-1}$.

Looking back, 2010 was a profoundly important year for a chastened and cautious accelerator sector that had to move carefully through a parameter space that demanded full awareness of the associated risks. The energy stored in the magnets has demonstrated its destructive power, and clearly the beam was to be treated with the utmost respect. Some key systems, such as the LHC Beam Dump System and Beam Interlock System, had been designed and implemented with safety very much taken into account; pulling everything else together to bring high intensity beams through the operational cycle was another matter and a culture of safe exploitation of the machine was bedded in. The LHC became magnetically and optically well understood (judging by standards at the time — which were impressively surpassed in later years). It was stunningly magnetically reproducible. Coupled with the revelatory performance of the collimation system, the machine was able to perform its dual role of cleaning and protection impeccably throughout the full cycle. The injectors were doing a great job throughout, reliably providing high intensity bunches with unforeseen low transverse emittances.

2010 ended with a switch from operations with protons to, for the first time, with lead ions. The operations team successfully leveraged the experience gained to rapidly push through the ion commissioning programme and declared Stable Beams for physics on 7 November.

### 1.2. *2011*

While beam energy remained at 3.5 TeV, 2011 saw combined exploitation and the exploration of performance limits. Following a ramp-up to around 200 bunches (75 ns bunch spacing) taking about 2 weeks, there was a scrubbing run of 10 days which included 50 ns injection commissioning. After an encouraging performance, decision was made to operate with 50 ns bunch spacing, and then a staged ramp-up in the number of bunches up to a maximum of 1380 bunches took place.

Having raised the number of bunches to 1380, performance was further increased by reducing the emittances of the beams delivered by the injectors and by gently increasing the bunch intensity. The result was a peak luminosity of $2.4 \times 10^{33}$ cm$^{-2}$s$^{-1}$ and some healthy delivery rates which topped 90 pb$^{-1}$ in 24 hours.

A reduction in $\beta^*$ in ATLAS and CMS from 1.5 m to 1 m delivered the next step up in peak luminosity. This step was made possible by careful measurements of the available aperture in the interaction regions concerned. These measurements revealed excellent aperture consistent with a very good alignment and close to design mechanical tolerances. The reduction in $\beta^*$ and further gentle increases in bunch intensity produced a peak luminosity of $3.8 \times 10^{33}$ cm$^{-2}$s$^{-1}$, well beyond expectations from the start of the year. Coupled with a concerted effort to improve availability, the machine went on to deliver a total of around 5.6 fb$^{-1}$ for the year to both ATLAS and CMS.

Meanwhile, excitement was building in the experiments. A colloquium at the end of the year showed a strengthening significance of an excess around 125 GeV. The possible discovery of the Higgs boson in 2012 was recognised, and corresponding LHC running scenarios were discussed in depth — first at the Evian workshop and then crystallized later at the

2012 Chamonix workshop, where CERN Director General Rolf Heuer firmly stated: as a top priority the LHC machine must produce enough integrated luminosity to allow the ATLAS and CMS experiments an independent discovery of the Higgs before the start of LS1. Soon after the workshop, Council president Michel Spiro sent a message to CERN's member stating: After a brilliant year in 2011, 2012 should be historic, with either the discovery of the Standard Model Higgs boson or its exclusion.

### 1.3. *2012 and 2013*

2012 was a production year at an increased beam energy of 4 TeV. The increase in beam energy came, following efforts to better understand and quantify the problem of the interconnect splice resistances. The conclusion was that the risks in moving to 4 TeV were acceptable.

The choice was made to continue to exploit 50 ns bunch spacing and run with a total number of bunches of around 1380. Based on experience from 2011, the decision was taken to operate with tight collimator settings, which allowed for a more aggressive squeeze to a $\beta^*$ of 0.6 m. The injectors continued to provide exceptional quality beam and routinely delivered $1.7 \times 10^{11}$ protons per bunch. Peak luminosity increased to near its maximum pretty quickly. This was followed by determined and long running attempts to improve peak performance. This was successful to a certain extent, revealing some interesting issues at high bunch and total beam intensity, but had little effect on integrated rates. Beam instabilities, although never debilitating, were a reoccurring problem and there were phases when they cut into operational efficiency. Nonetheless by the middle of 2012, another 6 $\text{fb}^{-1}$ had been delivered to both ATLAS and CMS. Combined with the 2011 dataset, this paved the way for the announcement of the Higgs boson discovery on $4^{\text{th}}$ July 2012.

It was a very long operational year and included the extension of the proton-proton run until December, resulting in the shift of a four week proton-lead run to 2013. Integrated rates were healthy at around the 1 $\text{fb}^{-1}$ per week level and this allowed a total of around 23 $\text{fb}^{-1}$ to be delivered to both ATLAS and CMS in 2012.

### 1.4. *Long Shutdown 1 (LS1)*

The primary aim of LS1 (2013 to 2014) was the consolidation of the superconducting splices in the magnet interconnects following the incident of 2008. The successful completion of this work allowed, in principle, the current in the main dipole and quadrupole circuits to be increased to the nominal value for 7 TeV operation. The subsequent main dipole magnet training campaign confirmed systematic de-training and the need for a very long training programme to get to 7 TeV, and the decision was made to operate the machine at a beam energy of 6.5 TeV during Run 2. However, a new problem with metallic debris in the diode boxes surfaced during the magnet training after LS1 and led to the decision to operate the machine at a beam energy of 6.5 TeV in order to minimize the number of training quenches. Besides splice consolidation, a significant amount of maintenance and other consolidation work was performed on all accelerator systems.

## 2. Overview of Run 2

Important milestones were reached by the LHC during Run 2 and these included the demonstration of reliable operation with 6.5 TeV beams and exploitation with 25 ns bunch spacing, with over 2500 bunches. Luminosity achieved surpassed the design luminosity of $1 \times 10^{34}$ cm$^{-2}$s$^{-1}$, reaching a peak of $2.1 \times 10^{33}$ cm$^{-2}$s$^{-1}$. Around 160 fb$^{-1}$ was delivered to ATLAS and CMS, along with 6.7 fb$^{-1}$ to LHCb and 33 pb$^{-1}$ to ALICE.

### 2.1. *2015*

The principle aims were to re-commission the machine without beam following the major consolidation and upgrades that took place during LS1, and, from a beam perspective, to safely establish operations at 6.5 TeV with 25 ns bunch spacing. The beam configuration targeted was close to nominal i.e. 25 ns bunch spacing with around 2800 bunches of near nominal bunch intensity ($1.15 \times 10^{11}$ protons per bunch). A relatively relaxed $\beta^*$ of 80 cm in ATLAS and CMS was chosen to provide some aperture margin in the Inner Triplets and thereby less rigorous demands on the collimator settings were required to protect said aperture.

Recommissioning at 6.5 TeV with a bunch spacing of 25 ns was anticipated to be more of a challenge than previous operations at 4 TeV with 50 ns beams. The increased energy implies lower quench margins and thus lower tolerance to beam loss. The hardware (beam dumps, power converters, magnets) is pushed closer to maximum with potential knock-on effects to availability. 25 ns beam was anticipated to have significantly higher electron-cloud than that experienced with 50 ns. It also implies higher total beam current and also higher intensity per injection. UFOs

("Unidentified Falling Objects") are micrometer sized dust particles that lead to fast, localized beam losses when they interact with the beam. The phenomenon had already appeared during Run 1 and they were expected to become more of an issue at higher energy. All of these factors came into play, making 2015 a challenging year.

The LHC suffered from electron cloud (see below). This led to degradation of beam quality, and power deposition onto the arc beam screens. Provoked, beam-induced, high levels of electron cloud is able to mitigate the issue to a large extent via a process known as scrubbing.

Two scrubbing runs delivered good beam conditions for around 1500 bunches per beam after a concerted campaign to re-condition the beam vacuum. However, electron cloud, as anticipated, was still significant at the end of the scrubbing campaign.

The initial 50 ns and 25 ns intensity ramp-up phases were tough, having to contend with a number of issues, including magnet circuit earth faults, UFOs, an unidentified aperture restriction in a main dipole, and radiation affecting specific electronic components in the tunnel. Combined, these problems made operations difficult during this phase but nonetheless the LHC was able to operate with up to 460 bunches and to deliver some luminosity to the experiments, albeit with poor efficiency.

The second phase of the ramp-up, following a technical stop at the start of September, was dominated by the electron cloud generated heat load and the subsequent challenge for the cryogenics system, which had to wrestle with transients and operation close to their cooling power limits. Consequently, the ramp-up in number of bunches was slow but steadily culminating, until the final figure of 2244 bunches per beam for 2015.

The overall machine availability was respectable with around 32% of the scheduled time spent in Stable Beams during the final period of proton-proton physics from September to November. By the end of the 2015 proton run, 2244 bunches per beam were giving peak luminosities of $5.5 \times 10^{33}$ cm$^{-2}$s$^{-1}$ in the high luminosity experiments with a total delivered integrated luminosity of around 4 fb$^{-1}$ delivered to both ATLAS and CMS. Levelled luminosity of $3 \times 10^{32}$ cm$^{-2}$s$^{-1}$ in LHCb and $5 \times 10^{30}$ cm$^{-2}$s$^{-1}$ in ALICE was provided throughout the run.

### 2.2. *2016–2018*

**2016** started with four weeks of relatively smooth commissioning with beam with the machine fully validated for $\beta^* = 40$ cm. The first part of the operating period was hit by a number of serious problems in both the LHC and the injectors — in particular a leak from a cooling circuit to the beam vacuum in the SPS beam dump which limited the beam intensity to the LHC. However, after recovering from the main LHC's problems, things progressed well. The number of bunches was increased to 2240 per beam — the maximum with the SPS limit of 72 bunches per injection. A bunch population of $1.1 \times 10^{11}$ gave a peak luminosity of $\approx 8 \times 10^{33}$ cm$^{-2}$s$^{-1}$. Design luminosity was reached on the 26$^{th}$ June thanks to the reduced $\beta^*$ and lower transverse beam sizes from the injectors, following significant effort to optimise beam brightness via: continuous optimisation; the change of the PS Booster's working point; and the deployment of the batch compression, merging and splitting (BCMS) scheme in the PS.[1] As a result, peak luminosity increased by around 20%, reaching a new record of $1.2 \times 10^{34}$ cm$^{-2}$s$^{-1}$.

The smaller emittances allowed the reduction of the crossing angle from 370 $\mu$rad to 280 $\mu$rad and a concomitant increase in the geometrical reduction factor from around 0.59 to 0.7. Performance was also helped by the use of reduced bunch length in Stable Beams. Thus, despite the limit in the number of bunches and a limit in bunch intensity from injection kicker vacuum issues, a peak performance of 40–50% over nominal was obtained.

2016 was also blessed by unprecedented machine availability: the machine was available for operation for 72% of the time scheduled for physics. Overall Stable Beam efficiency was around 49% (compared to 36% in 2012, and 30% during the short production period in 2015).

**2017** saw a further reduction in beam size at the interaction point ($\beta^*$ = 30 cm), which, together with small beams from the injectors, gave a peak luminosity of $2.1 \times 10^{34}$ cm$^{-2}$s$^{-1}$, the maximum possible given the cryogenic limit for cooling the triplet magnets from the heat load of the luminosity debris. Despite the effects of an accidental ingress of air into the beam vacuum during the winter technical stop, referred to as "16L2" after the location of the contamination, around 50 fb$^{-1}$ was still delivered to ATLAS and CMS.

**2018** essentially followed the set-up of 2017 with a squeeze with ATS optics[2] to 30 cm in ATLAS and CMS. Soon after the intensity ramped up, the debilitating effects of 16L2 returned, limiting the maximum bunch intensity to approximately $1.2 \times 10^{11}$ protons per bunch.

Despite the limitation by 16L2, the peak luminosity was systematically close to $2 \times 10^{34}$ cm$^{-2}$s$^{-1}$ and somewhat more integrated luminosity was possible thanks to the levelling strategy pursued:

- continuous crossing angle reduction ("anti-levelling") in Stable Beams, from an initial 160 $\mu$rad smoothly to 130 $\mu$rad as a function of the beam current;
- $\beta^*$ levelling: for the first time the LHC was operated with a dynamically changed optics in Stable Beams, with the $\beta^*$ in ATLAS and CMS being reduced from 30 cm to 27 cm to 25 cm while colliding.

## 3. Performance

### 3.1. *Run 1*

One of the main features of operations in Run 1 was the use of the high bunch intensity with 50 ns bunch spacing offered by the injectors. The injector complex has succeeded in delivering beam with significantly higher bunch intensities with lower emittances than nominal. This is particularly significant for the 50 ns beam. Happily the LHC was capable of absorbing these brighter beams, notably from a beam-beam perspective. This also resulted in increased pile-up for the high luminosity experiments, which they successfully dealt with.

The corresponding values for the main luminosity related parameters at the peak performance of the LHC through the years are shown in Table 1. The design report values are shown for comparison. Remembering that the beam size is naturally larger at lower energy, it can be seen that the LHC has achieved 77% of design luminosity at 4 sevenths of the design energy with a $\beta^*$ of 0.6 m (cf. design value of 0.55 m) with half the nominal number of bunches.

Table 1. Run 1: Proton performance related parameter overview.

| Parameter | 2010 | 2011 | 2012 | Design |
|---|---|---|---|---|
| Energy [TeV] | 3.5 | 3.5 | 4 | 7 |
| Bunch spacing [ns] | 150 | 75/50 | 50 | 25 |
| Number of bunches | 368 | 1380 | 1380 | 2808 |
| Bunch population ($1 \times 10^{11}$) | $1.2 \times 10^{11}$ | $1.45 \times 10^{11}$ | $1.7 \times 10^{11}$ | $1.15 \times 10^{11}$ |
| $\beta^*$ in IP 1 and 5 [m] | 3.5 | 1.0 | 0.6 | 0.55 |
| Normalised emittance ($\mu$m) | $\approx$2.0 | $\approx$2.4 | $\approx$2.5 | 3.75 |
| Peak luminosity [$\text{cm}^{-2}\text{s}^{-1}$] | $2.1 \times 10^{32}$ | $3.7 \times 10^{33}$ | $7.7 \times 10^{33}$ | $1 \times 10^{34}$ |
| Pileup | 4 | 17 | 37 | 19 |
| Stored beam energy [MJ] | $\approx$28 | $\approx$110 | $\approx$140 | 362 |

### 3.2. *Run 2*

Following a conservative and indeed difficult 2015, peak luminosity in ATLAS and CMS was resolutely pushed throughout the run, principally by:

- a staged reduction of the $\beta^*$ down to 30 cm at the start of Stable Beams;
- operational use of luminosity levelling via separation, crossing angle reduction and change of $\beta^*$ — all during Stable Beams;
- provision of high-brightness beams from the injectors (BCMS).

This resulted in a peak luminosity of over twice design and was in fact limited there by the cryogenic cooling capacity of the inner triplets.

Table 2. Run 2: Proton performance related parameter overview.

| Parameter | 2015 | 2016 | 2017 | 2018 |
|---|---|---|---|---|
| Energy (TeV) | 6.5 | 6.5 | 6.5 | 6.5 |
| No. of bunches | 2244 | 2220 | 2556 - 1868 | 2556 |
| No. of bunches per train | 144 | 96 | 144 - 128 | 144 |
| Bunch population ($1 \times 10^{11}$) | 1.2 | 1.25 | 1.25 | 1.1 |
| $\beta^*$ [cm] in IP 1 and 5 [cm] | 80 | | 40 | 40→25 |
| Normalised emittance [$\mu$m] | 2.6–3.5 | 1.8–2 | 1.8–2.2 | 1.8–2.2 |
| Peak Luminosity [$cm^{-2}s^{-1}$] | $0.6 \times 10^{34}$ | $1.5 \times 10^{34}$ | $2.0 \times 10^{34}$ | $2.1 \times 10^{34}$ |
| Half Crossing Angle ($\mu$rad) | 185 | 185 → 140 | 150 → 120 | 160 → 130 |

This peak performance was accompanied by impressive availability and a low level of premature dumps following a concerted program of measures outlined in more detail below (5).

The resultant integrated luminosity delivered to ATLAS and CMS is shown in Fig. 1.

CMS's peak luminosity by day is shown in Fig. 2. This illustrates nicely the results of all the measures outlined above.

An interesting snapshot of the LHC's overall performance during Run 1 and Run 2 is given by ATLAS's collection of performance records as of the end of 2018 — see Fig. 3.

### 3.3. *Other users*

Throughout Run 1 and Run 2, the operational flexibility of the LHC has allowed the pursuit of a rich variety of physics programmes ranging through lead-lead, lead-proton, xenon-xenon, and an interesting, and sometimes demanding, forward physics programme.

The time limited ion programme inevitably represents a challenge for LHC operations.[3] The team has to commission new configurations and

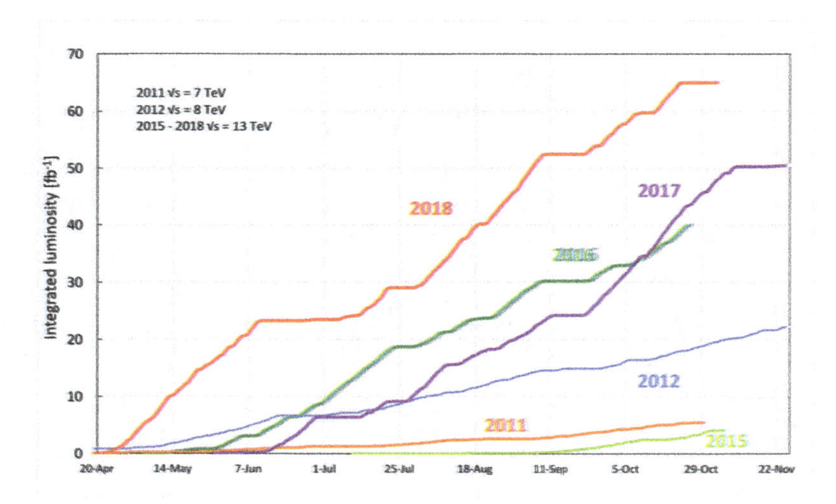


Fig. 1. Average integrated luminosity delivered to ATLAS and CMS during Run 1 and 2.

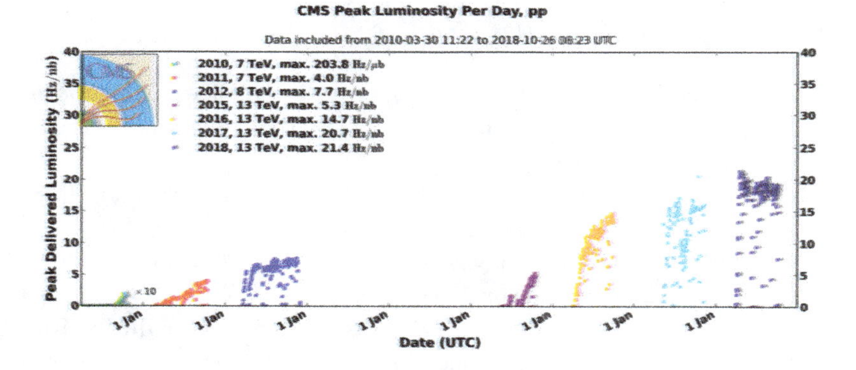


Fig. 2. CMS peak luminosity by day 2010–2018.

provide stable physics operation within a month, and still meet demanding requirements from the experiments which include multiple changes of beam conditions (intensity ramp-up, solenoid polarity reversal, crossing angle reversal, low/high/levelled luminosity, special beam energies, luminosity scans). Nonetheless, heavy-ion operation of LHC has surpassed initial expectations, both quantitatively (3.5 times design luminosity after about 10 weeks of Pb-Pb operation since 2010) and qualitatively (asymmetric p-Pb

| Record | Value | Date |
|---|---|---|
| Peak Stable Luminosity Delivered | $2.10 \times 10^{34}$ cm$^{-2}$s$^{-1}$ | 06.05.18 04:19 |
| Maximum Average Events per Bunch Crossing* | 88.6 | 26.10.18 08:11 |
| Maximum Stable Luminosity Delivered in one fill | 766.8 pb$^{-1}$ | 06.11.17 03:35 |
| Maximum Stable Luminosity Delivered in one day | 912.4 pb$^{-1}$ | 22.07.18 |
| Maximum Stable Luminosity Delivered for 7 days | 5.182 fb$^{-1}$ | 2 to 8 September, 2018 |
| Longest Time in Stable Beams for one fill | 1 day, 6 hrs, 4 min | 09.07.18 22:59 |
| Longest Time in Stable Beams for one day | 1 day, 0 min | 10.07.18 |
| Longest Time in Stable Beams for 7 days | 4 days, 22 hrs, 27 min | 17 to 23 October, 2018 |
| Fastest Turnaround to Stable Beams | 1 hr, 46 min | 14.10.18 |
| Maximum Colliding Bunches | 2544 | 05.05.18 |
| Maximum Charge per Bunch Colliding* | $1.83 \times 10^{11}$ | 26.10.18 08:11 |
| Maximum Charge per Beam Colliding | $3.08 \times 10^{14}$ | 09.08.17 23:45 |
| Maximum Total Charge per Beam | $3.09 \times 10^{14}$ | 09.08.17 23:45 |
| Average Specific Luminosity | $6.94 \times 10^{30}$ cm$^{-2}$s$^{-1}$($10^{11}$ p)$^{-2}$ | 08.08.18 03:47 |

Fig. 3. LHC performance records at the end of 2018 as noted by ATLAS. * indicates a record achieved during machine development.

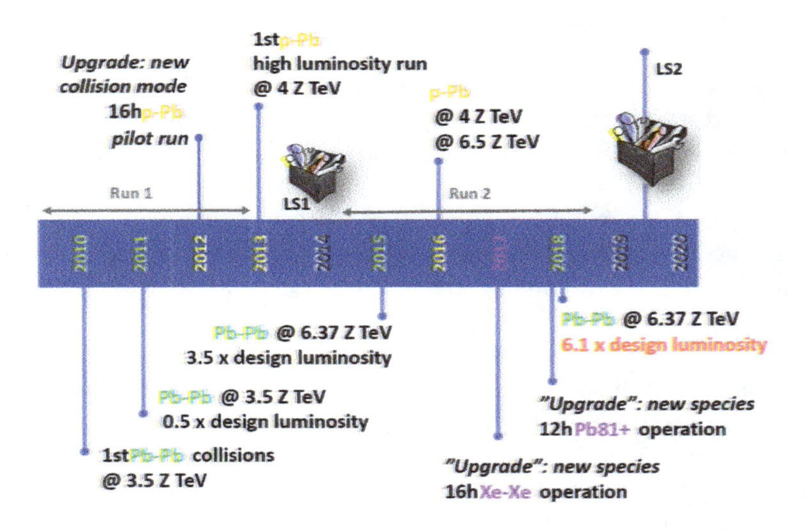


Fig. 4. Timeline of the heavy-ion runs during Run 1 and Run 2. Figure courtesy John Jowett and Michaela Schaumann.[4]

collisions, unforeseen in the design, have yielded almost 6 times their nominal luminosity, as well as a rich harvest of unexpected physics results). The fact that the LHC was able to be recomissioned rapidly and efficiently in multiple new configurations is testament to the understanding and level of control that has been established; the salient points are summarized below.

## 4. Overview of LHC operational characteristics

The performance described above is on the back of some excellent system performance and some fundamental operational characteristics of the LHC. Very good understanding of the beam physics and a good level of operational control has been established and the following features related to beam-based operation may be elucidated.

- The linear optics is well measured and is remarkably close to the machine model. The bare beta-beating is acceptable and has been corrected to excellence. The availability of multi-turn orbit measurements and impressive analysis tools should be noted.
- There is excellent single beam lifetime and on the whole the LHC enjoys very good vacuum conditions.
- Head-on beam-beam is not a limitation, although long-range beam-beam has to be taken seriously with enough separation at the long-range encounters guaranteed by sufficiently large crossing angles. The tolerance to high head-on beam-beam tune shifts can be partially attributed to: well-corrected lattice errors, via both an excellent magnet model and a superb optics measurement and correction programme; low external noise, and other perturbations. A full analysis may be found at.[5]
- Better than nominal beam intensity and beam emittance is delivered by the injectors and it has been proven that it is possible to collide nominal bunch currents with smaller than nominal emittances with no serious problems.
- Collective effects have been seen with high bunch intensities and with nominal bunch intensities in the presence of electron cloud. Single and coupled bunch instabilities have been suppressed using a range of tools (high chromaticity, Landau damping octupoles and transverse feedback).
- There is better than expected aperture due to good alignment and respect of mechanical tolerances.
- There is excellent field quality, as well as good correction of non-linearities. The magnetic machine is well understood and the modelling of all magnet types has delivered an excellent field description at all energies. This model includes persistent current effects which have been fully corrected throughout the cycle.
- A strict pre-cycling regime means that the magnetic machine is remarkably reproducible. This is reflected in the optics, orbit, collimator set-up, tune and chromaticity. Importantly, orbit stability (or the ability to

consistently correct back to a reference), means that collimator set-up remains good for a year's run.
- There is low tune modulation, low power converter ripple, and low RF noise. Power converters are delivering remarkably stable and accurate currents ranging from single digits to several thousand amps. Tracking between power converters in the ramp and squeeze is exceptional and the whole system is complemented by a very good front-end control system.
- Efficient, stable, operating procedures and supporting software are in place.

## 5. Operational cycle and availability

The nominal operation cycle provides the framework driving luminosity production. Given the high stored beam energy, the nominal cycle must be fully mastered for effective, safe operation. As of Run 2, the operational cycle has already been well established for both 50 and 25 ns, with bunch population exceeding nominal.

The turnaround time is defined as the time taken to go from the dump of a physics fill at top energy back into colliding beams following a refill. Following concerted efforts over the years, as well as numerous operational improvements, the minimum turnaround time had been reduced to around 110 minutes in 2018 from around 150 minutes in 2016.

Availability is defined as the overall percentage of the scheduled machine time left to execute the planned physics program after removing the total time dedicated to fault resolution. Faults cover an enormous range, from a simple front-end computer reboot to the loss of a cold compressor of the cryogenics system with a corresponding loss of time to operations from 10 minutes to potentially days. Availability has, in general, been excellent considering the size, complexity, and operating principles of the LHC. The percentage of scheduled proton-proton physics time spent delivering collisions to the experiments ("Stable Beams") was around 36% in 2012. Following a prolonged campaign of consolidation and targeted system improvements, the corresponding value in 2017 and 2018 was around 50%. By the end of Run 2, there was good overall system performance and availability based on solid foundations and vigorous follow-up of problems. This is the result of a sustained, targeted effort across the board by all teams, backed by effective fault tracking. Beam related issues such as radiation to electronics, UFOs, and beam induced heating have all been relentlessly tackled.

Operations also depend heavily on the superb performance of machine protection and associated systems. These include the beam interlock system, the beam dump system, the beam loss monitors, and the collimation system. There is rigorous machine protection follow-up, qualification, and monitoring; all non-conformities are carefully examined. The importance of this to the success of the LHC thus far cannot be overemphasized and due credit must be given to the teams involved for ensuring the safety of the machine during beam based operation over the two runs.

Remarkable operational flexibility has been demonstrated, and has allowed the team to handle, for example, the slower than expected electron cloud conditioning, and the effects of the accidental air ingress in Sector 12 — the infamous "16L2" that hampered operations in 2017 and 2018.

## 6. Issues

There have inevitably been a number of challenges during the utilization of the LHC. Initially, single event effects (SEEs) caused by beam induced radiation to tunnel electronics was a serious cause of inefficiency. However, this problem had been foreseen and its impact was considerably reduced, following sustained program of mitigation measures. There were several shielding campaigns prior to the 2011 run including relocation "on the fly" and equipment upgrades. The 2011/12 Christmas stop saw some "early" relocation and additional shielding and further equipment upgrades. Further improvement followed an extensive campaign of relocation, shielding, and hardware upgrades during LS1.

### *UFOs*

UFOs (Unidentified Falling Objects) are microscopic particles of the order of 10 microns across. These fall from the top of the vacuum chamber or beam screen, become ionised by collisions with circulating protons and are then repelled by the positively charged beam. While interacting with the circulating protons they generate localised beam loss which may be sufficient to dump the beam or, in the limit, cause a quench. They have now been very well studied and simulated. There were occasional dumps in 2012 following adjustment of BLM thresholds at the appropriate timescales (the beam loss spike caused by a UFO is typically of order 1 ms). With the increase in energy to 6.5 TeV and the move to 25 ns the UFOs become harder (energy) and more frequent (25 ns). Indeed, during the first half of 2015 they were a serious issue but happily there was conditioning and the UFO rate fell to acceptable levels as the year progressed. It should also be noted that it was fortunate that UFO rates have conditioned down, accompanied, as elsewhere, by excellent diagnostics, well thought through mitigation actions and understanding through simulation. The start of Run 3 in 2022 again saw high UFO levels with some loss of conditioning and high rates in magnets that were swapped in during LS2.

### *Beam induced heating*

Beam induced heating has been an issue and essentially all cases have been local and, in some way, due to non-conformities either in design or installation. The guilty parties have been clearly identified. Design problems have affected the injection protection devices and the mirror assemblies of the synchrotron radiation telescopes. Installation problems have occurred in a low number of vacuum assemblies. These singularities have all been addressed and the issue is not expected to be problematic in the long term.

### *Beam instabilities*

Beam instabilities were an interesting problem that dogged operations throughout 2012. It should be noted that this problem paralleled a gentle push in bunch intensity with the peak going into stable beams reaching around $1.7 \times 10^{11}$ protons per bunch i.e. ultimate bunch intensity. In 2015 operations with 25 ns bunch spacing and lower bunch population meant that intrinsically instabilities should have been less of an issue. However, high electron cloud proved to be a driver and defence mechanisms were deployed in the form of high chromaticity, high octupole field strength and the rigorous use of transverse damper system.

### *Electron Cloud*

Electron cloud is the result of an avalanche-like process in which electrons from gas ionisation or photo-emission are accelerated in the electromagnetic field of the beam and hit the beam chamber walls with energies of few hundred electronvolts, producing more electrons. The electron impact on the chamber wall causes gas desorption as well as heat load for the cryogenic system in the cold regions. High electron densities in the beam chamber can lead to beam oscillations and blow-up of the particle bunches due to the electromagnetic interaction between electrons and protons.[6] Electron bombardment of a surface has been proven to reduce drastically the secondary electron yield (SEY) of a material. In a process known as scrubbing, deliberate invocation of high electron cloud with beam provides a means to reduce or suppress subsequent electron cloud build-up.

Although electron cloud was not an issue with 50 ns beam, operating with 25 ns beam proved to be a challenge in 2015, and extensive scrubbing — both dedicated at low energy and while delivering collisions to the experiments — was required. Conditioning thereafter has been slow and the heat load from electron cloud to the cryogenics system remained a limitation in 2018.

## 7. Conclusions

After seven full years of operation, in the beam parameter regime concerned, the extended LHC team has managed to evolve an impressive mastery of the LHC and the delivery of the requisite beam from the injectors. A concise summary of the salient observations is attempted below.

- Good peak luminosity via exploitation of all available parameters ($\beta^*$, bunch population, bunch length, crossing angle, transverse emittance).
- Stunning availability following sustained effort from hardware groups accompanied by effective fault tracking.
- Few premature dumps allowing long fills: the UFO rate conditioned down and radiation to electronics effects have been largely mitigated, again after a sustained and successful campaigns.
- Excellent and improved system performance across the board, for example, the new developments of the transverse damper system; collimator alignment software; improved injection kicker performance via hardware modifications.
- The magnets, circuits and associated systems are behaving well at 6.5 TeV.
- Good beam lifetime through injection, ramp, and squeeze with tight control of tune and closed orbit, reflecting that operationally things are very well under control.
- Excellent luminosity lifetime in general with only moderate emittance blow-up in Stable Beams and minimal non-luminosity beam loss after the first hour or so.
- Well established and tuned magnet model, good compensation of persistent current decay and snapback, which, when coupled with a strict magnet cycling give excellent magnetic reproducibility.

- The optics of the machine has been measured and corrected to a impressive level, both linear and higher orders, and a superb level of understanding has been established.
- Aperture is fine and compatible with the collimation hierarchy.
- The collimation system has consistently demonstrated excellent performance and impressive robustness.
- A reliable and well designed machine protection system coupled with a disciplined regime has assured safe exploitation.

2016 was really the first year when it all came together: injectors; operational efficiency; system performance; understanding and control; and availability. In 2017, and 2018, the LHC was able to build on this to move into a full exploitation regime, accompanied, as always, by continued efforts to improve integrated luminosity delivery.

The LHC has moved haltingly from commissioning to full exploitation, and is now enjoying the benefits of the decades long international design, construction, and installation efforts — it's clear that the foundations and the fundamentals are solid. Its present performance is worthy reflection of this effort and the huge amount of experience and understanding gained and fed-forward over the last years. Remarkably, not only can a 27 km superconducting collider work, it can work well.

## References

1. H. Damerau, S. Hancock, A. Lasheen, and D. Perrelet, RF Manipulations for Special LHC-Type Beams in the CERN PS. In *9th International Particle Accelerator Conference* (6, 2018). doi: 10.18429/JACoW-IPAC2018-WEPAF063.
2. S. Fartoukh, M. Solfaroli, J. Coello de Portugal, A. Mereghetti, A. Poyet, and J. Wenninger, Achromatic telescopic squeezing scheme and by-products: From concept to validation, *Physical Review Accelerators and Beams*. **24** (02, 2021). doi: 10.1103/PhysRevAccelBeams.24.021002.
3. J. Jowett, Colliding Heavy Ions in the LHC. p. TUXGBD2. 6 p (2018). doi: 10.18429/JACoW-IPAC2018-TUXGBD2. URL `https://cds.cern.ch/record/2648704`.
4. J. M. Jowett and M. Schaumann, Overview of Heavy Ions in LHC Run 2. pp. 15–25. 11 p (2019). URL `https://cds.cern.ch/record/2750273`.
5. S. V. Furuseth and X. Buffat, Modeling of nonlinear effects due to head-on beam-beam interactions, *Physical Review Accelerators and Beams*. 21 (8): 081002 (Aug, 2018). doi: 10.1103/PhysRevAccelBeams.21.081002.
6. A. Romano, O. Boine-Frankenheim, X. Buffat, G. Iadarola, and G. Rumolo, Electron cloud buildup driving spontaneous vertical instabilities of stored

beams in the large hadron collider, *Phys. Rev. Accel. Beams*. **21**, 061002 (Jun, 2018). doi: 10.1103/PhysRevAccelBeams.21.061002. URL `https://link.aps.org/doi/10.1103/PhysRevAccelBeams.21.061002`.

© 2024 The Author(s)
https://doi.org/10.1142/9789811280184_0003

# Chapter 3

# The Higgs Boson Discovery

Christoph Paus* and Stefano Rosati†
*Massachusetts Institute of Technology (MIT)*
†*Istituto Nazionale di Fisica Nucleare (INFN) Sezione di Roma*

The Higgs boson discovery has been the highlight of particle physics in the last few decades and the crowning achievement of the LHC to date. It was accomplished in its initial running period, Run 1 (2010–2012). In this chapter we will review the status of the standard model Higgs boson searches at the start of LHC, and how the ATLAS and CMS experiments prepared. Analyses of the five main decay channels that lead to the simultaneous discovery, announced in the CERN seminar on July 4th, 2012, will be discussed. We will give an overview of the preparation for the data taking, the physics organization put in place, the analyses of the discovery channels, and finally discuss the results that lead to the discovery.

## 1. Introduction

The discovery of the Higgs boson, announced by the ATLAS and CMS experiments on July 4th 2012, marks an important milestone in particle physics. The scalar Higgs field and the spontaneous symmetry breaking mechanism[1–6] and the related particle were suggested to exist since the mid 1960's as a way to allow elementary particles to acquire their masses in a natural way by ensuring the standard model (SM) Lagrangian remain locally gauge invariant. At its discovery the Higgs boson was the last SM particle that had not yet been experimentally confirmed and its central role is further underlined by the fact that it is the only elementary *scalar* known to date.

The Higgs boson discovery had long been anticipated even before the start of Run 1, in fact the ATLAS and CMS experiments had detailed

This is an open access article published by World Scientific Publishing Company. It is distributed under the terms of the Creative Commons Attribution 4.0 (CC BY) License.

analyses performed on Monte Carlo simulations before their detectors were even completed. The sensitivity of the LHC experiments required to detect the Higgs boson served as the benchmark for the detector design. A number of adjustments to detectors and analyses were made when in the 2000's the electroweak precision data more and more clearly indicated that the expected Higgs boson mass should be rather small, below about 200 GeV.[7]

## 2. Getting ready for the analysis

At the start of Run 1, ATLAS and CMS carefully reviewed the portfolio of analyses to be performed urgently and decided on a set of high priority analyses. Apart from obvious new physics searches, the Higgs boson search with five main channels were at the top of the priority list. The most relevant decay channels for the Higgs boson are the decays: $H \to ZZ^* \to 4\ell$, $H \to \gamma\gamma$, $H \to WW \to \ell\nu\ell\nu$, $H \to \tau\tau$, and $H \to b\bar{b}$. The decay channels are in descending order of their expected significance at 125 GeV, but the sensitivity depends on the mass value of the Higgs boson and the level of sophistication of the various analyses implementations. The expected significance of these analyses is depicted in Fig. 1 for the CMS experiment and ATLAS had very similar expected sensitivities.

In preparation for the Higgs boson search SM processes like $J/\psi$, $Z$ boson and $W$ boson decays to leptons were measured and then used to tune the lepton selections and efficiencies. It follows a short description of triggers, object reconstructions and the related key performances.

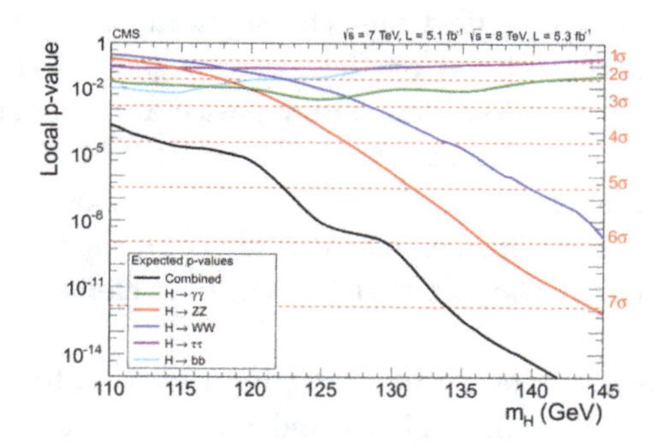


Fig. 1. Expected sensitivity for the five main Higgs boson decay channels for the CMS experiment.

### 2.1. *Trigger*

The online trigger selection was based on the identification of candidate muons, electrons, and photons with the lowest possible $p_T$ threshold given the Run 1 LHC instantaneous luminosity. Either single lepton or dilepton triggers were used for the $H \to ZZ^* \to 4\ell$ and $H \to WW \to \ell\nu\ell\nu$ channels, while diphoton calorimetric triggers were used for the $H \to \gamma\gamma$ channel.

### 2.2. *Reconstruction of the input objects*

The basic reconstructed objects used in the Higgs discovery channels were leptons (electrons or muons), photons, and missing energy $E_T^{miss}$. The Higgs boson decay channels to tau leptons and bottom quarks were also published by the CMS collaboration but did not significantly contribute to the observation.

In ATLAS muons were identified and reconstructed combining a muon spectrometer track or segment with a matching track in the inner detector (ID).[8,9] Photon and electron candidates were reconstructed from EM calorimeter energy clusters. Electrons were reconstructed by matching ID tracks to clusters satisfying a set of criteria aiming at the identification of electromagnetic showers.[10] Photons identification was based on shower shapes in the EM calorimeter and on energy leakage into the hadronic calorimeter.[11,12] A cut-based and a neural network selection was used for photon identification in 7 TeV and 8 TeV data, respectively. In CMS, all physics objects were reconstructed with the "particle-flow" event description algorithm,[13,14] which uses an optimized combination of all subdetector information to best reconstruct each particle (muons, electrons, photons, charged and neutral hadrons). Multivariate approaches were used to refine the initial loose selections of muons, electrons and photons. Both experiments were reconstructing jets using the anti-$k_t$ algorithm[15] to cluster the reconstructed objects. The missing energy was defined as the negative vector sum of the transverse momenta of the reconstructed objects, including muons, electrons, photons, jets and clusters of calorimeter cells not associated to these objects. Energy depositions and tracks from overlapping proton-proton collisions ("pileup") and the underlying event were carefully accounted for to ensure optimal selection efficiencies. In particular, lepton or photon isolation requirements are very sensitive to pileup if not designed properly.

For both ATLAS and CMS, data-driven methods were used to assess the reconstruction performances of all the objects used in the analyses.

Scaling factors were applied to the MC to achieve good representation of the actual performance in the data. For leptons, the reference channels were the $Z$ and $J/\psi$ decays to pair of muons or electrons. Energy scales and resolutions were derived from fits of the $Z$ peak.[9,16] The so-called "tag and probe" method was applied to Z bosons to determine reconstruction and identification efficiencies for electrons and muons. An example of the measured reconstruction and identification efficiencies is shown in Fig. 2(a) for muons in the ATLAS experiment and in Fig. 2(b) for electrons in the CMS experiment.

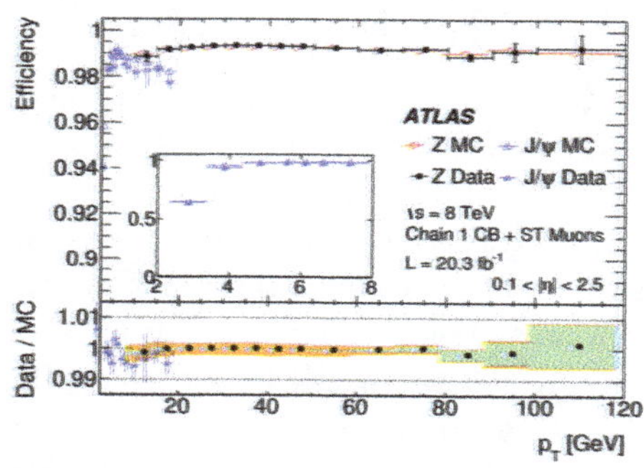


(a) ATLAS muon reconstruction efficiency, as a function of the muon $p_{\mathrm{T}}$. The inset shows the efficiency in the low $p_{\mathrm{T}}$ region. The bottom panel shows the ratio between the efficiencies in data and those expected from the MC simulation.

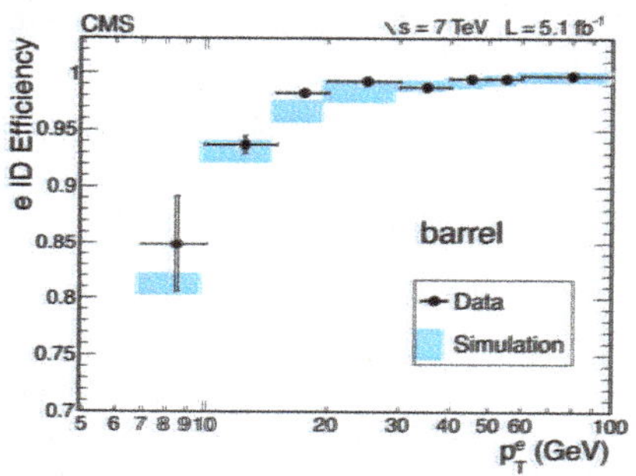


(b) CMS reconstruction efficiencies for electrons in the barrel, at 7 TeV center-of-mass energy. The points with error bars represent the measurements from data, while the histogram shows the efficiency obtained from MC. The shaded region represents the combined statistical and systematic uncertainties.

Fig. 2. An example of the lepton reconstruction efficiencies obtained from data-driven methods, compared to those expected from the MC simulation.

## 3. Analysis Organization

The Higgs search program was performed in a range consistent with the sensitivity of the individual decay channels but covering the entire region from about 100 GeV up to about 1 TeV, where the Higgs boson mass was not yet excluded by previous experiments. The analyses in ATLAS and CMS were optimized in an unbiased fashion, i.e., not looking at the signal region but only using MC simulation samples. The MC predictions were carefully normalized and constrained using background control regions, to estimate the background contamination in the signal region of each selection and to validate the MC description and relevant detector effects of the background processes. Data in the signal regions were only analyzed once the analysis optimization was completed and widely discussed within the working groups, and frozen after an approval procedure. In both experiments, the analysis was organized in working groups, each focusing on one of the highest sensitivity decay channels. Working group meetings were very frequent as well as general meetings for common discussions. Both experiments worked on the analysis optimization and validation until the last few weeks before the discovery. Signal regions remained hidden until the analysis procedures were finalized and approved.

In the following we begin with the two most significant channels for the observation. These channels also allow the full reconstruction of the final state and of the mass of the Higgs boson: the $H \to ZZ^* \to 4\ell$ and the $H \to \gamma\gamma$ channel. We finish with the $H \to WW \to \ell\nu\ell\nu$ channel which cannot reconstruct the full Higgs boson due the undetected neutrinos, but due to its large rate improves the combined sensitivity in the low mass region.

### 3.1. *The* $H \to ZZ^* \to 4\ell$ *channel*

In the $H \to ZZ^* \to 4\ell$ channel the experimental signature is a narrow four-lepton mass peak on top of a small background. The main background is the irreducible $Z\,Z^*$ contribution from direct production via $q\,\overline{q}$ and gluon-gluon interaction. Other backgrounds, relevant in particular in the low mass region, are $Z+$jets and $t\bar{t}$ production, where charged lepton candidates arise from leptonic decays of hadrons with $b$- and $c$-quarks, and from jets misidentified as leptons. The analysis selection starts from events with two same-flavor opposite-charge lepton pairs, with all four lepton tracks associated to the same interaction vertex.

Four independent sub-channels, $4e$, $2e\,2\mu$, $2\mu\,2e$, $4\mu$, with different mass resolutions and background compositions, were considered for the discovery analyses. In ATLAS, each electron (muon) had to satisfy $p_{\mathrm{T}} > 7$ GeV ($p_{\mathrm{T}} > 6$ GeV) and be within a fiducial region $|\eta| < 2.47$ ($|\eta| < 2.7$). The corresponding requirements in CMS are $p_{\mathrm{T}} > 7$ GeV and $p_{\mathrm{T}} > 5$ GeV with $|\eta| < 2.5$ ($|\eta| < 2.4$) for electrons and muons, respectively. The two lepton pairs masses corresponding the two $Z$ bosons are important variables to separate signal from background. To reject reducible backgrounds, the lepton isolation and the impact parameter significance were used by both experiments. The irreducible $ZZ^*$ background was determined from the MC simulation, and normalized to the theoretical cross section. Reducible backgrounds were determined from data control regions built by relaxing or reverting some of the identification, isolation or impact parameter requirements.

In ATLAS, a $Z$ boson mass constrained kinematic fit was applied to the lepton pair with the mass closest to the $Z$ mass to improve the four lepton mass resolution. In CMS, a kinematic discriminant was built based on five angles and the masses of the two leptons pairs, which fully describe the kinematics of the final state in its center-of-mass frame. In the hypothesis of a SM-like scalar Higgs boson, the discriminant was defined as the likelihood ratio $K_D = P_{\text{sig}}/(P_{\text{sig}} + P_{\text{bkg}})$.[17]

The distribution of the four-lepton mass in data and in MC, is shown in Fig. 3 for ATLAS (Fig. 3(a)) and CMS (Fig. 3(b)). The peak at the $Z$ boson mass is clearly visible for both experiments, it is however more pronounced in the case of CMS due to the looser requirements on the lepton momenta and on the subleading lepton pair mass.

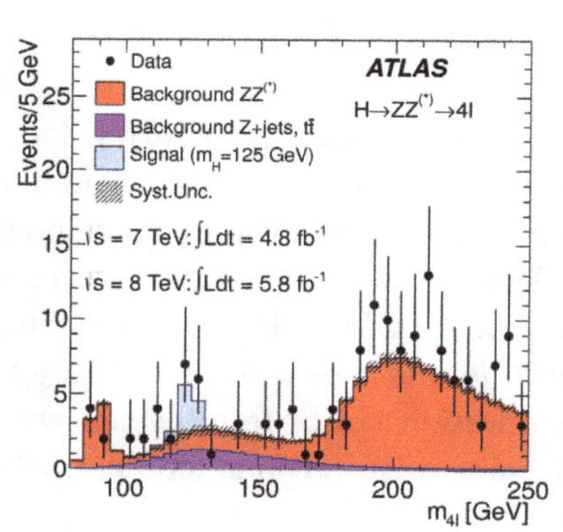


(a) Distribution of the four-lepton mass for the candidates selected by the ATLAS experiment. Points are the data, compared to the histograms that represent the background expectation. The expected signal for a SM Higgs with $m_H = 125$ GeV is also shown.

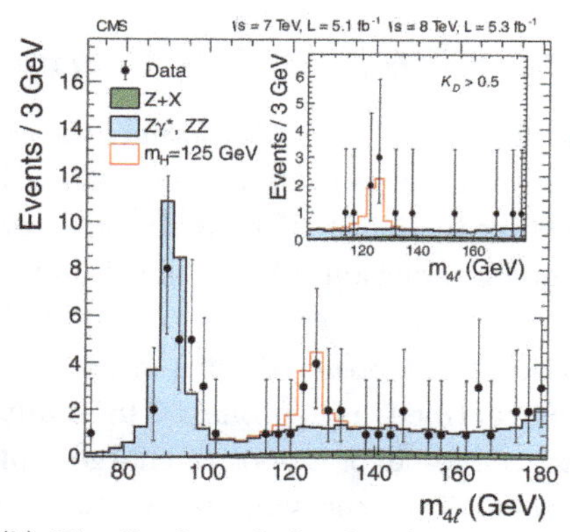


(b) Distribution of the four-lepton mass for the CMS experiment. The inset shows the mass distribution satisfying the requirement on the kinematic discriminant $K_D > 0.5$.

Fig. 3. Distribution of the four-lepton mass for the (a) ATLAS and (b) CMS experiment. The points represent the data, the filled histograms show the backgrounds and the expected signal for $m_H = 125$ GeV.

### 3.2. *The $H \to \gamma\gamma$ channel*

In the $H \to \gamma\gamma$ search channel, the signature is a narrow peak in the diphoton mass distribution. The main background is the irreducible background from SM diphoton production; additional contributions come from gluon plus jet and di-jet production with one or two jets misidentified as photons. In the analyses of both experiments, the events are separated into mutually exclusive categories based on the characteristics of the reconstructed photons and on the additional presence of two jets. In particular, a two-jets category aims to identify events in which the Higgs production has happened through the vector boson fusion process.

The identification of the interaction vertex is critical to keep an optimal resolution for the two photons invariant mass. In the dominant gluon fusion production process of the Higgs boson, it is hard to identify the correct vertex because photons have no tracks. To address this issue, the ATLAS analysis identifies the primary vertex by combining the flight directions of the two photons as reconstructed exploiting the longitudinal segmentation of the electromagnetic calorimeter and its pointing direction measurement, the parameters of the beam spot and the $\Sigma p_{\mathrm{T}}^2$ of the tracks associated to each reconstructed vertex. In the CMS analysis, the primary vertex is identified using a multivariate discriminant which uses, as input, the kinematic properties of the tracks associated to each vertex and the properties of the diphoton kinematics.[18]

The background in each category was estimated from data, by fitting the diphoton mass spectrum with a model selected for each category. Models were chosen to have a good statistical power while minimizing potential biases. The distribution of the diphoton mass for the ATLAS and the CMS experiment are shown in Figs. 4(a) and 4(b), respectively.

### 3.3. *The $H \to WW \to \ell\nu\ell\nu$ channel*

This channel is very sensitive in the Higgs mass region around 160 GeV, just above the threshold for the production of a pair of $W$ bosons, but its sensitivity extends downwards to the lower mass region. Only leptonic $W$ decays are considered because hadronic $W$ decays have a large background. Therefore the final state reconstructed in the detector is characterized by two opposite-charged leptons with high $p_{\mathrm{T}}$ and large $E_{\mathrm{T}}^{\mathrm{miss}}$ due to the presence of the two neutrinos, which cannot be seen in the detector. The signal topology and the background composition depend on the number of jets present in the final state. In order to optimize the signal sensitivity, the

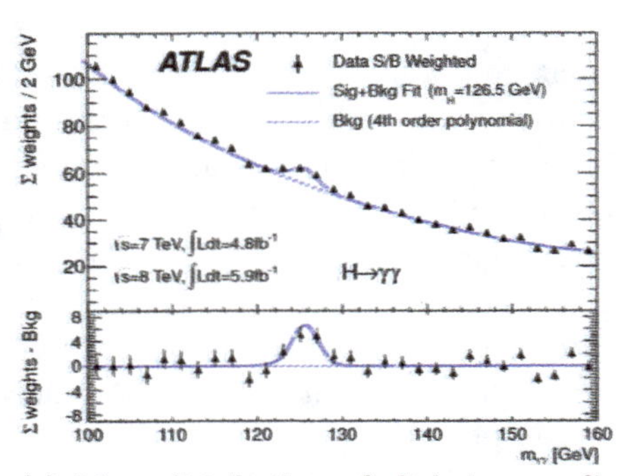


(a) Mass distribution of diphoton candidates of the ATLAS experiment. Events are weighted as a function of the signal over background ratios of their corresponding categories, as described in the text. The bottom inset shows the residual of the data with respect to the fitted background.

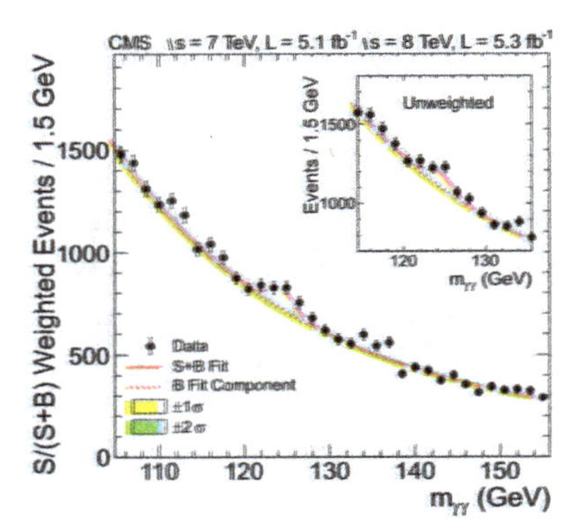


(b) Mass distribution of diphoton candidates of the CMS experiment. Each event is weighted by the signal over signal plus background value of its corresponding category as described in the text. The inset shows the unweighted mass distribution in the region around 125 GeV.

Fig. 4. Distributions of the diphoton mass for the (a) ATLAS and (b) CMS experiment.

event selection criteria were optimized separately for the zero-jet, one-jet and two-jet categories, where the two-jet category includes all events with two or more reconstructed jets.

The main backgrounds are the non-resonant $W\ W$, $t\bar{t}$ and $W\ t$. Drell-Yan lepton pair production, with $E_{\mathrm{T}}^{\mathrm{miss}}$ arising from mis-reconstructed leptons and jets, constitutes a significant and dominant background, for the same-flavor channel. This channel is included in the CMS analysis, while only the different-flavor channels $e\nu\mu\nu$ have been considered in ATLAS, due to the larger Drell-Yan contribution to the same-flavor decay. The main backgrounds were estimated using partially data-driven methods, i.e., normalizing the MC predictions to the data in control regions dominated by each background source. For all jet multiplicities, the distributions considered to test the presence of a signal were the transverse mass $m_T$ defined as: $m_T = \sqrt{(E_{\mathrm{T}}^{\ell\ell} + E_{\mathrm{T}}^{\mathrm{miss}})^2 - |\mathbf{p}_{\mathrm{T}}^{\ell\ell} + \mathbf{E}_{\mathrm{T}}^{\ell\ell}|^2}$ for ATLAS and the dilepton mass $m_{\ell\ell}$ for CMS. Figure 5(a) shows the transverse mass distribution for the zero-jet and one-jet channels together, for the ATLAS experiment. The distribution of $m_{\ell\ell}$ for the CMS experiment is shown in Fig. 5(b) for data and MC.

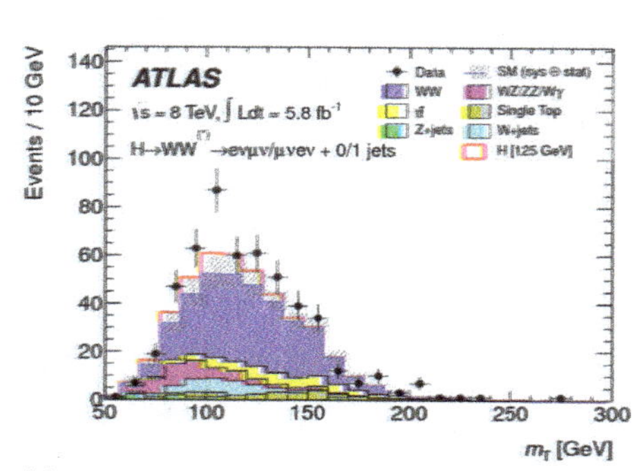


(a) ATLAS distribution of the transverse mass $m_{\mathrm{T}}$ for the $e\mu$ and $\mu e$ events selected by the zero- and one-jet analyses. The signal prediction for a SM Higgs boson with mass 125 GeV is shown as the red histogram stacked on top of the backgrounds. The hashed area indicates the total uncertainty on the background.

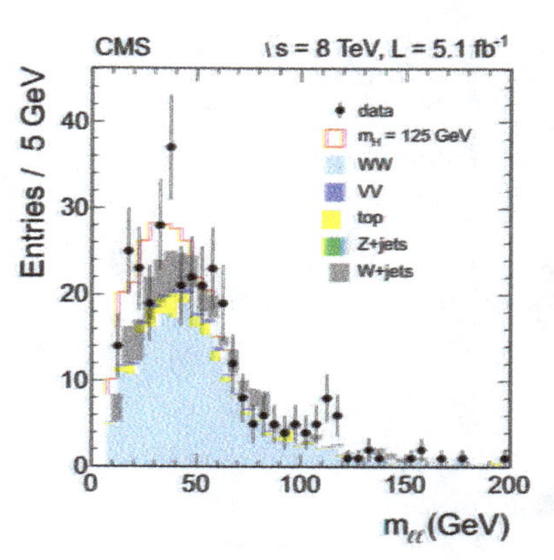


(b) Dilepton mass $m_{\ell\ell}$ distribution for the CMS experiment. The expected signal for a SM Higgs boson with mass 125 GeV is shown as the red histogram stacked on top of the backgrounds.

Fig. 5. Mass distributions for the $H \to WW \to \ell\nu\ell\nu$ Higgs decay channels for the (a) ATLAS and (b) CMS experiments.

## 4. Combination and results

### 4.1. *Statistical procedure*

The statistical procedure used to interpret the analysis results was developed by the ATLAS and CMS collaborations within the LHC Higgs Combination Group.[19–22] The parameter of interest is the cross section times the relevant branching fraction, denoted as signal strength $\mu = \sigma/\sigma_{\mathrm{SM}}$. This means that $\mu = 0$ corresponds to no Higgs boson signal, the background-only hypothesis, while $\mu = 1$ corresponds to the SM Higgs boson signal on top of the background. Exclusion limits are derived based on the $CL_s$ criterion.[23,24]

### 4.2. *Observing a narrow resonance*

The local $p$-values obtained from the combination of all search channels are shown in Figs. 6(a) and 6(b) for the ATLAS and CMS experiment, respectively. The ATLAS experiment combined the most sensitive channels $H \to ZZ^* \to 4\ell$, $H \to \gamma\gamma$ and $H \to WW \to \ell\nu\ell\nu$. The CMS experiment

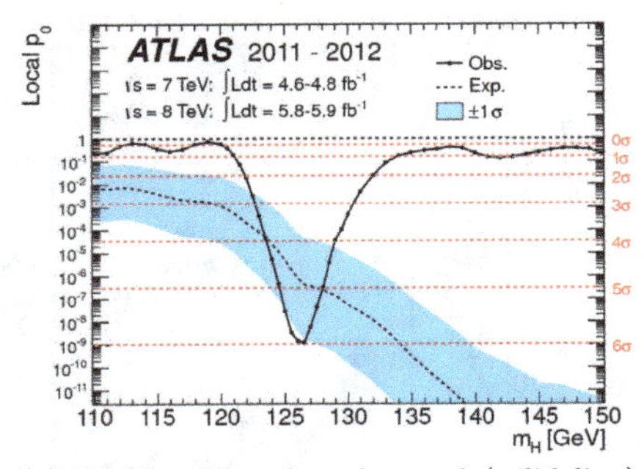


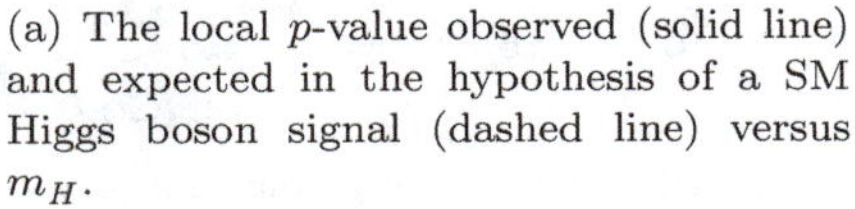
(a) The local $p$-value observed (solid line) and expected in the hypothesis of a SM Higgs boson signal (dashed line) versus $m_H$.

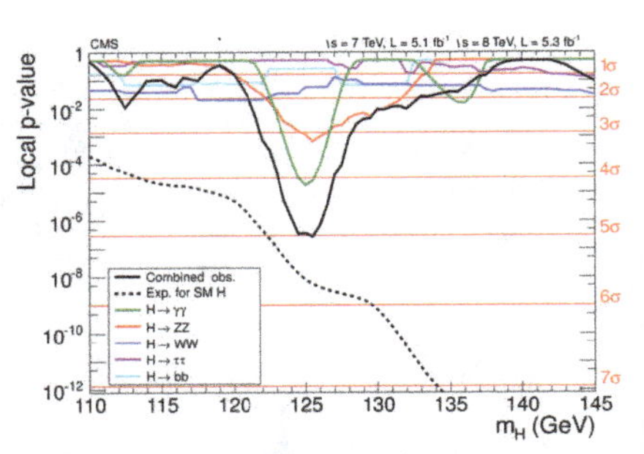


(b) The observed (expected) local $p$-values versus $m_H$ are shown as the black solid (dashed) line. The solid colored lines also show the observed local $p$-values per decay channel.

Fig. 6. Observed and expected local $p$-values in the low $m_H$ region. The horizontal lines indicate the $p$-values corresponding to significances of one to seven standard deviations.

added to those three channels also the $H \to b\bar{b}$ and $H \to \tau\tau$ decay channels for the final combination.

Both experiments observed a clear excess in the region close to $m_H =$ 125 GeV. The excess was dominated by the two high sensitivity and high mass resolution channels and was confirmed by the low mass resolution channels, in particular the $H \to WW \to \ell\nu\ell\nu$ channel. The leading contributions to the discovery came from the $H \to ZZ^* \to 4\ell$ channel for the ATLAS analysis, while the $H \to \gamma\gamma$ channel was most significant for the CMS analysis. From the combination of the $H \to ZZ^* \to 4\ell$, $H \to \gamma\gamma$ and $H \to WW \to \ell\nu\ell\nu$ channels, the ATLAS observed (expected) significance was 6.0 (4.9) standard deviations, while the CMS experiment's corresponding values were 5.0 (5.8) standard deviations for the combination of all five channels. The observed and expected local $p$-values are shown in Fig. 6(a) for the ATLAS experiment and in Fig. 6(b) for the CMS experiment.

While the excess was observed within the context of the SM Higgs boson search, the experiments were rather careful to state what they had found and not jump to conclusions. What was clear was that there was a new particle materializing as a narrow — consistent with detector resolution — resonance. Its observed decay to two photons excluded it from being a spin one particle and thus, it is most probably a spin zero boson. In addition more detailed tests were immediately performed. The compatibility of the observed excess with the expectation from the SM was evaluated by measuring the signal strength $\sigma/\sigma_{SM}$ in each decay channel. The results are shown in Figs. 7(a) and 7(b) for the two experiments. The signal strength depends on the Higgs mass considered at the time, which was $m_H = 126$ GeV for the ATLAS experiment and $m_H = 125.3$ GeV for the CMS experiment. The signal strength values measured by both experiments were compatible with the SM expectation.

These findings lead both experiments to announce the discovery of a new particle, a boson compatible with the expectations of a SM Higgs boson at around 125 GeV.

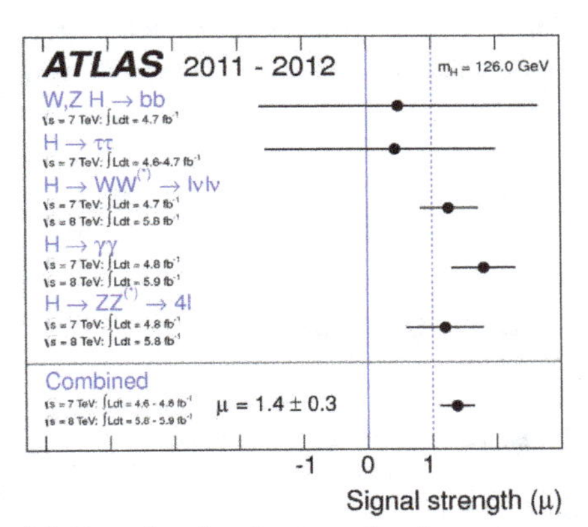


(a) Best-fit signal strengths for $m_H = 126$ GeV observed by the ATLAS experiment for each of the decay channels analyzed, and combined.

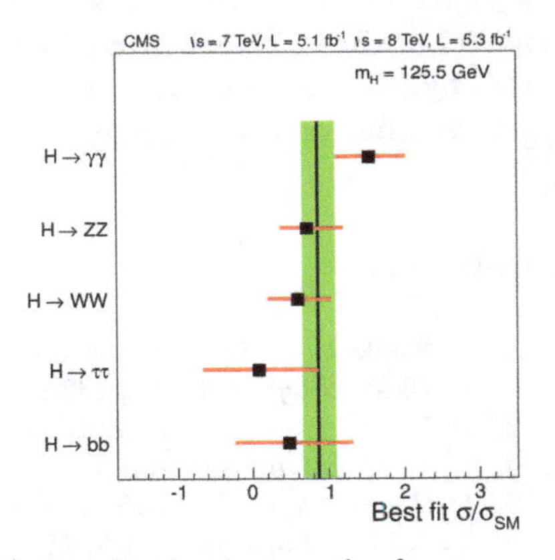


(b) Best-fit signal strengths for $m_H = 125$ GeV observed by the CMS experiment for each of the decay channels analyzed, and combined.

Fig. 7. Summary of the signal strengths for the various channels and the combined analyses per experiment. Both CMS and ATLAS observe a clear signal at five or more standard deviations consistent with the expectations of a SM Higgs boson at a mass around 125 GeV.

## 5. Conclusions

On July 4th 2012, the Large Hadron Collider experiments ATLAS and CMS announced the discovery of a new boson within the context of their standard model Higgs searches, a particle that behaved much like the Higgs bosons which had been hypothesized almost fifty years before. They did it within the first data taking period of the LHC, called Run 1, which operated at 7 and 8 TeV, about half the final planned center-of-mass energy. This new boson for all we know now is the Higgs boson and completed the standard model because at that time it was the last missing particle that had not yet been observed.

Completing the SM might seem like a final step for the outside observer, but explaining it and also addressing all fundamental questions it leaves unexplained are more like a new beginning. Questions at the top of the priority list for particle physicists are: is the Higgs boson really the Higgs boson and could it be a portal to another world of physics? What is the nature of dark matter, which we observe to exist? Why is the matter-antimatter asymmetry in the universe so large? How can gravity be included into our model of the universe? And there are many more questions which particle physicists hope to answer with the extension of the LHC program and future colliders at the energy frontier.

## References


1. F. Englert and R. Brout, *Phys. Rev. Lett.* **13**, 321 (1964).
2. P. Higgs, *Phys. Lett.* **12**, 132 (1964).
3. P. Higgs, *Phys. Rev. Lett.* **13**, 508 (1964).
4. C. H. G.S. Guralnik and T. Kibble, *Phys. Rev. Lett.* **13**, 585 (1964).
5. P. Higgs, *Phys. Rev.* **145**, 1156 (1966).
6. T. Kibble, *Phys. Rev.* **155**, 1554 (1967).
7. t. T. E. W. G. ALEPH CDF D0 DELPHI L3 OPAL SLD Collaborations, the LEP Electroweak Working Group, the SLD Electroweak, and H. F. Groups, *CERN-PH-EP-2010-095*. **arXiv:1012.2367 [hep-ex]** (2010).
8. ATLAS Collaboration, *Phys. Lett. B.* **710**, 383 (2012).
9. ATLAS Collaboration, *Eur. Phys. J. C.* **74**, 3130 (2014).
10. ATLAS Collaboration, *Eur. Phys. J. C.* **72**, 1909 (2012).
11. ATLAS Collaboration, *Phys. Rev. D.* **83**, 052005 (2011).
12. ATLAS Collaboration, *Phys. Rev. Lett.* **108**, 111803 (2012).
13. CMS Collaboration, *CMS Physics Analysis Summary.* **CMS-PAS-PFT-09-001** (2009). `http://cdsweb.cern.ch/record/1194487`.
14. CMS Collaboration, *CMS Physics Analysis Summary.* **CMS-PAS-PFT-10-001** (2010). `http://cdsweb.cern.ch/record/1247373`.
15. M. Cacciari, G.P. Salam and G. Soyez, *JHEP.* **0804**, 063 (2008).
16. CMS Collaboration, *JHEP.* **06**, 081 (2013).
17. CMS Collaboration, *JHEP.* **1204**, 036 (2012).
18. CMS Collaboration, *Phys. Lett. B.* **710**, 403 (2012).
19. ATLAS Collaboration and CMS Collaboration, *ATL-PHYS-PUB-2011-011, CERN-CMS-NOTE-2011-005* (2011).
20. L. Moneta, K. Belasco, K.S. Cranmer, S. Kreiss, A. Lazzaro et al., *PoS ACAT2010.* p. 057 (2010). arxiv:1009.1003 [physics.data-an].

21. K. Cranmer, G. Lewis, L. Moneta, A. Shibata, W. Verkerke, *CERN-OPEN-2012-016* (2012). `http://cdsweb.cern.ch/record/1456844`.
22. W. Verkerke, D. Kirby, *Tech. Rep., SLAC, Stanford, CA* (2003). arXiv:physics/0306116 [physics.data-an].
23. T. Junk, *Nucl. Instrum. Meth. A.* **434**, 435 (1999).
24. A.L. Read, *J. Phys. G.* **28**, 2693 (2002).

© 2024 The Author(s)
https://doi.org/10.1142/9789811280184_0004

# Chapter 4

# Physics Results

Albert De Roeck*, Monica Pepe Altarelli* and Pierre Savard†
*CERN, Geneva, Switzerland
†University of Toronto, Toronto, Canada and TRIUMF, Vancouver, Canada

## 1. Introduction

On the 23rd November 2009, the LHC came alive for the experiments with first proton-proton collisions delivered for physics at the beam injection energy, i.e., for a centre of mass (CM) energy, $\sqrt{s}$, of 900 GeV. On the 30th of November the beam energy was ramped-up to 1.18 TeV, thus setting the world record for highest energy particle collider in the world at $\sqrt{s} =$ 2.36 TeV, beating the Tevatron at FNAL by a narrow margin.

The LHC started its first real high-energy run on March 30th 2010 with the beam energy ramped-up to 3.5 TeV, leading to proton-proton collisions with $\sqrt{s}$ = 7 TeV, and which was the start of the exploration of a new energy regime for fundamental physics. The event was extensively covered by the popular media, and the particle physics community was particularly excited by this groundbreaking event. In particular the experiments demonstrated they were ready for the much anticipated physics data run. Experiments at the LHC collider that have collected data since then are the "high luminosity" general purpose detectors ATLAS[1] and CMS,[2] the heavy-flavor experiment LHCb,[3] the heavy ion experiment ALICE,[4] two forward detectors TOTEM[5] and LHCf[6] and the Monopole and Exotica search detector MoEDAL.[7]

The LHC continued its operation at $\sqrt{s}$= 7 TeV in 2011 and increased the CM energy to 8 TeV in 2012, the last year of run 1. The integrated

This is an open access article published by World Scientific Publishing Company. It is distributed under the terms of the Creative Commons Attribution 4.0 (CC BY) License.

luminosity collected by the experiments increased rapidly in the these first years. E.g., CMS collected 36 $\text{pb}^{-1}$, 5 $\text{fb}^{-1}$, and about 20 $\text{fb}^{-1}$ in 2010, 2011 and 2012, respectively.

Run 2 started in 2015 after a two-year shut-down and partial upgrade of the LHC machine. The beam energy for this run was raised to 6.5 TeV, close to the maximum nominal beam energy value of 7 TeV. Hence, $\sqrt{s}$ in run 2 was 13 TeV, and opened a new window for searches of new physics (NP) and precision measurements at the energy frontier. After the exploratory 2015 run, the main part of the integrated luminosity was collected in the years 2016–2018 with a total of about 140 $\text{fb}^{-1}$ each, for ATLAS and CMS. Collecting data at increasingly higher instantaneous luminosity led to having more proton-proton collisions within one bunch crossing of the beams, so called pileup collisions. The average number of pileup interactions was 23 (32) in 2016 (2017–2018). LHCb takes data at a lower instantaneous luminosity to reduce the impact of pileup events, and has collected a total integrated luminosity of 3 $\text{fb}^{-1}$ (6 $\text{fb}^{-1}$) in run 1 (run 2).

## 2. QCD measurements

Hadrons consist of partons, which carry the charge of the strong force. Collisions involving high $p_T$ phenomena can be treated in a perturbative way. But by far most hadronic collisions at the LHC are so called soft collisions in the non-perturbative regime.

The TOTEM collaboration, using forward detectors and a proton spectrometer integrated with the beampipe, has measured elastic, inelastic, diffractive and total cross-section at the available LHC CM energies to date.[8] The pp total cross section is measured in a range from $\sqrt{s}$ = 2.76 TeV to 13 TeV, obtaining a variation from $(84.7 \pm 3.3)$ mb to $(110.6 \pm 3.4)$ mb, and the nuclear slope $B$ of the elastic scattering is measured to be $(20.36 \pm 19)$ $\text{GeV}^2$ at the highest collision energy. During run 2 TOTEM and CMS jointly built a new proton spectrometer and now produce common physics studies (see, e.g. Ref. 9).

The precision of TOTEM allowed showing evidence for a non-exponential elastic proton–proton differential cross-section at low $|t|$ at 8 TeV.[10] Another important result of TOTEM, made recently by combining results with the D0 experiment at FNAL, is the first observation of Odderon exchange, from the elastic scattering differences between proton-proton and proton–antiproton data.[11]

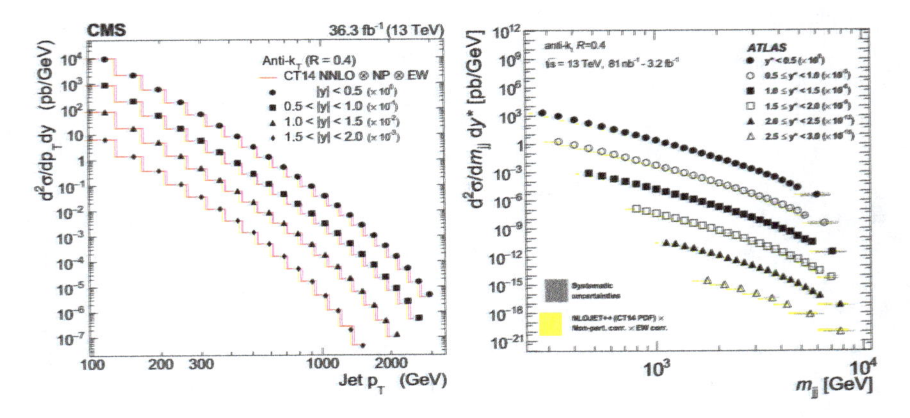


Fig. 1. (Left) The inclusive jet production cross sections as a function of the jet transverse momentum $p_T$ measured in intervals of the absolute rapidity $|y|$. The cross section obtained for jets clustered using the anti-$k_T$ algorithm with R = 0.4 is shown. The measurements are compared with fixed-order NNLO QCD predictions (solid line) using CT14nnlo PDF and corrected for electroweak and non-perturbativce effects. From Ref. 18. (Right) Dijet cross-sections as a function of $m_{jj}$ and $y^* = |y_1 - y_2|/2$, for anti-$k_T$ jets with R = 0.4. The dark gray shaded areas indicate the experimental systematic uncertainties. The data are compared to NLO pQCD predictions calculated using NLOJET++ with $p_T^{max}\exp(0.3y^*)$ as the QCD scale and the CT14 NLO PDF set, to which non-perturbative and electroweak corrections are applied. The light yellow shaded areas indicate the predictions with their uncertainties. From Ref. 19.

The LHCf experiment is designed to measures the very forward, at zero degrees, emitted neutral particles, namely neutrons and $\pi^0$s/photons, with the goal to test Monte Carlo (MC) models that are used for cosmic ray interaction studies. The LHCf data showed important deficiencies of the presently used MC models,[12,13] and was used to improve them. ALICE, ATLAS, CMS and LHCb have all made detailed measurements on particle multiplicities in soft collisions. In fact the first physics papers produced at the LHC were soft QCD measurements with the data from the initial short run at 900 GeV.[14–17]

The success of perturbative QCD has been demonstrated most convincingly with the measurements of inclusive jet cross sections for jet $p_T$ up to and above 2 TeV. Results for the double-differential jet $p_T$ and di-jet mass cross sections are shown in Fig. 1.[18,19] The QCD calculations shown in the figures describe the cross sections of the data over more than nine orders of magnitude.

Perturbative QCD measurements can be used to derive information on the strong force e.g., the strong coupling constant $\alpha_s$ and on the parton distributions functions (PDFs) of the proton. PDFs are key ingredients in predicting cross sections of all processes at a hadron collider, and are traditionally extracted from experimental data such as lepton-hadron scattering. In recent years perturbative QCD data from hadron colliders have emerged as important information to extract PDFs, as, e.g., demonstrated in Ref. 18. Most recent studies of PDFs use Tevatron and LHC jet data in general PDF fits (see, e.g. Ref. 20), demonstrating their power to constrain, e.g, the gluon distribution in the proton.

The strong coupling constant has been measured using various methods: inclusive jet measurements, ratios from 3-jet to 2-jet events (R32), 3-jet mass measurements, top production, transverse energy correlations and more. As an example, the R32 method gives $\alpha_s(m_Z) = 0.1148 \pm 0.0014_{\text{exp}} \pm 0.0018_{\text{PDF}} \pm 0.0050_{\text{theory}}$. Currently, the most precise $\alpha_s(m_Z)$ derivation from LHC data is that from inclusive $W$ and $Z$ cross sections[21] with about 1.6% uncertainty.

QCD data at the LHC are now used to make detailed jet and multi-jet, forward jet, and photoproduction studies and to address more physics topics, probing with increasing accuracy the perturbative QCD regime.

## 3. Electroweak measurements

The ATLAS and CMS detectors were designed to make high precision measurements of electrons, muons, and missing transverse momentum in the harsh high luminosity environment of the LHC. This level of performance coupled with the large data samples available after a decade of running, has resulted in measurements of electroweak observables at a precision that is now surpassing that of previous lepton colliders. In addition, the study of multi-boson final states now includes previously unmeasured processes that involve the quartic couplings of the SM. These measurements, in concert with those associated with the Higgs boson, provide numerous opportunities to test the electroweak sector of the SM.

The precision of $W$ mass measurements at hadron colliders has now surpassed that achieved at LEP. The ATLAS collaboration has reported results with an uncertainty of 19 MeV[22] and the LHCb collaboration has achieved a total uncertainty of 32 MeV,[23] dominated by statistics, based on about a third of the collected data. The fact that the LHCb experiment has reported a precision measurement of the $W$ boson mass is remarkable, considering the limited angular acceptance and lower instantaneous luminosity

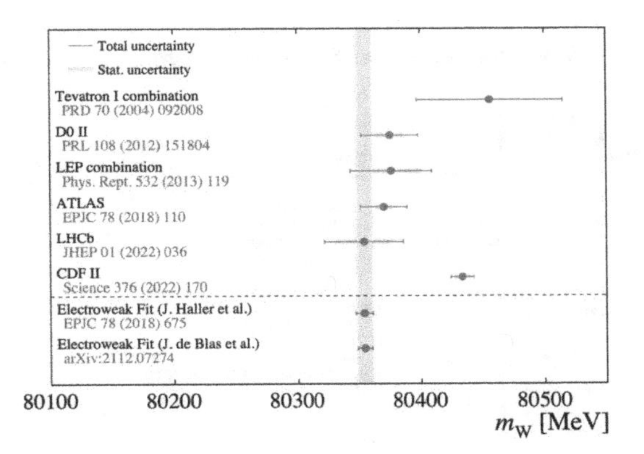


Fig. 2. $W$ boson mass measurements from LEP, LHC, and Tevatron experiments compared with the SM expectation.[24]

of the experiment. The LHCb measurement features many systematic uncertainties that are mostly uncorrelated with those of ATLAS and CMS, which will be helpful in future combinations. A comparison with other experiments, including the latest result from CDF, is shown in Fig. 2. Improved results using run 2 data are expected from all experiments in the future.

The Weinberg angle ($\theta_W$) is another central electroweak parameter that has been measured at the LHC. ATLAS, CMS, and LHCb have measured the value of $\sin^2(\theta_W)$ with a precision of 0.00036,[25] 0.00053,[26] and 0.00106,[27] respectively. The largest source of systematic uncertainty is associated with PDFs.

Weak boson production has been extensively studied by the LHC experiments over more than a decade. Precision measurements of single $W$ and $Z$ production, and of diboson production were performed during run 1 and run 2 at CM energies of 7, 8, and 13 TeV. With the larger datasets made available in run 2, more rare multi-boson production processes could also be investigated. This includes vector boson scattering measurements and triboson production measurements (see Fig. 3). Vector boson scattering has now been observed at the LHC in the $WW$, $WZ$, $ZZ$, $Z\gamma$ final states.

## 4. Flavour physics

The comprehensive program of flavour studies realized at the LHC, in particular by the LHCb experiment, has demonstrated that the LHC is an ideal

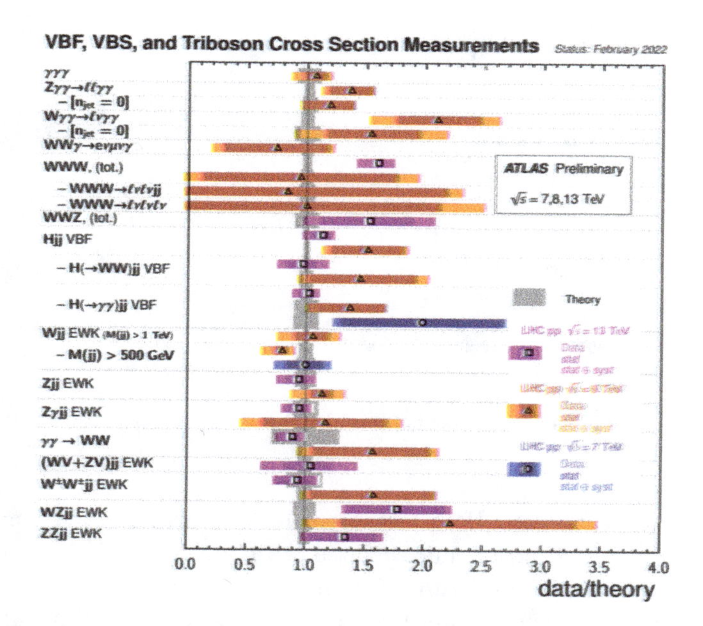


Fig. 3. Vector boson scattering, vector boson fusion, and triboson production measurements performed by the ATLAS experiment[28] and compared with the SM expectation.

laboratory for flavour physics. Much interest is devoted to rare $b$-hadron decays mediated by flavour-changing neutral currents (FCNC). In the SM these decays are forbidden at tree level but can occur at loop level; a new particle too heavy to be produced at the LHC can still give sizeable effects when exchanged in a loop, thus providing sensitivity to shorter distances or effectively higher energies. This is an indirect strategy to NP searches, which nicely complements direct searches performed by ATLAS and CMS and is particularly relevant in the absence of direct collider production of new particles. The baryon asymmetry of the universe needs $CP$ violation far beyond that provided by the SM. Precision measurements of the Cabibbo-Kobayashi-Maskawa (CKM) matrix elements and searches for possible inconsistencies in measurements of the angles and sides of the unitarity triangles probe the CKM description of flavour-changing processes and the mechanism of $CP$ violation through the phase in the quark-mixing matrix. Heavy-flavour physics also provides a unique laboratory for studying the strong interaction. Many hadrons with nonstandard quantum numbers that contain charm or beauty quarks have been discovered at the LHC, opening up a very active area of hadronic-physics research.

### 4.1. *Rare B-meson decays and flavour anomalies*

The study of rare decays is central to the LHC flavour program providing a wealth of opportunities, including in charm and strange decays. The decay $B^0_{(s)} \to \mu^+\mu^-$ has been identified as a very interesting potential constraint on the parameter space of NP models and is one of the milestones of the flavour program. Within the SM, this decay is very rare as it is a FCNC process with helicity and CKM suppression. Theoretically, it is reliably and precisely predicted,[29,30] with branching fractions $\mathcal{B}(B^0_s \to \mu^+\mu^-) = (3.66 \pm 0.14) \times 10^{-9}$ and $\mathcal{B}(B^0 \to \mu^+\mu^-) = (1.03 \pm 0.05) \times 10^{-10}$. This decay is also characterized by a very clean experimental signature and it has been studied by all high-energy collider experiments in an effort that lasted almost forty years. Experiments at the LHC have taken the lead in the analysis of these decays, profiting from the very large $B$-meson cross-section and their excellent muon reconstruction and identification. The combination of the ATLAS, CMS and LHCb measurements using data collected between 2011 and 2016 gives $\mathcal{B}(B^0_s \to \mu^+\mu^-) = (2.69^{+0.37}_{-0.35}) \times 10^{-9}$ and no significant signal for $B^0 \to \mu^+\mu^-$, leading to an upper limit of $1.9 \times 10^{-10}$ at 95% confidence level (CL).[31] The two-dimensional compatibility with the SM point is 2.1 $\sigma$ (see Fig. 4, left). A recent CMS update[32] based on the full run 1 and run 2 data, provides the most precise measurements to date, which are fully consistent with the SM, thus reducing the overall tension. The CMS dimuon invariant mass distribution of the selected $B^0_{(s)} \to \mu^+\mu^-$ candidates in a region of high signal purity is shown in Fig. 4, right.

Much interest is devoted to exclusive semileptonic $b \to s\ell^+\ell^-$ transitions of the type $B \to H_s\mu^+\mu^-$, with $H_s$ either a pseudoscalar or a vector meson. In many extensions of the SM, new particles can contribute to their amplitudes modifying the rates or angular distributions of the final-state particles. These decays are characterised by branching fractions of typically $\mathcal{O}(10^{-7})$, i.e., much larger than that of $B^0_{(s)} \to \mu^+\mu^-$. Contrary to the $B$-factory experiments, which exhibit similar efficiencies for muons and electrons, at the LHC measurements are performed in the dimuon final state, as muons can be triggered and reconstructed much more efficiently than electrons in the high-multiplicity hadronic environment. Measurements of the branching fractions are performed as a function of $q^2 = m^2_{\mu^+\mu^-}$, the dimuon invariant mass squared, to be able to exclude the regions around the $J/\psi$ or the $\Psi(2S)$ resonances, where the rates are dominated by the tree-level decays $B \to H_s J/\Psi(\Psi(2S))$. The differential branching ratios $d\mathcal{B}/dq^2$ are measured in bins of $q^2$ and are compared to theoretical calculations

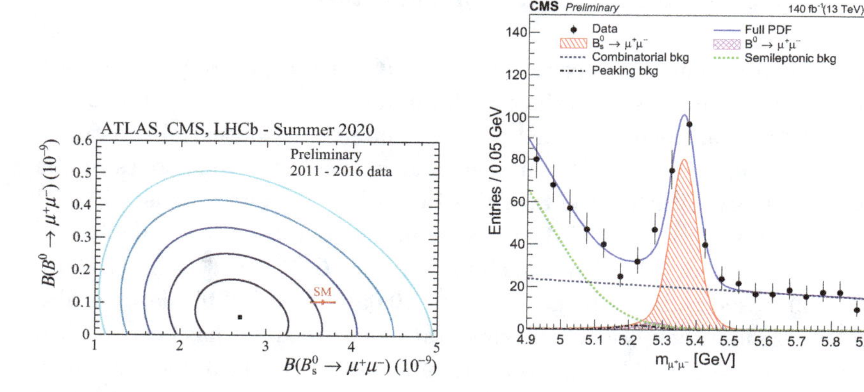


Fig. 4. (Left) Likelihood contours for the combination of ATLAS, CMS and LHCb in the plane $\mathcal{B}(B^0 \to \mu^+\mu^-)$ vs. $\mathcal{B}(B_s^0 \to \mu^+\mu^-)$ corresponding to 1 to 5 $\sigma$ levels. The red point shows the SM prediction with its uncertainties. (Right) CMS invariant dimuon mass distribution of the selected $B^0_{(s)} \to \mu^+\mu^-$ candidates in a region of high signal purity (multivariate classifier$>$ 0.99); the result of the fit is overlaid (blue solid line) and the different background components are detailed in the caption.

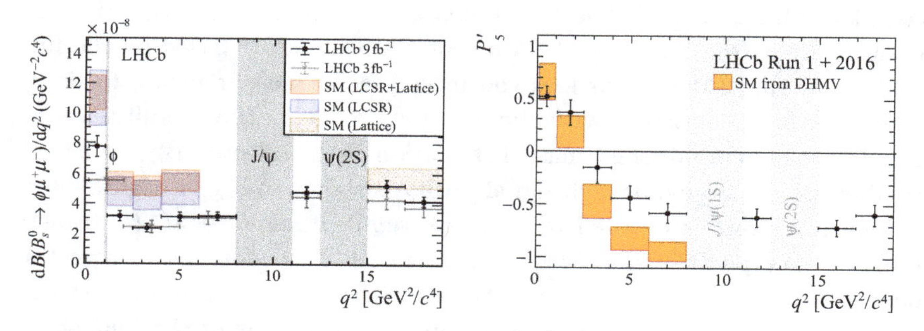


Fig. 5. (Left) Differential branching fraction $\mathrm{d}\mathcal{B}(B_s^0 \to \phi\mu^+\mu^-)/\mathrm{d}q^2$, overlaid with SM predictions using LCSR at low $q^2$,[37,38] and LQCD at high $q^2$.[39] The $J/\psi$ and $\psi(2S)$ regions are vetoed in the selection of the signal mode. (Right) $P_5'$ observable as a function of $q^2$ measured by LHCb[40] and compared with SM calculations from Refs. 41,42.

generally based on Lattice Cone Sum Rules (LCSR) or Lattice QCD (LQCD). LHCb measured the differential branching fractions in several exclusive transitions.[33–36] As an example, Fig. 5, left, shows the differential branching fraction $\mathrm{d}\mathcal{B}(B_s^0 \to \phi\mu^+\mu^-)/\mathrm{d}q^2$,[36] overlaid with SM predictions using LCSR at low $q^2$,[37,38] and LQCD at high $q^2$.[39] In the $q^2$ region between 1.1 and 6.0 $\mathrm{GeV}^2/c^4$, the measurement is found to lie 3.6 $\sigma$ below a SM prediction based on a combination of LCSR and LQCD calculations.

Also for other exclusive decays, such as $B \to K^{(*)}\mu^+\mu^-$,[33,34] the branching fractions at low $q^2$ lie generally below the SM predictions, which are however affected by significant hadronic uncertainties.

Searches for NP can also be conducted from the analysis of the angular distributions of the final state particles in semileptonic $b \to s\ell^+\ell^-$ decays. Measuring angular distributions is particularly interesting because it is possible to construct optimised variables[43] that depend less on hadronic uncertainties associated with the transition form factors. A global fit performed by LHCb to several angular observables in $B^0 \to K^{*0}\mu^+\mu^-$ decays results in an overall tension with the SM of about 3 $\sigma$.[40] Figure 5, right, shows a comparison of the data with one such optimized observable, the so-called $P_5'$ variable. A local discrepancy of about $2.5\,\sigma$ is measured in two bins in $q^2$. This discrepancy, however, should be taken with a grain of salt, given that there is no complete consensus about the theoretical uncertainty of the SM prediction.

LHCb has also performed tests of lepton flavour universality (LFU), which is a central property of the SM, looking for deviations from predictions in the ratios $R_H = \frac{B \to H\mu^+\mu^-}{B \to He^+e^-}$. Here, $H$ denotes a $K^+$ or a $K^{*0}$ (the so-called $R_K$ and $R_{K^*}$ ratios). These ratios are clean probes of NP because they are sensitive to new interactions that couple in a non-universal way to electrons and muons. They are also precisely calculated because hadronic uncertainties cancel in the ratios. In the SM, $R_H = 1$ neglecting lepton masses. The major challenge for this measurement stems from the markedly different detector response to electrons and muons, leading, e.g., to different trigger and reconstruction efficiencies, and background levels. Based on the total collected data, LHCb found deviations on $R_K$ from LFU at $3.1\,\sigma$ for $q^2 \in [1.1, 6.0]\,\mathrm{GeV}^2/c^4$.[44] Deviations were also measured for $R_{K^*}$, based on the analysis of about one third of the sample,[45] namely $\sim 2.2\,\sigma$ for $q^2 \in [0.045, 1.1]\,\mathrm{GeV}^2/c^4$ and $\sim 2.5\,\sigma$ for $q^2 \in [1.1, 6.0]\,\mathrm{GeV}^2/c^4$.

Lepton universality tests are also performed using tree-level $b \to c$ transitions from the measurement of $R(D^{(*)}) = \frac{B(B \to D^{(*)}\tau\nu_\tau)}{B(B \to D^{(*)}\ell\nu_\mu)}$, with $\ell = \mu, e$. This enables a comparison between muons (or electrons) and tau leptons. A combination of $R(D)$ and $R(D^*)$ results from Belle, BaBar and LHCb is in tension with the SM at the $\sim 3.2\,\sigma$ level.[46]

Although not significant individually, these deviations have generated immense interest in the community because they can be interpreted coherently, leading to the exploration of new interesting theoretical avenues.

However, more recently, LHCb performed an improved, simultaneous analysis of both $R_K$ and $R_{K^*}$ using the full dataset.[47,48] The two ratios were computed in two $q^2$ bins, thereby producing four independent measurements. This new analysis did not confirm the previous tensions and found results fully in line with the SM predictions. The main differences arise from a better understanding of misidentified hadronic backgrounds to the electron decay mode, determined at a tighter electron particle identification working point, and from the modelling of the residual hadronic backgrounds.

The search for deviations from the SM continues!

### 4.2. *Precision tests of the CKM framework*

The CKM matrix is able to describe a large range of phenomena with only four parameters, three angles and one phase, which is the only source of $CP$ violation in the SM. The physics impact of the precise determination of these parameters is not so much in the determination of their absolute values, given that the CKM matrix is not predicted, but rather in testing the (in)consistency of the "ensemble" of measurements and how precisely the SM description of flavour and $CP$ violation holds. Extensive measurements in $K$, $D$ and $B$ meson decays have been performed over the years by many different experiments, mostly at $B$ factories, at the Tevatron and at the LHC, mainly by LHCb. At the current level of precision, all measurements are consistent and intersect in the apex of the unitary triangle, which geometrically describes the unitarity of the CKM matrix, indicating that NP effects can appear at most as small corrections to the CKM description.

The golden SM benchmark is the CKM angle $\gamma$, which can be determined with negligible theoretical uncertainty entirely from tree-level processes. Deviations between direct measurements of $\gamma$ and the value derived from global CKM fits would be a clear indication of NP. Most measurements of $\gamma$ utilize the fact that interference of $B^\pm \to D^0 K^\pm$ and $B^\pm \to \overline{D}^0 K^\pm$ can be studied in final states accessible in both $D^0$ and $\overline{D}^0$ decays. LHCb performed a combination of several complementary $\gamma$ measurements that involve different intermediate neutral $D$-meson decays, including the LHCb run 2 update from the highly sensitive $B^\pm \to Dh^\pm$[49] with $D \to K^0_s h^+ h^-$, where $h^\pm$ is either a charged kaon or pion. From this combination, which also includes measurements sensitive to charm mixing (see also Sec. 4.3), LHCb finds[50] $\gamma = (65.4^{+3.8}_{-4.2})^\circ$ (see Fig. 6, left), which is the most precise determination from a single experiment and is in excellent agreement with the global CKM fit results.[51,52]

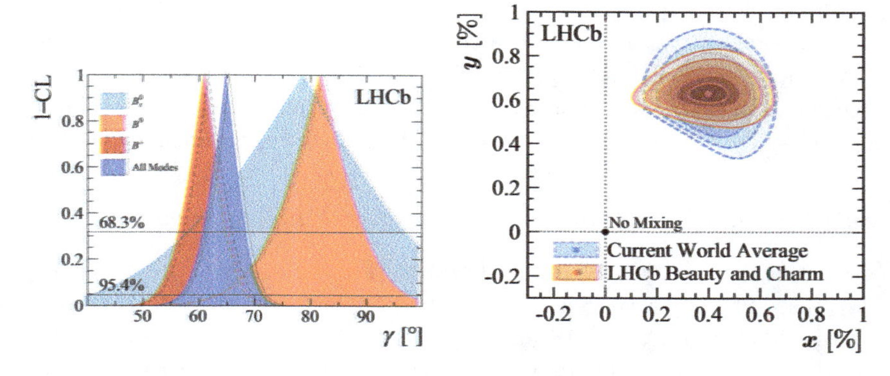


Fig. 6. (Left) One dimensional 1-CL profiles for $\gamma$ from combinations using inputs from $B_s^0$, $B^0$, $B^+$ mesons and all species together. (Right) Constraints for the charm mixing parameters $x$ and $y$. Contours are drawn out from 1 to 5 standard deviations.

## 4.3. *Mixing and CP violation in charm*

At the LHC, the charm cross-section is enormous; e.g., $\sigma(pp \to c\bar{c}X) = (2,369 \pm 192)\,\mu\text{b}$ at $\sqrt{s} = 13\,\text{TeV}$[53] in the LHCb acceptance. This has allowed LHCb to record the world's largest dataset of charm hadrons to date and to perform numerous high-precision measurements of their production and decay properties. By now, the existence of $D^0$-$\overline{D}^0$ mixing is well established. The mixing of charm flavour states can be described by two dimensionless parameters, $x \equiv (m_1 - m_2)/\Gamma$ and $y \equiv (\Gamma_1 - \Gamma_2)/2\Gamma$, where $m_i(\Gamma_i)$ is the mass (width) of the appropriate $D$ mass state, and $\Gamma$ their average decay width. LHCb determined $x$ and $y$ by performing a combination of measurements sensitive to $\gamma$ and to the charm mixing parameters,[50] thus including, for the first time, inputs from both $B$- and $D$-meson decays. Figure 6, right, shows the results for $x$ and $y$, which are the most precise determinations to date. The precision on $y$ is improved by a factor of two over the current world average.

The discovery of $CP$ violation in charm decays was another important milestone of the LHC flavour program. The size of $CP$ violation in charm decays is expected to be tiny in the SM. However, this might be altered by NP, even if theoretical predictions are difficult to compute reliably due to the presence of low-energy strong-interaction effects. LHCb measured the difference $\Delta A_{CP}$ of the time-integrated $CP$ asymmetries in $D^0 \to K^+K^-$ and $D^0 \to \pi^+\pi^-$ decays.[54] The measured value was found to differ from zero by more than five $\sigma$, providing the first observation of $CP$ violation in the decay of charm hadrons and prompting renewed interest in the charm-physics community. The result is consistent with, although in magnitude at the upper end of, SM expectations. More recently, at the 41$^{st}$ ICHEP conference in Bologna, LHCb announced the first evidence for $CP$ violation in a specific charm hadron decay, $D^0 \to \pi^+\pi^-$, with a significance of $3.8\,\sigma$.

### 4.4. *Exotic hadrons and spectroscopy with heavy flavour*

The LHC is an extremely rich laboratory for the study of heavy-flavour spectroscopy. In 11 years of LHC operation, over 60 hadrons have been discovered by ATLAS, CMS and LHCb; on average about one every two months. This includes exotic states, such as tetraquarks and pentaquarks, as well as many conventional hadrons. A major turning point in exotic baryon spectroscopy was achieved when the LHCb collaboration reported the observation of three significant pentaquark states, $P_c(4312)^+$, $P_c(4440)^+$, and $P_c(4457)^+$, decaying to $J/\psi p$ in $\Lambda_b^0 \to J/\psi p K^-$ decays.[55] Since then many other exotic hadrons have been observed by LHCb. These include, for example, fully heavy tetraquarks with hidden flavour, such as the $X(6900)$ discovered in the $J/\psi J/\psi$ mass distribution,[56] or doubly flavoured tetraquarks, such as the narrow doubly charmed $T_{cc}^+$ state decaying to $D^0 D^0 \pi^+$.[57] These discoveries have generated a lot of interest and the development of QCD-motivated models. However, no single theoretical model (e.g., based on loosely bound molecules or on tightly bound compact objects) can accommodate all of them, exposing our lack of understanding of hadronic states.

## 5. Top physics

The top quark is the heaviest known elementary particle and the only quark that decays before it hadronizes. The LHC, with its large collision energy and very high instantaneous luminosity makes it a top quark factory: over half a billion top quarks will have been produced during run 2. This large sample of top quarks allows for high precision measurements of many of its properties, for the measurement of rare production modes, and for searches for very rare decays.

The precise measurement of the top quark mass is an important goal of the LHC experiments as it allows for self-consistency tests of the SM when combined with other precision electroweak measurements. Direct and indirect measurements of the top quark mass have been performed: direct

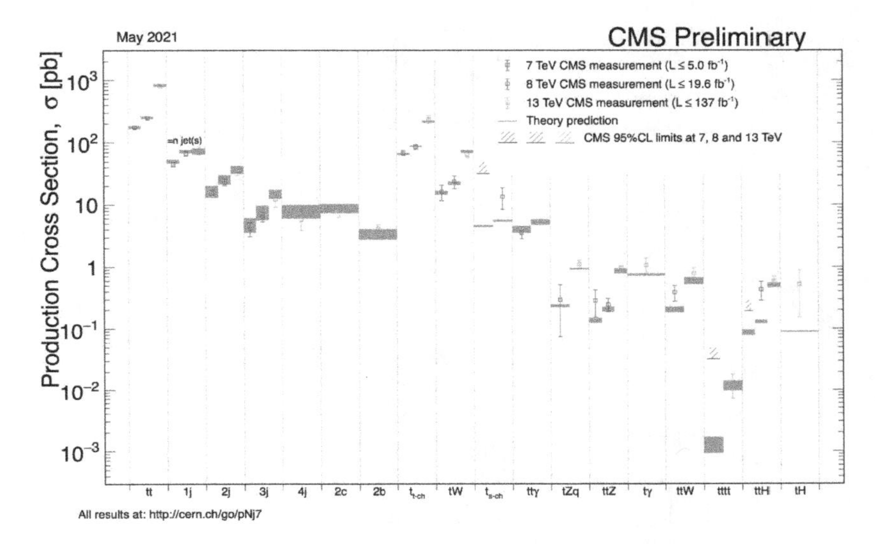


Fig. 7. Summary of several top-quark related production cross section measurements by CMS, compared to the corresponding theoretical expectations.[60]

measurements reconstruct the top quark mass from its decay products while the mass can also be measured indirectly using other measurements that exhibit a mass dependence, like the production cross section. The best single measurement of the top quark mass has been obtained by the CMS experiment and yields a value of $171.77 \pm 0.38$ GeV.[58]

The main top quark production process at the LHC is pair production. Top quarks can also be produced singly, in association with other particles like photons, $W$ bosons, or $Z$ bosons. Evidence of the rare production of four top quarks has also been reported by the ATLAS Experiment.[59] Figure 7 shows various measured top production processes compared to SM predictions. Overall, very good agreement is observed with theoretical predictions.

The very large sample of top quarks produced by the LHC allows for unprecedented sensitivity to extremely rare decays of the top quark, including FCNC decays. Given that some of these rare decays are not predicted by the SM to be observable at the LHC, an observation would signal NP. Figure 8 displays the limits in the branching fraction obtained for a wide variety of FCNC searches.[61] No significant excess of events has been observed up until now and some of the branching ratio limits extend to $10^{-5}$ at the end of run 2.

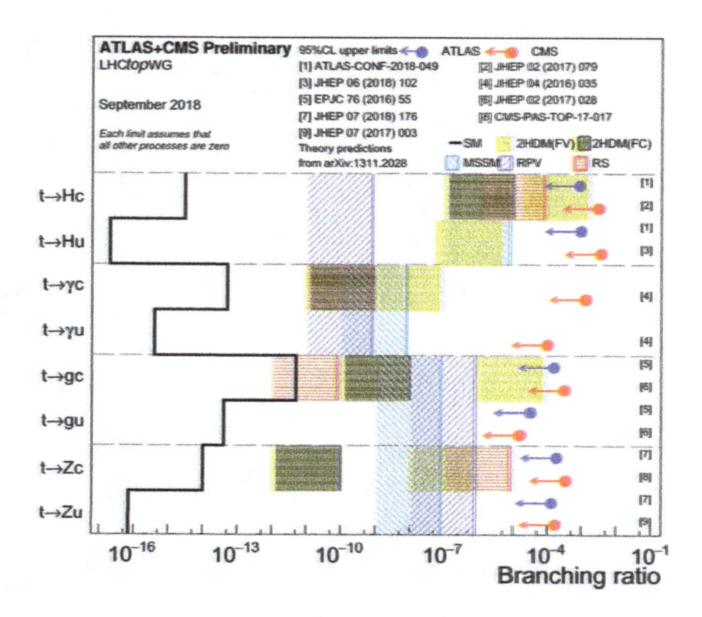


Fig. 8. Summary of the current 95% confidence level observed limits on the branching ratios of the top quark decays via FCNC to a quark and a neutral boson $t \to X_q$ ($X = g$, $Z$, $\gamma$ or $H$; $q = u$ or $c$) by the ATLAS and CMS collaborations compared to several NP models.[61] Each limit assumes that all other FCNC processes vanish.

## 6. The Higgs boson

Since the 2012 discovery by the ATLAS and CMS experiments of a new particle with properties consistent with those of the SM Higgs boson, our understanding of this particle has improved significantly. Using the data collected during run 1, the experiments demonstrated that the spin of the discovered particle is consistent with zero. Alternate spin-1 and spin-2 hypotheses were also tested and were excluded at a high level of confidence. In addition, studies of its $CP$ properties were consistent with a $CP$-even state as predicted by the SM. These detailed studies of the Higgs boson continued during run 2, where 30 times more Higgs bosons are predicted to have been produced compared to the dataset available at the time of discovery. This much larger dataset allowed for precision measurements of the Higgs boson mass, of its production and decay rates, and of its couplings. The Higgs boson mass is not predicted in the SM and must be measured to obtain production and decay rate predictions. The mass has now been measured by the ATLAS and CMS experiments with an uncertainty of

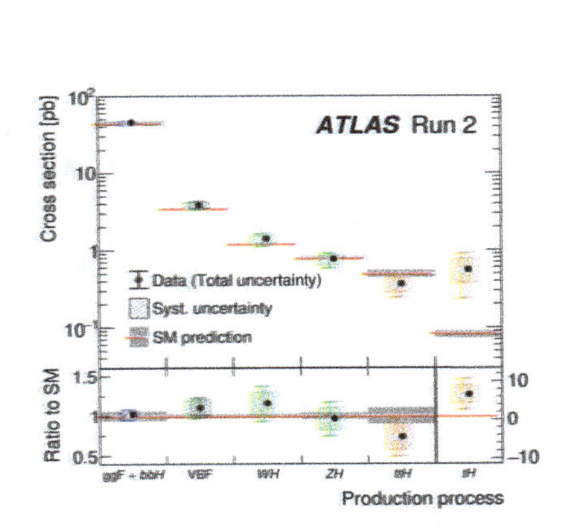


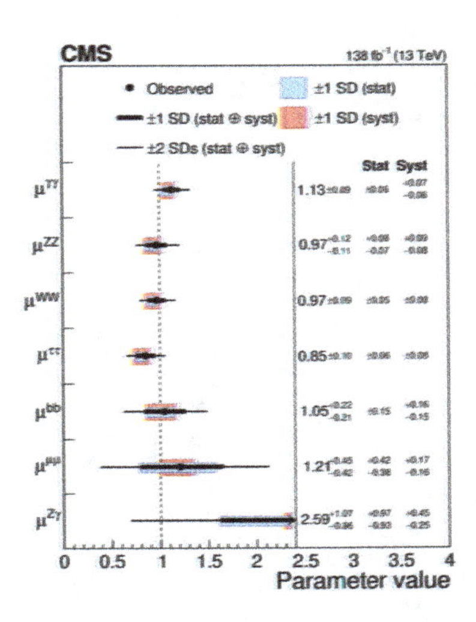


Fig. 9. (Left) Observed and predicted Higgs boson production cross sections from the ATLAS experiment.[65] Higgs boson production processes are measured assuming SM values for the decay branching fractions. (Right) CMS signal strength parameters extracted for decay channels ($\mu_f$) assuming the SM production cross sections.[66] The thick (thin) black lines indicate the 1 (2) $\sigma$ confidence intervals, with the systematic and statistical components of the 1 $\sigma$ interval indicated by the red and blue bands, respectively. The vertical dashed line at unity represents the predicted SM values.

approximately 0.1%. The most precise measurement of the Higgs boson mass has been obtained by CMS and yields a value of $125.38 \pm 0.14$ GeV.[62] The natural width of the Higgs boson at a mass of 125 GeV is predicted to be 4.1 MeV[63] and while a direct measurement of that width is not possible at the LHC, indirect measurements performed under certain assumptions have been carried out. The most precise published measurement is obtained by using the ratio of on-shell to off-shell production cross sections and yields a value of $3.2^{+2.4}_{-1.7}$ MeV.[64]

The main production and decay processes of the Higgs boson have been investigated using the run 2 dataset and measurements are shown in Fig. 9 for the production[65] and decay rates,[66] respectively. Overall, a very good agreement with SM predictions is observed. The results of these measurements can be combined in order to extract the couplings of individual particles to the Higgs boson. The results of such a fit by the ATLAS experiment[65] is shown in Fig. 10, left.

One of the main physics goals of the future High-Luminosity LHC will be to study the shape of the Higgs field potential through the measurement

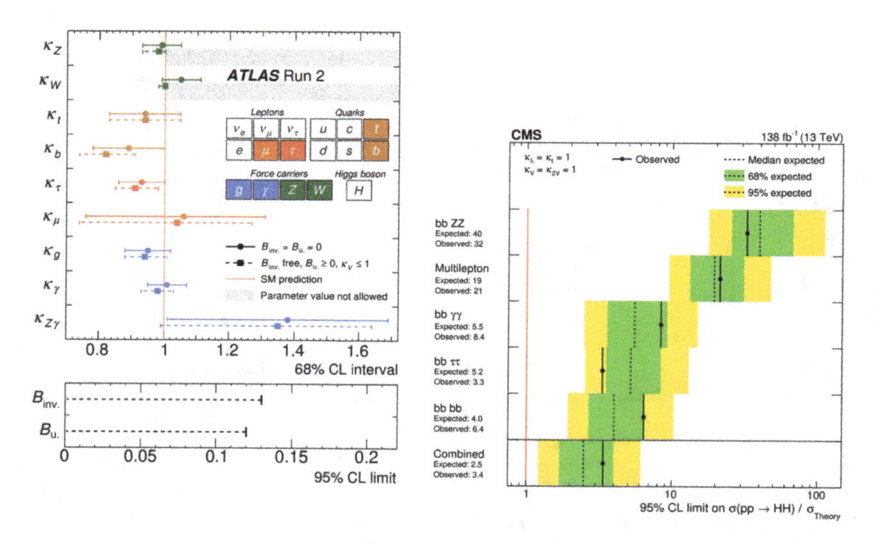


Fig. 10. (Left) Reduced coupling strength modifiers and their uncertainties per particle type with effective photon, $Z\gamma$ and gluon couplings.[65] The horizontal bars on each point denote the 68% confidence interval. The scenario where $B_{\mathrm{inv.}} = B_{\mathrm{u.}} = 0$ is assumed is shown as solid lines with circle markers. ($B_{\mathrm{inv.}}$ and $B_{\mathrm{u.}}$ denote branching fractions for decays to invisible and other undetected particles.) The $p$-value for compatibility with the SM prediction is 61% in this case. The scenario where $B_{\mathrm{inv.}}$ and $B_{\mathrm{u.}}$ are allowed to contribute to the total Higgs boson decay width while assuming that $\kappa_V \leq 1$ and $B_{\mathrm{u.}} \leq 0$ is shown as dashed lines with square markers. The lower panel shows the 95% CL upper limits on $B_{\mathrm{inv.}}$ and $B_{\mathrm{u.}}$ (Right) The expected and observed limits on the ratio of experimentally estimated production cross section and the expectation from the SM in searches using different final states and their combination.[66] The search modes are ordered, from upper to lower, by their expected sensitivities from the least to the most sensitive. The overall combination of all searches is shown by the lowest entry.

of the self-coupling of the Higgs boson. This is done by searching for double Higgs production in a variety of final states determined by the Higgs decay modes. With the run 2 dataset, the sensitivity that has been achieved by ATLAS and CMS has exceeded previous projections. The current limits at 95% confidence level from the CMS collaboration[66] are shown in Fig. 10 (right), and the associated 95% limits on the Higgs self-coupling parameter $\kappa_\lambda$ bound the range between $-1.24$ and 6.49.

## 7. Direct Searches for New Physics

Searches for NP is one of the main topics of the LHC experimental program. In this section focus is on the direct observation of new particles and interactions at the LHC. The palette of proposed theoretical scenarios for NP is very rich. Examples are supersymmetry, extra dimensions, quantum black holes, vector-like fermions, new gauge bosons, hidden valley phenomena, leptoquarks, heavy neutrinos, particles from the dark side,... Over the last years the community got, at times, confronted with suggestions of potential signals, creating excitement for a short time. One such occasion was a putative new resonance at a mass of 750 GeV in early 2016, observed with a significance of about $3\sigma$ in the first data collected at 13 TeV. This led to a flurry of about 500 theoretical papers within less than six months, up and until that it became clear that additional collected data did not confirm it. To date no evidence for any new particles or new interactions has been conclusively established, alas. In the following we will discuss some examples of conducted searches.

### 7.1. *Supersymmetry*

Supersymmetry (SUSY) is a concept proposed to solve the hierarchy problem, provide DM candidates, set the stage for the grand unification of all forces, and more. At the start of the LHC it was anticipated that SUSY was just around the corner waiting to deliver spectacular signals of decays of heavy new particles in the detectors. Interesting events were found, but not more than what could be explained by SM background processes. If realized in Nature, the LHC could have found ample evidence by now, unless the mass scales involved are larger than present sensitivities or the signatures difficult to isolate.

SUSY searches are conducted in ATLAS and CMS by exploring regions of signatures that correspond to decays of heavy new (SUSY) particles. There is a large variety of possible searches, which typically include signatures like large missing transverse momentum, corresponding to an escaping invisible stable lightest SUSY particles (LSP), isolated leptons and a number of jets. Backgrounds are estimated with data-driven techniques, and an optimal sensitivity to a signal is often obtained making use of machine learning techniques. So far no significant excess has been found in any of the SUSY analyses, and the results are mostly interpreted in terms of limits on so called Simplified ModelS (SMS),[67,68] a language designed to facilitate the interpretation of the experimental results. In Fig. 11, SMS limits are shown for searches for gluinos and for light quark SUSY partners versus the neutralino mass, excluding masses up to about 2 TeV.

Despite the null results so far, SUSY will remain a strong candidate for

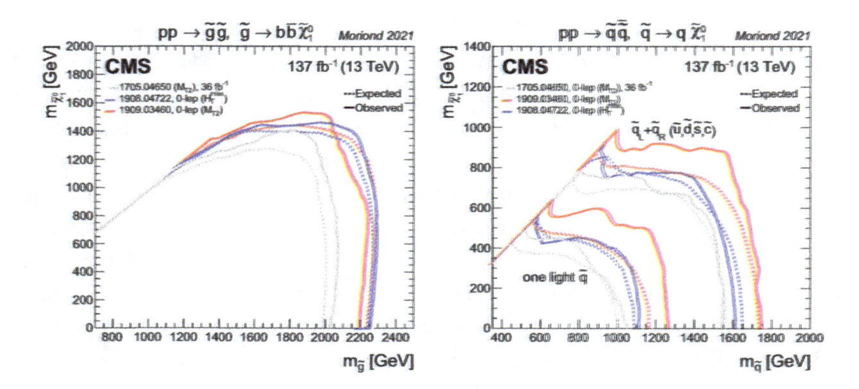


Fig. 11. (Left) Mass limits for a simplified model of gluino pair production with gluino decays to pairs of bottom quarks and the LSP. (Right) Mass limits for a simplified model of first or second generation squark pair production with squark decays to a quark and the LSP. The solid (dashed) lines correspond to the observed (median expected) limits. Limits are shown for two scenarios: production of eight degenerate squarks, or of a single squark. From Ref. 69.

searches for NP, but the preferred region is starting to narrow down.

## 7.2. *Dark Matter*

The nature of dark matter is one of the most intriguing open questions in fundamental physics today. If dark matter is caused by a particle, the SM does not deliver a candidate to play that role. If dark matter has more than only gravitational interactions, e.g., a weak interaction with the SM particles, it can be produced in the high energy collisions at the LHC. Produced dark matter would not be detected directly in the experiments, but via a large missing transverse momentum. These searches rely on the presence of additional initial state radiation, such as jets, gauge bosons, heavy quarks, etc. If the new interaction with the SM particles is as expected mediated by a new boson also interactions with only incoming and outgoing SM particles contain information for the search. To date, no evidence of dark matter production at the LHC has been observed. Figure 12 shows present limits in interaction strength and dark matter mass, compared to underground direct dark matter experiment limits.

Searches for dark matter are also conducted in other channels such as invisible Higgs decays, light dark matter ($m_{\rm DM} < 1$ GeV), axion/axion-like particles and more. Interestingly, a decommissioned LHC magnet has been used by the CAST experiment to track the sun for axion to photon

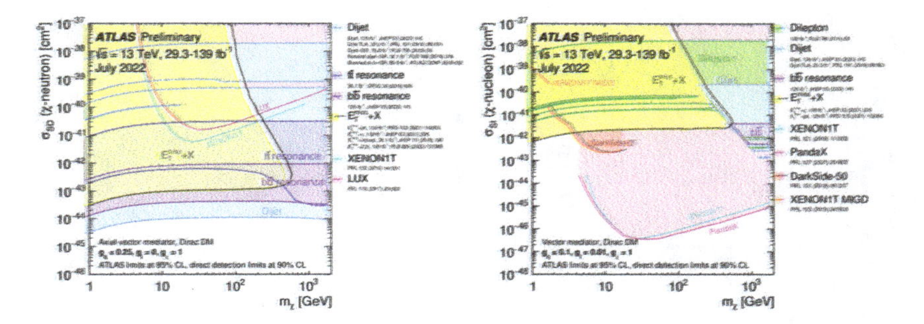


Fig. 12. A comparison of the inferred limits with the constraints from direct-detection experiments on the (Left) spin-dependent WIMP-neutron cross section and (Right) spin-independent WIMP-nucleus cross section in the context of the (Left) leptophobic axial-vector or (Right) vector mediator simplified model. Each shaded region represents the union of the exclusion contours of the individual analyses listed in the legend, where more than one result contributes. The results from this analysis are compared with limits from direct-detection experiments. LHC limits are shown at 95% CL and direct-detection limits at 90% CL. The comparison is valid solely in the context of this model, assuming a mediator width fixed by the dark matter mass, a DM coupling $g_\chi = 1$, quark coupling $g_q = 0.25$, and no coupling to leptons. LHC searches and direct-detection experiments exclude the shaded areas. Exclusions of smaller scattering cross sections do not imply that larger scattering cross sections are also excluded. The resonance and $E_T^{miss}$+X exclusion regions represent the union of exclusions from all analyses of that type. From Ref. 71.

conversions in the strong $\sim$ 9T magnetic field over a length of 9.26 m, and has produced world-leading limits.[70] The hunt for dark matter will continue in future LHC runs!

### 7.3. *Long Lived Particles*

The absence of clear evidence for any sign of New Physics at the LHC so far has recently led to an expansion in new search directions. One such new direction is the search for particles that are stable long enough to allow them to travel macroscopic distances, from at least a few mm to several meters or even much further. These so called long-lived particles (LLPs) require new reconstruction methods and search techniques.[72] Many searches for LLPs have been conducted in recent years. A few examples from ATLAS, CMS and LHCb are searches for heavy neutral leptons,[73] displaced muons,[74] displaced jets,[75,76] multicharged particles,[77] disappearing tracks[78] and R-Parity Violating SUSY decays.[79] Figure 13, left, shows limits from searches for a Higgs boson decaying into low mass LLPs in various decay channels.

A particular search is that for magnetic monopoles, i.e., particles that

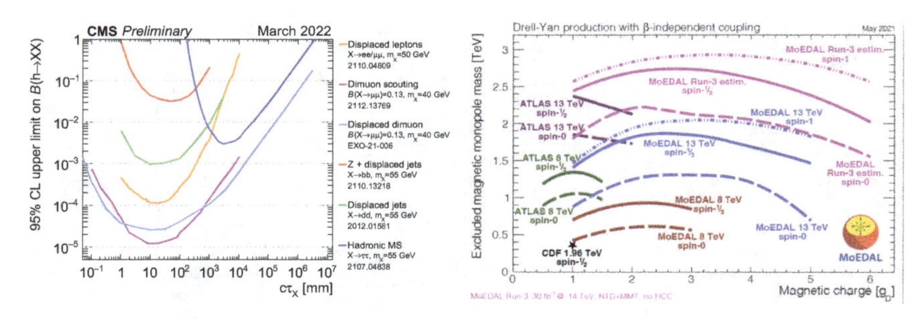


Fig. 13. (Left) The 95% CL observed exclusion limit from different CMS hadronic long-lived particle analyses on the branching fraction of the SM Higgs boson, h, to two neutral long-lived particles, X, shown as a function of the long-lived particle's proper lifetime. (Right) Magnetic monopole mass limits from CDF,[80] ATLAS[81,82] and MoEDAL searches[83–85] as a function of magnetic charge for various spins, assuming Drell-Yan pair- production mechanism and a beta-independent coupling. The MoEDAL projection for LHC run 3 assuming a 30 $\text{fb}^{-1}$ integrated luminosity and combined NTD and MMT data is superimposed.

carry a hypothetical magnetic charge, which is quantized and, according to Dirac, one magnetic charge unit corresponds to 137/2 times the electric charge unit of the electron. Such studies can be conducted by the LHC general detectors. However, the largest span in the magnetic charge versus mass space is covered by the MoEDAL experiment. MoEDAL is specially tailored for searches of exotic long-lived particles with a larger than elementary charge. This experiment consists of layers of plastic sheets through which the highly charged particles will literally "burn their way" (Nuclear Track Detector NTD), and aluminum rods that will slow down and stop the stable monopoles and can be unveiled by tracking the rods through a precision SQUID (Magnetic Monopole Tagger MMT). Figure 13, right, shows a summary of the limits set by CDF, ATLAS and MoEDAL.

Direct searches for BSM physics will continue to be a high activity of research at the LHC in the next 10–20 years; it needs really only one convincing significant deviation from the SM or a new observed particle to lead the way in the New Physics world.

## References


1. G. Aad et al., The ATLAS Experiment at the CERN Large Hadron Collider, *JINST*. **3**, S08003 (2008). doi: 10.1088/1748-0221/3/08/S08003.
2. S. Chatrchyan et al., The CMS Experiment at the CERN LHC, *JINST*. **3**, S08004 (2008). doi: 10.1088/1748-0221/3/08/S08004.

3. LHCb collaboration, A.A. Alves Jr. et al., The LHCb Detector at the LHC, *JINST.* **3**, S08005 (2008). doi: 10.1088/1748-0221/3/08/S08005.
4. K. Aamodt et al., The ALICE experiment at the CERN LHC, *JINST.* **3**, S08002 (2008). doi: 10.1088/1748-0221/3/08/S08002.
5. G. Anelli et al., The TOTEM experiment at the CERN Large Hadron Collider, *JINST.* **3**, S08007 (2008). doi: 10.1088/1748-0221/3/08/S08007.
6. O. Adriani et al., The LHCf detector at the CERN Large Hadron Collider, *JINST.* **3**, S08006 (2008). doi: 10.1088/1748-0221/3/08/S08006.
7. B. Acharya et al., The Physics Programme Of The MoEDAL Experiment At The LHC, *Int. J. Mod. Phys. A.* **29**, 1430050 (2014). doi: 10.1142/S0217751X14300506.
8. G. Antchev et al., First measurement of elastic, inelastic and total cross-section at $\sqrt{s} = 13$ TeV by TOTEM and overview of cross-section data at LHC energies, *Eur. Phys. J. C.* **79** (2), 103 (2019). doi: 10.1140/epjc/s10052-019-6567-0.
9. A. Tumasyan et al., First search for exclusive diphoton production at high mass with tagged protons in proton-proton collisions at $\sqrt{s} = 13$ TeV, *Phys. Rev. Lett.* **129**, 011801 (2022). doi: 10.1103/PhysRevLett.129.011801.
10. G. Antchev et al., Evidence for non-exponential elastic proton–proton differential cross-section at low —t— and $\sqrt{s} = 8$ TeV by TOTEM, *Nucl. Phys. B.* **899**, 527–546 (2015). doi: 10.1016/j.nuclphysb.2015.08.010.
11. V. M. Abazov et al., Odderon Exchange from Elastic Scattering Differences between $pp$ and $p\bar{p}$ Data at 1.96 TeV and from pp Forward Scattering Measurements, *Phys. Rev. Lett.* **127** (6), 062003 (2021). doi: 10.1103/PhysRevLett.127.062003.
12. A. Tiberio et al., Very-forward $\pi^0$ production cross section in proton-proton collisions at $\sqrt{s} = 13$ TeV measured with the LHCf experiment, *PoS.* **ICRC2021**, 386 (2021). doi: 10.22323/1.395.0386.
13. O. Adriani et al., Measurement of energy flow, cross section and average inelasticity of forward neutrons produced in $\sqrt{s} = 13$ TeV proton-proton collisions with the LHCf Arm2 detector, *JHEP.* **07**, 016 (2020). doi: 10.1007/JHEP07(2020)016.
14. K. Aamodt et al., First proton-proton collisions at the LHC as observed with the ALICE detector: Measurement of the charged particle pseudorapidity density at s**(1/2) = 900-GeV, *Eur. Phys. J. C.* **65**, 111–125 (2010). doi: 10.1140/epjc/s10052-009-1227-4.
15. V. Khachatryan et al., Transverse Momentum and Pseudorapidity Distributions of Charged Hadrons in pp Collisions at $\sqrt{s} = 0.9$ and 2.36 TeV, *JHEP.* **02**, 041 (2010). doi: 10.1007/JHEP02(2010)041.
16. G. Aad et al., Charged-particle multiplicities in $pp$ interactions at $\sqrt{s} =$ 900 GeV measured with the ATLAS detector at the LHC, *Phys. Lett. B.* **688**, 21–42 (2010). doi: 10.1016/j.physletb.2010.03.064.
17. LHCb collaboration, R. Aaij et al., Prompt $K_s^0$ production in $pp$ collisions at $\sqrt{s} = 0.9$ TeV, *Phys. Lett. B.* **693**, 69–80 (2010). doi: 10.1016/j.physletb.2010.08.055.
18. A. Tumasyan et al., Measurement and QCD analysis of double-differential

inclusive jet cross sections in proton-proton collisions at $\sqrt{s} = 13$ TeV, *JHEP*. **02**, 142 (2022). doi: 10.1007/JHEP02(2022)142.
19. M. Aaboud et al., Measurement of inclusive jet and dijet cross-sections in proton-proton collisions at $\sqrt{s} = 13$ TeV with the ATLAS detector, *JHEP*. **05**, 195 (2018). doi: 10.1007/JHEP05(2018)195.
20. R. D. Ball et al., The PDF4LHC21 combination of global PDF fits for the LHC Run III (3, 2022).
21. D. d'Enterria and A. Poldaru, Extraction of the strong coupling $\alpha_s(\mathrm{m}_Z)$ from a combined NNLO analysis of inclusive electroweak boson cross sections at hadron colliders, *JHEP*. **06**, 016 (2020). doi: 10.1007/JHEP06(2020)016.
22. M. Aaboud et al., Measurement of the $W$-boson mass in pp collisions at $\sqrt{s} = 7$ TeV with the ATLAS detector, *Eur. Phys. J. C*. **78** (2), 110 (2018). doi: 10.1140/epjc/s10052-017-5475-4. [Erratum: Eur. Phys. J. C 78, 898 (2018)].
23. LHCb collaboration, R. Aaij et al., Measurement of the $W$ boson mass, *JHEP*. **01**, 036 (2022). doi: 10.1007/JHEP01(2022)036.
24. Status of $W$ Mass Measurements. URL `https://cds.cern.ch/record/2806574`.
25. ATLAS Collaboration. Measurement of the effective leptonic weak mixing angle using electron and muon pairs from $Z$-boson decay in the ATLAS experiment at $\sqrt{s} = 8$ TeV. ATLAS-CONF-2018-037 (2018). URL `https://cds.cern.ch/record/2630340`.
26. A. M. Sirunyan et al., Measurement of the weak mixing angle using the forward-backward asymmetry of Drell-Yan events in pp collisions at 8 TeV, *Eur. Phys. J. C*. **78** (9), 701 (2018). doi: 10.1140/epjc/s10052-018-6148-7.
27. LHCb collaboration, R. Aaij et al., Measurement of the forward-backward asymmetry in $Z/\gamma^* \to \mu^+\mu^-$ decays and determination of the effective weak mixing angle, *JHEP*. **11**, 190 (2015). doi: 10.1007/JHEP11(2015)190.
28. ATLAS SM Public Results. URL `https://atlas.web.cern.ch/Atlas/GROUPS/PHYSICS/PUBNOTES/ATL-PHYS-PUB-2022-009/`.
29. C. Bobeth, M. Gorbahn, T. Hermann, M. Misiak, E. Stamou, and M. Steinhauser, $B_{s,d} \to l^+l^-$ in the Standard Model with reduced theoretical uncertainty, *Phys. Rev. Lett.* **112**, 101801 (2014). doi: 10.1103/PhysRevLett.112.101801.
30. M. Beneke, C. Bobeth, and R. Szafron, Power-enhanced leading-logarithmic QED corrections to $B_q \to \mu^+\mu^-$, *JHEP*. **10**, 232 (2019). doi: 10.1007/JHEP10(2019)232.
31. ATLAS, CMS, LHCb collaborations, Combination of the ATLAS, CMS and LHCb results on the $B^0_{(s)} \to \mu^+\mu^-$ decays (2020). LHCb-CONF-2020-002, ATLAS-CONF-2020-049, CMS-PAS-BPH-20-003.
32. CMS collaboration, Measurement of $B^0_s \to \mu^+\mu^-$ decay properties and search for the $B^0 \to \mu^+\mu^-$ decay in proton-proton collisions at $\sqrt{s} = 13$ TeV (2022). URL `https://cds.cern.ch/record/2815334`.
33. LHCb collaboration, R. Aaij et al., Differential branching fractions and isospin asymmetries of $B \to K^{(*)}\mu^+\mu^-$ decays, *JHEP*. **06**, 133 (2014). doi: 10.1007/JHEP06(2014)133.

34. LHCb collaboration, R. Aaij et al., Measurements of the S-wave fraction in $B^0 \to K^+\pi^-\mu^+\mu^-$ decays and the $B^0 \to K^*(892)^0\mu^+\mu^-$ differential branching fraction, *JHEP.* **11**, 047 (2016). doi: 10.1007/JHEP11(2016)047.
35. LHCb collaboration, R. Aaij et al., Differential branching fraction and angular analysis of $\Lambda_b \to \Lambda\mu^+\mu^-$ decays, *JHEP.* **06**, 115 (2015). doi: 10.1007/JHEP06(2015)115.
36. LHCb collaboration, R. Aaij et al., Branching fraction measurements of the rare $B_s^0 \to \phi\mu^+\mu^-$ and $B_s^0 \to f_2'(1525)\mu^+\mu^-$ decays, *Phys. Rev. Lett.* **127**, 151801 (2021). doi: 10.1103/PhysRevLett.127.151801.
37. A. Bharucha, D. M. Straub, and R. Zwicky, $B \to V\ell^+\ell^-$ in the Standard Model from light-cone sum rules, *JHEP.* **2016** (8), 1–64 (2016).
38. W. Altmannshofer and D. M. Straub, New physics in $b \to s$ transitions after LHC run 1, *The European Physical Journal C.* **75** (8), 1–30 (2015).
39. R. R. Horgan, Z. Liu, S. Meinel and M. Wingate, Calculation of $B^0 \to K^{*0}\mu^+\mu^-$ and $B_s^0 \to \phi\mu^+\mu^-$ observables using form factors from lattice QCD, *Phys. Rev. Lett.* **112** (21) (May, 2014). doi: 10.1103/physrevlett.112.212003.
40. LHCb collaboration, R. Aaij et al., Measurement of *CP*-averaged observables in the $B^0 \to K^{*0}\mu^+\mu^-$ decays decay, *Phys. Rev. Lett.* **125**, 011802 (2020). doi: 10.1103/PhysRevLett.125.011802.
41. S. Descotes-Genon, L. Hofer, J. Matias and J. Virto, On the impact of power corrections in the prediction of $B \to K^*\mu^+\mu^-$ observables, *JHEP.* **2014** (12), 125 (2014). doi: 10.1007/JHEP12(2014)125.
42. T. M. A. Khodjamirian, A. Pivovarov, and Y.-M. Wang, Charm-loop effect in $B \to K^{(*)}\ell^+\ell^-$ and $B \to K^*\gamma$, *JHEP.* **2010** (9) (Sep, 2010). doi: 10.1007/jhep09(2010)089.
43. S. Descotes-Genon, J. Matias, M. Ramon and J. Virto, Implications from clean observables for the binned analysis of $B \to K^*\ell\ell$ - at large recoil, *JHEP.* **2013** (1) (Jan, 2013). doi: 10.1007/jhep01(2013)048.
44. LHCb collaboration, R. Aaij et al., Test of lepton universality in beauty-quark decays, *Nature Physics.* **18** (3), 277–282 (2022).
45. LHCb collaboration, R. Aaij et al., Test of lepton universality with $B^0 \to K^{*0}\ell^+\ell^-$ decays, *JHEP.* **08**, 055 (2017). doi: 10.1007/JHEP08(2017)055.
46. Y. Amhis et al., Averages of *b*-hadron, *c*-hadron, and $\tau$-lepton properties as of 2021. URL `https://arxiv.org/abs/2206.07501` (2022).
47. LHCb collaboration, R. Aaij et al., Measurement of lepton universality parameters in $B^+ \to K^+\ell^+\ell^-$ and $B^0 \to K^{*0}\ell^+\ell^-$ decays (2022). Submitted to Phys. Rev. D.
48. LHCb collaboration, R. Aaij et al., Test of lepton universality in $b \to s\ell^+\ell^-$ decays (2022). Submitted to Phys. Rev. Lett.
49. LHCb collaboration, R. Aaij et al., Measurement of the CKM angle $\gamma$ in $B^\pm \to DK^\pm$ and $B^\pm \to D\pi^\pm$ decays with $D \to K_S^0 h^+h^-$, *JHEP.* **2021** (2) (Feb, 2021). doi: 10.1007/jhep02(2021)169.
50. LHCb collaboration, R. Aaij et al., Simultaneous determination of CKM angle $\gamma$ and charm mixing parameters, *JHEP.* **2021** (12) (Dec, 2021). doi: 10.1007/jhep12(2021)141.

51. CKMfitter group, J. Charles et al., Current status of the standard model CKM fit and constraints on $\Delta F = 2$ new physics, *Phys. Rev. D.* **91** (7) (Apr, 2015). doi: 10.1103/physrevd.91.073007. Updated results and plots available at http://ckmfitter.in2p3.fr/.
52. UTfit collaboration, M. Bona et al., The unitarity triangle fit in the standard model and hadronic parameters from lattice QCD: a reappraisal after the measurements of $\Delta m_s$ and $BR(B \to \tau\nu_\tau)$, *JHEP.* **2006** (10), 081–081 (Oct, 2006). doi: 10.1088/1126-6708/2006/10/081. Updated results and plots available at http://www.utfit.org/.
53. LHCb collaboration, R. Aaij et al., Measurements of prompt charm production cross-sections in *pp* collisions at $\sqrt{s} = 13$ TeV, *JHEP.* **03**, 159 (2016). doi: 10.1007/JHEP03(2016)159.
54. LHCb collaboration, R. Aaij et al., Observation of $CP$ Violation in Charm Decays, *Phys. Rev. Lett.* **122** (21) (May, 2019). doi: 10.1103/physrevlett.122.211803.
55. LHCb collaboration, R. Aaij et al., Observation of a Narrow Pentaquark State, $P_c(4312)^+$, and of two-peak structure of the $P_c(4450)^+$, *Phys. Rev. Lett.* **122** (22) (Jun, 2019). doi: 10.1103/physrevlett.122.222001.
56. LHCb collaboration, R. Aaij et al., Observation of structure in the $J/\psi$-pair mass spectrum, *Science Bulletin.* **65** (23), 1983–1993 (2020). ISSN 2095-9273. doi: https://doi.org/10.1016/j.scib.2020.08.032.
57. LHCb collaboration, R. Aaij et al., Study of the doubly charmed tetraquark $T_{cc}^+$, *Nature Communications.* **13** (1), 3351 (2022).
58. CMS collaboration, A profile likelihood approach to measure the top quark mass in the lepton+jets channel at $\sqrt{s} = 13$ TeV (2022). URL `https://cds.cern.ch/record/2806509`.
59. G. Aad et al., Measurement of the $t\bar{t}t\bar{t}$ production cross section in *pp* collisions at $\sqrt{s} = 13$ TeV with the ATLAS detector, *JHEP.* **11**, 118 (2021). doi: 10.1007/JHEP11(2021)118.
60. CMS Top Quark Cross Sections Measurements. URL `https://twiki.cern.ch/twiki/pub/CMSPublic/PhysicsResultsCombined/`.
61. Flavour Changing Neutral Current Decays. URL `https://twiki.cern.ch/twiki/bin/view/LHCPhysics/LHCTopWGSummaryPlots/`.
62. CMS Collaboration, A measurement of the Higgs boson mass in the diphoton decay channel, *Phys. Lett. B.* **805**, 135425 (Jun, 2020). doi: 10.1016/j.physletb.2020.135425. URL `https://doi.org/10.1016%2Fj.physletb.2020.135425`.
63. CERN. Cern yellow reports: Monographs, vol. 2 (2017): Handbook of lhc higgs cross sections: 4. deciphering the nature of the higgs sector. URL `https://e-publishing.cern.ch/index.php/CYRM/issue/view/32` (2017).
64. CMS Collaboration, Measurement of the higgs boson width and evidence of its off-shell contributions to zz production, *Nature Physics.* **19**, 1329–1334 (2022).
65. A detailed map of Higgs boson interactions by the ATLAS experiment ten years after the discovery, *Nature.* **607** (7917), 52–59 (Jul, 2022). doi: 10.1038/s41586-022-04893-w.

66. A portrait of the Higgs boson by the CMS experiment ten years after the discovery, *Nature.* **607** (7917), 60–68 (Jul, 2022). doi: 10.1038/s41586-022-04892-x.
67. J. Alwall, P. Schuster, and N. Toro, Simplified Models for a First Characterization of New Physics at the LHC, *Phys. Rev. D.* **79**, 075020 (2009). doi: 10.1103/PhysRevD.79.075020.
68. D. Alves, Simplified Models for LHC New Physics Searches, *J. Phys. G.* **39**, 105005 (2012). doi: 10.1088/0954-3899/39/10/105005.
69. CMS SUSY Public Results. URL `https://twiki.cern.ch/twiki/bin/view/CMSPublic/PhysicsResultsSUS`.
70. V. Anastassopoulos et al., New CAST Limit on the Axion-Photon Interaction, *Nature Phys.* **13**, 584–590 (2017). doi: 10.1038/nphys4109.
71. ATLAS EXO Public Results. URL `https://atlas.web.cern.ch/Atlas/GROUPS/PHYSICS/PUBNOTES/ATL-PHYS-PUB-2021-045/`.
72. J. Alimena et al., Searching for long-lived particles beyond the Standard Model at the Large Hadron Collider, *J. Phys. G.* **47** (9), 090501 (2020). doi: 10.1088/1361-6471/ab4574.
73. A. Tumasyan et al., Search for long-lived heavy neutral leptons with displaced vertices in proton-proton collisions at $\sqrt{\mathrm{s}}$ =13 TeV, *JHEP.* **07**, 081 (2022). doi: 10.1007/JHEP07(2022)081.
74. A. Tumasyan et al., Search for long-lived particles decaying into muon pairs in proton-proton collisions at $\sqrt{s} = 13$ TeV collected with a dedicated high-rate data stream, *JHEP.* **04**, 062 (2022). doi: 10.1007/JHEP04(2022)062.
75. A. M. Sirunyan et al., Search for long-lived particles decaying to jets with displaced vertices in proton-proton collisions at $\sqrt{s} = 13$ TeV, *Phys. Rev. D.* **104** (5), 052011 (2021). doi: 10.1103/PhysRevD.104.052011.
76. G. Aad et al., Search for neutral long-lived particles in $pp$ collisions at $\sqrt{s} =$ 13 TeV that decay into displaced hadronic jets in the ATLAS calorimeter (3, 2022).
77. M. Aaboud et al., Search for heavy long-lived multicharged particles in proton-proton collisions at $\sqrt{s} = 13$ TeV using the ATLAS detector, *Phys. Rev. D.* **99** (5), 052003 (2019). doi: 10.1103/PhysRevD.99.052003.
78. G. Aad et al., Search for long-lived charginos based on a disappearing-track signature using 136 $\mathrm{fb}^{-1}$ of $pp$ collisions at $\sqrt{s} = 13$ TeV with the ATLAS detector (1, 2022).
79. R. Aaij et al., Search for long-lived particles decaying to $e^{\pm}\mu^{\mp}\nu$, *Eur. Phys. J. C.* **81** (3), 261 (2021). doi: 10.1140/epjc/s10052-021-08994-0.
80. A. Abulencia et al., Direct search for Dirac magnetic monopoles in $p\bar{p}$ collisions at $\sqrt{s} = 1.96$ TeV, *Phys. Rev. Lett.* **96**, 201801 (2006). doi: 10.1103/PhysRevLett.96.201801.
81. G. Aad et al., Search for magnetic monopoles and stable particles with high electric charges in 8 TeV $pp$ collisions with the ATLAS detector, *Phys. Rev. D.* **93** (5), 052009 (2016). doi: 10.1103/PhysRevD.93.052009.
82. G. Aad et al., Search for Magnetic Monopoles and Stable High-Electric-Charge Objects in 13 Tev Proton-Proton Collisions with the ATLAS Detector, *Phys. Rev. Lett.* **124** (3), 031802 (2020). doi: 10.1103/PhysRevLett.124.031802.

83. B. Acharya et al., Search for magnetic monopoles with the MoEDAL prototype trapping detector in 8 TeV proton-proton collisions at the LHC, *JHEP.* **08**, 067 (2016). doi: 10.1007/JHEP08(2016)067.
84. B. Acharya et al., Search for magnetic monopoles with the MoEDAL forward trapping detector in 2.11 $fb^{-1}$ of 13 TeV proton-proton collisions at the LHC, *Phys. Lett. B.* **782**, 510–516 (2018). doi: 10.1016/j.physletb.2018.05.069.
85. B. Acharya et al., Magnetic Monopole Search with the Full MoEDAL Trapping Detector in 13 TeV pp Collisions Interpreted in Photon-Fusion and Drell-Yan Production, *Phys. Rev. Lett.* **123** (2), 021802 (2019). doi: 10.1103/PhysRevLett.123.021802.

© 2024 The Author(s)
https://doi.org/10.1142/9789811280184_0005

# Chapter 5

# Heavy-ion physics at the LHC

Benjamin Audurier*, Brian Cole†, Andrea Dainese‡, Yen-Jie Lee§

*Laboratoire Leprince-Ringuet, École Polytechnique, Paris, France
†Nevis Laboratories, Columbia University, New York, NY, USA
‡Istituto Nazionale di Fisica Nucleare, Padova, Italy
§Massachusetts Institute of Technology, Cambridge, MA, USA

Heavy-ion collisions have been part of the programme of the LHC since its conception. The main nucleus–nucleus system is ${}^{208}_{82+}$Pb+${}^{208}_{82+}$Pb. During the Runs 1 and 2, the LHC provided Pb+Pb collisions with centre-of-mass energy per nucleon pair $\sqrt{s_{\rm NN}}$ = 2.76 and 5.02 TeV, Xe+Xe with $\sqrt{s_{\rm NN}}$ = 5.44 TeV, and p+Pb with $\sqrt{s_{\rm NN}}$ = 5.02 and 8.16 TeV. The LHC performance with heavy-ion beams is described in Ref. 1. The ALICE detector was conceived to be dedicated to the study of heavy-ion collisions. The ATLAS and CMS experiments participated in the programme since the very beginning, and the LHCb experiment participated since the end of Run 1. Figure 1 shows event displays of Pb+Pb collisions from the four experiments.

## 1. The quark-gluon plasma and heavy-ion collisions

Collisions of heavy ions at ultra-relativistic energies are a unique tool to produce in the laboratory a hot and dense state of strongly interacting matter. Strongly interacting matter is expected to exist in different states. At low temperatures and for an energy density $\varepsilon \sim 0.1$ GeV/fm$^3$ we have ordinary atomic nuclei. When the energy density is increased beyond a critical value, the confinement of colour charges (quarks and gluons) into hadrons vanishes and a quark-gluon plasma (QGP) is formed. Lattice QCD calculations indicate that near zero baryon density — relevant for heavy-ion collisions at the LHC — the transition from confined matter to QGP is not a true phase transition, but a cross-over transition. The critical

This is an open access article published by World Scientific Publishing Company. It is distributed under the terms of the Creative Commons Attribution 4.0 (CC BY) License.

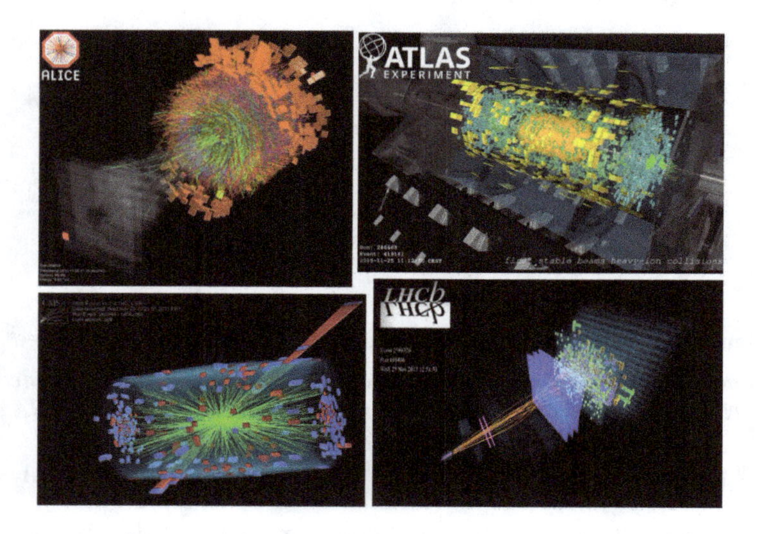


Fig. 1. Event displays of Pb+Pb collisions at $\sqrt{s_{\rm NN}} = 5.02$ TeV.

temperature at small baryon density is predicted by lattice calculations: $T_c = (155 \pm 5)$ MeV,[2,3] corresponding to $\varepsilon_c \approx 0.6$ GeV/fm$^3$.

The space-time evolution of a nuclear collision is shown in Fig. 2. After a short equilibration time $\tau_0 \sim 0.1$–$1$ fm/$c$, a QGP is formed and rapidly expands under large pressure gradients. As the expansion reduces the energy density, the system hadronises and hadron abundances are defined (chemical freeze-out). After a dense hadron gas phase, hadrons cease to interact with each other (kinetic freeze-out).

Before the startup of the LHC, experiments at the CERN-SPS and BNL-RHIC accelerators at $\sqrt{s_{\rm NN}}$ of 17.3 and 200 GeV, respectively, had provided evidence of the QGP formation and indications on its properties. In a nutshell, the QGP as studied at RHIC is deconfined, partonic, opaque to high-energy partons, and expands hydrodynamically like a low-viscosity liquid. The LHC deemed to bring a big step forward in the exploration of these new territories of QCD, because the QGP was expected be hotter, denser, and longer-lived, because the probes for its characterization would become abundantly available, and because the new-generation detectors would provide unprecedented experimental performance.

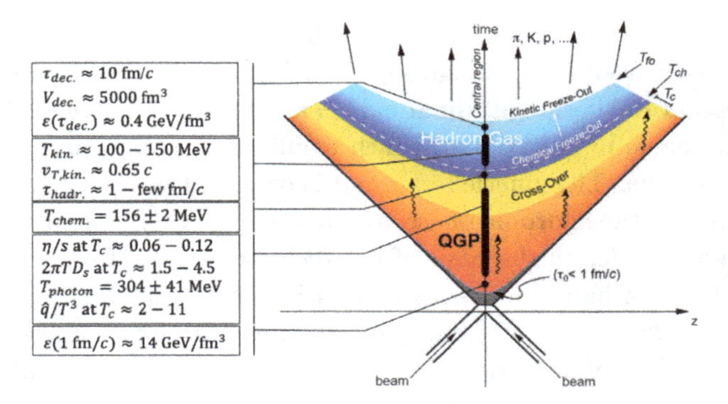


Fig. 2. Space-time evolution of a heavy-ion collision at LHC energies, with estimates from LHC measurements or their model descriptions (see text).

## 2. Macroscopic and thermodynamical QGP properties

The distributions of the produced charged hadrons and of their energy provide a first characterisation of nucleus–nucleus collisions. These quantities are related to the geometry of the collision: a larger multiplicity is connected with more central collisions. For the most central collisions, the number of nucleons participating in the collisions ($N_{\rm part}$) and the number of nucleon–nucleon collisions ($N_{\rm coll}$) are maximum. The centrality classes are denoted in terms of percentiles of the hadronic nucleus–nucleus cross section, with the lower percentages corresponding to more central events. The charged-hadron multiplicity $\mathrm{d}N_{\rm ch}/\mathrm{d}\eta$ or transverse energy $\mathrm{d}E_{\rm T}/\mathrm{d}\eta$ at mid-rapidity can be related with the energy density reached at the formation time $\tau_0$ of the partons in the collision, via the Bjorken formula $\varepsilon_{\rm Bjorken} = \mathrm{d}E_{\rm T}/\mathrm{d}\eta/(\tau_0 \cdot S)$, where $S$ is the transverse area of nuclear overlap. The CMS measurement[4] of $\mathrm{d}E_{\rm T}/\mathrm{d}\eta \approx 2$ TeV yields the estimate $\varepsilon_{\rm Bjorken} \approx 14$ GeV/fm$^3$ at $\tau_0 = 1$ fm/$c$ and for a Pb nucleus transverse area of about 160 fm$^2$. This energy density estimate is three times larger than that at RHIC energies and about ten times larger than the critical value.

The Bose-Einstein enhancement of identical-pion pairs at low relative momentum allows to assess the spatial scale of their emitting source. This technique, denoted femtoscopy, was used by ALICE to estimate the radii in three perpendicular directions:[5] the longitudinal and lateral radii can be connected to the volume at pion decoupling, approximately corresponding to the kinetic freeze-out time, by the relation $V_{\rm dec.} \approx (2\pi)^{3/2} R_{\rm out} \cdot R_{\rm side} \cdot R_{\rm long}$. In central Pb+Pb collisions at $\sqrt{s_{\rm NN}} = 2.76$ TeV this gives a volume of about 5000 fm$^3$, which is about two times larger than at RHIC energy. The measurement of $R_{\rm long}$ (longitudinal radius) can be used to estimate the total duration of the longitudinal expansion, i.e. of the kinetic freeze-out time of the pions in the system, which results in $\tau_{\rm dec} \approx 10$ fm/$c$ in central Pb+Pb collisions. The kinetic freeze-out is preceded by the hadronic phase. The duration of the hadronic phase is estimated by ALICE to range between one and a few fm/$c$, on the basis of the rescattering effect observed for the decay particles of hadronic resonances with lifetimes in the fm/$c$ range.

The temperature of hot strongly-interacting matter can be assessed from the emitted thermal radiation. Photons are produced during all stages of the system evolution. The photon energy distribution measured by ALICE in central Pb+Pb collisions[6] exhibits a soft exponential component with $T_{\rm slope} = 304 \pm 41$ MeV. The temperature measurement with virtual photons (dilepton pairs), which is free from Doppler effect, is a major goal for the upcoming LHC runs. The temperature of the system at chemical freeze-out is precisely estimated within statistical hadronization models (see Sec. 5) to be $T_{\rm chem} = 156 \pm 2$ MeV, a value close to the QCD critical temperature.

Figure 2 shows the evolution of a heavy-ion collision, with the macroscopic and thermodynamical properties, as well as transport coefficients, estimated on the basis of LHC measurements (see also next sections).

## 3. Expansion and hydrodynamical properties of the QGP

The experimental observable that is most sensitive to the expansion of the QGP is the azimuthal distribution of particles in the plane perpendicular to the beam direction. In semi-central collisions, the geometrical overlap region and therefore the initial matter distribution is anisotropic (almond shaped). The spatial asymmetry is converted via multiple collisions to an anisotropic momentum distribution. This anisotropy is quantified via a Fourier harmonic decomposition of the hadron azimuthal distribution, $dN/d\phi$, with respect to the reaction plane. The dominant second-order harmonic in non-central collisions is called elliptic flow coefficient and indicated with $v_2$. Large values of $v_2$ were measured at RHIC and described by hydrodynamic models of the QGP expansion. It was later demonstrated with LHC data that event-by-event fluctuations in the positions of the nucleons of the colliding nuclei generate higher-order terms ($v_3$ describes a triangular mode, $v_4$ a quadrupole one, etc.). The connection between initial spatial anisotropies and particles momentum anisotropies depends on the strength and frequency of the interactions among the QGP constituents.

These properties are encoded in QGP transport coefficients, such as the shear-viscosity-to-entropy-density ratio $\eta/s$, and their temperature dependence. Several theory groups have recently carried out multi-parametric Bayesian analyses that use a broad range of LHC (and RHIC) measurements and provide estimates of QGP properties. Figure 3 shows an example of bayesian analysis of measured flow coefficients and the resulting $\eta/s$ of the QGP: this form of matter has a ten times lower viscosity than any other fluid.[7]

The transverse-momentum-differential elliptic flow coefficients for the various species of light-flavour hadrons show a clear mass ordering with the $v_2$ trends shifted to higher $p_T$ for heavier hadrons. The ordering is consistent with collective radial expansion: between chemical and kinetic freeze-out, hadrons acquire additional momentum proportional to their mass times the transverse expansion velocity $\beta_{T,kin}$. A phenomenological fit of the mass ordering of the $p_T$ distributions provides an estimate of $\beta_{T,kin}$ and of the kinetic freeze-out temperature $T_{kin}$. In central Pb+Pb collisions ALICE obtained $\beta_{T,kin} \approx 0.65$ (2/3 of the speed of light) and $T_{kin} \approx 100$–$150$ MeV.[8]

The flow coefficients for charmed hadrons are smaller than for light-flavour hadrons, indicating that charm quarks take part in the QGP expansion but their large mass limits the momentum deflections. The comparison with models provides an estimate of the charm-quark spatial diffusion coefficient $D_s$ of the QGP, which is reported in the adimensional form $2\pi D_s \cdot T$, where $T$ is the temperature. ALICE recently reported $2\pi D_s \cdot T = 1.5$–$4.5$ for $T = T_c$.[9]

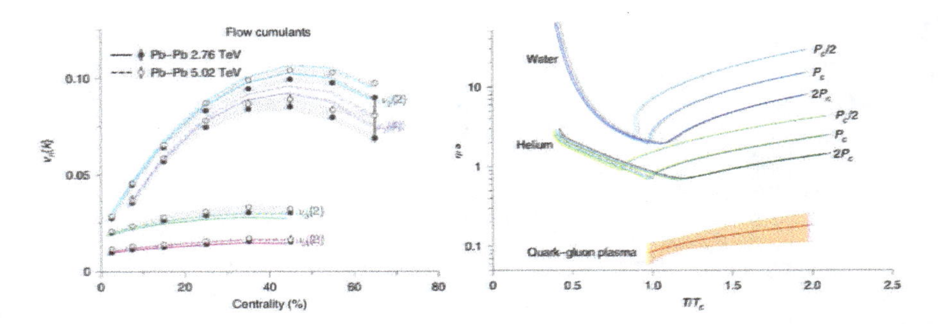


Fig. 3. Example of bayesian analysis of flow coefficients and estimated shear viscosity of the QGP, in comparison with other fluids.[7]

## 4. Effects of the QGP on colour charges

High momentum-transfer interactions between partons in the nuclei produce hard probes with QGP. One can study the impact of QGP on colour charges with fast-moving partons[10] and heavy quarks.[11] Medium-induced gluon radiations and elastic scatterings could transfer the parton energy to a large angle and may not be included in the jet. Partons could also excite a QGP wake. This "jet quenching" effect can be observed as the attenuation of the jets, and their substructure modification. At the LHC, jet quenching was observed as dijet $p_{\mathrm{T}}$ asymmetry[12,13] (Fig. 4, left), boson-jet $p_{\mathrm{T}}$ imbalance[14] and jet[15] and hadron suppression[16] in Pb+Pb collisions. Dijet missing transverse momentum and jet-hadron correlation measurements show that the medium efficiently transfers parton energy to large angles, and then emerges as low $p_{\mathrm{T}}$ particles.[13] The jet's hardening and narrowing are observed in inclusive jet fragmentation function, radial shape and substructures. Those results show that narrow jets with fewer constituents are less quenched and are more likely to pass the jet $p_{\mathrm{T}}$ threshold. In addition, QGP enhances low $p_{\mathrm{T}}$ particles inside the jet cone and broadens photon-tagged jet shape. Those results indicate the presence of medium-induced radiations and medium response. Jet transport coefficient $\hat{q}$ of QGP has been extracted from charged hadron spectra with $\frac{\hat{q}}{T^3}$ ranges in 2–11 at $T = T_c$, significantly higher than that of cold nuclear matter.[17]

Parton flavour dependence of jet quenching has been studied with jets and hadrons. High $p_{\mathrm{T}}$ gluons lose more energy than quarks due to colour factors. This effect has been studied with jet and hadron $R_{\mathrm{AA}}$ in different $\eta$ intervals, where the quark fractions vary. Further investigations are performed with photon-tagged jets and heavy quark jets. Except for jet

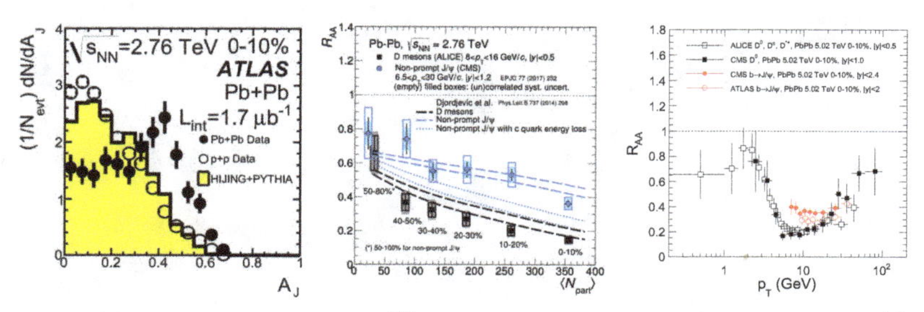


Fig. 4. Left: Dijet $p_{\mathrm{T}}$ asymmetry.[12] (Middle panel) $R_{\mathrm{AA}}$ of D from ALICE[18] and nonprompt $J/\psi$ from CMS[19] as a function of centrality. Right: $R_{\mathrm{AA}}$ of D and nonprompt $J/\psi$ as a function of hadron $p_{\mathrm{T}}$.

charge, results are consistent with the expectation of larger gluon jet suppression than quark jets. Heavy flavour hadron spectra have been studied extensively for the mass dependence of energy loss. Heavy quarks are expected to lose less energy through radiation due to the dead-cone effect,[20] the suppression of small-angle gluon radiation induced by the lower heavy quark velocity at the same kinetic energy as light quarks. By comparing the $R_{\mathrm{AA}}$ of beauty and charm mesons shown in Fig. 4, a significant flavour dependence is observed at low $p_{\mathrm{T}}$ and the effect disappears at high $p_{\mathrm{T}}$.

Quarkonia, with different hadron sizes and binding strengths, are probes of the Debye screening lengths in the medium and the medium-induced decays since finite temperature modifications produce different surviving probabilities in the QGP.[24] Measurements of inclusive bottomonia spectra show that the ground states are less suppressed than excited states (Fig. 5, left). Together with charmonia measurements, the results are consistent with the theoretical expectation. Quarkonia spectra are also sensitive to parton recombination involving two heavy quarks. At low hadron $p_{\mathrm{T}}$, the $R_{\mathrm{AA}}$ of inclusive $J/\psi$ is much larger at the LHC[23] than at RHIC[22] (Fig. 5, right). This observation can only be described in terms of recombination of charm and anticharm pairs that are deconfined in the QGP: the $c\bar{c}$ production cross section is larger by an order of magnitude at LHC than at RHIC energy, thus leading to a much larger recombination probability. The large elliptic flow of low-$p_{\mathrm{T}}$ $J/\psi$ also supports a scenario of recombining charm quarks from an expanding QGP. The recombination may happen at the phase boundary or even in the hadron phase, which may be studied further using $B_c$[25] and exotic hadrons such as $\chi_{c1}(3872)$.[26] Such measurements will become feasible with Run 3+4 data.

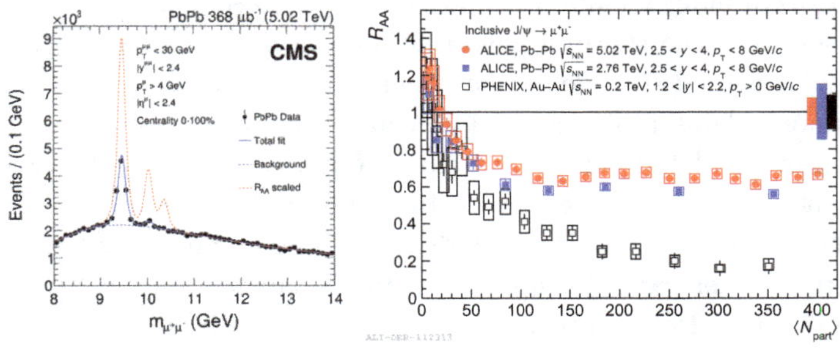


Fig. 5. Left: Sequential suppression pattern of $\Upsilon$-states observed by CMS.[21] Right: $R_{\mathrm{AA}}$ of $J/\psi$ at RHIC[22] and LHC with ALICE.[23]

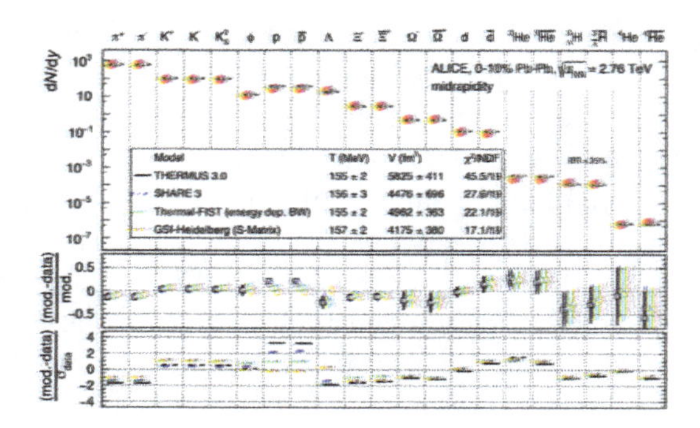


Fig. 6. Hadron yields in central Pb+Pb collisions compared to statistical hadronization model fits. Figure from Ref. 28.

## 5. Formation of hadrons from the QGP

The transition from the QGP to a hadronic system has been studied by ALICE with measurements of the yields and of the $p_{\mathrm{T}}$ distributions of a wide range of light-flavour, strange, charm hadrons, as well as light nuclei up to helium-4. The yields of light and strange mesons and baryons and of light nuclei, which span nine orders of magnitude, can be described by models that implement the concept of statistical hadronization. The hadron gas formed when the temperature falls below the chemical freeze-out temperature $T_{\mathrm{chem}}$ is described as a grand-canonical ensemble using a partition function with hadron abundances determined by their masses and quantum numbers and by a few system parameters, namely temperature, volume and chemical potentials.[27] Figure 6 shows the comparison of the fits of statistical hadronization models to ALICE measurements.[28] The fits estimate $T_{\mathrm{chem}} \approx 156 \pm 2$ MeV and a volume of about 5000 $\mathrm{fm}^3$, in agreement with the pion decoupling volume estimated with femtoscopy. The excellent fit quality supports the thermal nature of hadron distributions at freeze-out. The large measured strange baryon yields can only be described within the grand-canonical ensemble scenario, in which the strangeness quantum number is conserved globally in the system, as opposed to the local conservation in the canonical ensemble scenario. In addition, in a deconfined state, the abundances of parton species, including the strange quark, are expected to quickly reach their equilibrium values due to the low energy threshold to produce $s\bar{s}$ pairs. Hence, the observed validity of a grand-canonical description of strange-hadron production can be seen as a natural consequence of the formation of a QGP.

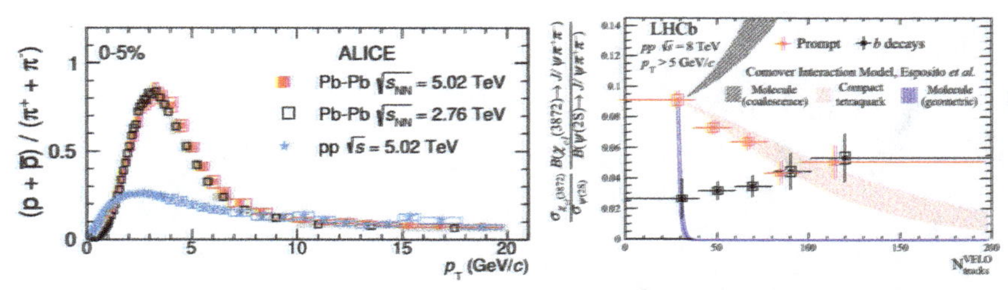


Fig. 7. Left: enhancement at intermediate $p_T$ of proton-to-pion ratio in Pb+Pb compared pp collisions.[8] Right: study on the nature of the $\chi_{c1}(3872)$ exotic hadron in high-multiplicity pp collisions.[29]

Insight on the mechanisms of hadron formation from the QGP is provided by the comparison of the $p_T$ distributions of various hadron species with pp collisions. As shown in Fig. 7(left), the hadron ratios at high $p_T$ (above 8–10 GeV/$c$) are independent of centrality and similar to pp collisions,[8] indicating that hadron formation dominantly occurs by quark and gluon fragmentation out the QGP. In the $p_T$ range below 8 GeV/$c$, instead, a large enhancement of the baryon-to-meson ratio is observed, which is qualitatively similar for light-flavour, strange and charm hadrons. This pattern is quantitatively described by models that include hadronisation via recombination of effective constituent quarks.

The abundant production of light nuclei and antinuclei can, as well, be described by modelling the coalescence of protons, neutrons and $\Lambda$ baryons or by using the statistical hadronization model. In analogy with the case of light nuclei and of charmonium, the statistical hadronization or coalescence ansatz can be used to gain unique insight on the structure (tetraquark or molecular state) of exotic hadrons, like $\chi_{c1}(3872)$ studied by LHCb in high-multiplicity pp[29] (see Fig. 7, right) and by CMS in Pb+Pb collisions.[26]

## 6. Emergence of high-density QCD effects in small collision systems

Among the more surprising results obtained from the LHC physics programme are the observations, in nearly every kind of collision studied at the LHC — even $\gamma + A$ collisions — of angular correlations between soft particles and of hadronization patterns similar to those observed in Pb+Pb collisions. The first indication of these unexpected phenomena was the observation of the "ridge", in 7 TeV pp collisions[31] and 5.02 TeV p+Pb collisions.[30] Figure 8 shows two-charged particle correlation functions measured

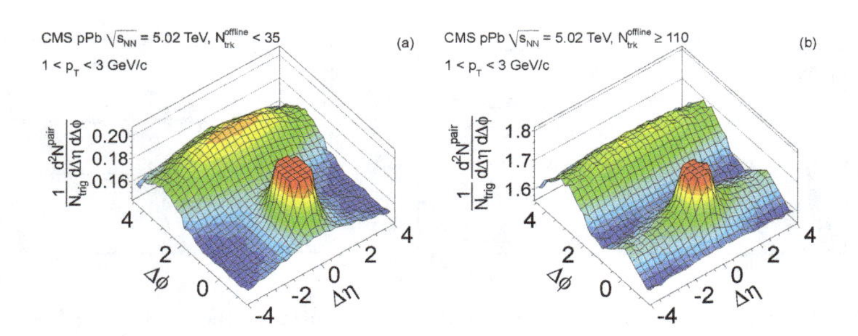


Fig. 8. Two-particle correlation functions[30] in p+Pb collisions as a function of $\Delta\phi$ and $\Delta\eta$ for low (left) and high (right) multiplicity collisions.

in low- and high-multiplicity p+Pb collisions as a function of the azimuthal angle ($\Delta\phi$) and pseudorapidity ($\Delta\eta$) separation between the two particles. The ridge — the enhancement seen in the high-multiplicity correlation function for small $\Delta\phi$ values that extends over full measured $\Delta\eta$ range — does not naturally arise in pQCD but arises in Pb+Pb collisions due to QGP expansion.

The presence of "non-flow" correlations arising from hard-scattering processes initially made interpreting the ridge difficult, especially in pp collisions where the non-flow correlations dominate. However, with the advent of higher-multiplicity pp collisions at 13 TeV and through the introduction of methods to account for non-flow correlations,[34] it was found that the ridge in both pp and p+Pb collisions results from a combination of elliptic and higher harmonics in the azimuthal distribution of produced particles similar to those observed in Pb+Pb collisions. One striking feature of the pp data is that the Fourier coefficients of the elliptic modulation, $v_2$, are independent of both $\sqrt{s}$ and multiplicity (see left panel of Fig. 9). The p+Pb $v_2$ values are similar to those in pp collisions at low multiplicity but grow with increasing multiplicity while the pp values remain essentially constant with the same value at the two energies. The $p_{\mathrm{T}}$ dependence of the pp and p+Pb $v_2$ values differ only by a multiplicative scale factor. The $p_{\mathrm{T}}$ dependence of the $v_2$ values in both pp and p+Pb collisions is found to vary with particle mass in the same manner as in Pb+Pb collisions where the mass-dependence arises from the velocity boost associated with the collective expansion of the plasma.

In addition to the angular correlations, measurements of strange hadron yields[33] and hadron flavor composition in pp and p+Pb collisions also show features similar to that observed in Pb+Pb collisions. For example, Fig. 9

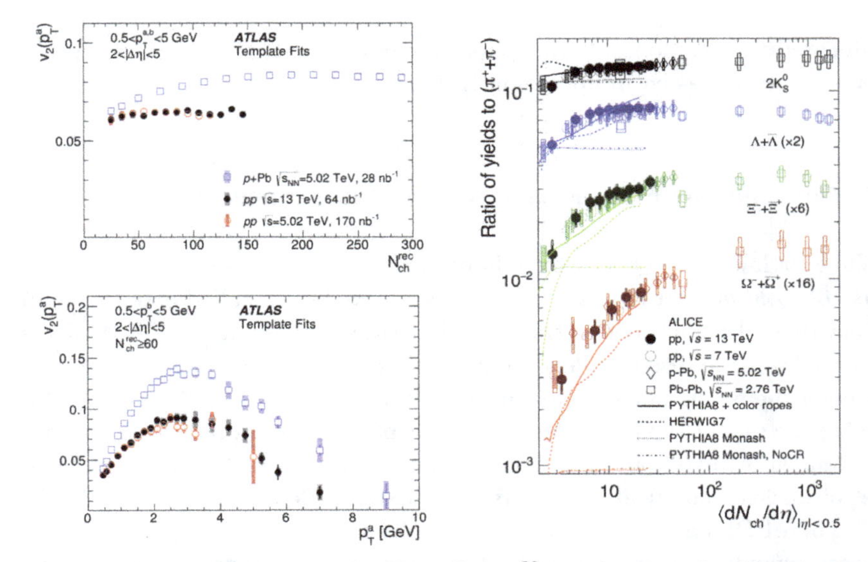


Fig. 9. Left: $v_2$ values in pp and p+Pb collisions[32] as a function of multiplicity (top) and $p_{\mathrm{T}}$ (bottom). Right: Ratios of strange particle and pion yields in pp, p+Pb, and Pb+Pb collisions versus multiplicity.[33]

shows ratios of different strange particle yields to pion yields in pp, p+Pb, and Pb+Pb collisions as a function of multiplicity. These ratios follow a common trend for pp and p+Pb data, increasing monotonically with multiplicity to match the Pb+Pb values.

The above observations have generated a great deal of theoretical interest and controversy. Viscous hydrodynamic models can simultaneously describe the $v_n$ values measured in pp, p+Pb, and Pb+Pb collisions,[35] though there still remains some theoretical disagreements regarding the applicability of hydrodynamics to small systems. The success of hydrodynamic models combined with the phenomenological similarities between the results obtained in small systems and Pb+Pb collisions has led to claims that small droplets of strongly-coupled QGP are produced in these smaller colliding systems. However, an alternative mechanism,[36] based on non-linear evolution of the partons in the initial state (saturation), combined with quantum interference in the production of final-state gluons can qualitatively reproduce some of the features observed in the pp and p+Pb data. Separately, it has been argued[37] that non-perturbative interactions between longitudinal gluon fields, i.e. "colour strings", could be responsible for the observed angular correlations in small systems and for the enhanced strange baryon yields. Resolving the theoretical ambiguities regarding the interpretation of the small-system data remains one of the main open problems to be addressed in upcoming LHC Runs.

## 7. Cold nuclear matter studies in proton-nucleus collisions

While high-density QCD effects have been observed in small systems (see Sec. 6), cold nuclear matter (CNM) effects have been studied extensively in inelastic p+Pb collisions at the LHC. Indeed, constraining the CNM effects is mandatory to disentangle a large variety of effects from QGP signatures. Among them, the nuclear parton distribution functions (nPDFs) (Fig. 10, right) are one of the main ingredients for pQCD-based predictions when a nucleus is involved, and are usually constrained via global fits on a large set of nuclear modification factors $R_{\mathrm{pPb}}$ (Fig. 10, left).

The nPDFs are not the only CNM effects. For instance, hard probes measurements such as charmonia states (e.g. $J/\psi$ and excited $\psi(2S)$ states, see Fig. 11) provide strong constraints for effects such as coherent energy loss (CEL). In particular, studies have outlined limits of the CEL model, unable to explain the difference between the two charmonia states. New final-state effects (e.g. interactions with co-moving hadronic or pre-hadronic particles) mimicking the colour screening mechanism (see Sec. 4), are needed to explain the p+Pb data. Future LHC runs and the search for jet quenching in small systems will further address these questions.

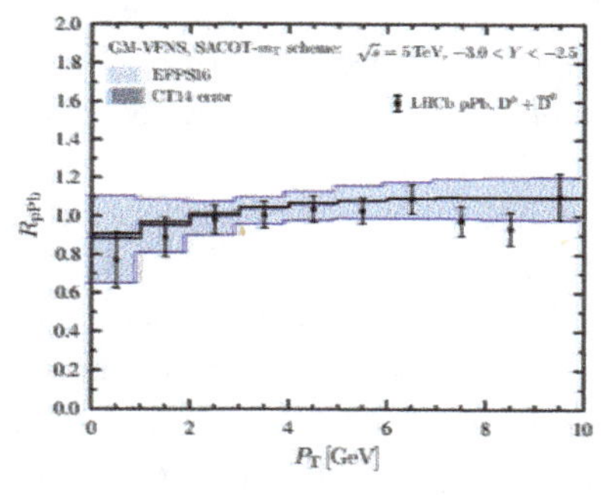


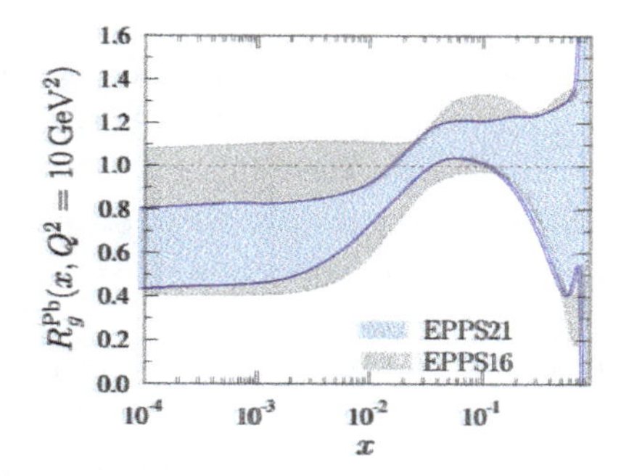


Fig. 10. Left: Comparison of nPDFs predictions with prompt $D^0$ nuclear modification factor measured by LHCb.[38] Right: improvement of the nuclear modification $R_g^{\mathrm{Pb}}$ of the gluon PDF versus the Bjorken-$x$ variable in the Pb nucleus[39] thanks to the latest LHC data.

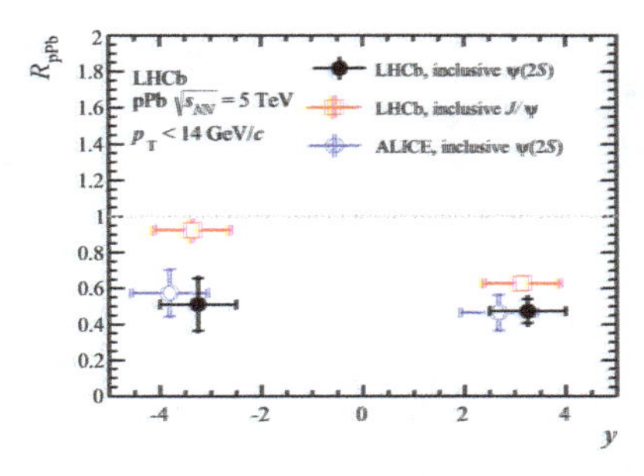


Fig. 11. Nuclear modification factor $R_{\mathrm{pPb}}$ for inclusive $J/\psi$ and $\psi(2S)$ measured by LHCb[40] and ALICE.[41]

## 8. Photon–photon collisions with LHC lead beams

Photon-photon scattering enables the study of fundamental problems in particle physics such as inherent non-linearities in QED, properties of hadronic final states, and searches for BSM particles and non-SM couplings. The large fluxes of photons associated with ultra-relativistic nuclei make it possible to perform measurements of high-energy $\gamma + \gamma$ scattering in ultra-peripheral collisions (UPC) and, even, non-ultraperipheral nuclear collisions at the LHC. One of the simplest $\gamma + \gamma$ processes that can be studied in UPC collisions is the production of dileptons. High-statistics measurements of exclusive UPC production of $\mu^+\mu^-$ and $e^+e^-$ pairs provide crucial tests of the theoretical descriptions of $\gamma+\gamma$ scattering processes needed for tests of the standard model and searches for BSM physics.[42]

Measurements of dilepton production in non-ultra-peripheral Pb+Pb collisions are interesting as the small transverse momenta of the initial-state photons makes the angular alignment of the outgoing leptons potentially sensitive to electromagnetic interactions of the leptons with the quark-gluon plasma. ATLAS has observed[43] a centrality-dependent broadening of dimuon pairs in Pb+Pb collisions that might have indicated such interactions, but recent theoretical studies[44,45] suggest the broadening results from impact parameter dependence of the initial-state photon transverse momenta.

The elastic scattering of two photons or "light-by-light" (L-by-L) scattering is interesting because it is classically forbidden and can only occur via the exchange of virtual charged particles. It could also occur via the BSM production of axions or "axion-like particles" (ALPs). ATLAS and CMS have provided a larger-than-8$\sigma$ observation of L-by-L scattering at the

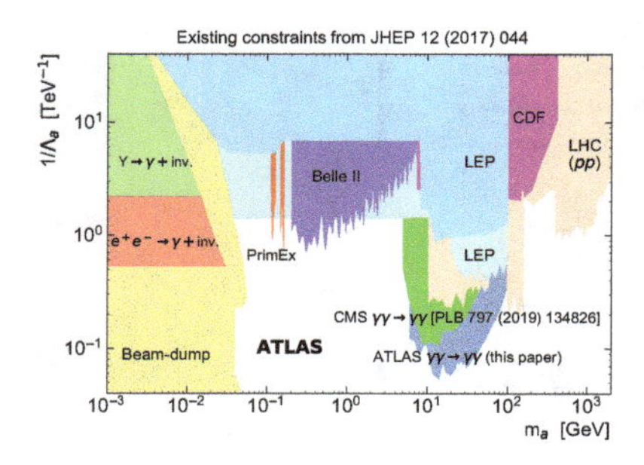


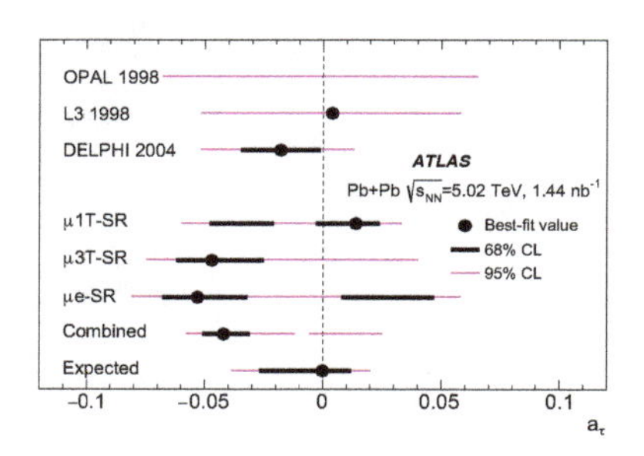


Fig. 12. Left: Constraints[46] on the mass and photon coupling strength of an hypothetical ALP from L-by-L data. Right: Constraints on $a_\tau$ from a measurement of $\gamma + \gamma \to \tau^+\tau^-$[47] compared to previous results.

LHC. Limits, shown in the left panel of Fig. 12, were placed[46] on the coupling of photons to a hypothetical ALP and on its mass, with a significant improvement over previous results in the mass range 6–100 GeV.

Another example of the potential of $\gamma + \gamma$ collisions for probing BSM physics is the gyromagnetic ratio of the $\tau$ lepton. Because of its larger mass, the $\tau$ $g-2$ and, in turn, the $\gamma + \gamma \to \tau^+\tau^-$ cross section, is uniquely sensitive to BSM contributions. ATLAS reported[47] an observation of $\tau^+\tau^-$ production. The constraints on $a_\tau \equiv \frac{1}{2}(g_\tau - 2)$ are shown in the right-hand panel of the figure and compared to LEP results. The new results are a significant improvement, though interference between SM and possible BSM contributions causes the 95% confidence region to allow both signs for $a_\tau$.

## 9. Prospects for future LHC heavy-ion runs

The next decade will represent a precision era for high-density QCD with small and large colliding systems.[48] With the LHC Runs 3-4, an increase of the delivered luminosity by a factor of about 10 compared to Run 2 is foreseen for all experiments in p+Pb and Pb+Pb collisions. The emergence of QGP-like effects and possibly the formation of small QGP droplets will also be explored with oxygen–oxygen collisions. Proton–oxygen collisions will constrain models of primary cosmic-ray interactions.

For the LHC Runs 5-6, a next-generation detector, ALICE 3, is proposed (see Chapter 14), to enable new measurements in the heavy-flavour sector and multi-differential measurements of dielectron emission. A further LHCb upgrade (see Chapter 13) will grant access to central Pb+Pb collisions and to fixed-target collisions on a polarized target.

## References


1. R. Bruce et al., Performance and luminosity models for heavy-ion operation at the CERN Large Hadron Collider, *Eur. Phys. J. Plus.* **136** (7), 745 (2021). doi: 10.1140/epjp/s13360-021-01685-5.
2. S. Borsanyi et al., Full result for the QCD equation of state with 2+1 flavors, *Phys. Lett. B.* **730**, 99–104 (2014). doi: 10.1016/j.physletb.2014.01.007.
3. A. Bazavov et al., Equation of state in (2+1)-flavor QCD, *Phys. Rev. D.* **90**, 094503 (2014). doi: 10.1103/PhysRevD.90.094503.
4. CMS Collaboration, Measurement of the pseudorapidity and centrality dependence of the transverse energy density in PbPb collisions at $\sqrt{s_{NN}}$ = 2.76 TeV, *Phys. Rev. Lett.* **109**, 152303 (2012). doi: 10.1103/PhysRevLett.109.152303.
5. ALICE Collaboration, Two-pion Bose-Einstein correlations in central Pb-Pb collisions at $\sqrt{s_{NN}}$ = 2.76 TeV, *Phys. Lett. B.* **696**, 328–337 (2011). doi: 10.1016/j.physletb.2010.12.053.
6. ALICE Collaboration, Direct photon production in Pb-Pb collisions at $\sqrt{s_{NN}}$ = 2.76 TeV, *Phys. Lett. B.* **754**, 235–248 (2016). doi: 10.1016/j.physletb.2016.01.020.
7. J. E. Bernhard et al., Bayesian estimation of the specific shear and bulk viscosity of quark–gluon plasma, *Nature Phys.* **15** (11), 1113–1117 (2019). doi: 10.1038/s41567-019-0611-8.
8. ALICE Collaboration, Production of charged pions, kaons, and (anti-) protons in Pb-Pb and inelastic *pp* collisions at $\sqrt{s_{NN}}$ = 5.02 TeV, *Phys. Rev. C.* **101** (4), 044907 (2020). doi: 10.1103/PhysRevC.101.044907.
9. ALICE Collaboration, Transverse-momentum and event-shape dependence of D-meson flow harmonics in Pb–Pb collisions at $\sqrt{s_{NN}}$ = 5.02 TeV, *Phys. Lett. B.* **813**, 136054 (2021). doi: 10.1016/j.physletb.2020.136054.
10. J. D. Bjorken. FERMILAB-PUB-82-059-THY. Technical report (1982).
11. X. Dong et al., Open Heavy-Flavor Production in Heavy-Ion Collisions, *Ann. Rev. Nucl. Part. Sci.* **69**, 417–445 (2019). doi: 10.1146/annurev-nucl-101918-023806.
12. ATLAS Collaboration, Observation of a Centrality-Dependent Dijet Asymmetry in Lead-Lead Collisions at $\sqrt{s_{NN}}$ = 2.77 TeV with the ATLAS Detector at the LHC, *Phys. Rev. Lett.* **105**, 252303 (2010). doi: 10.1103/PhysRevLett.105.252303.
13. CMS Collaboration, Observation and studies of jet quenching in PbPb collisions at nucleon-nucleon center-of-mass energy = 2.76 TeV, *Phys. Rev. C.* **84**, 024906 (2011). doi: 10.1103/PhysRevC.84.024906.
14. CMS Collaboration, Studies of jet quenching using isolated-photon+jet correlations in PbPb and *pp* collisions at $\sqrt{s_{NN}}$ = 2.76 TeV, *Phys. Lett. B.* **718**, 773–794 (2013). doi: 10.1016/j.physletb.2012.11.003.
15. ATLAS Collaboration, Measurement of the jet radius and transverse momentum dependence of inclusive jet suppression in lead-lead collisions at $\sqrt{s_{NN}}$= 2.76 TeV with the ATLAS detector, *Phys. Lett. B.* **719**, 220–241 (2013). doi: 10.1016/j.physletb.2013.01.024.

16. ALICE Collaboration, Suppression of Charged Particle Production at Large Transverse Momentum in Central Pb-Pb Collisions at $\sqrt{s_{NN}} = 2.76$ TeV, *Phys. Lett. B.* **696**, 30–39 (2011). doi: 10.1016/j.physletb.2010.12.020.
17. JET Collaboration, Extracting the jet transport coefficient from jet quenching in high-energy heavy-ion collisions, *Phys. Rev. C.* **90** (1), 014909 (2014). doi: 10.1103/PhysRevC.90.014909.
18. ALICE Collaboration, Centrality dependence of high-$p_T$ D meson suppression in Pb-Pb collisions at $\sqrt{s_{\mathrm{NN}}} = 2.76$ TeV, *JHEP.* **11**, 205 (2015). doi: 10.1007/JHEP11(2015)205. [Addendum: JHEP 06, 032 (2017)].
19. CMS Collaboration, Suppression and azimuthal anisotropy of prompt and nonprompt J/$\psi$ production in PbPb collisions at $\sqrt{s_{\mathrm{NN}}} = 2.76$ TeV, *Eur. Phys. J. C.* **77** (4), 252 (2017). doi: 10.1140/epjc/s10052-017-4781-1.
20. Y. L. Dokshitzer and D. E. Kharzeev, Heavy quark colorimetry of QCD matter, *Phys. Lett. B.* **519**, 199–206 (2001). doi: 10.1016/S0370-2693(01) 01130-3.
21. CMS Collaboration, Measurement of nuclear modification factors of $\Upsilon$(1S), $\Upsilon$(2S), and $\Upsilon$(3S) mesons in PbPb collisions at $\sqrt{s_{\mathrm{NN}}} = 5.02$ TeV, *Phys. Lett. B.* **790**, 270–293 (2019). doi: 10.1016/j.physletb.2019.01.006.
22. PHENIX Collaboration, $J/\psi$ suppression at forward rapidity in Au+Au collisions at $\sqrt{s_{NN}} = 200$ GeV, *Phys. Rev. C.* **84**, 054912 (2011). doi: 10.1103/PhysRevC.84.054912.
23. ALICE Collaboration, J/$\psi$ suppression at forward rapidity in Pb-Pb collisions at $\sqrt{s_{\mathrm{NN}}} = \mathbf{5.02}$ TeV, *Phys. Lett. B.* **766**, 212–224 (2017). doi: 10.1016/j.physletb.2016.12.064.
24. T. Matsui and H. Satz, $J/\psi$ Suppression by Quark-Gluon Plasma Formation, *Phys. Lett. B.* **178**, 416–422 (1986). doi: 10.1016/0370-2693(86)91404-8.
25. CMS Collaboration, Observation of the $B_c^+$ Meson in Pb-Pb and pp Collisions at $\sqrt{s_{NN}} = 5.02$ TeV and Measurement of its Nuclear Modification Factor, *Phys. Rev. Lett.* **128** (25), 252301 (2022). doi: 10.1103/PhysRevLett.128. 252301.
26. CMS Collaboration, Evidence for X(3872) in Pb-Pb Collisions and Studies of its Prompt Production at $\sqrt{s_{NN}} = 5.02$ TeV, *Phys. Rev. Lett.* **128** (3), 032001 (2022). doi: 10.1103/PhysRevLett.128.032001.
27. A. Andronic et al., Decoding the phase structure of QCD via particle production at high energy, *Nature.* **561** (7723), 321–330 (2018). doi: 10.1038/ s41586-018-0491-6.
28. ALICE Collaboration, Production of $^4$He and $^4\overline{\mathrm{He}}$ in Pb-Pb collisions at $\sqrt{s_{\mathrm{NN}}} = 2.76$ TeV at the LHC, *Nucl. Phys. A.* **971**, 1–20 (2018). doi: 10. 1016/j.nuclphysa.2017.12.004.
29. LHCb Collaboration, Performance of the LHCb Vertex Locator, *JINST.* **9**, P09007 (2014). doi: 10.1088/1748-0221/9/09/P09007.
30. CMS Collaboration, Observation of Long-Range Near-Side Angular Correlations in Proton-Lead Collisions at the LHC, *Phys. Lett. B.* **718**, 795–814 (2013). doi: 10.1016/j.physletb.2012.11.025.

31. CMS Collaboration, Observation of Long-Range Near-Side Angular Correlations in Proton-Proton Collisions at the LHC, *JHEP.* **09**, 091 (2010). doi: 10.1007/JHEP09(2010)091.
32. ATLAS Collaboration, Measurements of long-range azimuthal anisotropies and associated Fourier coefficients for *pp* collisions at $\sqrt{s} = 5.02$ and 13 TeV and $p$+Pb collisions at $\sqrt{s_{\rm NN}} = 5.02$ TeV with the ATLAS detector, *Phys. Rev. C.* **96** (2), 024908 (2017). doi: 10.1103/PhysRevC.96.024908.
33. ALICE Collaboration, Enhanced production of multi-strange hadrons in high-multiplicity proton-proton collisions, *Nature Phys.* **13**, 535–539 (2017). doi: 10.1038/nphys4111.
34. ATLAS Collaboration, Observation of Long-Range Elliptic Azimuthal Anisotropies in $\sqrt{s} = 13$ and 2.76 TeV *pp* Collisions with the ATLAS Detector, *Phys. Rev. Lett.* **116** (17), 172301 (2016). doi: 10.1103/PhysRevLett.116.172301.
35. R. D. Weller and P. Romatschke, One fluid to rule them all: viscous hydrodynamic description of event-by-event central p+p, p+Pb and Pb+Pb collisions at $\sqrt{s} = 5.02$ TeV, *Phys. Lett. B.* **774**, 351–356 (2017). doi: 10.1016/j.physletb.2017.09.077.
36. K. Dusling et al., Novel collective phenomena in high-energy proton–proton and proton–nucleus collisions, *Int. J. Mod. Phys. E.* **25** (01), 1630002 (2016). doi: 10.1142/S0218301316300022.
37. C. Bierlich et al., arXiv:2203.11601. Technical report (2022).
38. LHCb Collaboration, Prompt $\Lambda_c^+$ production in *p*Pb collisions at $\sqrt{s_{NN}} = 5.02$ TeV, *JHEP.* **02**, 102 (2019). doi: 10.1007/JHEP02(2019)102.
39. K. J. Eskola et al., EPPS21: a global QCD analysis of nuclear PDFs, *Eur. Phys. J. C.* **82** (5), 413 (2022). doi: 10.1140/epjc/s10052-022-10359-0.
40. LHCb Collaboration, Study of $\psi(2S)$ production and cold nuclear matter effects in pPb collisions at $\sqrt{s_{NN}} = 5$TeV, *JHEP.* **03**, 133 (2016). doi: 10.1007/JHEP03(2016)133.
41. ALICE Collaboration, Suppression of $\psi$(2S) production in p-Pb collisions at $\sqrt{s_{\rm NN}} = 5.02$ TeV, *JHEP.* **12**, 073 (2014). doi: 10.1007/JHEP12(2014)073.
42. D. d'Enterria et al., (2022). arXiv: 2203.05939.
43. ATLAS Collaboration, Observation of centrality-dependent acoplanarity for muon pairs produced via two-photon scattering in Pb+Pb collisions at $\sqrt{s_{\rm NN}} = 5.02$ TeV with the ATLAS detector, *Phys. Rev. Lett.* **121** (21), 212301 (2018). doi: 10.1103/PhysRevLett.121.212301.
44. W. Zha, J. D. Brandenburg, Z. Tang, and Z. Xu, Initial transverse-momentum broadening of Breit-Wheeler process in relativistic heavy-ion collisions, *Phys. Lett. B.* **800**, 135089 (2020). doi: 10.1016/j.physletb.2019.135089.
45. B.-W. Xiao, F. Yuan, and J. Zhou, Momentum Anisotropy of Leptons from Two Photon Processes in Heavy Ion Collisions, *Phys. Rev. Lett.* **125** (23), 232301 (2020). doi: 10.1103/PhysRevLett.125.232301.
46. ATLAS Collaboration, Measurement of light-by-light scattering and search for axion-like particles with 2.2 nb$^{-1}$ of Pb+Pb data with the ATLAS detector, *JHEP.* **11**, 050 (2021). doi: 10.1007/JHEP11(2021)050.

47. ATLAS Collaboration (2022). accepted by PRL, arXiv: 2204.13478.
48. Z. Citron et al., Report from Working Group 5: Future physics opportunities for high-density QCD at the LHC with heavy-ion and proton beams, *CERN Yellow Rep. Monogr.* **7**, 1159–1410 (2019). doi: 10.23731/CYRM-2019-007.1159.

© 2024 The Author(s)
https://doi.org/10.1142/9789811280184_0006

# Chapter 6

# HL-LHC configuration and operational challenges

Andrea Apollonio, Xavier Buffat, Roderik Bruce, Riccardo De Maria,
Massimo Giovannozzi, Giovanni Iadarola, Anton Lechner, Elias Métral,
Guido Sterbini, Rogelio Tomás and Markus Zerlauth

*CERN,*
*Esplanade des Particules 1, 1211 Meyrin, Switzerland*

Recently, the operational configuration of the HL-LHC along its first Run has been carefully established to reach nominal performance. This chapter presents the key beam and machine parameters together with the most critical operational challenges. The contents of this chapter reflect the project situation prior to experience from the LHC Run 3.

## 1. Proton parameters and machine optics

The HL–LHC operational scenario for Run 4 is continuously evolving following changes in the hardware configuration, findings in beam dynamics, and schedule updates.[1–3] The latest HL-LHC Interaction Region (IR) layout is shown in Fig. 1 compared to the current LHC. The larger aperture quadrupoles next to the Interaction Point (IP) and the Crab Cavities (CC) allow to reduce the IP beam size while compensating for the loss of geometric overlap during collisions, hence maximizing luminosity. Other key HL-LHC hardware upgrades are described in the following.

*The installation of new sextupoles (MS10) in the dispersion suppressor.* The loss of Dynamic Aperture (DA, see Sec. 2.3) of about 0.5 $\sigma$ for not having MS10 is acceptable for optics with $\beta^* \geq 20$ cm in IP1 and IP5, which is the current assumption for Run 4. However, at $\beta^* = 15$ cm, considered for possible new Run 4 scenarios, the DA loss is larger and these sextupoles are mandatory.[4–6] Figure 2 shows the machine optics at $\beta^* = 15$ cm.

This is an open access article published by World Scientific Publishing Company. It is distributed under the terms of the Creative Commons Attribution 4.0 (CC BY) License.

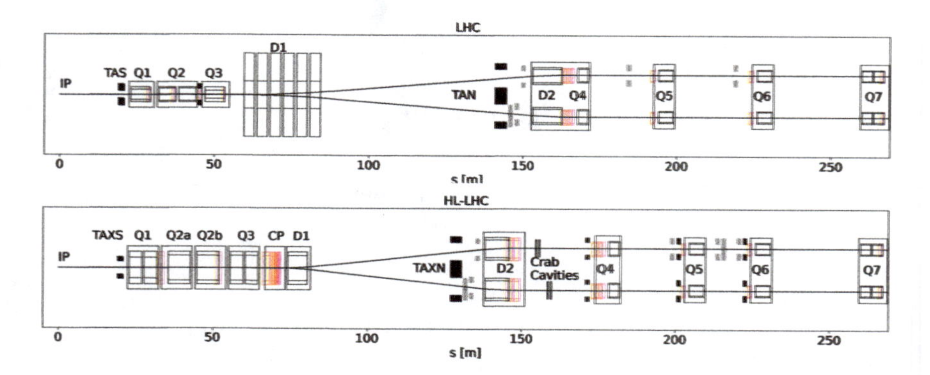


Fig. 1. A schematic comparison of the right side of the interaction region in Points 1 and 5 between LHC (top) and the HL-LHC (bottom). Black boxes represent main magnets and red boxes linear and non-linear correctors. Black filled boxes represent fixed absorbers while gray filled boxes represent collimators. The HL-LHC has large aperture magnets up to D2, a cold D1 and separated D2-Q4 to host the crab cavities.

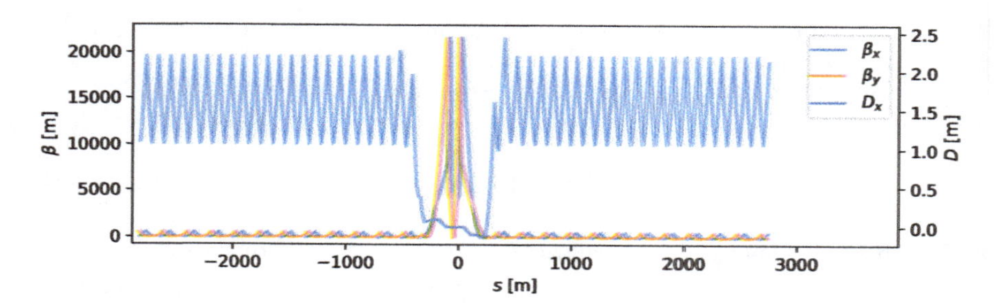


Fig. 2. Plot of the Twiss parameters of the HL-LHC at $\beta^* = 15$ cm. Peak $\beta$-function reaches 20 km in the triplet and about 600 m in the arcs around Point 1 and 5. This enhances the sensitivity to field imperfections.

*Low impedance upgrade of the secondary Carbon Fiber Composite (CFC) collimators.* Most of these collimators are being replaced with Mo-coated collimators, made of Molybdenum-graphite, to guarantee beam stability. Beam experiments in Run 3 should determine if the low impedance collimator upgrade should be carried out in full, whether a reduction in the number of upgraded units would be feasible.

*The Hollow Electron Lens (HEL).* The HEL is an advanced tool for active control of the diffusion speed of halo particles, which will serve to mitigate losses from fast processes. Due to resource limitations the HEL will not be ready for Run 4. A primary collimator gap at 8.5 $\sigma$ is considered the main mitigation against halo issues in the absence of the HEL. The need for additional mitigation measures in Run 4, such as reducing the bunch charge, will need to be evaluated with dedicated measurements in Run 3.

Figure 3 shows a schematic view of the Run 4 operational cycle including key beam parameters and luminosity. The abrupt jumps in bunch intensity and emittances during the collision adjustment process, just before 2.5 h, correspond to the intensity loss and emittance growth budgets assigned for the interval between injection and the start of collisions; however, here they are pessimistically lumped when collisions are established. The slow horizontal emittance growth at injection is due to intra-beam scattering

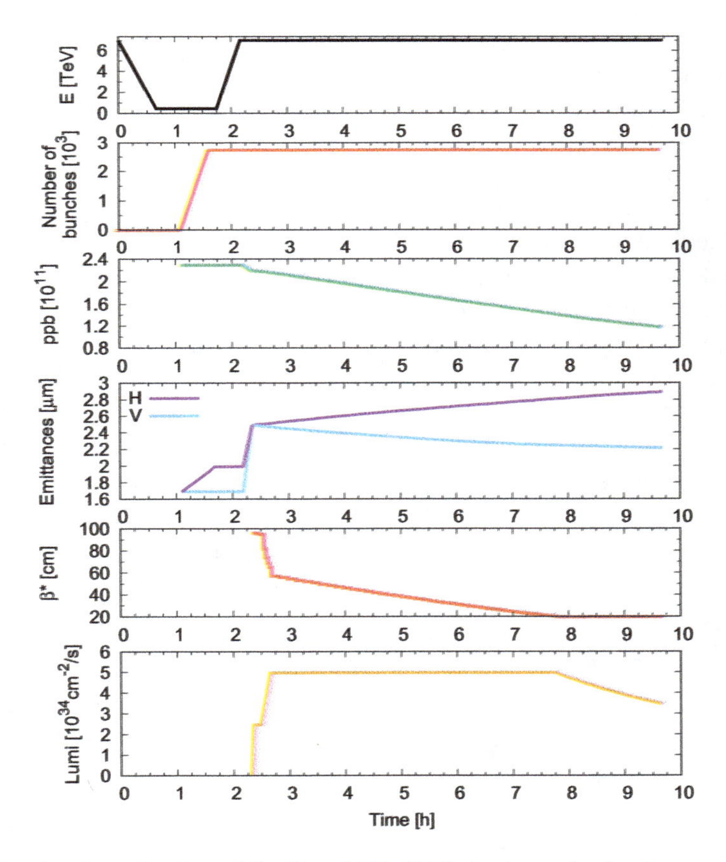


Fig. 3. A schematic view of the Run 4 HL–LHC physics cycle showing energy cycle, number of bunches, protons per bunch (ppb), transverse emittances (Batch Compression Merging and Splitting beam case), and luminosity (top to bottom) versus time until the beam dump.

(IBS). The luminosity starts with a step at $2.5\times10^{34}$ cm$^{-2}$s$^{-1}$ followed by a linear ramp to meet cryogenic requests. The bunch intensity and emittance evolution during physics include burn-off, IBS, synchrotron radiation (SR) damping, and emittance growth from CC noise. Emittance growth from luminosity burn-off has a small effect on the integrated luminosity of HL–LHC of approximately 1%, not included here, and should be further studied for more accurate predictions.[7]

Assuming no limitations to the beam parameters in Run 4, the potential HL–LHC performance ramp-up is given in Fig. 4, allowing to integrate above 715 fb$^{-1}$ over the 4 years of operation. The minimum $\beta^*$ in Run 4 is tentatively kept to 20 cm but 15 cm is being considered.

In the first year of operation, bunch intensity is assumed to match the updated Run 3 expectation of $1.8\times10^{11}$ ppb with minimum $\beta^* = 30$ cm and without CCs in physics (but commissioning them with dedicated machine time). Crossing angle is assumed to be about 450 $\mu$rad as validated with DA simulations. It is foreseen to steadily reduce the crossing angle during the physics fill as the bunch population decays to maximize performance and reduce the peak radiation dose to the triplet magnets.

The beam-based IR non-linear corrections are expected to require considerable commissioning time and iterations between the different magnet types. Therefore, it is assumed that optics commissioning in the first years will only include correction magnets up to the octupolar order, leaving the commissioning of the decapolar and dodecapolar correctors for the years with lower $\beta^*$. Simulations have confirmed that DA is sufficient at $\beta^* = 30$ cm without decapolar and dodecapolar IR corrections. Moreover,

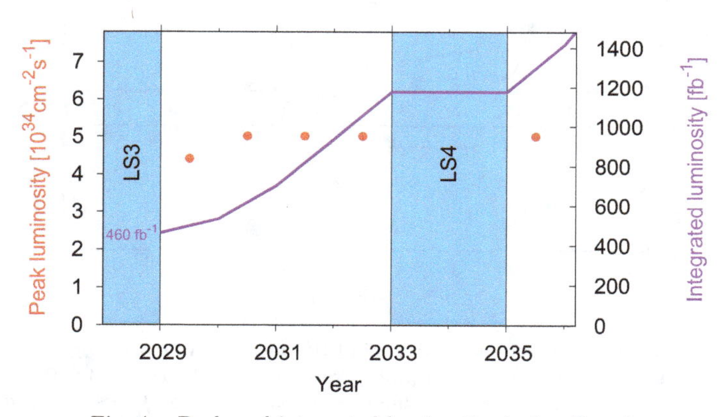


Fig. 4. Peak and integrated luminosity during Run 4.

techniques to speed-up the optics commissioning including high-order corrections are being developed and will require dedicated machine experiments in Run 3.

Bunch intensity could be limited in Run 4 due to the absence of the HEL or if RF voltage limitations are encountered at injection. Taking these aspects into account, a minimum bunch intensity of $1.8 \times 10^{11}$ ppb is estimated to be easily achievable. Further studies are ongoing to investigate the maximum bunch intensity feasible with the current RF system. If bunch charge is limited to $1.8 \times 10^{11}$ ppb at injection, the fill shortens by more than 2 hours and the levelling time by more than 3 hours with respect to the baseline shown in Fig. 3. Annual integrated luminosity (assuming 160 days) is reduced from 242 to 194 $\mathrm{fb}^{-1}$, reducing the Run 4 expected integrated luminosity by about 20% for the case with $\epsilon = 2.5$ $\mu$m. Figure 5 shows the annual integrated luminosity versus bunch charge at injection in the range between $1.8 \times 10^{11}$ ppb and the baseline value. Mitigation measures imply reducing $\beta^*$ or the crossing angle. The first requires that the MS10 sextupoles are installed during Long Shutdown 3 (LS) to guarantee sufficient lifetime. The latter requires that long-range beam-beam compensators, not yet in the baseline, are installed.

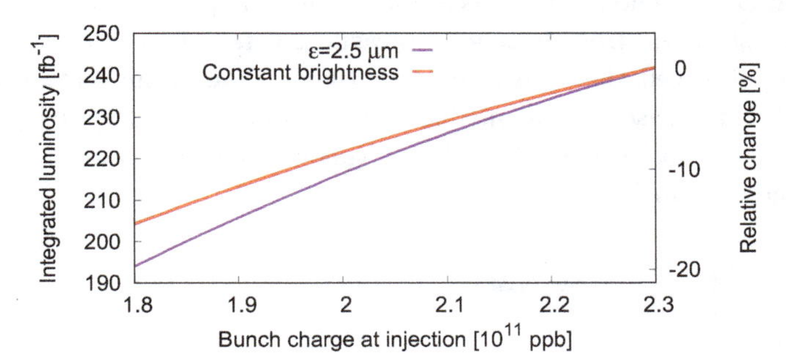


Fig. 5. Annual integrated luminosity versus bunch charge at injection in Run 4.

In the following the most critical operational challenges are reviewed.

## 2. Proton operational challenges

### 2.1. *Heat load and e-cloud*

The LHC and HL-LHC cryogenic magnets are equipped with actively cooled beam screens, which intercept beam induced heating mainly due to synchrotron radiation, impedance and e-cloud effects. During the LHC Run 2 large heat loads were observed on the beam screens during operation with the nominal bunch spacing of 25 ns. In particular, the heat loads in some of the arcs reached levels close to the design cooling capacity of 8 kW/arc. In all sectors, the heat-loads were significantly larger than expected from impedance and synchrotron radiation.[8–10]

By analyzing the heat load data collected during Run 2 and comparing them against models and simulations, it was possible to conclude that a dominant fraction of the observed heat loads is due to electron cloud effects, as a result of a larger than expected Secondary Electron Yield (SEY) of the beam screen surfaces. During the LS2 (2019–2022) surface analyses were conducted of beam screens extracted from the accelerator, which identified specific surface modifications associated with the magnets showing the highest heat load, namely the presence of cupric oxide (CuO) and a very low carbon concentration. These modifications are associated with a larger SEY and therefore with a stronger e-cloud.[11]

Numerical simulations can be used to estimate the arc heat loads expected for the HL-LHC beam parameters. Figure 6 shows the arc heat loads expected for the most critical LHC arc (S81) as a function of the bunch intensity. The predictions are made assuming for each cryogenic cell the SEY estimated from heat load measurements collected during Run 2. It can be observed that the heat load contributions from e-cloud are not expected to increase significantly for intensities above $1.8 \times 10^{11}$ p/bunch. Such a feature has been confirmed experimentally using short bunch trains at the end of Run 2.[12]

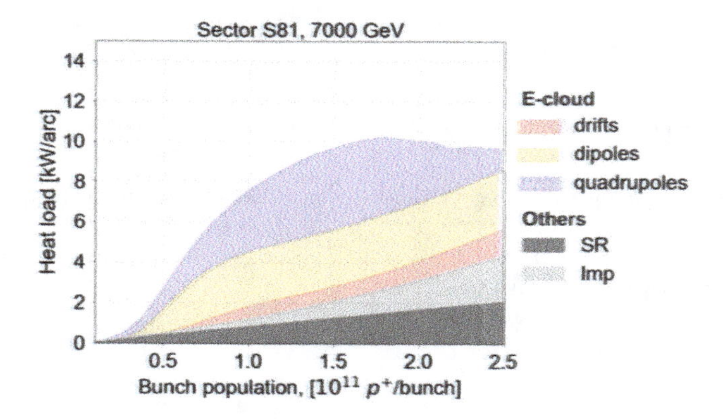


Fig. 6. Expected heat loads at 7 TeV as a function of the bunch intensity for the most critical arc of the LHC (to be compared to the available cooling capacity of 10 kW/arc).

During Run 2, the LHC cryogenics has been operated in an optimized configuration (using one cold-compressor unit to serve two consecutive sectors). The cryoplants feeding the high-load sectors have been recently characterized by the cryogenics team, and they were found to perform better than their design specifications, being able to deliver 10 kW/arc. Assuming that the cryoplants can reliably provide such a cooling capacity and that no degradation of the SEY will take place after Run 2, the HL-LHC nominal beam configuration is expected to be compatible with the limits defined by the cryogenic system. In case these conditions are not met, surface treatments will need to be performed in order to reduce the SEY of the beam-screen surfaces. Alternatively, hybrid filling patterns could be used to partially mitigate the performance loss.[13]

### 2.2. *Beam instabilities*

Once the machine is well scrubbed, the main limitation due to beam instabilities occurs at flat top, mostly because of the impedance of the collimators. Indeed, to maintain the cleaning efficiency, the collimators move closer to the beam during the energy ramp, following the adiabatic damping of the transverse emittances. This significantly increases their impedance. At the same time the reduction of the physical emittance of the beam reduces the effectiveness of the Landau octupoles. These magnets located in the arcs generate a spread in the transverse tunes, thus stabilising the so-called weak head-tail instability driven by the machine impedance via the mechanism of Landau damping. The strength of the damping is however limited by the maximum strength of the octupole magnets and by their detrimental impact on the beam lifetime.[14] In order to allow for the highest beam brightness without exceeding those limits, new collimators were designed with the goal of maintaining the robustness to radiation, while reducing their resistivity and consequently their impedance. While the LHC collimators are based on CFC jaws, the new primary and secondary collimators are based on Molybdenum-Graphite blocks. The secondary collimators feature in addition a Molybdenum coating[15] in order to optimise their impedance. Along with this technological improvement, it is necessary to master the various mechanisms that affect Landau damping taking into account the operational constraints.

Non-linear magnetic components in the final focusing quadrupoles can significantly impact the tune spread when the beams are squeezed at the IP. Since this contribution may enhance or cancel the tune spread driven by the Landau octupoles in an uncontrolled manner, the measurement and correction of these non-linearities is critical to maintain the beam stability.[16] Linear coupling can severely reduce the tune spread driven by the octupoles.[17] This mechanism caused several instabilities leading to beam aborts during the first run of the LHC. With the implementation of an online coupling measurement and correction tool, this instability was no longer observed. A tight control of linear coupling will be needed for the HL–LHC.[14]

The beam-beam interactions also affect the beam stability, mainly through their impact on the tune spread.[14] The polarity of the octupoles is chosen to interact constructively with the long-range beam-beam interactions which dominate in the phase from the end of the ramp to the start of collision. This phase is indeed the most critical for beam stability, since afterwards the head-on collisions generate a much larger tune spread which efficiently suppresses all expected instabilities. Configurations featuring an offset between the beams in the order of the transverse rms beam size require a special attention due to the very specific non-linear behaviour of the beam-beam force in this range. Yet no performance limitations are expected due to this effect in the HL–LHC.

The relation between instabilities arising with a latency of several minutes and noise acting on the beam was both observed and explained.[18] This mechanism of loss of Landau damping puts tight constraints on the noise of existing and new equipment. In addition, this mechanism favours shortening of the most critical phase, i.e. between the end of the ramp and the establishment of collisional orbits. The operation with a combined ramp and squeeze as well as $\beta^*$ levelling is therefore greatly beneficial, by reducing to the minimum the phase of the $\beta^*$ squeeze with non-colliding beams.

### 2.3. *Beam lifetime*

The expected performance of the HL–LHC relies not only on very challenging beam parameters, e.g. beam intensity and emittance, to be achieved at the start of the luminosity production, but also on preserving those parameters throughout the luminosity production period of the fill. The beam lifetime is the figure of merit to quantify the time constant of the beam intensity decay. In an ideal collider, the beam lifetime should be dominated by the burn-off losses induced by luminosity.

The approach assumed at the time of the LHC design[19] evolved quite strongly after the beginning of the operations. Indeed, the original paradigm based on well-separated, quasi-static, and sequential changes to the ring optics has been replaced, following also the implementation of the so-called Achromatic Telescopic Squeeze (ATS) optics,[20] by rapid and parallel changes to the machine configuration, encompassing, as an example, squeeze of the insertion optics and variation of the strength of the Landau octupoles to fight collective instabilities. A similar approach has been extended to the stage when the beams are put in collision, when several luminosity levelling options have been already implemented, such as levelling by varying the parallel separation and the crossing angle of the beams at the interaction points.

This dynamic environment is applied to a system in which nonlinear effects perturb the beam motion, be them generated by the unavoidable magnetic field errors stemming from the superconducting magnets or the strong beam-beam interactions. Nonlinear effects induce resonance excitation, which, combined with IBS, synchrotron radiation, luminosity effect,[21,22] and with other time-dependent perturbations, lead to emittance growth or degradation of the beam lifetime, finally affecting the collider performance in terms of luminosity evolution.

Possible sources of time-dependent perturbations are the ripples in the power converters of the various magnet families, perturbations coming from the UPS connected to the machine electronics, or from specific devices, e.g. CC,[23] transverse damper,[24] HEL,[25,26] etc. In the HL–LHC, the revolution frequency is about 11.24 kHz, and a noise spectrum with frequency larger than about 1 kHz can affect the beam lifetime. As an example, the power converter ripples may introduce pseudo-random effects in the beam dynamics, thus creating a diffusive behaviour of the beam distribution leading to emittance growth and losses that affect the beam lifetime. For these reasons, important efforts are devoted to scrutinize the limiting circuits of the machine[27–29] with the goal of devising mitigation measures to the ripples on existing and new power converters. As an example, in Fig. 7 we report, starting from the measured LHC transverse noise spectra of the beams, the simulated impact on Beam 1 (B1 - the clock-wise beam) and Beam 2 (B2 - the counter-clock-wise beam) intensity decay. The simulations are in agreement with the observed lifetime difference of the two beams in LHC.

In this light, the design studies for the HL–LHC have tackled a series of new challenges. Since the LHC design, the concept of DA, i.e. the extent of the phase-space region in which bounded motion occurs, has been used as the key figure of merit to scrutinize the suitability of the field quality

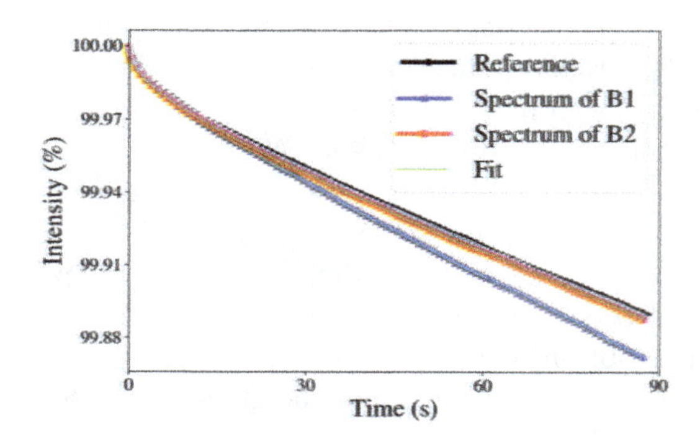


Fig. 7. Intensity evolution in the absence of power supply ripple (black), including the measured power supply ripple spectrum of Beam 1 (blue) and Beam 2 (red) (the parameters used for the numerical simulations are listed in Ref. 28).

of the various magnet families.[19,30] To make such a rather abstract, i.e. not directly observable using the machine diagnostic, figure of merit better suited to the analysis of the actual collider performance, a link between DA and beam lifetime has been established[31] and used also to derive models for the luminosity lifetime.[32,33] Experimental methods to determine the DA in a circular accelerator[34] were developed (note that a qualitative DA estimate was obtained by displacing the beam close to the DA border,[35]) together with improved tracking capability leveraging on the potential of the Graphical Processing Unit hardware.[36,37]

It is worth noting that a consistent approach to the determination of the beam lifetime has to be complemented by means and tools to determine the evolution of the beam distribution, which is of paramount importance to assess emittance growth and beam losses. Therefore, new emphasis has been put on the long-term numerical tracking (e.g. about 30 min, corresponding to $2 \times 10^7$ turns in the LHC) and the derivation of diffusive models, basing this framework on a fundamental theorem of the theory of dynamical systems, i.e. Nekhoroshev theorem.[38–40] This led to a number of successful analyses of beam measurement taken at top energy in the LHC[41,42] and to a possible revision[43] of the so-called collimator scans that are used to determine the diffusion coefficient for the LHC beam dynamics.[44,45] This research is currently in full development and future steps include the development of diffusive models for systems with two degrees of freedom and the possibility of extracting information on the diffusion coefficient from dedicated tracking simulations. Accurate information on the evolution of the beam distribution over realistic time scales would then be obtained by solving the Fokker-Planck equation with the diffusion coefficient determined by direct tracking. This approach would have also the advantage that solving Fokker-Plank equations over a time scale compatible with the actual machine cycle is certainly more affordable, in terms of CPU power required, than carrying out element-by-element tracking simulations over the same time scale.

### 2.4. *Machine availability*

Large-scale research infrastructures and in particular circular colliders such as the LHC represent a major challenge in terms of equipment reliability, as many ten thousand accelerator and infrastructure components must operate simultaneously and continuously for many hours to produce the desired physics output. For HL-LHC, the nominal fill length (including the leveling period) will be in the order of 7.5 hours, which will be interleaved with a turn-around time of at least 2.5 hours to bring back the beams into collisions after a machine failure or a deliberate termination of the prior physics fill. Therefore, in addition to the accurate and reliable control of proton and ion beams, with twice and five times the stored beam energy of the nominal LHC design respectively, machine operation in the HL-LHC era will also require further improvements of the already outstanding machine availability that was steadily improved during the first two operational runs.[46] During the three final years of its second operational run, the LHC managed to produce particle collisions during almost half the time devoted to high-intensity proton operation, while the remaining time was equally shared between equipment failures and regular operational time (such as the injection of beams from the injectors or the energy ramps). This is an unprecedented achievement for such a complex machine, which in addition is using many novel technologies that were never used at such industrial scales before. One of the main reasons for this achievement is that dependability considerations were a fundamental part of every equipment design from the very beginning. This is in particular the case for the backbone of the machine protection system, for which state of the art reliability engineering methods were employed to guarantee meeting both, the challenging reliability as well as availability targets. The second, equally important ingredient is a continuous identification and documentation of the root causes of down-time arising during the operational periods of the accelerator equipment. A dedicated tool, the so-called Accelerator Fault Tracker (AFT) has been developed to this end, allowing to identify and quantify the impact of recurring equipment failures on machine operation and trigger targeted consolidation activities to mitigate these weaknesses. Examples are major consolidation or displacement activities for electronics installed close to the accelerator which have shown weaknesses to radiation induced effects, the optimization of interlock levels across numerous protection systems based on beam operation experience and the preventive replacement of several thousands of local power supplies with unsatisfactory reliability. Another important ingredient is the development of more and more powerful operational tools, both for the diagnostic of the machine state as well as for the automation of recurrent operations and adjustments, which ensure repeatability while avoiding as much as possible human errors in the execution of the complex operational sequences. Following these continued efforts, LHC availability has today reached a level where it is dominated on one hand by the availability of its injector complex, and on the other hand by a few, but often long stops in the infrastructure systems necessary for the operation of the large superconducting magnet system as shown in Fig. 8. The full injector complex underwent a major upgrade program during the second long shutdown (LS2),[47] the impact of which on overall availability will only become visible during the upcoming Run 3. Failures in particular in the cryogenic system on the other hand will, despite often minor root causes, require many hours to recover nominal operating conditions as employing redundancy techniques is only possible to a very limited extent in such large-scale industrial systems.

As the complexity of the LHC will further increase with the deployment of the HL-LHC upgrade, it is therefore important to maintain and even further improve the availability for HL-LHC operation as shown in Fig. 9. This is not only true for the newly installed machine components, but also for the remaining parts of the machine which are based on components that will approach the end-of-life at the time of the HL-LHC era. Pursuing preventive maintenance and consolidation activities as well as further improvements of intervention procedures are therefore a necessity, limiting as much as possible the need of physical access to the tunnel to perform corrective actions.

### 2.5. *Energy choice and beam-induced magnet quenches*

The LHC has been designed for a center of mass collision energy of 14 TeV, and all superconducting main dipole magnets have been individually

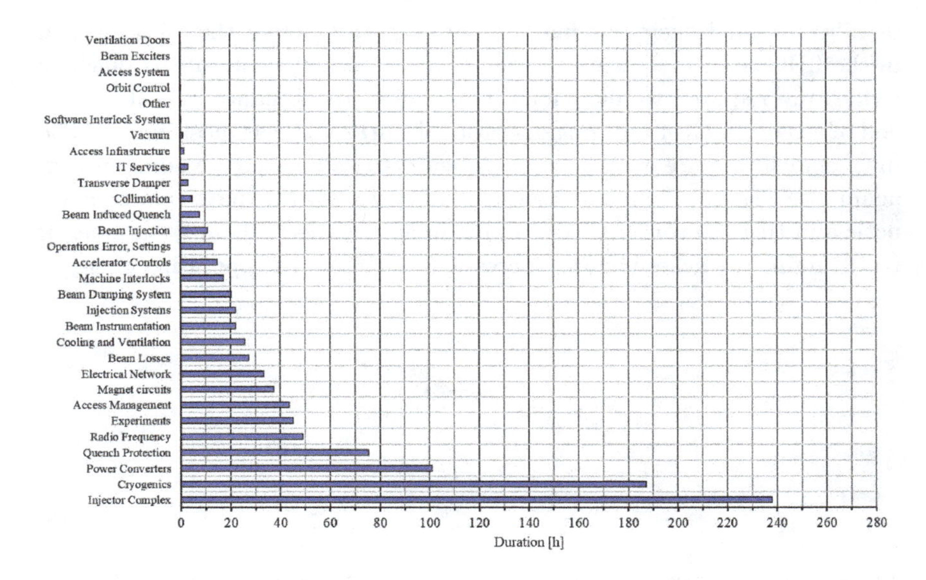


Fig. 8. Downtime of the LHC as a function of the root cause failure during the last year of Run 2 in 2018.

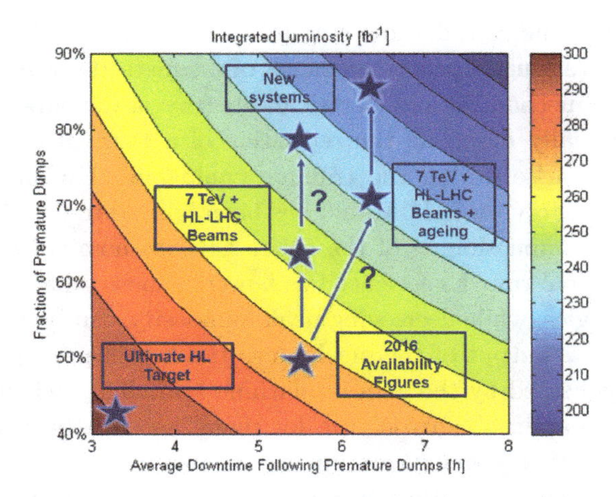


Fig. 9. Yearly luminosity reach of the LHC as a function of premature beam dumps and the average downtime caused. The availability figures reached during a typical Run 2 year (2016) are highlighted along with the HL-LHC target and possible negative performance impacts.

qualified beyond their nominal energy of 7 TeV before their installation in the LHC tunnel. Recent experience after thermal cycles of the machine (which are required to allow for the extended maintenance periods at the end of a typically 3–4 yearlong operational run) show that most main dipole magnets will require a (re-)training quench to reach their nominal current again (see Fig. 10). Combined with secondary quenches due to electromagnetic and thermal coupling this results in an unexpectedly large number of (re-)training quenches that are necessary to restore the operating energy.

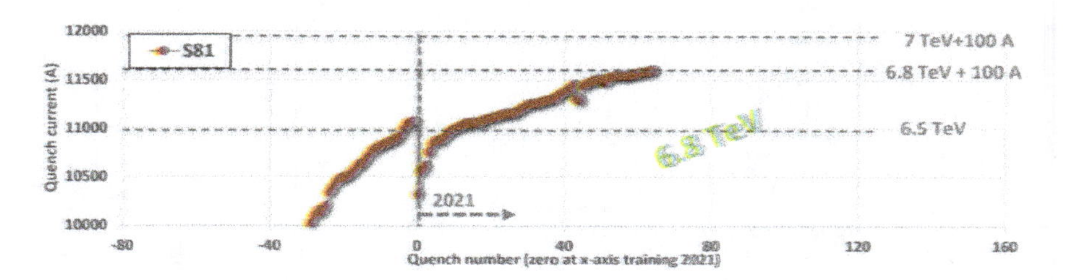


Fig. 10. Circuit current as a function of number of (re-)training quenches in LHC sector 81. During Run 2, the machine was operated at 6.5 TeV, while the 2021 commissioning campaign pushed the sector to achieve the 6.8 TeV energy equivalent.

In addition to the considerable amount of time that needs to be spent for the magnet training, repeated quenches of magnets also represent a certain risk for the mechanical and electrical integrity of the superconducting elements, potentially requiring the repetition of a thermal cycle if major non-conformities arise during the training campaign. For these reasons, the operational energy for LHC Run 3 has been limited to 6.8 TeV, further postponing the commissioning of the entire LHC to nominal energy until a future operational run. Experience has however shown that operation at lower beam energies, while decreasing peak luminosity, can be beneficial for the overall physics output of the machine as the loss in peak performance is largely compensated by the gain in machine availability when operating with increased margins. This is true for many accelerator systems such as power converters, cryogenic equipment, etc., but also for their interplay with the high intensity particle beams, such as the increased likelihood of beam induced quenches of superconducting magnets when operating at higher beam energies. Controlling the losses of highly energetic particles in a superconducting accelerator is a challenging task, especially for localized loss events which can be caused by fast beam instabilities or interactions of the proton beams with dust particles (UFOs). The latter were the main reason for beam-induced quenches in Run 2,[48] and they are expected to remain the primary source of transient beam loss events in future runs. When entering the beam, dust particles get rapidly ionized and are repelled from the circulating protons within a few turns of the beams. While most events are harmless, a small fraction of dust particles can still induce sufficient beam losses to perturb beam operation. This fraction will, however, increase in future runs due to more challenging operational conditions.

An efficient protection of the magnets against beam-induced quenches requires an in-depth understanding of the underlying physics of the energy deposition mechanisms as well as the quench limits of the different superconducting magnets. Both have been extensively simulated and empirically studied during the first two operational runs of the LHC, allowing a good compromise between beam loss protection settings and beam-induced magnet quenches to be defined. Figure 11 shows the maximum energy density in dipole coils for different dust events observed at 6.5 TeV in Run 2 (left plot). The energy density values were obtained by means of particle shower simulations. Events, which resulted in a quench, are displayed as crosses, whereas events without quench are represented by dots. As illustrated in the right plot, the number of quenches in Run 2 would have been higher, had the operational energy been 6.8 TeV as in Run 3. The energy increase reduces the quench limit of the main dipole magnets by 20%, while the same number of lost protons will lead to 7–8% higher energy densities in the magnet coils. This increased likelihood of beam induced quenches and in general reduced operational margins will very likely require further optimizations of the protection thresholds and strategies as a function of the

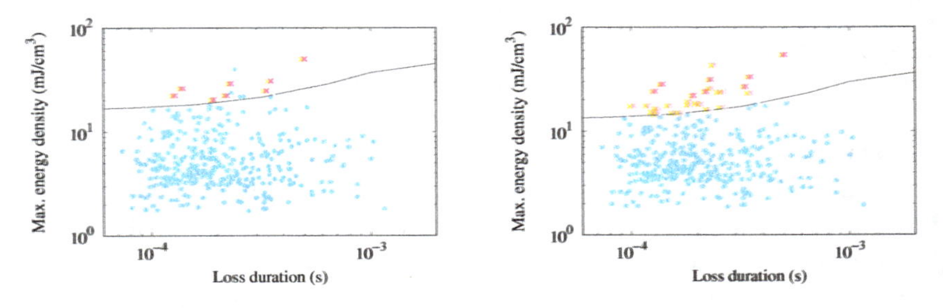


Fig. 11. Maximum energy density in dipole coils for different dust-induced loss events. Events without quench are shown as blue dots, whereas events with quench are shown as crosses. The solid line represents the quench level. The left figure illustrates the actual situation in Run 2 (6.5 TeV), while the right plot shows the expected number of quenches if the beam energy would have been 6.8 TeV.

experience gained when operating at increased beam energies approaching the nominal energy of 7 TeV.

## 3. Ion operation and challenges

Apart from the main physics programme with proton collisions, the LHC has also been designed to collide heavy ions. So far, the LHC has typically operated for about one month per year with heavy-ion beams, mainly fully stripped Pb nuclei. Initially, the goal was mainly to provide Pb-Pb collisions to ALICE, which is specialised in heavy-ion physics, but over time all the LHC experiments have joined the heavy-ion programme.

Operation with Pb ions entails different challenges and limitations from proton operation. The magnetic fields are the same, however, since the charge-to-mass ratio is lower, the energy per mass is also lower (2.76 TeV per nucleon for Pb and 7 TeV per proton in the LHC design scenario). The Pb ion bunch charge is only about 15% of the proton bunch charge, however, each $^{208}Pb^{82+}$ ion has 82 times higher charge and about 208 times higher mass than a proton. Because of this, limitations from beam-beam effects and machine impedances are more relaxed, although IBS and radiation damping are stronger. Furthermore, the total interaction cross section is more than 6000 times larger for Pb ions than for protons. This results in a much larger fraction of ions colliding at every passage through the collision point, and hence a very rapid burn-off of the beam.

It should also be noted that the interaction cross section is strongly dominated by ultraperipheral electromagnetic interactions, which occur in about 98.5% of all collision events. These interactions take place when two colliding ions pass close to each other without a direct nuclear overlap, as opposed to the hadronic nuclear interactions where the nuclei physically overlaps. The hadronic nuclear interactions, occurring only in about 1.5% of the events, are usually the main object of study of the experiments.

So far four one-month runs have been carried out with Pb-Pb,[49–51] and two runs with proton-Pb collisions. So far, the integrated luminosity for Pb-Pb collisions (corresponding to the total amount of data collected by the experiments) is $2.5\,\mathrm{nb}^{-1}$ at ATLAS and CMS, $1.5\,\mathrm{nb}^{-1}$ at ALICE and $0.25\,\mathrm{nb}^{-1}$ at LHCb. It should be noted that an integrated luminosity of $1\,\mathrm{nb}^{-1}$ corresponds to about $5.15\times10^{11}$ collision events for Pb-Pb. For p-Pb collisions, an integrated luminosity of about $250\,\mathrm{nb}^{-1}$ has been collected at ATLAS and CMS, $75\,\mathrm{nb}^{-1}$ at ALICE and $36\,\mathrm{nb}^{-1}$ at LHCb, where $100\,\mathrm{nb}^{-1}$ corresponds to $2.2 \times 10^{11}$ p-Pb collision events.

It has been planned for the LHC heavy-ion programme to continue in the future LHC and HL-LHC operation in Run 3 and Run 4, still using Pb-Pb and p-Pb collisions. The target for the ALICE physics programme after the proposed upgrade have been set to $13\,\mathrm{nb}^{-1}$ of integrated Pb–Pb luminosity during the next eight years of operation,[52,53] which requires producing almost nine times more collision events than what has been produced so far in the first eight years (in about the same amount of time). A similar increase in p-Pb collision events is also required.

A detailed machine scenario for future heavy-ion operation has already been devised,[54,55] containing the configuration of all relevant accelerator subsystems. Numerical simulation models of the ion beam evolution, including a range of physical effects, are used to study the development of the luminosity, intensity, and emittance.[55] These studies show that the physics goals are just within reach, however, there is a significant error bar on the input conditions, such as the achievable intensity and the machine availability. Therefore, various performance improvements are being studied, such as decreased beam size at the collision points, or smaller crossing angles, which would boost the luminosity.[55]

Dedicated upgrades of the LHC and the injectors are necessary in order to reach these ambitious goals. Firstly, a new method of stacking the bunches in the Pb beam more closely together will be implemented in the SPS. This method, called slip-stacking,[47] relies on an upgrade of the RF system in the SPS and allows shortening of the spacing between the bunches from 75 ns to 50 ns. Hence about 70% more bunches can be fitted in the LHC, following also optimizations in the pattern of injected bunches into the LHC. This allows significant increasing of the luminosity, most notably by more than a factor 6 at ALICE.[55] A first demonstration of the slip-stacking in the SPS was done in 2021.

The higher beam intensity and luminosity entail significant operational challenges connected to beam losses, where the most serious one is connected to the ultraperipheral electromagnetic interactions. The most common one is called bound-free pair production (BFPP). It causes the creation of an electron-positron pair, where the electron is caught in a bound state at one of the colliding nuclei, hence changing its charge. Therefore, the magnetic force on the affected ions is reduced, so they follow a dispersive trajectory and eventually impact on the machine aperture a few hundred metres downstream of the collision point.[56,57] Such beam losses due to BFPP have been observed both at RHIC[58] and at the LHC.[59]

Losses from BFPP, carrying a total power of up to about 165 W at

HL-LHC, would heat the impacted magnet such that it loses its superconducting state (a so-called quench). Quenches must be avoided by all means during collider operation as the recovery is a lengthy process that reduces the available time for physics operation. Therefore, these losses will be mitigated by redirecting them with a local orbit bump into a safe location: either an empty connection cryostat, or newly installed collimators. This alleviates the risk of quenches at the luminosities considered for Run 3 and HL-LHC.[59]

Electromagnetic dissociation is another ultraperipheral interaction, where one nucleus emits one or several nucleons, hence changing the charge-to-mass ratio and the trajectory of the affected ions. These processes are, however, less hindering for operation, since they either carry only a small amount of power and can be intercepted by collimators.

Another serious operational challenge comes from the higher beam intensity, and hence higher stored beam energy. It will exceed 20 MJ for Pb beams, which makes the LHC beams highly destructive, since even minor beam losses could cause magnet quenches or even damage. Therefore, beam losses need to be tightly controlled and safely intercepted by a collimation system, consisting of several stages of massive absorbers.[60–64] The same collimation system was used during Run 1 and Run 2 for protons and heavy ions, but the cleaning process is about two orders of magnitude less efficient for ions, due to nuclear fragmentation processes inside the primary collimators. Affected ions, with altered charge-to-mass ratio, can exit the collimator material and continue through the beam pipe on dispersive trajectories until they are lost.[65,66] Therefore, the collimation performance is more critical for heavy ions than for protons, in spite of lower stored beam energy, and the inefficiency of the present collimation system risks limiting the machine availability due to frequent beam aborts.

To overcome this limitation, the future collimation system for heavy ions is based on a fundamentally different principle, called crystal channeling. A silicon crystal, only a few millimeters long and bent with curvature of about 50 $\mu$ rad, replaces the role of the present massive primary collimators.[67,68] Inside the crystal, any intercepted ions are captured in a potential well between the crystalline planes, a “channel”, where it can propagate with a strongly reduced probability of interacting with the atoms of the crystal. The bending of the channel causes affected ions to exit the crystal with an angular kick, strong enough to make it hit so deeply inside the standard secondary collimators (that are used as absorbers), that the risk of any secondary ion fragments leaking out is greatly reduced. With this technique, the efficiency of the collimation system can be improved sufficiently to safely store the future Pb beams. Following successful tests in previous LHC runs, new crystal collimators have been installed in the LHC for operational deployment in Run 3.[69]

The present heavy-ion physics programme, based on Pb beams, is scheduled to continue until the end of Run 4. Starting from Run 5, a new programme is proposed and under study,[53] which relies on upgraded experimental detectors that can digest a much higher luminosity. In order to provide this luminosity, the possibility of operating the LHC with ion species other than Pb is being investigated. Such beams have the potential of a significantly increased intensity and nucleon-nucleon luminosity, however, limitations in the injector complex as well as in the LHC require further studies.

## References

1. G. Arduini et al., HL–LHC Run 4 proton operational scenario. Technical Report CERN-ACC-2022-0001, CERN (2022).
2. E. Métral et al., Update of the HL–LHC operational scenarios for proton operation. Technical Report CERN-ACC-NOTE-2018-0002, CERN (January, 2018). URL `https://cds.cern.ch/record/2301292/files/CERN-ACC-NOTE-2018-0002.pdf`.
3. R. Tomás et al., Operational scenario for the first HL–LHC run. In *Proc. 13th Int. Particle Accelerator Conf. (IPAC'22)*, JACoW Publishing (2022).
4. S. Kostoglou and R. De Maria, Reviewing the "No MS10" scenario during collisions for Run 4 with DA beam-beam simulations (2021). 132$^{nd}$ HL-LHC TCC meeting, `https://edms.cern.ch/document/2569782/1`.
5. R. De Maria and S. Kostoglou, Review of the situation without MS10 at the beginning of collisions and at the end of leveling (2021). 189$^{th}$ HL-LHC WP2 meeting, `https://indico.cern.ch/event/1014471/`.
6. S. Kostoglou, H. Bartosik, R. D. Maria, Y. Papaphilippou, G. Sterbini, and R. Tomás, Dynamic aperture studies for the Run 4 High-Luminosity LHC operational scenario, in preparation.
7. R. Tomás, J. Keintzel, and S. Papadopoulou, Emittance growth from luminosity burn-off in future hadron colliders, *Phys. Rev. Accel. Beams.* **23**, 031002 (Mar, 2020). doi: https://doi.org/10.1103/PhysRevAccelBeams.23.031002. URL `https://journals.aps.org/prab/abstract/10.1103/PhysRevAccelBeams.23.031002`.
8. G. Iadarola, B. Bradu, P. Dijkstal, L. Mether, A. Romano, G. Rumolo, G. Skripka, and L. Tavian, Electron Cloud and Heat Loads in Run 2. In *9th LHC Operations Evian Workshop*, pp. 221–232, Geneva, Switzerland (2019).
9. G. Skripka, P. Dijkstal, G. Iadarola, L. Mether, G. Rumolo, and E. G. T. Wulff, Comparison of electron cloud build-up simulations

against heat load measurements for the LHC arcs with different beam configurations. In *Proc. 10th Int. Particle Accelerator Conf. (IPAC'19)*, pp. 3232–3235, JACoW Publishing (May, 2019). doi: doi:10.18429/JACoW-IPAC2019-WEPTS051. URL http://accelconf.web.cern.ch/ipac2019/papers/WEPTS051.pdf.
https://doi.org/10.18429/JACoW-IPAC2019-WEPTS051.
10. G. Iadarola et al., Progress in mastering electron clouds at the large hadron collider. In *Proc. 12th Int. Particle Accelerator Conf. (IPAC'21)*, pp. 1273–1278, JACoW Publishing (May, 2021). doi: doi:10.18429/JACoW-IPAC2021-TUXA03. URL https://jacow.org/ipac2021/papers/TUXA03.pdf. https://doi.org/10.18429/JACoW-IPAC2021-TUXA03.
11. V. Petit, M. Taborelli, D. A. Zanin, M. Himmerlich, H. Neupert, P. Chiggiato, and G. Iadarola, Beam-induced surface modifications as a critical source of heat loads in the Large Hadron Collider, *Communications Physics*. **4** (1), 1–10 (2021). ISSN 23993650. doi: 10.1038/s42005-021-00698-x. URL http://dx.doi.org/10.1038/s42005-021-00698-x.
12. G. Iadarola et al., Beam-induced heat loads on the LHC arc beam screens with different beam and machine configurations: experiments and comparison against simulations. Technical Report CERN-ACC-NOTE-2019-0057, CERN (Dec, 2019). URL https://cds.cern.ch/record/2705513.
13. G. Skripka and G. Iadarola, Beam-induced heat loads on the beam screens of the HL-LHC arcs. Technical Report CERN-ACC-NOTE-2019-0041, CERN (Oct, 2019). URL https://cds.cern.ch/record/2692753.
14. X. Buffat, S. Antipov, G. Arduini, R. De Maria, N. Karastathis, S. Kostoglou, A. Koval, E. H. Maclean, N. Mounet, Y. Papaphilippou, T. H. B. Persson, and R. Tomas Garcia, Strategy for Landau damping of head-tail instabilities at top energy in the HL-LHC. Technical Report CERN-ACC-NOTE-2020-0059, CERN (Nov, 2020). URL http://cds.cern.ch/record/2745703/files/CERN-ACC-NOTE-2020-0059.pdf.
15. S. Redaelli, S. Antipov, N. Biancacci, A. Bertarelli, R. Bruce, F. Carra, G. Gobbi, A. Lechner, A. Mereghetti, and E. Metral, Staged implementation of low-impedance collimation in IR7: plans for LS2. Technical Report CERN-ACC-NOTE-2019-0001, CERN (Jan, 2019). URL http://cds.cern.ch/record/2654779/files/CERN-ACC-NOTE-2019-0001.pdf.
16. E. H. Maclean, R. Tomás, F. S. Carlier, M. S. Camillocci, J. W. Dilly, J. Coello de Portugal, E. Fol, K. Fuchsberger, A. Garcia-Tabares Valdivieso, M. Giovannozzi, M. Hofer, L. Malina, T. H. B. Persson, P. K. Skowronski, and A. Wegscheider, New approach to LHC optics commissioning for the nonlinear era, *Phys. Rev. Accel. Beams*. **22**, 061004 (Jun, 2019). doi: 10.1103/PhysRevAccelBeams.22.061004. URL https://link.aps.org/doi/10.1103/PhysRevAccelBeams.22.061004.
17. L. R. Carver, X. Buffat, K. Li, E. Métral, and M. Schenk, Transverse beam instabilities in the presence of linear coupling in the large hadron collider, *Phys. Rev. Accel. Beams*. **21**, 044401 (Apr, 2018). doi: 10.1103/PhysRevAccelBeams.21.044401. URL https://link.aps.org/doi/10.1103/PhysRevAccelBeams.21.044401.

18. S. V. Furuseth and X. Buffat, Loss of transverse Landau damping by noise and wakefield driven diffusion, *Phys. Rev. Accel. Beams.* **23**, 114401 (Nov, 2020). doi: 10.1103/PhysRevAccelBeams.23.114401. URL `https://link.aps.org/doi/10.1103/PhysRevAccelBeams.23.114401`.
19. O. S. Brüning, P. Collier, P. Lebrun, S. Myers, R. Ostojic, J. Poole, and P. Proudlock, *LHC Design Report*. CERN Yellow Rep. Monogr., CERN, Geneva (2004). doi: 10.5170/CERN-2004-003-V-1.
20. S. Fartoukh, Achromatic telescopic squeezing scheme and application to the LHC and its luminosity upgrade, *Phys. Rev. ST Accel. Beams.* **16**, 111002 (2013).
21. S. Papadopoulou, F. Antoniou, T. Argyropoulos, M. Hostettler, Y. Papaphilippou, and G. Trad, Impact of non-Gaussian beam profiles in the performance of hadron colliders, *Phys. Rev. Accel. Beams.* **23**, 101004 (Oct, 2020). doi: 10.1103/PhysRevAccelBeams.23.101004. URL `https://link.aps.org/doi/10.1103/PhysRevAccelBeams.23.101004`.
22. R. Tomás, J. Keintzel, and S. Papadopoulou, Emittance growth from luminosity burn-off in future hadron colliders, *Phys. Rev. Accel. Beams.* **23**, 031002 (Mar, 2020). doi: 10.1103/PhysRevAccelBeams.23.031002. URL `https://link.aps.org/doi/10.1103/PhysRevAccelBeams.23.031002`.
23. P. Baudrenghien and T. Mastoridis, Transverse emittance growth due to rf noise in the high-luminosity lhc crab cavities, *Phys. Rev. ST Accel. Beams.* **18**, 101001 (Oct, 2015). doi: 10.1103/PhysRevSTAB.18.101001. URL `https://link.aps.org/doi/10.1103/PhysRevSTAB.18.101001`.
24. X. Buffat, W. Herr, T. Pieloni, and D. Valuch, Modeling of the emittance growth due to decoherence in collision at the Large Hadron Collider, *Phys. Rev. Accel. Beams.* **23**, 021002 (Feb, 2020). doi: 10.1103/PhysRevAccelBeams.23.021002. URL `https://link.aps.org/doi/10.1103/PhysRevAccelBeams.23.021002`.
25. S. Redaelli, R. Appleby, R. Bruce, O. Brüning, A. Kolehmainen, G. Ferlin, A. Foussat, M. Giovannozzi, P. Hermes, D. Mirarchi, D. Perini, A. Rossi, and G. Stancari, Hollow electron lenses for beam collimation at the high-luminosity large hadron collider (HL-LHC), *Journal of Instrumentation.* **16** (03), P03042 (Mar, 2021). doi: 10.1088/1748-0221/16/03/p03042. URL `https://doi.org/10.1088/1748-0221/16/03/p03042`.
26. D. Mirarchi, R. B. Appleby, R. Bruce, M. Giovannozzi, A. Mereghetti, S. Redaelli, and G. Stancari, Nonlinear dynamics of proton beams with hollow electron lens in the cern high-luminosity lhc, *The European Physical Journal Plus.* **137** (1), 7 (2021). doi: 10.1140/epjp/s13360-021-02201-5. URL `https://doi.org/10.1140/epjp/s13360-021-02201-5`.
27. S. Kostoglou, G. Arduini, Y. Papaphilippou, G. Sterbini, and L. Intelisano, Origin of the 50 Hz harmonics in the transverse beam spectrum of the large hadron collider, *Phys. Rev. Accel. Beams.* **24**, 034001 (Mar, 2021). doi: 10.1103/PhysRevAccelBeams.24.034001. URL `https://link.aps.org/doi/10.1103/PhysRevAccelBeams.24.034001`.
28. S. Kostoglou, G. Arduini, Y. Papaphilippou, G. Sterbini, and L. Intelisano, Impact of the 50 Hz harmonics on the beam evolution of the

large hadron collider, *Phys. Rev. Accel. Beams.* **24**, 034002 (Mar, 2021). doi: 10.1103/PhysRevAccelBeams.24.034002. URL https://link.aps.org/doi/10.1103/PhysRevAccelBeams.24.034002.
29. S. Kostoglou, H. Bartosik, Y. Papaphilippou, G. Sterbini, and N. Triantafyllou, Tune modulation effects for colliding beams in the High Luminosity Large Hadron Collider, *Phys. Rev. Accel. Beams.* **23**, 121001 (Dec, 2020). doi: 10.1103/PhysRevAccelBeams.23.121001. URL https://link.aps.org/doi/10.1103/PhysRevAccelBeams.23.121001.
30. S. Fartoukh and M. Giovannozzi, Dynamic aperture computation for the as-built CERN Large Hadron Collider and impact of main dipoles sorting, *Nucl. Instrum. Methods Phys. Res., A.* **671**, 10–23 (2011). doi: 10.1016/j.nima.2011.12.052.
31. M. Giovannozzi, A proposed scaling law for intensity evolution in hadron storage rings based on dynamic aperture variation with time, *Phys. Rev. ST Accel. Beams.* **15**, 024001 (2012). doi: 10.1103/PhysRevSTAB.15.024001.
32. M. Giovannozzi and F. Van der Veken, Description of the luminosity evolution for the CERN LHC including dynamic aperture effects. Part I: the model, *Nucl. Instrum. Methods Phys. Res.* **A905**, 171–179 (2018). doi: 10.1016/j.nima.2019.01.072. [Erratum: Nucl. Instrum. Methods Phys. Res. A927, 471(2019)].
33. M. Giovannozzi and F. F. Van der Veken, Description of the luminosity evolution for the CERN LHC including dynamic aperture effects. Part II: application to Run 1 data, *Nucl. Instrum. Methods Phys. Res.* **A908**, 1–9 (2018). doi: 10.1016/j.nima.2018.08.019.
34. E. H. Maclean, M. Giovannozzi, and R. B. Appleby, Innovative method to measure the extent of the stable phase-space region of proton synchrotrons, *Phys. Rev. Accel. Beams.* **22**, 034002 (Mar, 2019). doi: 10.1103/PhysRevAccelBeams.22.034002. URL https://link.aps.org/doi/10.1103/PhysRevAccelBeams.22.034002.
35. E. H. Maclean, R. Tomás, F. Schmidt, and T. H. B. Persson, Measurement of nonlinear observables in the Large Hadron Collider using kicked beams, *Phys. Rev. Spec. Top. Accel. Beams.* **17**, 081002. 19 p (2014). doi: 10.1103/PhysRevSTAB.17.081002. URL https://cds.cern.ch/record/2135840.
36. R. De Maria, J. Andersson, L. Field, M. Giovannozzi, P. Hermes, N. Hoimyr, G. Iadarola, S. Kostoglou, E. Maclean, E. McIntosh, A. Mereghetti, J. Molson, V. Berglyd Olsen, D. Pellegrini, T. Persson, M. Schwinzerl, S. Singh, K. Sjobak, and I. Zacharov, SixTrack project: Status, runtime environment, and new developments. p. TUPAF02. 7 p (2019). doi: 10.18429/JACoW-ICAP2018-TUPAF02. URL https://cds.cern.ch/record/2697441.
37. R. De Maria and G. Iadarola, Xsuite (2022). https://xsuite.readthedocs.io.
38. N. Nekhoroshev, An exponential estimate of the time of stability of nearly-integrable Hamiltonian systems, *Russ. Math. Surv.* **32**, 1 (1977).
39. A. Bazzani, S. Marmi, and G. Turchetti, Nekhoroshev estimate for isochronous non resonant symplectic maps, *Cel. Mech.* **47**, 333 (1990).

40. G. Turchetti, Nekhoroshev stability estimates for symplectic maps and physical applications. In *Proc. of the Winter School*, vol. 47, *Springer Proceedings in Physics*, p. 223, Les Houches, France ('89) (1990).
41. A. Bazzani, O. Mazzarisi, M. Giovannozzi, and E. Maclean, Diffusion in stochastically perturbed Hamiltonian systems with applications to the recent LHC dynamic aperture experiments. In eds. S. Chattopadhyay, M. Cornacchia, and S. Di Mitri, *Proceedings, 2017 Nonlinear Dynamics and Collective Effects (NOCE) workshop on Particle Beam Physics: Arcidosso, Italy, 19–22 September 2017*, pp. 70–85, WSP (2019).
42. A. Bazzani, M. Giovannozzi, and E. Maclean, Analysis of the non-linear beam dynamics at top energy for the CERN Large Hadron Collider by means of a diffusion model, *Eur. Phys. J. Plus.* **135** (1), 77 (2020). doi: 10.1140/epjp/s13360-020-00123-2.
43. A. Bazzani, M. Giovannozzi, and C. Montanari, Probing the diffusive behaviour of beam-halo dynamics in circular accelerators. *Eur. Phys. J. Plus* **137**, 1264 (2022). https://doi.org/10.1140/epjp/s13360-022-03478-w.
44. G. Valentino, R. Aßmann, R. Bruce, F. Burkart, V. Previtali, S. Redaelli, B. Salvachua, G. Stancari, and A. Valishev, Beam diffusion measurements using collimator scans in the LHC, *Phys. Rev. ST Accel. Beams.* **16**, 021003 (Feb, 2013). doi: 10.1103/PhysRevSTAB.16.021003. URL https://link.aps.org/doi/10.1103/PhysRevSTAB.16.021003.
45. A. Gorzawski, R. B. Appleby, M. Giovannozzi, A. Mereghetti, D. Mirarchi, S. Redaelli, B. Salvachua, G. Stancari, G. Valentino, and J. F. Wagner, Probing LHC halo dynamics using collimator loss rates at 6.5 TeV, *Phys. Rev. Accel. Beams.* **23**, 044802 (Apr, 2020). doi: 10.1103/PhysRevAccelBeams.23.044802. URL https://link.aps.org/doi/10.1103/PhysRevAccelBeams.23.044802.
46. A. Apollonio, M. Brugger, L. Rossi, R. Schmidt, B. Todd, D. Wollmann, and M. Zerlauth, Roadmap towards high accelerator availability for the CERN HL-LHC era, *Proceedings of IPAC15, Richmond, US* (2015). URL https://cds.cern.ch/record/2141848/files/tupty053.pdf.
47. J. Coupard, H. Damerau, A. Funken, R. Garoby, S. Gilardoni, B. Goddard, K. Hanke, D. Manglunki, M. Meddahi, G. Rumolo, R. Scrivens, and E. Chapochnikova, LHC Injectors Upgrade, Technical Design Report, Vol. II: Ions. Technical Report CERN-ACC-2016-0041, CERN, Geneva (Apr, 2016). URL https://cds.cern.ch/record/2153863.
48. A. Lechner, P. Bélanger, I. Efthymiopoulos, L. Grob, B. Lindstrom, R. Schmidt, and D. Wollmann, Dust-induced beam losses in the cryogenic arcs of the CERN Large Hadron Collider, *Phys. Rev. Accel. Beams.* **25**, 041001 (2022). doi: 10.1103/PhysRevAccelBeams.25.041001.
49. J. Jowett, G. Arduini, R. Assmann, P. Baudrenghien, C. Carli, M. Lamont, M. S. Camillocci, J. Uythoven, W. Venturini, and J. Wenninger, First run of the LHC as a heavy-ion collider, *Proceedings of IPAC11, San Sebastian, Spain.* p. 1837 (2011). URL http://accelconf.web.cern.ch/AccelConf/IPAC2011/papers/tupz016.pdf.

50. J.M. Jowett, Colliding heavy ions in the LHC, *Proceedings of the International Particle Accelerator Conference 2018, Vancouver, Canada.* p. 584 (2018). URL `https://accelconf.web.cern.ch/ipac2018/papers/tuxgbd2.pdf`.
51. J. Jowett, C. B. Castro, W. Bartmann, C. Bracco, R. Bruce, J. Dilly, S. Fartoukh, A. Garcia-Tabares, M. Hofer, M. Jebramcik, J. Keintzel, A. Lechner, E. Maclean, L. Malina, T. Medvedeva, D. Mirarchi, T. Persson, B. Petersen, S. Redaelli, M. Schaumann, M. Solfaroli, R. Tomas, J. Wenninger, J. Coello, E. Fol, N. Fuster-Martinez, E. Holzer, A. Mereghetti, B. Salvachua, C. Schwick, M. Spitznagel, H. Timko, A. Wegscheider, and D. Wollmann, The 2018 Heavy-Ion Run of the LHC, *Proceedings of the 10th International Particle Accelerator Conference (IPAC2019): Melbourne, Australia, May 19–24, 2019.* p. 2258 (2019). doi: https://doi.org/10.18429/JACoW-IPAC2019-WEYYPLM2.
52. B. Abelev et al., Upgrade of the ALICE experiment: Letter of intent, *Journal of Physics G: Nuclear and Particle Physics.* **41** (8), 087001 (Jul, 2014). doi: 10.1088/0954-3899/41/8/087001. URL `https://doi.org/10.1088%2F0954-3899%2F41%2F8%2F087001`.
53. Z. Citron et al., Future physics opportunities for high-density QCD at the LHC with heavy-ion and proton beams, *HL/HE-LHC Workshop: Workshop on the Physics of HL-LHC, and Perspectives at HE-LHC Geneva, Switzerland, June 18–20, 2018, CERN-LPCC-2018-07* (2018). URL `https://cds.cern.ch/record/2650176?ln=en`.
54. R. Bruce, T. Argyropoulos, H. Bartosik, R. D. Maria, N. Fuster-Martinez, M. Jebramcik, J. Jowett, N. Mounet, S. Redaelli, G. Rumolo, M. Schaumann, and H. Timko, HL-LHC operational scenario for Pb-Pb and p-Pb operation, *CERN-ACC-2020-0011* (2020). URL `https://cds.cern.ch/record/2722753`.
55. R. Bruce, M. Jebramcik, J. Jowett, T. Mertens, and M. Schaumann, Performance and luminosity models for heavy-ion operation at the CERN Large Hadron Collider, *Eur. Phys. J. Plus.* **136**, 745 (2021). doi: 10.1140/epjp/s13360-021-01685-5. URL `https://doi.org/10.1140/epjp/s13360-021-01685-5`.
56. S. R. Klein, Localized beampipe heating due to e- capture and nuclear excitation in heavy ion colliders, *Nucl. Inst. & Methods A.* **459**, 51 (2001).
57. R. Bruce, D. Bocian, S. Gilardoni, and J. M. Jowett, Beam losses from ultra-peripheral nuclear collisions between Pb ions in the Large Hadron Collider and their alleviation, *Phys. Rev. ST Accel. Beams.* **12** (7), 071002 (Jul, 2009). doi: 10.1103/PhysRevSTAB.12.071002.
58. R. Bruce, J. M. Jowett, S. Gilardoni, A. Drees, W. Fischer, S. Tepikian, and S. R. Klein, Observations of beam losses due to bound-free pair production in a heavy-ion collider, *Phys. Rev. Letters.* **99** (14), 144801 (2007). doi: 10.1103/PhysRevLett.99.144801. URL `https://link.aps.org/doi/10.1103/PhysRevLett.99.144801`.
59. M. Schaumann, J. M. Jowett, C. Bahamonde Castro, R. Bruce, A. Lechner, and T. Mertens, Bound-free pair production from nuclear collisions and the

steady-state quench limit of the main dipole magnets of the CERN Large Hadron Collider, *Phys. Rev. Accel. Beams.* **23**, 121003 (Dec, 2020). doi: 10.1103/PhysRevAccelBeams.23.121003. URL `https://link.aps.org/doi/10.1103/PhysRevAccelBeams.23.121003`.

60. R.W. Assmann, Collimators and Beam Absorbers for Cleaning and Machine Protection, *Proceedings of the LHC Project Workshop - Chamonix XIV, Chamonix, France.* p. 261 (2005).
61. R.W. Assmann et al., The Final Collimation System for the LHC, *Proc. of the European Particle Accelerator Conference 2006, Edinburgh, Scotland.* p. 986 (2006).
62. R. Bruce, R. W. Assmann, V. Boccone, C. Bracco, M. Brugger, M. Cauchi, F. Cerutti, D. Deboy, A. Ferrari, L. Lari, A. Marsili, A. Mereghetti, D. Mirarchi, E. Quaranta, S. Redaelli, G. Robert-Demolaize, A. Rossi, B. Salvachua, E. Skordis, C. Tambasco, G. Valentino, T. Weiler, V. Vlachoudis, and D. Wollmann, Simulations and measurements of beam loss patterns at the CERN Large Hadron Collider, *Phys. Rev. ST Accel. Beams.* **17**, 081004 (Aug, 2014). doi: 10.1103/PhysRevSTAB.17.081004. URL `http://link.aps.org/doi/10.1103/PhysRevSTAB.17.081004`.
63. R. Bruce, R. W. Assmann, and S. Redaelli, Calculations of safe collimator settings and $\beta^*$ at the CERN Large Hadron Collider, *Phys. Rev. ST Accel. Beams.* **18**, 061001 (Jun, 2015). doi: 10.1103/PhysRevSTAB.18.061001. URL `http://link.aps.org/doi/10.1103/PhysRevSTAB.18.061001`.
64. R. Bruce, C. Bracco, R. D. Maria, M. Giovannozzi, A. Mereghetti, D. Mirarchi, S. Redaelli, E. Quaranta, and B. Salvachua, Reaching record-low $\beta^*$ at the CERN Large Hadron Collider using a novel scheme of collimator settings and optics, *Nucl. Instrum. Methods Phys. Res. A.* **848**, 19–30 (Jan, 2017). doi: http://dx.doi.org/10.1016/j.nima.2016.12.039. URL `http://www.sciencedirect.com/science/article/pii/S0168900216313092`.
65. P. Hermes, R. Bruce, J. Jowett, S. Redaelli, B. S. Ferrando, G. Valentino, and D. Wollmann, Measured and simulated heavy-ion beam loss patterns at the CERN Large Hadron Collider, *Nucl. Instrum. Methods Phys. Res. A.* **819**, 73–83 (Feb, 2016). doi: http://dx.doi.org/10.1016/j.nima.2016.02.050. URL `https://www.sciencedirect.com/science/article/pii/S0168900216002175?via%3Dihub`.
66. N. Fuster-Martínez, R. Bruce, F. Cerutti, R. De Maria, P. Hermes, A. Lechner, A. Mereghetti, J. Molson, S. Redaelli, E. Skordis, A. Abramov, and L. Nevay, Simulations of heavy-ion halo collimation at the CERN Large Hadron Collider: Benchmark with measurements and cleaning performance evaluation, *Phys. Rev. Accel. Beams.* **23**, 111002 (Nov, 2020). doi: 10.1103/PhysRevAccelBeams.23.111002. URL `https://link.aps.org/doi/10.1103/PhysRevAccelBeams.23.111002`.
67. V. Previtali, *Performance evaluation of a crystal-enhanced collimation system for the LHC*. PhD thesis, EPFL, Lausanne (2010). URL `https://cds.cern.ch/record/1302274`. Presented on 07 Oct 2010.
68. D. Mirarchi, *Crystal Collimation for LHC*. PhD thesis, Imperial College, London (Aug, 2015). URL `http://cds.cern.ch/record/2036210`.

69. M. D'Andrea, *Applications of Crystal Collimation to the CERN Large Hadron Collider (LHC) and its High Luminosity Upgrade Project (HL-LHC)*. PhD thesis, University of Padova (Feb, 2021). URL `http://cds.cern.ch/record/2758839`. Presented 23 Feb 2021.

© 2024 The Author(s)
https://doi.org/10.1142/9789811280184_0007

# Chapter 7

# Large-Aperture High-Field $Nb_3Sn$ Quadrupole Magnets for HiLumi

Giorgio Ambrosio* and Paolo Ferracin†

**Fermi National Accelerator Laboratory, Batavia, IL 60510*

*†Lawrence Berkeley National Laboratory, Berkeley, CA 94720*

The HL-LHC Inner Triplet quadrupole magnets use $Nb_3Sn$ superconductor for the first time in a particle accelerator. Coil peak field, forces, and stored energy density are significantly higher than in previous accelerator magnets. The original design addressed all these challenges. Lessons learned in the early phases made the final design and specifications for series magnets more robust.

## 1. Introduction

The HL-LHC Inner Triplet (IT) quadrupole magnets (MQXF)[1] are first-of-their-kind accelerator magnets. With a 150-mm aperture and 132.2 T/m magnetic field gradient, they operate at 11.3 T peak field on the coils and use $Nb_3Sn$ superconductor. Although the superconducting properties of $Nb_3Sn$ exceed those of Nb-Ti, no high-field magnet with this conductor has ever been used in a particle accelerator. This is due to the strain sensitivity and brittleness of $Nb_3Sn$,[2] which requires very careful handling at all steps, from coil fabrication to magnet assembly, pre-load, and transportation.

The beginning of HL-LHC's operation will mark the successful conclusion of a long journey, initiated in the 1960s, after many contributions from several institutions.[3] Over the past two decades the US LHC Accelerator Research Program (LARP[4,5]) took the lead in this effort and demonstrated a 120-mm aperture quadrupole magnet (HQ) with accelerator-quality features[6] and a scale-up of a 90-mm aperture quadrupole magnet from 1 to 3 m in magnetic length.[7] CERN and LARP developed together the MQXF design, beginning with the HQ magnet. In this chapter we describe

This is an open access article published by World Scientific Publishing Company. It is distributed under the terms of the Creative Commons Attribution 4.0 (CC BY) License.

the bare magnets (i.e., without the stainless-steel skin that makes up the coldmass helium vessel[8]).

The magnets (MQXFB) in the Q2a/Q2b Inner Triplet elements are 7.15 m in length and are fabricated by CERN. The US HL-LHC Accelerator Upgrade Project (AUP)[9] is fabricating 4.2-m-long magnets (MQXFA), two of which are housed within each Q1/Q3 Inner Triplet element.[10] AUP decided to fabricate two magnets for each Q1/Q3 element because of available infrastructure, and because of a need to reduce risks for the fabrication of these magnets. In fact, the 4.2-m-long MQXFA can be tested in a vertical cryostat before cryo-assembly fabrication, and their length was a small scale-up from the 3-m length previously demonstrated by LARP.[7] MQXFA and MQXFB magnets have identical cross-sections but different lengths.

## 2. Requirements and Challenges

All of the requirements for the bare magnets are presented in Ref. 11. The major requirements are summarized here:

(1) The magnet aperture is 150 mm. This is more than double the aperture (70 mm) of the LHC IT quadrupole magnets (MQXA/B), in order to accommodate larger beams and a tungsten absorber with a thickness of 16 mm in Q1 IT elements and of 6 mm in Q2/Q3.
(2) The operating magnetic field gradient is 132.2 T/m, with an 11.3 T peak field on the coils and a 24% margin on the magnet load line.
(3) The field harmonics shall be optimized at nominal operating current. The expected values for integral field harmonics at a reference radius of 50 mm are less than 3.3 and 4.4 units for $b_3$ and $b_6$, respectively; less than ~2 units for other low-order harmonics; and less than 1 unit for high-order harmonics.
(4) After a thermal cycle to room temperature, MQXF magnets shall attain the nominal operating current with no more than three quenches. The desired target is no more than one quench.
(5) The magnets must survive 50 quenches and 10,000 power cycles.

The main challenges for the design and operation of the MQXF magnets are:

(1) The electromagnetic forces at operating current in the straight section and in the ends are four and six times higher, respectively, than in the LHC Inner Triplet magnets.

(2) The stored energy per unit length is 1.2 MJ/m, which is more than double the energy per unit length of the LHC main dipole magnets.
(3) Despite the tungsten liner in the aperture, heat loads and integrated dose are similar to those in the LHC IT magnets because of higher luminosity.

## 3. Magnet Design

The design of the MQXFA/B quadrupole magnet is described in detail in Ref. 12 and is shown in Fig. 1(top), where the different assembly steps are shown from left to right. The main coil and magnet parameters are provided in Table 1. The shell-yoke sub-assembly (left) comprises four iron yoke stacks locked with four gap keys inside an aluminum shell. The coil-pack sub-assembly (center) includes four double-layer coils wound with a single-piece cable that is 18.150 mm wide and composed of 40 strands with a 0.85 mm diameter. The strand architecture is the 108/127 Rod Restack Process (RRP) by Bruker-OST.[13] A second strand architecture, the Powder In Tube (PIT) by Bruker-EAS, with and without bundle barrier,[14] was utilized on three CERN short models.[10]

In each coil, 50 turns are placed around a Ti-alloy winding pole, which incorporates a G11 pole-key for alignment. The coils are surrounded by aluminum collars and bolted iron pads. The two sub-assemblies are combined in the magnet (top right in Fig. 1), where two trapezoidal master keys, containing two loading keys and one central alignment key, provide room for the water-pressurized bladders that are used to pre-compress the coil and pre-tension the shell at room temperature. After magnet preload, the holes for the bladders can house magnetic shims used to correct low-order harmonics up to ~4 units. The 3D design is characterized by segmented aluminum shells and two end-plates, connected by four full-length axial rods for coil axial support (see Fig. 1, bottom).

Quench heaters, composed of 25-$\mu$m stainless steel strips with copper-plated sections to lower the resistance between heating stations, and with a 50-$\mu$m polyimide insulation layer, are placed on the outer layer of the superconducting coils to protect them in the event of a quench. The heaters work together with a Coupling-Loss Induced Quench (CLIQ) system, which heats up the coils by injecting an oscillating current into the magnet.[15] This protection scheme allows maintenance of the hot spot temperature after a spontaneous quench at nominal current below 280 K, despite the large energy density of the coil (~0.08 J/mm$^3$, in consideration of the volume of

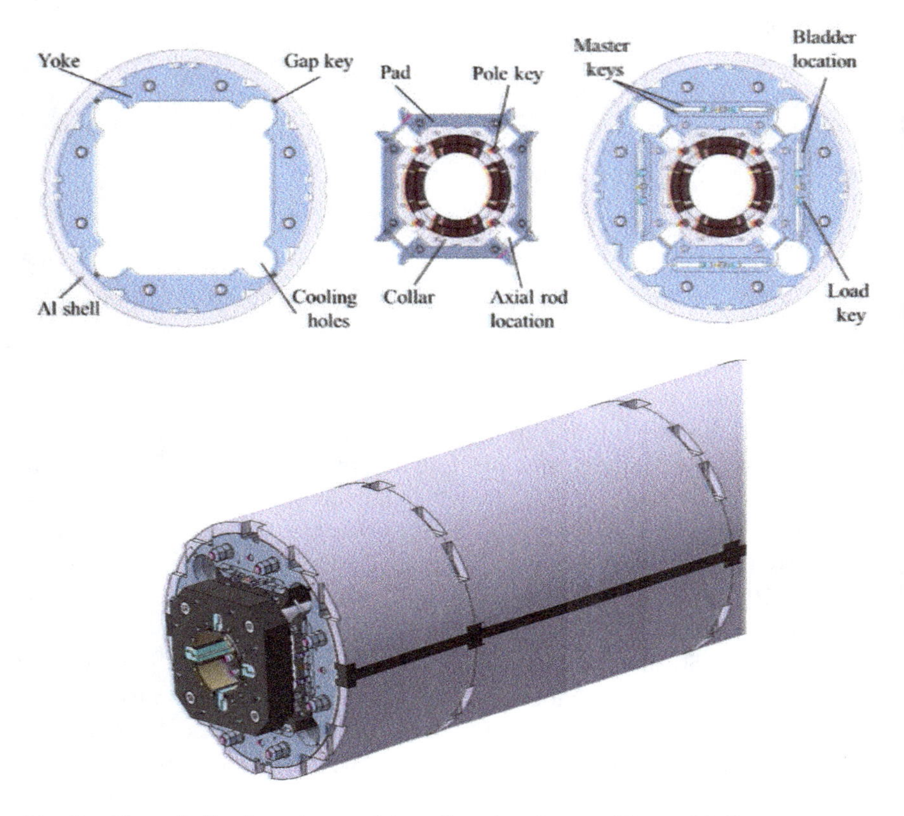


Fig. 1. Top: shell-yoke sub-assembly, coil-pack sub-assembly, and full magnet cross-sections. Bottom: 3D view of the magnet's end region.

the insulated cable). No dump resistor is included in the protection system due to the high inductance of the circuit.

Cooling is obtained by maintaining the magnet in an HeII bath contained within the coldmass stainless steel skin. Two heat exchangers are inserted into the two upper 77-mm diameter holes of the iron yoke. In order to remove the heat deposited in the superconducting coils, a 1.5-mm gap between the coil inner layer and the cold bore tube is filled with liquid He. In addition, 8-mm diameter holes in the winding poles and in the alignment pole-keys allow the free passage of He up to the heat exchanger. Inside the cold bore, a beam screen, on which 40-cm-long tungsten alloy blocks are mounted, shields the magnet from particle collision debris and beam-induced heating.

Table 1. Coil and magnet parameters.

| Parameters | Units | |
|---|---|---|
| Coil aperture diameter | mm | 150 |
| Magnet outer diameter | mm | 614 |
| Strand diameter | mm | 0.85 |
| Magnetic length MQXFA/MQXFB | m | 4.20/7.15 |
| Cu/SC | | 1.2±0.1 |
| Number of strands in cable | | 40 |
| Cable bare width (before/after HT) | mm | 18.150/18.363 |
| Cable bare mid-thickn. (before/after HT) | mm | 1.525/1.594 |
| Keystone angle | Deg. | 0.40 |
| No. turns in layers 1/2 (octant) | | 22/28 |
| Nominal gradient $G_{nom}$ | T/m | 132.2 |
| Nominal current $I_{nom}$ | kA | 16.23 |
| Nominal conductor peak field $B_{op}$ | T | 11.3 |
| $I_{nom}/I_{ss}$ at 1.9 K for RRP (specs.) | % | 76 |
| Stored energy density at $I_{nom}$ | MJ/m | 1.15 |
| Differential inductance at $I_{nom}$ | mH/m | 8.26 |
| Stored energy at $I_{nom}$ (Q1-Q3)/(Q2) | MJ | 4.83/8.22 |
| $F_x/F_y$ (per octant) at $I_{nom}$ | MN/m | +2.41/−3.41 |
| $F_\vartheta$ layers 1/2 (per octant) at $I_{nom}$ | MN/m | −1.80/−2.08 |
| $F_z$ (whole magnet) at $I_{nom}$ | MN | 1.15 |

## 4. Short Models and Prototypes

As part of the MQXF short model program carried out both at CERN and in the US through an exceptional collaboration, six magnets (MQXFS) with a magnetic length of 1.1 m were fabricated and tested.[10] The first magnet (MQXFS1) reached ultimate current (for 7.5 TeV operation) but it exhibited some detraining after a pre-stress change. The second magnet (MQXFS3) was limited to below nominal current by conductor degradation.

The experience gained in the first two short models allowed for fine tuning of the assembly procedure and pre-load level: all of the subsequent short model magnets (MQXFS4-7) managed to reach ultimate current, both at 1.9 K and 4.5 K,[16] with perfect memory after thermal cycles. A maximum conductor peak field of 13.4 T was reached in magnet MQXFS6. In addition, both MQXFS6 and MQXFS7 reached 97% of the short-sample limit at 4.5 K (based on conductor properties) on the magnet load line.

Finally, an endurance test performed on magnet MQXFS4[17] did not reveal any sign of performance degradation up to 19.15 kA (Fig. 2), well above ultimate current (17.49 kA), after 1,000 current cycles to nominal current, a high Quench-Integral test that reached a hot-spot temperature of $\approx 400$ K (performed during the MQXFS4c test), and a total of 11 thermal cycles.

The short model program was followed by the fabrication and testing of two long prototypes, both of the MQXFA and MQXFB magnet type. The two "A" prototypes were limited by, respectively, a coil-to-ground short (AP1) and a fracture of one of the aluminum shells (AP2), which triggered an update to both the electrical and structural design criteria applied to subsequent magnets.[18,19] The two "B" prototypes were limited at 15.2 kA (BP1) and 16.0 kA (BP2) by conductor degradation in the magnet's straight section. As a result, a full revision of the fabrication and assembly procedure has been carried out, aimed at minimizing the risk of conductor

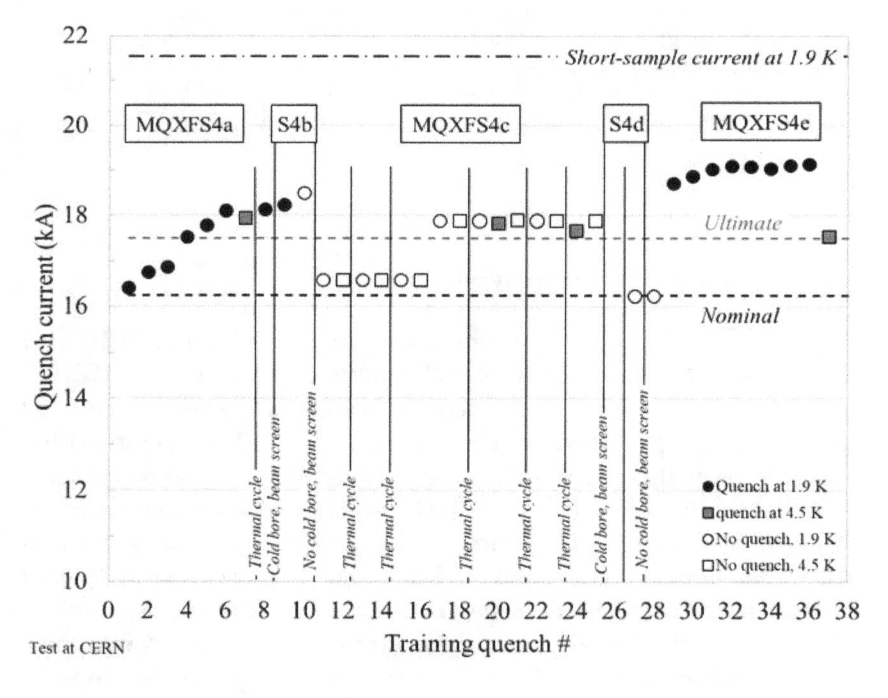


Fig. 2. Test results of short model magnet MQXFS4: the vertical lines indicate either a thermal cycle or a thermal cycle with a mechanical change (described at the bottom of the lines). After the initial test ("a"), the quench performance was verified with (test "b" and "d") and without (test "c" and "e") two different types of beam screens inside the cold bore. The maximum current reached at the end of MQXFS4e training (19.15 kA) generated a coil peak field of 13.2 T.

damage caused by excessive strain during coil fabrication, magnet assembly and loading, stainless steel shell welding, cool-down, and testing.[20,21]

Short models and prototypes showed field quality consistent with expectations, although b6 was on average close to the lower range. Therefore, it was decided to move a 125-$\mu$m-thick shim from midplane to pole in all future coils, using an option that was included in the coil design.[22]

## 5. Production Specifications

Following the experience gained during the short model and prototype phase, a set of specifications was established for the fabrication of the series magnets.[23] The specifications are aimed in particular at guaranteeing a uniform and controlled stress in the brittle $Nb_3Sn$ superconducting coils, thus minimizing the risk of conductor degradation, but applying at the same time sufficient pre-compressions to the coils to withstand the electromagnetic forces. In terms of coil size after impregnation, a key parameter that has significant impact on the magnet assembly, the acceptable arc length excess (the difference between the measured and nominal coil arc length) along the straight section is set to within $+0.150/-0.250$ mm. In addition, when combined in a coil pack, the average size variation of the four coils shall be kept between $\pm 100$ $\mu$m to maintain the resulting coil stress variations within $\pm 13$ MPa (according to a sensitivity analysis carried out on a Finite Element model of the magnet).[24–27]

The pre-load operation must be conducted in steps, alternating between azimuthal and axial pre-stress. The average azimuthal coil pre-load among the four coils after bladder operation is $-80 \pm 8$ MPa. With $-20$ to $-30$ MPa added to the coils during cool-down, the expected value of coil pre-load at cold is $-100$ to $-110$ MPa. With this level of pre-stress at cold, the coils are expected to remain compressed against the winding pole at up to 80–90% of the nominal current.[10,16] Finally, to minimize the risk of conductor degradation due to stress, it was decided to keep the maximum coil peak stress during the assembly and pre-load operation below $-110$ MPa, a conservative value that accounts for a possible increase in coil stress due to coil size variations and the fabrication tolerance of the different magnet components. At nominal current, the accumulated electromagnetic force is expected to generate a compressive stress of $-110$ MPa on the coil midplane (outer layer).

## 6. MQXFA Pre-Series and First Series Magnet Production

The US AUP project assembled and tested five MQXFA pre-series magnets (MQXFA03-07) between May 2019 and August 2021. Pre-series magnets were fabricated to meet all of the requirements[11] for use in the cryo-assemblies intended for the String Test[28] or the HL-LHC. Feedback from fabrication and testing of these magnets was used to complete and update the Series Magnet Production Specifications.[23]

All MQXFA magnets tested so far were tested at the BNL vertical test facility (VTF),[18,29] and were subjected to two or three thermal cycles. All magnets were trained at 1.9 K except for MQXFA06, which was initially trained at 4.5 K and after quench #12 at 1.9 K. The first four pre-series magnets (MQXFA03-06) met all of the requirements that could be tested at BNL VTF. They were able to hold acceptance current (nominal current + 300 A) for several hours and demonstrated excellent training memory and temperature margin up to 4.5 K. The test history during their first thermal cycle is shown in Fig. 3.

In contrast, the last pre-series magnet (MQXFA07) and the first series magnet (MQXFA08) showed detraining after they reached 16.1 kA and 16.5 kA, respectively.[30] These magnets were assembled during the COVID-19 pandemic and COVID-19 restrictions contributed to the issue that caused their limitation. The disassembly, inspection, and 3D finite element analysis of both magnets revealed that the gaps between the collars and the pole-key of the limiting coil were closed during assembly. As a result, the ends of that coil were not sufficiently axially restrained during magnet energization,[30] and they developed high tensional strain in the end region, in particular at the junction between wedge and end-spacers in the coil inner layer. Metallurgical inspections of these areas performed at CERN[31] confirmed the presence of broken filaments in the turns close to the wedge/end-spacer interface.[32] In order to prevent this issue from re-occurring, new assembly procedures and tighter specifications aimed at avoiding the closure of the collars on the pole keys, were implemented in magnets MQXFA10 and MQXFA11. They were successfully tested at BNL (Fig. 3), where they met all acceptance criteria. In order to demonstrate the endurance properties of MQXFA production magnets, MQXFA05 was re-tested at the BNL vertical test facility. As shown in Fig. 4, MQXFA05 was able to reach acceptance current at 1.9 K and nominal current at 4.5 K at the end of the fifth thermal cycle, after 52 quenches and 79 powering cycles.[30]

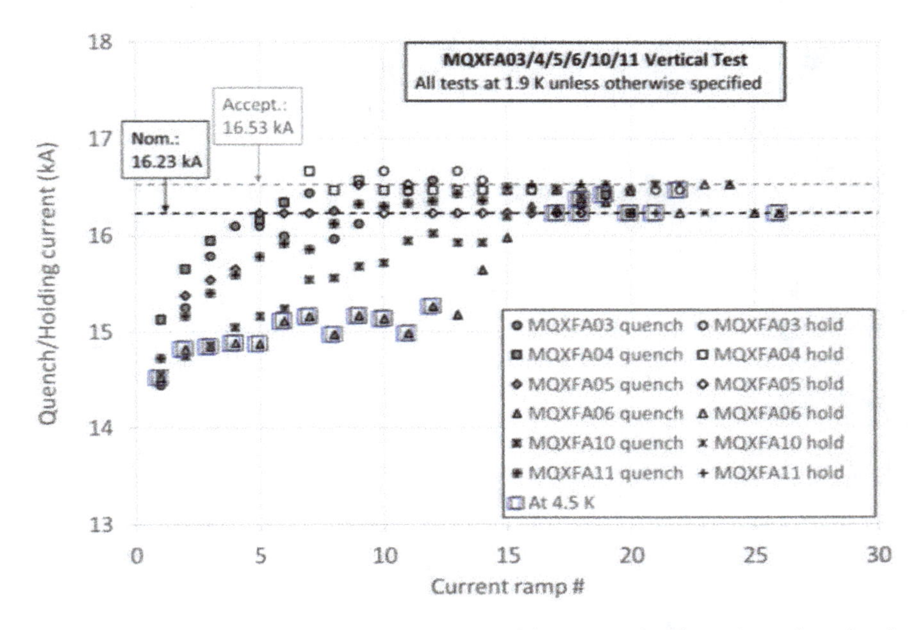


Fig. 3. Sequence of current ramps above 10 kA during the first thermal cycle of MQXFA03/4/5/6/10/11 magnets. Some ramps ended with magnet quench (markers with grey background). In all other ramps the magnet was able to hold current without quench (markers with white background); only some of these are shown in this plot. The vertical scale covers a small current range (13–18 kA) for plot readability.

The field quality of all pre-series MQXFA magnets is consistent with expectations. Magnetic shims inserted in selected bladder locations (Fig. 1) have been used to correct some low-order harmonics in MQXFA05 (+4 and −4 units of $b_3$ and $a_3$, respectively) and MQXFA06 (+4 units of $a_3$).

## 7. Plans and Conclusions

Fabrication of the large-aperture quadrupole magnets for the HL-LHC Inner Triplets is in progress at AUP and CERN. Short model magnets have demonstrated (i) reproducibility, with five out of six magnets reaching required performance, (ii) large margin, with five magnets able to operate at current providing a 1 T higher peak field on the conductor than nominal current, (iii) excellent memory with no retraining after thermal cycle, even at 12–13 T peak field, (iv) large temperature margin, with a proven ability to operate in the HL-LHC even at 4.5 K, and (v) achievement of this performance within a large range of pre-load assembly parameters. The first

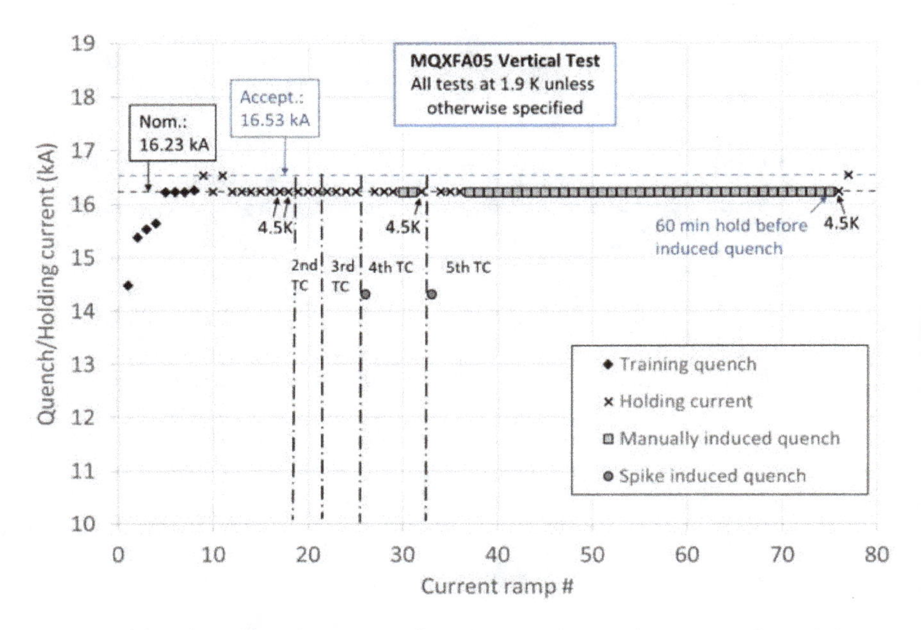


Fig. 4. MQXFA05 vertical tests: the first three thermal cycles were performed during the first test campaign; the fourth and fifth thermal cycles were performed during the "endurance test" that included 41 manually induced quenches and two quenches induced by a voltage spike that triggered the magnet protection system. Current ramps for magnetic measurements are not shown.

four MQXFA pre-series magnets, 4.2 m long, met all of the requirements that could be tested in a vertical cryostat. MQXFA07 and MQXFA08 exhibited a limitation. Their disassembly and subsequent analysis offered an opportunity to learn a useful lesson for MQXF assembly, which resulted in tightened specifications. Fabrication of the 7-m-long MQXFB magnets is beginning after implementation of the lessons learned from the prototypes.

All of the MQXF magnets that met the vertical test requirements demonstrated a large temperature margin, very good memory, and field quality for accelerators. An extensive test campaign performed on MQXFA05 demonstrated the endurance quality of MQXF magnets. The lessons learned in the early phases highlighted the importance of tight specifications and quality control in order to avoid conductor damage. These lessons have been implemented in the design and specifications for series magnet production, making them more robust. The first set of Inner Triplet elements is planned to be tested in a String Test at CERN before the start of LHC Long Shutdown 3 (LS3).

Although still far from completion, production of the first series of $Nb_3Sn$ accelerator magnets is demonstrating that $Nb_3Sn$ conductor can be used in accelerator magnets to achieve magnetic fields in the 11–13 T conductor peak field range, unattainable with the standard Nb-Ti conductor. In addition, the experience and the lessons learned during this production will be extremely valuable for the design and production of future $Nb_3Sn$ accelerator magnets, in particular for 12 T dipoles to be used in future colliders, which would have very similar peak fields and accumulated stresses to MQXF magnets.

During the final editing of this chapter, valuable test results and analyses became available and were presented at the 2022 Applied Superconductivity Conference. These are listed below, together with the main references:

- MQXFA11 was involved in a truck accident during transportation to the BNL vertical test facility. The crate was displaced by about 2 m and the shipping frame experienced shocks up to 6–10 g as recorded by different devices.[30] Nonetheless, MQXFA11 met all acceptance criteria for vertical testing (its quench history is shown in Fig. 3), demonstrating the resilience of MQXF magnets.
- MQXFBP3, the third prototype fabricated at CERN, was able to achieve acceptance current at 1.9 K. At 4.5 K it was limited at 15.9 kA, i.e. 2% below nominal current, resulting in a "2 K temperature margin for operation at nominal current and 1.9 K".[33] Therefore, the "MQXFBP3 temperature margin is 6 times larger than the expected temperature rise in the conductor during HL-LHC operation",[33] which is 0.32 K in MQXF magnets.[34] The limiting mechanism is similar to the one that limited MQXFBP1 and MQXFBP2[33] and efforts are ongoing to obtain the full potential of the design as proven in short models and in MQXFA magnets.

## References

1. P. Ferracin et al., "Development of MQXF: the $Nb_3Sn$ low-beta quadrupole for the HiLumi LHC", *IEEE Trans. Appl. Supercond.* vol. 26 (2016) 4000207.
2. L. F. Goodrich et al., "Method for determining the irreversible strain limit of $Nb_3Sn$ wires", *2011 Supercond. Sci. Technol.* **24** 075022.
3. $Nb_3Sn$ Accelerator Magnets, D. Schoerling and A. Zlobin, Eds., Springer Open, 2019.
4. S. Gourlay et al., "Magnet R&D for the US LHC Research Accelerator Program (LARP)", *IEEE Trans. Appl. Supercond.* 2006, vol. 16, pp. 234–327.

5. P. Ferracin, "LARP $Nb_3Sn$ Quadrupole Magnets for the LHC Luminosity Upgrade", in *AIP Conf. Proc.* 1218, 1291–1300.
6. J. DiMarco et al., "Test Results of the LARP $Nb_3Sn$ Quadrupole HQ03a", *IEEE Trans. Appl. Supercond.* Vol. 26, No. 4, June 2016.
7. G. Ambrosio et al., "Test Results and Analysis of LQS03 Third Long $Nb_3Sn$ Quadrupole by LARP", *IEEE Trans. Appl. Supercond.* Vol. 23, No. 3, June 2013.
8. P. Ferracin et al., "Development of MQXF: The $Nb_3Sn$ Low-$\beta$ Quadrupole for the HiLumi LHC", *IEEE Trans. Appl. Supercond.* Vol. 26, No. 4, June 2016, Art. no. 4000207.
9. G. Apollinari, R. Carcagno, G. Ambrosio, S. Feher, L. Ristori, "US contribution to the High Luminosity LHC Upgrade: focusing quadrupoles and crab cavities", *J. Phys. Conf. Ser.* 1350 (2019) 1, 012005.
10. E. Todesco et al., "The High Luminosity LHC interaction region magnets towards series production", 2021 *Supercond. Sci. Technol.* 34 053001.
11. MQXFA Magnets Functional Requirements Specification, CERN EDMS No. 1535430.
12. P. Ferracin et al., "The HL-LHC low-b quadrupole magnet MQXF: from short model to long prototype", *IEEE Trans. Appl. Supercond.* vol. 29, no. 5, August 2019, Art. no. 4001309.
13. L. D. Cooley, A. K. Ghosh, D. R. Dietderich, and I. Pong, "Conductor Specification and Validation for High-Luminosity LHC Quadrupole Magnets", *IEEE Trans. Appl. Supercond.* vol. 27, no. 4, June 2017, Art. no. 6000505.
14. B. Bordini et al., "The Bundle-Barrier PIT Wire Developed for the HiLumi LHC Project", *IEEE Trans. Appl. Supercond.* vol. 27, no. 4, June 2017, Art. no. 6000706.
15. E. Ravaioli et al., "Protecting a Full-Scale $Nb_3Sn$ Magnet With CLIQ, the New Coupling-Loss-Induced Quench System", *IEEE Trans. Appl. Supercond.* vol. 25, no. 3, art. no. 4001305.
16. S. Izquierdo Bermudez et al., "Performance of a MQXF $Nb_3Sn$ Quadrupole Under Different Stress Level", *IEEE Trans. Appl. Supercond.*, submitted for publication.
17. F. J. Mangiarotti et al., "Powering Performance and Endurance Beyond Design Limits of HL-LHC Low-Beta Quadrupole Model Magnets", *IEEE Trans. Appl. Supercond.* vol. 31, no. 5, August 2021, Art. no. 4000805.
18. J. F. Muratore et al., "Test Results of the First Two Full-Length Prototype Quadrupole Magnets for the LHC Hi-Lumi Upgrade", *IEEE Trans. Appl. Supercond.* vol. 30, no. 4, June 2020, Art. no. 4004205.
19. G. Ambrosio et al., "Lessons Learned from the Prototypes of the MQXFA Low-Beta Quadrupoles for HL-LHC and Status of Production in the US", *IEEE Trans. Appl. Supercond.* vol. 31, no. 5, August 2021, Art. no. 4001105.
20. S. Izquierdo Bermudez et al., "Progress in the Development of the $Nb_3Sn$ MQXFB Quadrupole for the HiLumi Upgrade of the LHC", *IEEE Trans. Appl. Supercond.* vol. 31, no. 5, August 2021, Art. no. 4002007.
21. F. J. Mangiarotti et al., "Power Test of the First Two MQXFB Quadrupole Magnets Built at CERN for the HL-LHC Low-Beta Insertion", *IEEE Trans. Appl. Supercond.*, vol. 32, no. 6, September 2022, Art. no. 4003305.

22. S. Izquierdo Bermudez et al., "Second-Generation Coil Design of the $Nb_3Sn$ low-$\beta$ Quadrupole for the High Luminosity LHC", *IEEE Trans. Appl. Supercond.* vol. 26, no. 4, June 2016, Art. no. 4001105.
23. P. Ferracin et al., "Assembly and Pre-Loading Specifications for the Series Production of the $Nb_3Sn$ MQXFA Quadrupole Magnets for the HL-LHC", *IEEE Trans. Appl. Supercond.* vol. 32, no. 6, September 2022, Art. no. 4000306.
24. E. Takala et al., "Preload Characterization of Short Models of MQXF the $Nb_3Sn$ Low-$\beta$ Quadrupole for the Hi-Lumi LHC", *IEEE Trans. Appl. Supercond.* vol. 30, no. 4, June 2020, Art. no. 4002806.
25. G. Vallone et al., "Mechanical Analysis of the Short Model Magnets for the $Nb_3Sn$ Low-$\beta$ Quadrupole MQXF", *IEEE Trans. Appl. Supercond.* Vol. 28, no. 3, April 2018, Art. no. 4003106.
26. G. Vallone et al., "Summary of the Mechanical Performances of the 1.5 m Long Models of the $Nb_3Sn$ Low-$\beta$ Quadrupole MQXF", *IEEE Trans. Appl. Supercond.* vol. 29, no. 5, August 2019, Art. no. 4002805.
27. E. Takala et al., "On the mechanics of MQXFB - the low-beta quadrupole for the HL-LHC", 2021 *Supercond. Sci. Technol.* 34 095002.
28. M. Bajko and M. Pojer, "*IT String and Hardware Commissioning*", in "*High-Luminosity Large Hadron Collider: Technical design report*", edited by I. Bejar Alonso, O. Bruning, P. Fessia, M. Lamont, L. Rossi, L. Tavian, M. Zerlauth, CERN Yellow Reports: Monographs, CERN-2020-010, 2020.
29. J. F. Muratore et al., "Test Results of the First Pre-Series Quadrupole Magnets for the LHC Hi-Lumi Upgrade", *IEEE Trans. Appl. Supercond.* vol. 31, no. 5, August 2021, Art. no. 4001804.
30. G. Ambrosio et al., "Challenges and Lessons Learned from fabrication, testing and analysis of eight MQXFA Low Beta Quadrupole magnets for HL-LHC", *IEEE Trans. Appl. Supercond.*, vol. 33, no. 5, August 2023, Art. no. 4003508.
31. I. Aviles Santillana et al., "Advanced Examination of $Nb_3Sn$ Coils and Conductors for the LHC Luminosity Upgrade: A Methodology Based on Computed Tomography and Materialographic Analyses", *Superconductor Science and Technology*, submitted for publication.
32. A. Moros et al., "A Metallurgical Inspection Method to Assess the Damage in Performance-Limiting $Nb_3Sn$ Accelerator Magnet Coils", *IEEE Trans. Appl. Supercond.*, vol. 33, no. 5, August 2023, Art. no. 4000208.
33. S. Izquierdo Bermudez et al., "Status of the MQXFB $Nb_3Sn$ quadrupoles for the HL-LHC", *IEEE Trans. Appl. Supercond.*, vol. 33, no. 5, August 2023, Art. no. 4001209.
34. P. Borges de Sousa et al., "Numerical Assessment of the Inhomogeneous Temperature Field and the Quality of Heat Extraction of $Nb_3Sn$ Impregnated Magnets for the High Luminosity Upgrade of the LHC", *IEEE Trans. Appl. Supercond.*, vol. 33, no. 5, August 2023, Art. no. 4000705.

© 2024 The Author(s)
https://doi.org/10.1142/9789811280184_0008

# Chapter 8

# Radio Frequency systems

Rama Calaga and Frank Gerigk
*CERN*

## 1. Challenges for the existing LHC Accelerating System (ACS)

Superconducting cavities were introduced at CERN to boost the energy of the Large Electron Positron (LEP) collider and first prototypes of 4-cell 352 MHz cavities for LEP2 were produced in 1991. At that time and due to the relatively low frequency, copper cavities sputtered with a thin film of Niobium were shown to provide a reliable performance at the required gradients, and hence was an economical option compared to bulk Niobium. Other benefits include thermal stability due to the large copper substrate and insensitivity to stray magnetic fields. Hence, this technology was adopted to construct 288 cavities for LEP2. For the LHC, a proton machine with significantly lower synchrotron radiation losses per turn, the need for accelerating voltage is greatly reduced and therefore the LEP technology was adapted "as-is" without trying to push cavity performance. Furthermore the technology proved to be reliable and stable during LEP operation. The first LHC module with four single-cell 400.79 MHz cavities was successfully tested in 2000 and all four modules were installed and commissioned with beam by 2008 [Linnecar (2008)], in time for the LHC start-up.

**LHC cavities** Two cryomodules per beam are in operation (see Fig. 1). The cavities are connected with large aperture beam tubes (300 mm) to reduce the longitudinal impedance and thereby transient beam loading.

This is an open access article published by World Scientific Publishing Company. It is distributed under the terms of the Creative Commons Attribution 4.0 (CC BY) License.

Fig. 1. ACS cryomodules in the LHC tunnel.

Table 1. LHC ACS cavity parameters.

| | |
|---|---|
| RF frequency | 400.79 MHz |
| Accelerating gradient | 5.5 MV/m |
| Accelerating voltage | 2 MV |
| $Q_0$ | $2 \times 10^9$ |
| $R/Q$ | 88.1 Ω |
| Adjustable power coupler | $11000 < Q_{ex} < 200000$ |
| Operating temperature | 4.5 K |
| Tuning range | 180 kHz |

Two dipole mode couplers and two broadband couplers are mounted on the beam tubes adjacent to the cells to provide Higher Order Mode (HOM) damping [Haebel (1997)]. The main cavity parameters are summarised in Table 1.

**RF powering & HL-LHC requirements** Originally the LHC was designed for a maximum intensity of $1.15 \times 10^{11}$ p/b. A full detuning scheme was proposed to cope with the transient beam loading effects during the energy ramp and collisions and was expected to enable the HL-LHC's intensity of $2.3 \times 10^{11}$ p/b [Mastoridis (2017)]. With this scheme, the klystron power is independent of the beam current and maintained constant over one full turn at the expense of bunch-to-bunch phase modulation.

The cavities provide a voltage of up to 16 MV per beam. During Run I (2009–2012) and Run II (2015–2018) the voltage was adjusted to 6 MV at injection and to 12 MV at flat top, requiring between 80 and 120 kW per cavity. Each cavity is powered by a klystron with a maximum of 300 kW continuous wave (CW) forward power. Taking into account that the klystrons are operated today at $\approx 1.5\,dB$ below saturation (control margin) and that losses due to waveguides and reflections from circulators account for around 5–10%, the available klystron power at the cavity input is reduced to a maximum of around 200 kW.

During injection of the HL-LHC beams from the SPS into the LHC the half-detuning scheme is required to strictly preserve the bunch-to-bunch spacing. Recent studies [Timko (2018)] based on the operational experience during Run II and advanced beam dynamics simulations with CERN's BLOND code [BLOND (2022)] have concluded that 200 kW at the cavity input is likely too low to achieve a balance between reducing injection oscillations and keeping beam losses at acceptable limits. In the 2022 LHC Performance workshop [Chamonix (2022)] it was established that, using the half-detuning scheme, a minimum injection voltage of 7.8 MV is required, which translates to 265 kW at the cavity level. Additional margin may be needed as today's measurement of the klystron forward power has a 20% uncertainty and because the klystrons themselves have a $\approx 20\%$ spread in their power saturation values.

**Means to increase RF performance** The first measure to increase the power into the cavities is the replacement of the existing LHC klystrons with new high-efficiency plug-compatible klystrons. Their internal cavity structure, responsible for electron beam shaping and beam extraction, has been redesigned by CERN [Syratchev (2021)] and is presently under construction in European industry [Thales (2019)]. This modification is expected to increase the RF efficiency from 62% to 71%, raising the klystron forward power from 300 kW to 350 kW with the same input power. Plug-compatibility means that the existing high-voltage and modulator infrastructure can be used, limiting the needed investment. Automatic adjustment of the circulator settings to minimum reflection together with improved power measurements and optimised control algorithms is expected to reduce the present control margin from 1.5 dB to 1.0 dB below saturation, which means that 350 kW klystron forward power would result in 275 kW power in the cavities, just enough to cover the estimated minimum power of 265 kW needed.

Further measures to increase the available voltage can be: i) transient detuning of the cavities with ferro-electric fast reactive tuners (FE-FRT) [Shipman (2021)], currently under development at CERN, or; ii) the installation of additional cavities and RF power systems.

Along with any power increase a precise longitudinal impedance model of the LHC is needed to improve the simulations of injection, capture, ramp and flat top operation.

## 2. The HL-LHC Crab Cavities

For high luminosity operation of the LHC (HL-LHC), proton beams are squeezed to very small $\beta^*$ at IP1 and IP5 (well below the nominal 55 cm). A non-zero crossing angle is required to control the effect of the large number of parasitic collisions. A crossing angle in combination with small $\beta^*$ results in a geometric reduction of the luminosity by a factor of $R_\Phi = (1+\Phi^2)^{-1/2}$ due to imperfect overlap of the colliding bunches. Here, $\Phi = \frac{\sigma_z}{\sigma_x^*}\phi$ is referred to as the Piwinski angle and $\phi$ is the half crossing angle. In the HL-LHC, up to 70% of the peak luminosity is lost due to the geometric reduction factor, $R_\phi = 2.66$. The effect the crossing geometry is illustrated in Fig. 2 and compared to with the crab crossing scheme where the head and the tail of the bunches transported along different orbits maximize the overlap at the interaction point.

The HL-LHC upgrade will use a crab crossing scheme using superconducting crab cavities, also operating at 400.79 MHz, to compensate for geometric luminosity loss. The crab cavities placed on either side of the IP are used to generate a localized perturbation upstream of the IP where crabbing is required and compensates for it downstream, such that through the rest of the ring the bunches remain unperturbed. The local scheme requires up to 8 cavities per beam and per IP operating at 3.4 MV for a full compensation of the HL-LHC crossing angle. This scheme allows for the different crossing planes in IP1 and IP5 and exploits the large optical functions in this region to minimize the required voltage.

Fig. 2. Bunches colliding with a crossing angle without crab crossing (left); with the crab crossing (right).

**Superconducting Crab Cavities** In order to sustain the extremely high surface fields at a kick voltage of 3.4 MV per cavity for the HL-LHC in continuous wave (CW), superconducting technology is essential; space restrictions, voltage requirements, and impedance considerations strongly rule out a normal conducting option. The placement of the cavities in the interaction region requires unconventional cavities that are compact enough for the nominal beam pipe distance of 194 mm.

Fig. 3. Schematic view of the cavity with interfaces (left) DQW; (right) RFD.

Table 2. RF parameters for the DQW and RFD cavities.

| quantity | unit | value |
|---|---|---|
| Frequency | MHz | 400.79 |
| Bunch length | ns | 1.0-1.2 (4 $\sigma$) |
| Maximum cavity radius | mm | $\leq 145$ |
| Nominal kick voltage | MV | 3.4 |
| $R/Q$ (linac convention) | $\Omega$ | 430 |
| $Q_0$ | | $\geq 10^{10}$ |
| $Q_{ext}$ (fixed coupling) | | $5 \times 10^5$ |
| RF power | kW-CW | 40 (80 peak, 1 ms) |
| LLRF loop delay | µs | $\approx 1$ |
| Cavity detuning (if parked, optional) | kHz | $\approx 1.0$ |

As a result of an intense R&D within the FP7 HiLumi LHC, EuCARD and LARP programs and with other external collaborators, three compact designs at 400 MHz emerged as potential candidates of which two were retained after the prototyping phase [Verdu, Xiao (2015); De Silva (2013)]. The vertical crabbing is realized by the Double Quarter Wave (DQW) and the horizontal crabbing by the RF Dipole (RFD). The final design of the cavities including all external interfaces is shown in Fig. 3. The proposed designs are at least four times smaller in the plane of crossing compared to a conventional elliptical cavity with a ratio of the kick gradient to the peak surface fields lower by a factor of 2 or better. Table 2 summarises the main crab cavity parameters.

**RF powering of Crab Cavities** The longitudinal impedance of the operating mode of these cavities vanishes on axis, i.e. there is no beam loading for a centered beam; the RF generator does not exchange energy with the beam [Joachim (2011)]. For a beam circulating at an offset $\Delta x$, the beam-induced voltage is proportional to the offset and the average beam current. In deflecting cavities operated in the crabbing mode, kick voltage and beam current are in quadrature ($\phi_s = 0$, synchrotron convention). Restricting the unavoidable orbit offsets and drifts to a maximum of 1 mm, the required RF power is approximately $\leq$ 40 kW for an optimal $Q_L$ of $5 \times 10^5$ This $Q_L$ provides a good compromise between the required cavity bandwidth and the available RF power. For short excursions of the orbit, an input RF power of 80 kW up to 1 ms is feasible to cope with injection and other transients. An independent powering system for each cavity using Inductive Output Tubes (IOTs) at 400.79 MHz are used to produce the 40 kW CW power. This scheme allows for fast and independent control of the cavity set point voltage and phase to ensure accurate control of the closed orbit and the crossing angle in the multi-cavity scheme. Most importantly, fast control of the cavity fields will minimize the risk to the LHC during an abrupt failure of one of the cavities, ensuring machine protection before the beams can be safely extracted. For such fast and active feedback, a small loop delay between the RF system and the cavity is used for the RF infrastructure design.

**RF Feedback and Controls for Crab Cavities** The amplifier driven by a feedback system feeds a compensating current to cancel the beam current. The cavity impedance is then effectively reduced by the feedback gain with the round-turn-loop delay as the limiting factor. This delay is specified to be less than 1.5 µs for HL-LHC which allows a significant reduction of the cavity impedance seen by the beam. A rapid and unforeseen change of the field in one cavity should trigger the LHC Beam Dump System (LBDS) to extract the beam in a minimum time of three turns (270 µs). The RF controls should minimize the effect on the beam within the 3 turns to avoid abrupt displacements which can potentially damage the machine elements. Therefore, independent power systems of each cavity with a short delay cavity controller are used [Baudrenghien (2012)]. A central controller between the two systems across the IP makes the required corrections to adjust the cavity set points as necessary.

**Beam operation with crab cavities** The use of the crab cavities for HL-LHC also requires that during the injection, energy ramp or operation without crab cavities, the cavities remain transparent to the beam, known as "crabbing off". Since more than one cavity is used, counter-phasing (such that the relative cavity RF phase, $\phi_1 - \phi_2 = \pi$) reduces the effective kick voltage to zero while always keeping accurate control of the cavity field. This scheme is also most effective for beam stability. In the HL-LHC, a single frequency reference generated at the RF controls of the accelerating cavities at IP4 will be sent over phase-compensated links to the respective crab cavities at IP1 and IP5 to synchronize the crab cavities with the beam.

RF noise in the form of amplitude jitter in the crab cavity voltage introduces a residual crossing angle at the IP and a phase jitter results in a transverse offset at the IP. The long physics fills in the HL-LHC imply that amplitude or phase jitter of the crab cavity voltage leads to growth in the transverse phase space (emittance) and thereby reduces the luminosity [Calaga (2010); Baudrenghien (2012)]. This emittance growth is of particular concern with proton beams, which have very low synchrotron radiation damping. First performance estimates in HL-LHC with realistic crab cavity amplitude and phase noise yield about a 2% luminosity loss [Medina (2018)]. To achieve this extremely low level of the RF noise budget the noise floor of the RF controls has to be further reduced from the present state-of-the-art. A dedicated noise feedback system may be necessary to further reduce the impact on the luminosity loss [Baudrenghien (2019)]. This feedback system will work in conjunction with the existing transverse damping system to counter-act against the crab cavity RF noise.

**Beam impedance** For Higher Order Modes (HOMs) in HL-LHC, the total maximum allowed longitudinal impedance from each HOM, summing over all cavities in one beam, assuming the pessimistic case that the HOM falls exactly on a beam harmonic, is specified to be $\leq 200\,\mathrm{k}\Omega$ [HL-LHC TDR (2020); Shaposhnikova, Burov (2010)]. The same limit was imposed for higher frequencies although the allowed impedance has a quadratic behaviour. In the transverse plane, considering stability criteria for multi-bunches and assuming the pessimistic case, the maximum total transverse impedance in each plane is set to be $1\,\mathrm{M}\Omega/\mathrm{m}$ [Biancacci (2014)]. The crab cavities equipped with HOM couplers were carefully designed to keep the impedance within tight tolerances, and the system remains close to the limits [Mitchell (2019)]. Modes with frequencies above 2 GHz are expected to be Landau-damped due to natural frequency spread in the respective planes. The beam power deposited in the longitudinal HOMs can become significant when the frequencies coincide with bunch harmonics. The HOM couplers are designed to accept up to 1 kW per coupler.

The crabbing field of the cavity geometries contain higher order components (described using multipoles) due to the lack of azimuthal symmetry. As the cavities are placed at locations with large beam size, the higher order components of the main deflecting mode can affect long-term particle stability. RF multipole components $b_n$ of the RF deflecting field can be approximated and hence expressed in a similar fashion to magnets [Navarro (2013)].

**SPS test facility** The first proof of principle system with two superconducting DQW cavities was tested in the special test bench in the Super Proton Synchrotron (SPS) in 2018. The primary aim of these tests was to validate the technology with proton beams, demonstrate the ability to make the system transparent and establish a robust operational control of a multi-cavity system for the different modes of operation. A straight section (LSS6) of the SPS ring was equipped with a special bypass on a movable table and featuring Y-chambers with mechanical bellows that can be displaced horizontally (see Fig. 4). This allows for the crab cavity module to be placed out of the circulating beam during regular operation of the SPS and to be moved in only during dedicated machine development [Calaga (2018)].

A complete cryogenic system on the surface (SPS-BA6) and in the tunnel (SPS-LSS6) was installed to deliver 2 K helium for the test operation of the crab cavities. Two coaxial transmission lines are used to feed RF power of up to 40 kW from the amplifiers (IOTs) installed on the surface. Placement of the passive RF elements (circulators and RF loads) on the table was required to allow for the horizontal movement of the bypass remotely. All beam-pipes in this vacuum sector are coated with a thin film of amorphous carbon to reduce secondary electron yield and hence mitigate electron cloud effects.

A detailed campaign of dedicated experiments was carried out in the SPS with proton beams in 2018. Crabbing of the proton bunches was demonstrated (see Fig. 5) for the first time and several aspects related to the RF sychronization, cavity transparency, beam quality preservation,

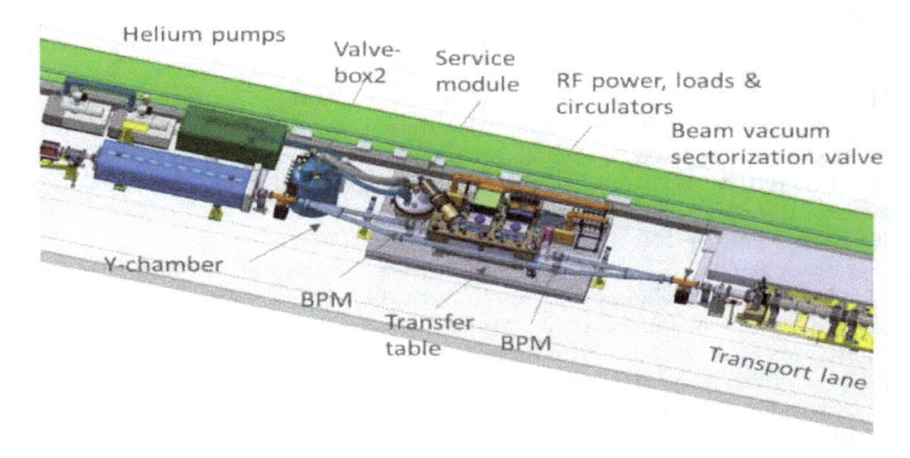


Fig. 4. SPS-LSS6 bypass for the installation of a 2-cavity crab cavity module for the first beam tests with protons [Calaga (2018)].

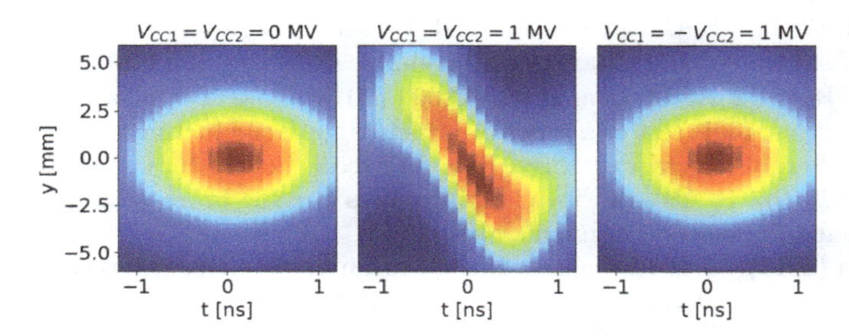


Fig. 5. Intra-bunch motion from three different cases measured with the HT monitor. Left: Crab cavities switched off (voltage = 0). Center: Synchronous crabbing with both cavities in phase corresponding to $V_{CC} \approx 2\,\text{MV}$ total voltage ($V_{CC1} = V_{CC2} = 1\,\text{MV}$). Right: Cavities in counter-phase, corresponding to residual $V_{CC} \approx 60\,\text{kV}$ total voltage.

transverse emittance growth and intensity related effects were demonstrated [Calaga (2021)].

## References

[Linnecar (2008)] T. Linnecar et al., "Hardware and initial beam commissioning of the LHC RF systems", Tech. Rep. LHC Project Report 1172, 2008.

[Haebel (1997)] E. Haebel et al., "Higher-Order Mode Dampers of the 400 MHz Superconducting LHC Cavities", 8th Workshop on RF Superconductivity, Albano Therme, Italy, 1997.

[Mastoridis (2017)] T. Mastoridis, P. Baudrenghien, J. Molendijk, "Cavity voltage phase modulation to reduce the high-luminosity Large Hadron Collider rf power requirements", Phys. Rev. Accel. Beams 20 101003 (2017).

[Timko (2018)] H. Timko, E. Shaposhnikova, K. Turaj, "Estimated LHC RF system performance reach at injection during Run III and beyond", CERN-ACC-NOTE-2019-0005, CERN, Switzerland, 2018.

[BLOND (2022)] H. Timko et al., "Beam Longitudinal Dynamics Simulation Suite BLonD", submitted to PRAB.

[Chamonix (2022)] https://indico.cern.ch/event/1097716/.

[Syratchev (2021)] I. Syratchev, "Efficiency frontiers of high-efficiency klystrons", SY Technical meeting, 15 April 2021, CERN, Geneva, Switzerland.

[Thales (2019)] A. Beunas et al., "Towards high efficiency klystrons for LHC", FCC week, 24–28 June 2019, Brussels, Belgium.

[Shipman (2021)] N. Shipman et al., "Ferro-Electric Fast Reactive Tuner applications for SRF cavities", IPAC2021, 24–28 May 2021, Campinas, Brazil.

[Verdu, Xiao (2015)] S. Verdu-Andres et al., Design and vertical tests of double-quarter wave cavity prototypes for the high-luminosity LHC crab cavity system, Phys. Rev. Accel. Beams 21, 082002; B. Xiao et al., Design, prototyping, and testing of a compact superconducting double-quarter wave crab cavity, Physical Review Special Topics - Accelerators and Beams 18, 041004 (2015).

[De Silva (2013)] S. U. De Silva and J. R. Delayen, "Design evolution and properties of superconducting parallel-bar rf-dipole deflecting and crabbing cavities", Phys. Rev. ST Accel. Beams 16, 012004 (2013); S. U. De Silva and J. R. Delayen, "Cryogenic test of a proof-of-principle superconducting rf-dipole deflecting and crabbing cavity", Phys. Rev. ST Accel. Beams 16, 002001 (2013).

[Calaga (2010)] R. Calaga, "LHC crab cavities", in the proceedings of LHC performance workshop, Chamonix, 2010.

[Joachim (2011)] J. Tuckmantel, Cavity beam transmitter interaction formula collection with derivation, CERN Internal Note, Geneva, CERN-ATS-Note-2011-002-TECH, Jan. 2011.

[Baudrenghien (2012)] P. Baudrenghien, LLRF for Crab Cavities, presented at the 2nd HiLumi-LHC Meeting, Frascati, 2012.

[Medina (2018)] L. Medina, R. Tomás, G. Arduini and M. Napsuciale, "Assessment of the performance of High Luminosity LHC operational scenarios: integrated luminosity and effective pile-up density", Canadian Journal of Physics **97** 498 (2018).

[Baudrenghien (2019)] P. Baudrenghien, T. Mastoridis, "Crab Cavity RF Noise Feedback and Transverse Damper Interaction", CERN-ACC-NOTE-2019-0006, 2019.

[HL-LHC TDR (2020)] "High-Luminosity Large Hadron Collider (HL-LHC): Technical design report", CERN-2020-010.

[Shaposhnikova, Burov (2010)] E. Shaposhnikova, "Impedance effects during injection, energy ramp & store", LHC-CC10 workshop, CERN, Geneva, 2010; A. Burov et al., Impedance Aspects, LHC-CC11 workshop, CERN, Geneva, 2011.

[Biancacci (2014)] N. Biancacci et al., "HL-LHC impedance and stability studies", presented at the 4th HiLumi-LHC workshop, KEK, 2014.

[Mitchell (2019)] J. Mitchell, Ph.D. Dissertation, University of Lancaster, https://doi.org/10.17635/lancaster/thesis/743, 2019.

[Navarro (2013)] M. Navarro-Tapia, R. Calaga, A. Grudiev, RF mulipoles from crab cavities, in the proceedings of IPAC13, Shanghai, 2013.

[Barranco (2016)] J. Barranco et al., Phys. Rev. Accel. Beams 19, 101003 (2016).

[Calaga (2018)] R. Calaga et al., SPS tests of HL-LHC crab cavities, in Proc. IPAC18, Vancouver, 2018.

[Calaga (2021)] R. Calaga et al., "First demonstration of the use of crab cavities on hadron beams", Phys. Rev. Accel. Beams 24, 062001 (2021).

© 2024 The Author(s)
https://doi.org/10.1142/9789811280184_0009

# Chapter 9

# Beam Collimation, Dump and Injection Systems

Chiara Bracco and Stefano Redaelli
*CERN*

High-performance beam collimation, injection and beam-disposal systems are essential for operating efficiently and safely modern hadron accelerators at the beam intensity frontier. In particular, the superconducting environment in colliders that work in the multi-TeV energy regime poses specific challenges that need to be addressed through optimized designs and operational schemes. The upgraded collimation system for HL-LHC is presented, addressing both the performance with protons and heavy-ion beams, and taking into account recent changes of the upgrade baseline. The upgrades applied to improve the reliability of the fast pulsed kickers of the injection and extraction systems are also addressed together with the new advanced designs and cutting-edge materials of the different collimators and dumps to cope with the unprecedented energy of the HL-LHC beams.

## 1. The HL-LHC beam stored-energy challenge

Figure 1 shows the stored beam energy for a selection of past, operating and future hadron and lepton machines. Handling beams at the beam stored-energy frontier poses obvious challenges in superconducting environments like the ones of the LHC and of its high-luminosity upgrade, HL-LHC. During its Run 2 (2015–2018) at 6.5 TeV, the LHC achieved successfully regular, high-efficiency operation close to its design value of 362 MJ [Brüning *et al.* (2004)]. This is about two orders of magnitude above the previous state-of-the-art achieved by the Tevatron (see Fig. 1). The LHC performance benefited from the high-performance multi-stage collimation system [R. W. Assmann *et al.* (2006)] but a further step is needed to meet the HL-LHC ambitious goals.

This is an open access article published by World Scientific Publishing Company. It is distributed under the terms of the Creative Commons Attribution 4.0 (CC BY) License.

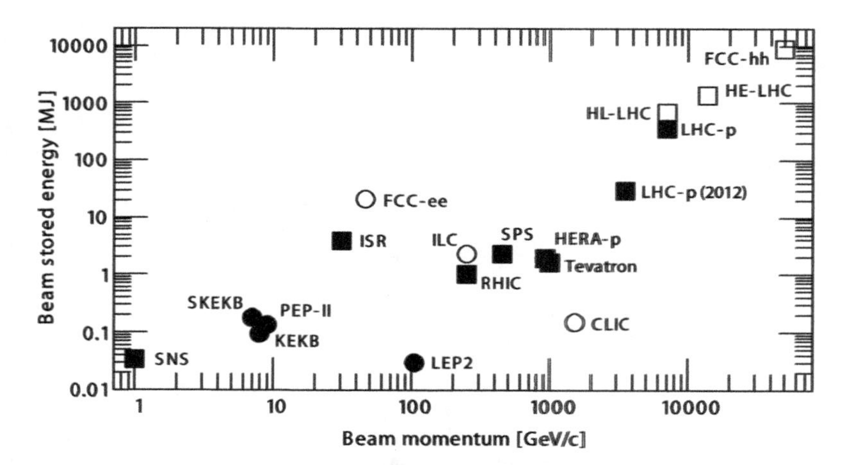


Fig. 1. Livingston-like plot of the stored beam energies for hadron (squares) and lepton (circles) accelerators. Filled symbols are used for past or operating machine and empty symbols indicate future accelerators. *Courtesy R. Aßmann.*

Injection and extraction are among the most critical systems in the LHC. The magnets used in these systems are fast-pulsed kickers operating at very high voltage. In case of a failure, high intensity beams can be mis-kicked and induce quenches in several superconducting magnets and, in the worst case, even damage the machine and the experiments. Passive protection elements are installed downstream of the injection and extraction kickers to intercept mis-kicked beams and shadow the machine aperture in order to reduce the energy deposited on the magnets and prevent any damage.

The HL-LHC upgrade project foresees doubling the stored beam energy with beams up to 5 times brighter, leading to unprecedented challenges for beam collimation and machine protection systems. A number of upgrades of the LHC systems were conceived in order to handle efficiently and safely beam stored energies up to 700 MJ at 7 TeV throughout the operational cycle, achieving the target performance with sufficient margins. Note that the injected stored energy per train increases from about 2.4 MJ to 4.8 MJ. The HL-LHC also operates as a heavy-ion collider with nearly double intensity compared to Run 2 [Coupard *et al.* (2016)].

## 2. The HL-LHC upgrade of the injection system

The LHC ring has an eight-fold symmetry with 8 insertion regions (IRs) dedicated to the four main experiments (ATLAS [ATLAS Collaboration

(2008)] in IR1, ALICE [The ALICE Collaboration (2008)] in IR2, CMS [CMS Collaboration (2008)] in IR5 and LHCb [The LHCb Collaboration (2008)] in IR8), to the off-momentum (IR3) and betatron (IR7) collimation systems, to the radio-frequency system (IR4), to the beam dumping system (IR6). IR2 and IR8 also house the injection systems for clock-wise Beam 1 and anti-clock-wise Beam 2, respectively. The injection takes place about 200 m upstream of ALICE and LHCb experiments [Brüning *et al.* (2004a)]. The beam to be injected into the LHC passes through five horizontally deflecting steel septum magnets (MSI) and four vertical deflecting kickers (MKI) steer the injected beam onto the LHC closed orbit. The kickers pulse at 25 kV and operate at $\sim 10^{-11}$ mbar to minimise the risk of flashovers. The full magnetic pulse consists of a 900 ns rise time, a flattop of 7.87 µs and a fall time of 3 µs. Uncontrolled beam losses resulting from MKI errors (missing pulses, erratic, partial, badly synchronized or wrong kick strength) could result in serious damage to the downstream equipment. In particular, the beam could directly hit the superconducting separation dipole D1, the triplet quadrupoles magnets near the ALICE and LHCb experiments or any other exposed machine aperture around the ring. In addition, particle showers, generated by proton losses, could damage components of the detectors, which are close to the beam pipe. Precautions must therefore be taken against damage and magnet quenches and, to that purpose, a dump (TDIS) is installed at about 90° phase advance from the injection kicker to protect the downstream components in case of MKI malfunctions and timing errors.

### 2.1. *Injection kicker*

The injection kicker magnets consist of U-core ferrite cells between two high voltage (HV) conducting plates. Extruded ceramic tubes (99.7% alumina), with 24 screen conductors lodged in its inner wall, are placed within the aperture of each MKI magnet [Ducimètiere *et al.* (2003)]. A set of toroidal ferrite rings is mounted around each end of the alumina tube, outside of the magnet aperture to damp low-frequency resonances. To ensure reliable operation of the MKI magnets, the temperature of the ferrite yokes must not exceed their Curie point, which is ~125°C. Above this temperature, the magnetic properties of the ferrite are compromised, and the beam cannot be injected. The MKI kickers installed in IR2 and IR8 for the first LHC run encountered a number of issues that affected operation [Barnes *et al.* (2018)]. These include beam-induced heating and electron cloud related vacuum pressure rise, which caused, in some cases, electrical breakdowns and surface flashovers. The conditioning process of the alumina tubes with beam was slow, requiring approximately 300 hours, and this could strongly affect beam operation in particular in case of replacement of a magnet in the middle of a run. As a mitigation to this problem, a 50 nm thick $Cr_2O_3$ coating [Barnes *et al.* (2017)], applied by magnetron sputtering to the inner part of the alumina tubes, allows a rapid reduction of the dynamic vacuum and a significantly faster conditioning with respect to the original design. The power deposition in the MKI for operation with HL-LHC beams is expected to be a factor of four higher than for the LHC, which would be unacceptably high for the original magnet design and would require cooling the ferrite yokes. Impedance studies show that it is possible to redistribute the beam induced power deposition, from the yoke to the upstream ferrite rings by modifying the geometry of the metallic cylinder that provides capacitive coupling to the open ends of the screen conductors. The rings can then be much more easily cooled down than the yokes since they are not pulsed at high voltage (Fig. 2). Studies shows that an active water cooling system just of the rings is sufficient to keep the temperature of the full magnet below 100°C also for HL-LHC beams [Vlachodimitropoulos *et al.* (2018)]. A complete prototype with $Cr_2O_3$ coated chambers, upgraded beam screen with active cooling of the ferrite rings, the so called "MKI-cool" will be installed and tested in the LHC during the 2022–2023 winter stop for the final validation before launching the upgrade of the full series.

Fig. 2. Detail of the cooling system installed around the upstream ferrite rings of the MKI-cool.

## 2.2. *Injection Protection Dump*

The original injection protection dump consisted in a movable two-sided vertical absorber composed by two, 4.185 m long jaws accommodating blocks of graphite, aluminum and copper alloy CuCr1Zr. This design proved to be affected by several anomalies including outgassing, vacuum spikes, structural damage of the beam screens and elastic deformation of the jaws due to beam induced RF heating during the first years of the LHC operation [Lechner *et al.* (2016)]. A new improved design in terms of mechanics, robustness, reliability, setup accuracy, impedance and operational aspects was conceived for HL-LHC. The new injection dump [Carbajo Perez *et al.* (2021)] (Fig. 3), called TDIS, consists of three modules of equal length (each 1.6 m long) hosting different materials. The first two modules are made of low-Z graphite absorbers blocks to dilute the beam while the last module is constituted of a sandwich of higher-Z materials ($Ti_6Al_4V$ and CuCr1Zr) to partially absorb and efficiently attenuate the particle showers from the low-density upstream blocks. The shorter jaws and the improved mechanics allows a more precise beam-based alignment, less prone to deformation due to beam induced heating and less sensitive to mechanical offsets and angles. The correct positioning of the TDIS jaws around the beam is indeed vital for machine protection. Each jaw of each module is independently movable in order to be aligned with respect to the circulating beam. Redundant position measurements are performed and checked via the Beam Interlock System. The upgraded systems were installed in both injection regions already during the Long Shutdown 2 (LS2); this will allow to probe their performance with higher intensity beams well in advance before the start of HL-LHC operations.

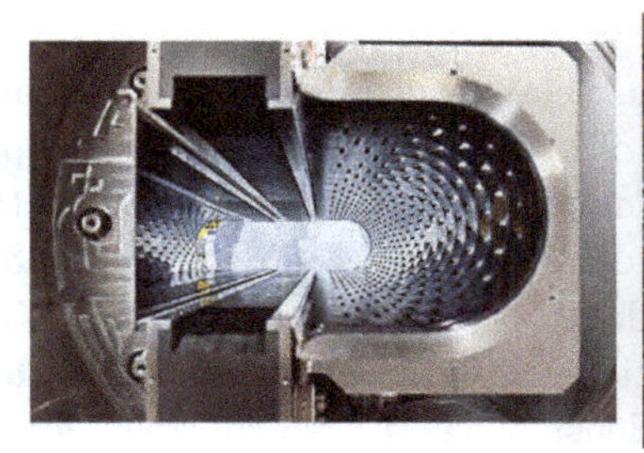

Fig. 3. The cross-section of one of the new TDIS modules is shown on the left: the graphite jaws have to intercept any vertically mis-kicked beam while the counter-rotating beam circulates unperturbed, in the same vacuum tank, to the side of the jaws. The TIDS installed in IR8 of the LHC is also shown on the right.

## 3. The HL-LHC Upgrade of the Beam Dump System

The LHC Beam Dump System (LBDS) [Brüning *et al.* (2004b)] is installed in IR6 and has to fast-extract the beam in a loss-free way and to transport it to an external dump (TDE), which is located approximately 650 m downstream of the extraction point. This requires a particles-free gap, in the circulating beam, for the rise time of the field of the fifteen fast-pulsed extraction kicker magnets (MKD) which deflect the beam horizontally into a set of fifteen Lambertson septum magnets (MSD). The septa deflect the beam vertically out of the LHC ring into the extraction channel up to the TDE. Ten dilution kickers (four horizontal MKBH and six vertical MKBV), installed ~80 m downstream of the MSD, dilute the extracted beam on the front face of the dump over a quasi-elliptical shaped area. The dilution and the long drift allow minimizing the local energy density on the dump. The synchronization between the beam-free abort gap and the field rise-time is ensured by a highly reliable timing system. Nevertheless, several failure modes exist in the synchronization system and the kicker switches which could led to an asynchronous dump, in which several hundreds of bunches could be swept across the LHC aperture by the rising kicker field. Two absorbers, a fixed block (TCDS) and a mobile diluter (TCDQ), are installed immediately upstream of the MSD and the Q4 superconducting quadrupole magnets in order to protect them from damage in the event of an asynchronous firing of the MKD kickers. No or minor modifications are required for both the TCDS and TCDQ to make them compatible with HL-LHC beams operations.

### 3.1. *Extraction and dilution kickers*

The fast-pulsed kicker magnets which are employed in the extraction and dilution system, operate at several tens of kV and can be subjected to spurious firings and flashovers. In case of loss of synchronisation between the RF system and the abort gap, or if one of the kickers undergoes a spontaneous firing, an asynchronous beam dump occurs. A fault of the MKBs would instead result in a reduced dilution at the dump front face with a consequently higher energy deposition density. The LBDS controls include a re-triggering system which, if one kicker pulses spontaneously, detects the pulse and fires all the remaining magnets. The detection of an erratic and a fast reaction of the system, within specified limits, is fundamental to minimise the effects of the failure. In order to deal with the new HL-LHC beam parameters and to ensure the highest machine availability, several consolidation items were put in place to reduce the occurrence of erratics and limit their consequences [Allonneau *et al.* (2018); Ducimètiere and Senaj (2018); Magnin *et al.* (2019)]. In particular, a third capacitor was added in the high voltage generator to lower the operational voltage and hence reduce the probability of self-trigger. New power trigger modules were implemented to increase the trigger peak current at the HV switches and reduce the stress on the switches. The re-triggering line length was shortened to reduce the reaction time to self-triggers events. The related diagnostics was also upgraded and a Spark Activity Monitoring (SAM) systems for early detection of sparking activity inside the HV generators, before degradation that could result in a self-trigger, was deployed.

### 3.2. *Dump*

Each LHC dump consists of a graphite core housed in a stainless-steel tube. The core is composed of different graphite segments, comprising six 70 cm-long isostatic Sigrafine© blocks (1.7 g/cm$^3$) and a 350 cm long segment made of 2 mm-thick Sigraflex© sheets (1.1–1.2 g/cm$^3$) which are supported by a few cm-thick extruded Graphite plates. The dump core has to be kept under a slight over pressure (100–200 mbar) of Nitrogen ($N_2$) to avoid oxidation and mass loss when heated up by the beam. During the first two LHC runs, a problem with beam induced vibrations and displacements was discovered which led to $N_2$ leaks at the different gaskets in the connection line. The vibrations were attributed to the lateral leakage of particle showers from the graphite core (mainly the Sigraflex© segment), which deposit a non-negligible amount of energy in the stainless-steel shell within the beam dump event. This rapid heating results in a highly dynamic response of the whole dump structure. In order to mitigate the vibration effects and avoid any risk of contaminating the LHC ultra-high vacuum (UHV), several modifications were put in place during LS2 [Martin *et al.* (2021)]. The core, which is kept in $N_2$ atmosphere, was disconnected from the LHC vacuum line and was suspended by a cradle-like support (Fig. 4) to absorb the vibrations and prevent as much as possible permanent longitudinal drifts of the system. Upgraded 3D forged Ti6V4Al alloy windows were installed at the entrance and exit of the dump and an additional UHV Ti window was placed at the end of the vacuum line. Despite the applied modifications, new upgraded dumps will be needed to cope with HL-LHC standard beams during nominal operation ($\sim$52% of initial intensity, i.e.

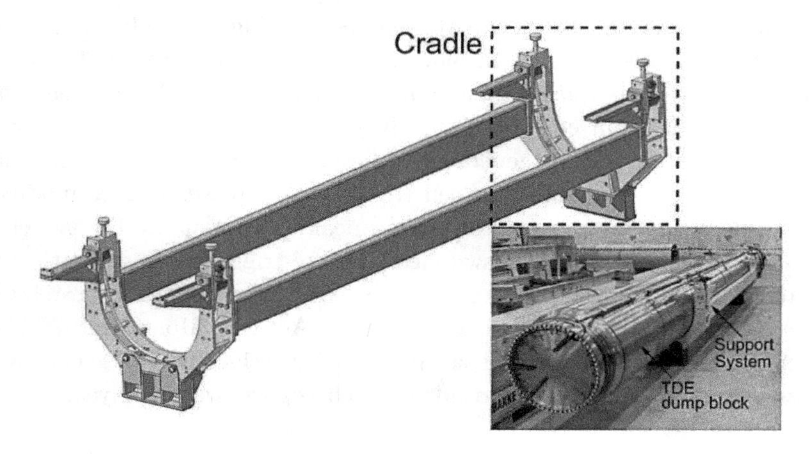


Fig. 4. 3D view of the new dump support frame with cradles at each end. The inset photo shows both upgraded dump blocks assembled in their support systems prior to installation. *Courtesy of J. Maestre*

$3.3 \times 10^{14}$ protons, dumped at the end of each fill) and in case of MKB failures (two missing horizontal MKBs as worst failure case scenario). The experience gained during the first years of beam operation in Run 3 with the present dumps will provide fundamental inputs and information for the design of the new dumps. Characterisation studies, the autopsy of the operational beam dumps and beam-impact tests in the HiRadMat facility allowed some preliminary information to be collected concerning possible materials for the dump core. In particular, Sigraflex© or C/C composites were identified as candidates for the low density sector while CFC should replace the extruded Graphite plates in the high density part. Further tests are needed to qualify the long-term resistance to beam impacts and studies are being performed to finalise the design and the choice of the vessel material (stainless steel or Ti).

## 4. The HL-LHC upgrade of the LHC collimation system

### 4.1. *Introduction*

The backbone of the HL-LHC collimation system will remain, as for the current LHC, the betatron and momentum multi-stage cleaning systems. Each system relies on the multi-stage transverse hierarchy illustrated in Fig. 5. In the following, circulating beam particles with transverse

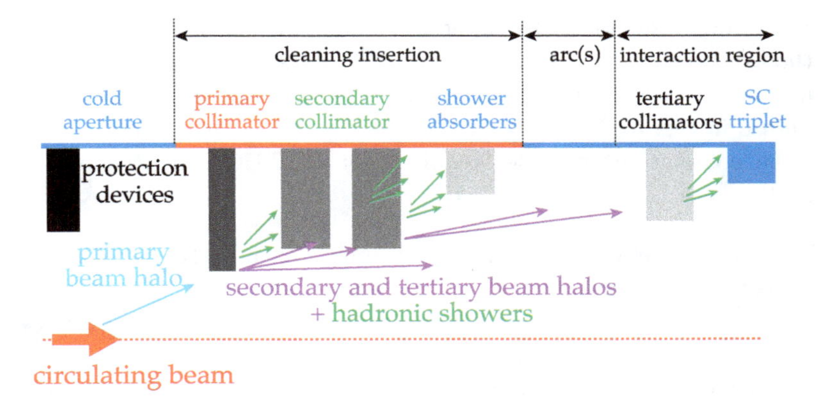


Fig. 5. Schematic illustration of the LHC multi-stage collimation cleaning system. Primary and secondary collimators (darkest grey) are closest to beam and are made of robust carbon-fibre-carbon composites. Shower absorbers and tertiary collimators (lighter grey) sit at larger apertures and are made of a tungsten alloy to improve absorption, in the shadow of protection devices (black). Collimators of different families are ordered in a pre-defined collimation hierarchy that must be respected to ensure the required system functionalities.

amplitudes within, say, 3–4 $\sigma$ of the RMS beam distribution are referred to as *beam core* as opposed to the ones above this value that are referred to as *beam halo* or *halo particles*. The border between these two regimes is somehow arbitrary. Core particle contribute significantly to the collider's luminosity production while halo particle do not and are subject to cause potentially detrimental beam losses. The multi-stage collimation system constrains the maximum betatronic and off-momentum amplitudes of halo particles with sufficient margins to the accelerator aperture to ensure a safe operation.

Primary collimators (TCPs) intercept the beam losses in case of diffusion, beam instabilities or failures experienced by the circulating beam. A very efficient halo cleaning at the energies of interest requires several secondary (TCS) collimators and shower absorbers (TCLA) to suppress the large-amplitude beam halos and to safely dispose of the energy deposited by the electromagnetic and hadronic showers produced by the interactions of halo particles with the collimator materials. Tertiary collimators (TCTs) are part of the betatron system and are located in front of the aperture bottlenecks at the super-conducting (SC) triplet magnets that provide the final focusing close to the experiments. In addition, the high-luminosity experimental regions need a physics debris collimation scheme to safely dispose of the collision products. These different collimation schemes are already part of the LHC systems and will be upgraded for HL-LHC. Injection protection and dump devices discussed above are also part of the transverse hierarchy as shown in Fig. 5.

The collimation upgrades required for HL-LHC are [Redaelli *et al.* (2020)]:

(i) Improved betatron collimation cleaning around IR7, particularly for heavy-ion operation.
(ii) Reduction of the collimator-induced impedance to allow operation with higher-brightness beams.
(iii) Improved protection of the dispersion suppressors (DSs) around ALICE in view of the luminosity upgrade for ion operation.
(iv) Improved physics debris collimation around ATLAS and CMS to enable a 5 times larger peak luminosity (levelled) than the LHC design.
(v) Improved tertiary-halo collimation around the high-luminosity experiments.

The active halo control of the high stored-energy beams, for example through hollow electron beams [Redaelli *et al.* (2021a)], is no longer part of the HL-LHC baseline however it is still pursued as a potential future upgrade beyond the scope of the HL-LHC project.

The upgrade strategy to fulfill these complex requirements was to stage the key collimation upgrades in two phases. The first phase started in the LHC second long shutdown (LS2, 2019–2021) with the deployment of:

- the collimation cleaning upgrades around IR2 (2 TCLD collimators);
- the first phase of the low-impedance upgrade of the system, involving the installation of 8 new secondary collimators (TCSPM) and 4 primary collimators (TCPPM) in IR7;
- the improvement of the protection of warm magnets in IR7 against radiation effects by means of two new passive absorbers (TCAPM);
- the crystal collimation for ion beams in IR7.

The completion of the upgrade is planned for the LS3 (2026–2029), with the installation in IR7 of 10 additional low-impedance secondary collimators and possibly 2 TCLD in the DS, with the deployment of all the IR1 and IR5 upgrades (28 movable collimators and 12 fixed-aperture masks). The complete list of collimators that will be part of the HL-LHC collimation system is given in Table 1. Including damage protection elements, details

Table 1. List of the HL-LHC collimators, including their abbreviated names, plane (H = horizontal, V = vertical, S = skew), the number of installed units, the material (CFC = carbon-fibre composite, W = heavy tungsten alloy (Inermet180), MoGr = molybdenum-graphite, CuCD = copper-diamond) and the operational openings in collision in units of beam $\sigma$ for p-p operation at $\beta^\star = 15$ cm.

| Functional type | Name | Plane | Num. | Mat. | Setting[a] |
|---|---|---|---|---|---|
| Primary IR3 | TCP | H | 2 | CFC | $17.7\sigma$ |
| Secondary IR3 | TCS | H | 8 | CFC | $21.3\sigma$ |
| Absorber IR3 | TCLA | H, V | 8 | W | $23.7\sigma$ |
| Passive absorber IR3[c] | TCAP | – | 4 | W | – |
| Primary IR7 | TCP | H, V, S | 2 | CFC | $6.7\sigma$ |
| Primary crystal IR7 | TCPC | H, V | 4 | Si | $6.5\sigma$ |
| Low-impedance primary IR7 | TCP | H,V | 4 | MoGr | $6.7\sigma$ |
| Secondary IR7 | TCS | H, V, S | 4 | CFC | $9.1\sigma$ |
| Low-impedance secondary IR7 | TCS | H,V,S | 18 | MoGr | $9.1\sigma$ |
| Absorber IR7 | TCLA | H, V | 10 | W | $12.7\sigma$ |
| Passive absorber IR7[c] | TCAP | – | 8 | W | – |
| Passive absorber mask IR7[c] | TCAPM | – | 2 | Steel | – |
| Dispersion suppressor IR7 | TCLD | H | 2 | W | $16.6\sigma$ |
| Dispersion suppressor IR2[c] | TCLD | H | 2 | W | $30\sigma$ |
| Tertiary IR8 | TCT | H, V | 8 | W | $43.8\sigma$ |
| Tertiary IR2 | TCT | H, V | 8 | W | $17.7\sigma$ |
| Tertiary IR1/IR5 | TCT | H | 8 | CuCD[d] or W | $10.4\sigma$ |
| Tertiary IR1/IR5 | TCT | V | 8 | W | $10.4\sigma$ |
| Physics debris IR1/IR5 | TCL | H | 12 | W | $14\sigma$ |
| Physics debris IR1/IR5 mask | TCLM | – | 12 | Cu / W | – |
| Dump protection IR6 | TCDQ | H | 2 | CFC | $10.1\sigma$ |
| | TCSP | H | 2 | CFC | $10.1\sigma$ |

[a] A reference proton emittance of 2.5 $\mu$m has been used for a 7 TeV beam energy.
[b] Only used for operation with heavy-ion beams.
[c] Non movable, fixed-aperture collimators or masks.
[d] CuCD is no longer part of the HL-LHC baseline following the decision in Sep. 2022 to build all TCTs in Inermet180. The possibility to deploy it as future upgrade of the HL-LHC is being considered.

of the collimator designs can be found in [Redaelli *et al.* (2020); Carra *et al.* (2014); Dallocchio *et al.* (2011); Valentino *et al.* (2017)].

### 4.2. *The new collimation in the high-luminosity regions*

The magnetic elements and collimator layout around the ATLAS experiment in IR1 are shown as a function of the longitudinal coordinate in Fig. 6 for the Beam 1. The IR5 layouts are equivalent. Two pairs of horizontal and vertical tertiary collimators are needed in front of the Q6 magnets (labelled "TCT6") and of the triplet magnets ("TCT4") in order to protect

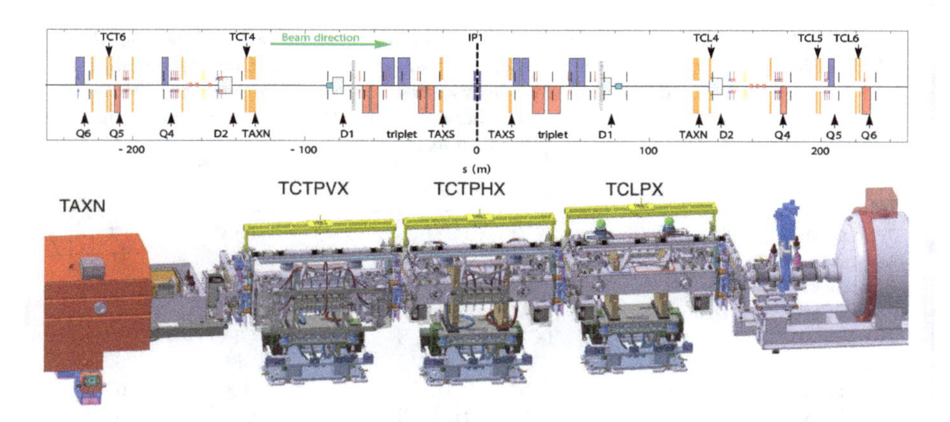


Fig. 6. Layout around ATLAS for the HL-LHC optics version 1.3 (top graph) and 3D drawing of the novel, 2-in-1 collimators called TCTPVX, TCTPHX and TCLPX (bottom). Courtesy of R. Bruce and J. Oliveira, CERN.

adequately the interaction region magnets from incoming beam losses. On the outgoing beam, three movable physics debris collimators (TCL) and three fixed-aperture masks are needed to avoid quenches and radiation damage from the collisional debris losses in high-luminosity proton collisions. A new, two-beam design concept has been conceived to install the required devices in the tight space of the beam recombination region around the experiments, where the beams share the same vacuum pipe (Fig. 6).

During LS2, the ALICE experiment was upgraded to be compatible with a peak luminosity up to 7 times higher than in Run 2, up to at least $7 \times 10^{27}$ cm$^{-2}$s$^{-1}$. This entails specific challenges that are discussed in Chapter WP2. Ultraperipheral electromagnetic interactions between the opposing ion beams, in particular the bound-free pair production (BFPP), limits the achievable peak luminosity in absence of improved collimation to dispose of the products that emerge from the collision points. During LS2, two TCLD collimators were added around IP2, in two connection cryostats in cell 11 at both sides of IR2, to efficiently remove this limitation. A photograph of the installed collimator in shown in Fig. 7.

### 4.3. *The new betatron collimation system*

The collimation cleaning upgrades are primarily required by the increased risk of quenches from off-momentum losses around IR7. These regions, called dispersion suppressors (DS), are the first part of the arcs downstream of IR7 where the dispersion function starts rising. Particles that

Fig. 7. Photograph of a TCLD collimator installed in the connection cryostat around IR2 (July 2020, Courtesy of M. Brice, CERN).

leak out of IR7 with modified rigidity after the interaction with the betatron collimators, follow perturbed trajectories — the dipole magnets act on them as powerful spectrometers — and risk to be lost locally, possibly causing quenches. This would result in costly downtime and reduced HL-LHC availability. These losses represent the primary collimation cleaning limitation of the present system, both for proton and heavy-ion beams, i.e. these are the locations of the accelerator with the largest losses in superconducting magnets from collimation leakage.

Various solutions were studied as part of HL-LHC to mitigate risk of quenches in the DS magnets. The baseline upgrade relied on adding collimators called TCLDs in the cold region. The space to add this collimator can be made by replacing a standard, 15 m-long LHC dipole with two shorter, 11 T dipoles. One TCLD per IR7 dispersion suppressor would be used, which requires a total of 4 new 11 T magnets. Losses simulated at 7 TeV for the configurations without and with TCLDs are shown in Fig. 8. This upgrade was initially planned to take place in LS2, driven by the upgrade of the ion beam parameters, however it was deferred because of delays with the 11 T dipoles. This is not an immediate concern for proton beam operation as Run 3 will not reach the design HL-LHC parameters yet. The HL-LHC target ion beam parameters will instead be achieved in

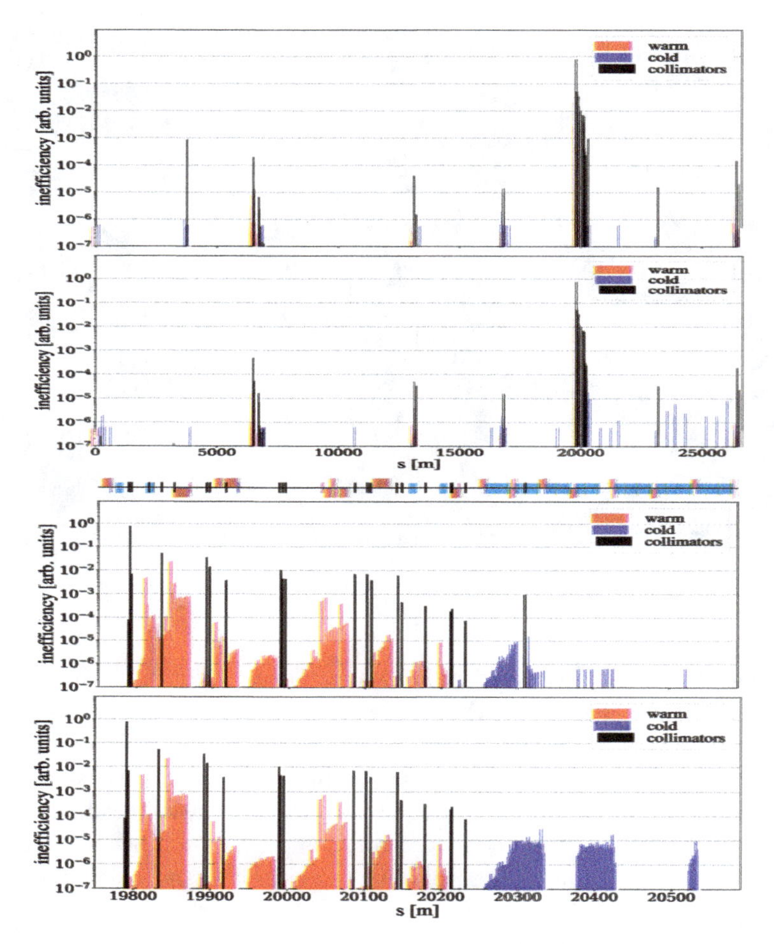


Fig. 8. Simulated loss maps for protons at the HL-LHC, at 7 TeV with 15 cm $\beta^*$. Top: full ring, with and without TCLD collimators; Bottom: IR7 and its DS, with and without TCLD collimators. Courtesy of B. Lindstrom, CERN.

Run 3. It is important to note that collimation cleaning for ion beams will be improved by the deployment of crystal collimation discussed in the next section that was pursued in WP5 as schedule-risk mitigation for the 11 T program.

Needs for the 11 T dipoles for proton operation in Run 4 will be studied during Run 3 to decide if this upgrade should be deployed. In particular, it is planned to test with beam the quench behaviour of the present dispersion suppressor magnets: this is identified as a critical input for future upgrades based on 11 T magnets. Other critical inputs are the decision on future beam energy, limited to 6.8 TeV in Run 3, and the assessment of the beam lifetime at the LHC during Run 3 with enhanced beam parameters enabled by the LIU upgrade. Alternative improvements of the proton cleaning performance are also under study, exploring for example changes of IR7 locally-generated dispersion or optics [Bruce *et al.* (2021)]. Key results are expected during 2023.

The betatron collimation system will additionally be upgraded to reduce its contribution to the machine impedance. New secondary collimators will replace 18 out of 22 TCS (see Table 1). The novel material molybdenum-graphite (MoGr) [Guardia-Valenzuela *et al.* (2018)], coated with a 6 $\mu$m layer of Mo, was chosen for the first upgrade during LS2. The improvement in surface resistivity from the present CFC secondary collimators reaches about a factor 100. The expected performance will be assessed in Run 3 thanks to the first-phase of the upgrade, where 8 new TCSs have already been installed during LS2. It is noted that the LS2 upgrade also involved the replacement of 4 primary collimators with a new design based on MoGr without coating, which would be at risk to be removed from primary beam losses impacting on the TCP surface. All new HL-LHC collimator embed in-jaw BPMs for faster alignment and local orbit monitoring [Dallocchio *et al.* (2011)].

### 4.4. *Crystal collimation of heavy-ion beams*

The collimation performance of heavy-ion beams in Run 3 relies on the crystal collimation scheme, shown schematically in Fig. 9. While the multi-stage cleaning requires several secondary collimators and absorbers, one single absorber per collimation plane would instead be sufficient, in theory, in a crystal-based collimation scheme. A bent crystal replaces the primary collimator and steers the impinging halo coherently on a single spot [Previtali (2010)]. Si crystals, 4 mm long and bent to a curvature radius of 80 m (producing a 50 $\mu$rad kick), are needed at the LHC [Mirarchi *et al.* (2017)]. Nuclear interactions are much reduced in this case [Redaelli *et al.* (2021b)], which translates into a reduction of dispersive losses downstream of the cleaning insertion. In practise, the crystal primary collimator will be inserted in the existing collimation hierarchy, slightly retracted with respect to the primary collimators that remain at their nominal positions, ensuring the passive machine protection in case of failures or orbit drifts opposite to the crystal [D'Andrea (2021)]. A minimum of 4 bent crystals is required

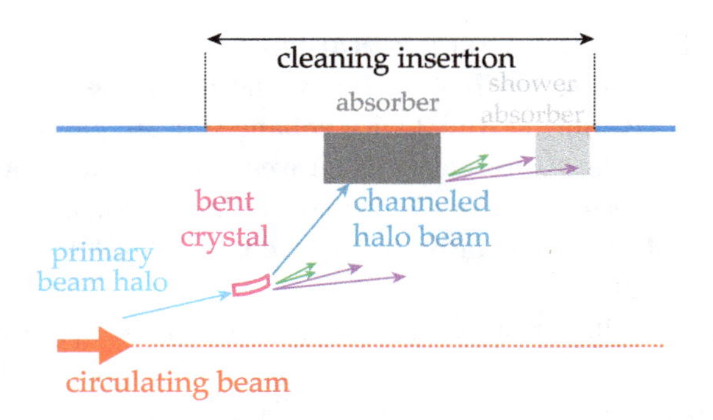


Fig. 9. The crystal collimation scheme.

for the horizontal and vertical collimation of both beams. Measurements performed in Run 2 demonstrated an improvement with Pb ion beams at 6.37 Z TeV by up to a factor 7, which is expected to be sufficient to achieve the HL-LHC goals if achieved in all beams and planes. The performance will be assessed in the first high-intensity Pb run in 2023.

### 4.5. *Active halo controls with hollow electron lenses*

Following the consistent observation of over-populated beam tails at the LHC [Valentino *et al.* (2013); Gorzawski *et al.* (2020)] and other colliders, hollow electron lenses (HELs) [Shiltsev (2016)] were integrated in the HL-LHC upgrade baseline to mitigate detrimental effects to the machine availability and safety from over-populated halos [Redaelli *et al.* (2021a)]. The HEL-based collimation scheme is shown in Fig. 10: a hollow electron beam, with inner radius below the aperture of the primary collimators, excited transverse particles outside the beam core. An ideal HEL generates zero field on the beam core, with no impact on the luminosity performance, while controlling the halo loss rates and populations over a broad range [Mirarchi *et al.* (2021)]. In reality, the field on the core cannot be null, and specific powering schemes are needed to drive the halo resonantly unstable. Various schemes are being studied and optimized for the HL-LHC operation, as shown for example in Fig. 11.

The construction of the HL-LHC lenses was planned as part of a Russian in-kind contribution. Following the critical international situation at the time of writing, it has became clear that the timeline for their deployment falls beyond the HL-LHC project duration, i.e. beyond the end of LS3. In these conditions, the lenses are no longer part of the HL-LHC baseline.

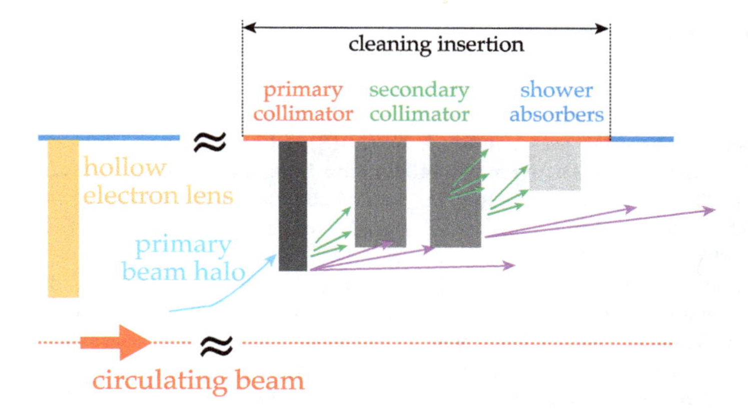


Fig. 10. Schematic view of the HEL collimation scheme: the HEL, typically located outside of the warm collimation region, actively controls the diffusion speed of particles at transverse amplitudes 1–2 $\sigma$ below the TCP, without perturbing the core. The standard collimation system remains responsible of the safe disposal of the beam halo.

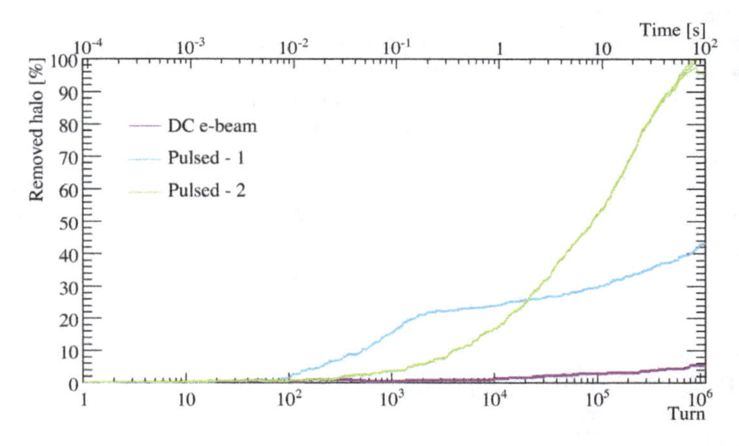


Fig. 11. Example of active halo depletion with HEL at the HL-LHC from [Mirarchi *et al.* (2021)] for different powering schemes for the electron beam: DC current (violet), random variations (green) and pulsed pattern with 9 turns ON and 14 turns OFF, as described in [Mirarchi *et al.* (2021)].

However, they are still pursued as a potential upgrade beyond the HL-LHC project, with earliest installation in LS4.

## References

[Allonneau *et al.* (2018)] Allonneau, L., Boucly, C., Carlier, E., Ducimetiere, L., Magnin, N., Mottram, T., Senaj, V., Strobino, L., Rodziewicz, J., Van Trappen, P., Vez, B., and Voumard, N. (2018). LBDS Control Upgrade, `https://indico.cern.ch/event/784431/contributions/3263918/attachments/1790346/2916731/LBDS_Review_2019_LBDS_Controls_Upgrade_v2.pdf`.

[ATLAS Collaboration (2008)] The ATLAS experiment at the CERN Large Hadron Collider, *JINST* **3**, p. S08003.

[Barnes *et al.* (2017)] Barnes, M., Adraktas, A., Bregliozzi, G., Goddard, B., Ducimètiere, L., Salvant, B., Sestak, J., Vega Cid, L., Weterings, W., and Yin Vallgren, C. (2017). Operational experience of the upgraded LHC injection kicker magnets during Run 2 and future plans, *Journal of Physics: Conference Series* **874**, p. 012101, `https://doi.org/10.1088/1742-6596/874/1/012101`.

[Barnes *et al.* (2018)] Barnes, M., Bracco, C., Bregliozzi, G., Chmielinska, Ducimètiere, L., Goddard, B., Kramer, T., Neupert, H., Vega Cid, L., Vlachodimitropoulos, V., Weterings, W., and Yin Vallgren, C. (2018). An Upgraded LHC Injection Kicker Magnet, *"Proceedings of 9th International Particle Accelerator Conference", Vancouver, BC, Canada*, pp. 2632–2635, `http://cds.cern.ch/record/2672230/files/wepmk003.pdf`.

[Bruce *et al.* (2021)] Bruce, R., Maria, R. D., Giovannozzi, M., Mounet, N., and Redaelli, S. (2021). Optics configurations for improved machine impedance and cleaning performance of a multi-stage collimation insertion, *Proceedings of the 12th International Particle Accelerator Conference (IPAC'21): Campinas, Brazil, May 2021*, pp. 57–60, doi:doi: 10.18429/JACoW-IPAC2021-MOPAB006, `https://jacow.org/ipac2021/papers/MOPAB006.pdf`.

[Brüning *et al.* (2004)] Brüning, O. S., Collier, P., Lebrun, P., Myers, S., Ostojic, R., Poole, J., and P. Proudlock (editors), (CERN, Geneva, Switzerland, 2004). LHC design report v. 1: The LHC main ring, *CERN-2004-003-V1*.

[Brüning *et al.* (2004a)] Brüning, O., Collier, P., Lebrun, P., Meyers, S., Ostojic, R., Poole, J., and Proudlock, P. (2004a). LHC Design Report, **I**, pp. 417–440, `https://cds.cern.ch/record/782076?ln=fr`.

[Brüning *et al.* (2004b)] Brüning, O., Collier, P., Lebrun, P., Meyers, S., Ostojic, R., Poole, J., and Proudlock, P. (2004b). LHC Design Report, **I**, pp. 441–466, `https://cds.cern.ch/record/782076?ln=fr`.

[Carbajo Perez *et al.* (2021)] Carbajo Perez, D., Perillo-Marcone, A., Berthome, E., Bertone, C., Biancacci, N., Bracco, C., Bregliozzi, G., Bulat, B., Cadiou, C., Calviani, M., Cattenoz, G., Cherif, A., Costa Pinto, P., Dallocchio, A., Di Castro, M., Fessia, P., Frankl, M., Fuchs, J. F., Garcia Gavela, H., Gentini, L., Geisser, J. M., Gilardoni, S., Gonzalez De La Aleja Cabana, M. A., Grenard, J. L., Joly, S., Lechner, A., Lendaro, J., Maestre, J., Page, E., Perez Ornedo, M., Pugnat, D., Rigutto, E., Seidenbinder, R., Salvant, B., Sapountzis, A., Scibor, K., Sola Merino, J., Taborelli, M., Urrutia, E., Franqueira Ximenes, R., Vieille, A., Vollinger,

C., and Yin Vallgren, C. (2021). New Generation CERN LHC Injection Dump – Assembly and Installation, *"Proceedings of 12th International Particle Accelerator Conference", Campinas, SP, Brasil*, pp. 3548–3551, `https://cdsweb.cern.ch/record/2809172/files/document.pdf`.

[Carra *et al.* (2014)] Carra, F. *et al.* (2014). Mechanical Engineering and Design of Novel Collimators for HL-LHC, in *5th International Particle Accelerator Conference*, p. MOPRO116, doi:10.18429/JACoW-IPAC2014-MOPRO116.

[CMS Collaboration (2008)] The CMS experiment at the CERN LHC, *JINST* **3**, p. S08004.

[Coupard *et al.* (2016)] Coupard, J., Damerau, H., Funken, A., Garoby, R., Gilardoni, S., Goddard, B., Hanke, K., Manglunki, D., Meddahi, M., Rumolo, G., Scrivens, R., and Chapochnikova, E. (2016). LHC Injectors Upgrade, Technical Design Report, Vol. II: Ions, Tech. Rep. CERN-ACC-2016-0041, CERN, Geneva, `https://cds.cern.ch/record/2153863`.

[Dallocchio *et al.* (2011)] Dallocchio, A., Boccard, C. B., Gasior, M., Carra, F., Bertarelli, A., Gentini, L., and Timmins, M. A. (2011). LHC Collimators with Embedded Beam Position Monitors: A New Advanced Mechanical Design, *Conf. Proc. C* **110904**, pp. 1611–1613.

[D'Andrea (2021)] *Applications of Crystal Collimation to the CERN Large Hadron Collider (LHC) and its High Luminosity Upgrade Project (HL-LHC)*, Ph.D. thesis, University of Padova, `http://cds.cern.ch/record/2758839`, presented 23 Feb 2021.

[Ducimètiere *et al.* (2003)] Ducimètiere, L., Garrel, N., and Barnes, M. (2003). The LHC Injection Kicker Magnet, *"Proceedings of the 2003 Particle Accelerator Conference", Portland, Oregon, US*, pp. 1162–1164, `https://accelconf.web.cern.ch/p03/PAPERS/TPAE036.pdf`.

[Ducimètiere and Senaj (2018)] Ducimètiere, L. and Senaj, V. (2018). LBDS Kickers: HW upgrade,
`https://indico.cern.ch/event/784431/contributions/3263915/attachments/1790226/2916321/HW_LBDS_review_05_Feb_2019.pdf`.

[Gorzawski *et al.* (2020)] Gorzawski, A., Appleby, R., Giovannozzi, M., Mereghetti, A., Mirarchi, D., Redaelli, S. *et al.* (2020). Probing LHC halo dynamics using collimator loss rates at 6.5 TeV, *Phys. Rev. ST Accel. Beams* **23**, p. 044802.

[Guardia-Valenzuela *et al.* (2018)] Guardia-Valenzuela, J., Bertarelli, A., Carra, F., Mariani, N., Bizzaro, S., and Arenal, R. (2018). Development and properties of high thermal conductivity molybdenum carbide - graphite composites, *Carbon* **135**, pp. 72–84, doi:https://doi.org/10.1016/j.carbon.2018.04.010, `http://www.sciencedirect.com/science/article/pii/S0008622318303555`.

[Lechner *et al.* (2016)] Lechner, A., Barnes, M., Bartmann, W., Biancacci, N., Bracco, C., Bregliozzi, G., Butcher, M., Esteban Muller, J., Folch, R., Gentini, L., Gilardoni, S., Grudiev, A., Lamas Garcia, I., Losito, R., Masi, A., Metral, E., Perillo Marcone, A., Salvant, B., Taborelli, M., Uythoven, J., Vollenberg, W., and Yin Vallgren, C. (2016). TDI – Past Observations and Improvements for 2016, *"Proceedings 6th Evian Work-*

*shop on LHC beam operation", Evian Les Bains, France*, pp. 123–130, http://cds.cern.ch/record/2294666/files/1637016_123-130.pdf.

[Magnin *et al.* (2019)] Magnin, N., Bartmann, W., Bracco, C., Carlier, E., Gräwer, G., Mottram, T., Renner, E., Rodziewicz, P., Senaj, V., and Wiesner, C. (2019). Consolidation of Re-Triggering System of LHC Beam Dumping System at CERN, *"Proceedings of 17th International Conference on Accelerators and Large Experimental Physics Control Systems"*, pp. 412–416, https://cds.cern.ch/record/2778521/files/10.18429_JACoW-ICALEPCS2019-MOPHA088.pdf.

[Martin *et al.* (2021)] Martin, R., Jose, M., Kershaw, K., Calviani, M., Grenier, D., Maestre Herdia, J., Torregrosa, M., Claudio, L., Buonocore, L., Serrano, G., Matheson, E., and Sola Merino, J. (2021). Practical Challenges of the LHC Main Beam Dump Upgrades, https://cds.cern.ch/record/2786788.

[Mirarchi *et al.* (2021)] Mirarchi, D., Appleby, R., Bruce, R., Giovannozzi, M., Mereghetti, A., Redaelli, S., and Stancari, G. (2021). Nonlinear dynamics of proton beams with hollow electron lens in the CERN high-luminosity LHC, *Eur. Phys. J. Plus* **137**, p. 7, https://doi.org/10.1140/epjp/s13360-021-02201-5.

[Mirarchi *et al.* (2017)] Mirarchi, D., Hall, G., Redaelli, S., and Scandale, W. (2017). Design and implementation of a crystal collimation test stand at the large hadron collider, *The European Physical Journal C* **77**, 6, p. 424, doi:10.1140/epjc/s10052-017-4985-4, https://doi.org/10.1140/epjc/s10052-017-4985-4.

[Previtali (2010)] Previtali, V. (2010). *Performance evaluation of a crystal-enhanced collimation system for the LHC*, Ph.D. thesis, Ecole Polytechnique, Lausanne.

[Redaelli *et al.* (2020)] Redaelli, S., Bruce, R., Lechner, A., and Mereghetti, A. (2020). Chapter 5: Collimation system, *CERN Yellow Rep. Monogr.* **10**, pp. 87–114, doi:10.23731/CYRM-2020-0010.87.

[Redaelli *et al.* (2021a)] Redaelli, S., Appleby, R., Bruce, R., Brüning, O., Kolehmainen, A., Ferlin, G., Foussat, A., Giovannozzi, M., Hermes, P., Mirarchi, D., Perini, D., Rossi, A., and Stancari, G. (2021a). Hollow electron lenses for beam collimation at the High-Luminosity Large Hadron Collider (HL-LHC), *Journal of Instrumentation* **16**, 03, p. P03042, doi:10.1088/1748-0221/16/03/p03042, https://doi.org/10.1088/1748-0221/16/03/p03042.

[Redaelli *et al.* (2021b)] Redaelli, S. *et al.* (2021b). First observation of ion beam channeling in bent crystals at multi-TeV energies, *Eur. Phys. J. C* **81**, 2, p. 142, doi:10.1140/epjc/s10052-021-08927-x.

[R. W. Assmann *et al.* (2006)] The Final Collimation System for the LHC, *Proc. of the European Particle Accelerator Conference 2006, Edinburgh, Scotland*, p. 986.

[Shiltsev (2016)] Shiltsev, V. D. (2016). *Electron Lenses for Super-Colliders*, Particle Acceleration and Detection (Springer), ISBN 978-1-4939-3315-0, 978-1-4939-3317-4, doi:10.1007/978-1-4939-3317-4.

[The ALICE Collaboration (2008)] The ALICE experiment at the CERN LHC, *Journal of Instrumentation* **3**, 08, p. S08002.

[The LHCb Collaboration (2008)] The LHCb Detector at the LHC, *Journal of Instrumentation* **3**, 08, p. S08005.

[Valentino *et al.* (2013)] Valentino, G., Aßmann, R., Bruce, R., Burkart, F., Previtali, V., Redaelli, S., Salvachua, B., Stancari, G., and Valishev, A. (2013). Beam diffusion measurements using collimator scans in the LHC, *Phys. Rev. ST Accel. Beams* **16**, p. 021003, doi:10.1103/PhysRevSTAB.16.021003.

[Valentino *et al.* (2017)] Valentino, G., Baud, G., Bruce, R., Gasior, M., Mereghetti, A., Mirarchi, D., Olexa, J., Redaelli, S., Salvachua, S., Valloni, A., and Wenninger, J. (2017). Final implementation, commissioning, and performance of embedded collimator beam position monitors in the large hadron collider, *Phys. Rev. Accel. Beams* **20**, p. 081002, doi:10.1103/PhysRevAccelBeams.20.081002, `https://link.aps.org/doi/10.1103/PhysRevAccelBeams.20.081002`.

[Vlachodimitropoulos *et al.* (2018)] Vlachodimitropoulos, V., Barnes, M., Ducimètiere, L., Vega Cid, L., and Weterings, W. (2018). Longitudinal Impedance Analysis of an Upgraded LHC Injection Kicker Magnet, *"Proceedings of 9th International Particle Accelerator Conference", Vancouver, BC, Canada*, pp. 2628–2631, `https://accelconf.web.cern.ch/ipac2018/papers/wepmk002.pdf`.

© 2024 The Author(s)
https://doi.org/10.1142/9789811280184_0010

# Chapter 10

# Machine Protection and Cold Powering

Amalia Ballarino and Daniel Wollmann
*CERN*
*Espl. des Particules 1*
*1211 Genève 23*
*Switzerland*

## 1. Introduction

In the HL-LHC era the bunch intensity will nearly double as compared to the design value of the LHC. Therefore, the stored energy per beam will also nearly double from 360 MJ to about 700 MJ [1]. This increases the criticality of existing beam failure cases and requires the upgrade of protection elements in the injection and dump regions of the LHC (see Chapter 9).

At the same time several optics parameters will change. To achieve the very small beam sizes in the interaction points of IR1 and IR5, the $\beta$ functions will nearly triple in the respective final focusing magnets, which requires the installation of novel large aperture quadruple magnets based on $Nb_3Sn$ superconductors. These magnets will be installed in a complex nested circuit arrangement and require highly reliable and redundant quench protection systems [1, Chapter 7]. Furthermore, the $\beta$ functions will increase by a factor of four and three, respectively, in the separation and recombination dipole magnets, the so-called D1 and D2 [1, Chapter 3]. These changes will significantly increase the effect of failures in these elements on the beam.

This is an open access article published by World Scientific Publishing Company. It is distributed under the terms of the Creative Commons Attribution 4.0 (CC BY) License.

The implementation of $Nb_3Sn$ based magnets for the first time in an accelerator raises the question of their damage limits in view of the impact of beam or high density particle showers in case of accepted failures and especially for failures beyond design.

The LHC will also see the installation of new accelerator equipment like crab cavities, coupling loss induced quench systems or a full remote alignment system for the majority of elements installed in the straight sections of IR1 and 5. These new systems will introduce new failure cases in the LHC, which require dedicated interlocks.

## 2. New Fast Failures and Interlocks

The new $Nb_3Sn$ final focusing quadrupole magnets, also known as Q1, Q2 and Q3 in IR1 and 5 will be equipped with quench heater strips and the novel Coupling Loss Induced Quench (CLIQ) system [1, Chapter 7] [2] as part of the quench protection system. Detailed studies have shown that, if they are triggered, when the beams are still circulating in the LHC, both systems can cause unacceptable beam losses within less than 1 ms, which is equivalent to about 10 LHC turns [3–5]. These losses could cause critical damage to accelerator equipment or the close by experimental detectors. In case of a magnet quench, the quench detection system, therefore, needs to first initiate a beam dump request via the LHC's Beam Interlock System (BIS) [6] before triggering the discharge of the quench heater power supplies and the CLIQ systems. These strict requirements are new for the LHC quench protection system, but can be fulfilled without any negative side effects. This mitigates the failure in case of a regular quench in one or more of the inner triplet quadrupole magnets.

However, a spurious discharge of a single quench heater power supply or a single CLIQ system cannot be fully excluded and can still cause critical losses [4, 5]. In case of a spurious discharge of one of the CLIQ units of the Q2, critical loss levels would be reached after only 450 $\mu$s from the beginning of the event. This leaves only 170 $\mu$s to detect the spurious discharge, as about 280 $\mu$s — slightly more than 3 turns — are required for the signal transmission via the beam interlock system and the extraction of the two beams [7]. Therefore, the quench detection system also has to provide a fast interlock to mitigate spurious discharges of quench heater power supplies and CLIQ systems in the new inner triplet circuits of the HL-LHC era.

Due to the large $\beta$ functions in the separation and recombination dipoles — D1 and D2 — quench heater discharges in these magnets can also have an important impact on the LHC beam. Therefore, the requirements described above for the inner triplet circuits also apply here, with the exception that these magnets are only protected by quench heaters.

The HL-LHC project also contributed to the design and integration studies of hollow e-lens systems in each beam for beam collimation. While this system is not part of the current HL-LHC baseline, an eventual installation after the start of the HL-LHC operation would allow partial depletion of the beam halo in a radius of about two beam $\sigma$ outside the cut of the primary collimators [8]. The partial depletion of the beam halo can reduce the criticality of fast failures, which cause a sudden movement of the beam, as it increases the time margin between the onset of the failure and reaching critical loss levels in the collimation system. This effect is relevant in the case of failures in magnet protection systems and crab cavities which feature dedicated fast interlocks. Also, the criticality of the missing beam-beam kick will be reduced by the partial depletion of the beam halo [4].

However, in case the failure detection relies solely on beam loss monitors, the depleted halo can increase the criticality of the failure, as it might reduce the time between reaching the interlock threshold of the beam loss monitors and reaching critical loss levels in the collimation system. This was carefully studied and discussed in [9]. The studies show that the beam halo should not be depleted by significantly more than 50%. Otherwise, in case of a symmetric quench in one of the new inner triplet magnets, the time between reaching the interlock threshold in the beam loss monitors and reaching critical loss levels in the collimation system will be too short to safely dump the beams before damage occurs. It has to be noted that the quench detection system will reliably detect the symmetric quench and initiate the timely firing of the quench protection elements. However, to avoid spurious dumps and allow for detection thresholds in the order of 100 mV, the quench detection system requires a discrimination time in the order of 10 ms, whereas critical loss levels will already be reached after less than 8 ms. Therefore the protection against the effect of symmetric quenches on the beam has to rely on the detection of losses by the beam loss monitors.

The situation is similar in the case of an accidental coherent excitation of the full HL-LHC beam by the transverse damper, as it does not have a dedicated interlock. Furthermore, similar beam losses can be caused by fast transverse beam instabilities.

## 3. Circuit Protection

The backbone of the magnet and circuit protection for the new superconducting circuits in the HL-LHC era is the highly reliable Universal Quench Detection System (UQDS) [10–12]. Due to its adaptability it can be used to protect such different circuit elements as Nb-Ti and $Nb_3Sn$ based superconducting magnets as well as $MgB_2$ based superconducting links, Nb-Ti bus-bars and HTS current leads. The UQDS features, among other things, current dependent thresholds — required to immunise the quench detection system against flux jumps in $Nb_3Sn$ magnets at low current levels, which would otherwise cause spurious triggers of the quench protection systems. Furthermore, the UQDS is equipped with a high bandwidth communication field bus which allows efficient transmission of high resolution data, in case of a quench with a time synchronisation significantly shorter than 1 ms. This is especially important to validate the correct functioning of the magnet and circuit protection systems in case of a quench in one element or during a fast power abort in the new inner triplet circuits [1, Chapter 7]. These will become the most complex circuits in the LHC, featuring four power converters, cold and warm by-pass diodes [13, 14], crowbars and the novel Coupling Loss Induced Quench Systems [1, Chapter 7] [2].

Several of the new HL-LHC circuits will be protected by a novel generation of energy extraction systems, which are equipped with in-vacuum circuit breakers, allowing for highly reliable extraction of the circuit energy and low maintenance [1, Chapter 7] [15].

## 4. Damage limits of Nb-Ti and $Nb_3Sn$ superconductors due to beam impact

With the increase in bunch intensity and stored energy in the HL-LHC era, failure cases, which lead to high levels of high energy particle showers into superconducting magnets like injection or dump failures, become more critical. To understand the criticality of such events for superconducting magnets, detailed studies have been performed to determine the damage limits of superconducting strands based on Nb-Ti and $Nb_3Sn$ due to the impact of high intensity proton beams [14, 16–19]. The hadronic showers developing during such an event lead to a sudden temperature rise in the strand material. The temperature rises to peak levels of several hundred Kelvin within one to a few micro seconds, depending on the length of the proton pulse. Due to the very localised energy deposition, the hot-spots are accompanied by high temperature gradients in the strand material. Both effects lead to high levels of stress, which can damage the superconductor.

For Nb-Ti strands no degradation of the critical current density $J_c$ has been observed up to hot spot temperatures of 1150 K. However, for peak temperatures above 800 K, the RRR at the beam impact reduces significantly to below 100. This reduction of the conductivity in the copper matrix leads to a reduction of the minimum quench energy, which can potentially cause thermal instabilities in the strand and make the operation of a superconducting magnet difficult [14].

For $Nb_3Sn$ strands a significant degradation of $J_c$ was observed for hotspot temperatures above 460 K and temperature gradients above 200 K/mm after the beam impact. Thermo-mechanical simulations allowed the identification of the two main damage mechanisms in $Nb_3Sn$ strands [19]. The first is the breaking of filaments in the strand due to high axial strain. This was also confirmed by microscopic analyses of the impacted samples. The second is the degradation of the second critical field $B_{c_2}$ due to residual strain between the copper matrix and other copper/bronze phases present in the strand. Although both mechanisms cause a reduction of $J_c$, filament breaking is by far the dominating effect [14].

## 5. Cold Powering Systems

The power converters for the HL-LHC superconducting magnets will be installed in new radiation-free galleries about 10 m above the LHC tunnel. The electrical connection between the power converters and the magnets will be provided by novel cold powering systems, eight in total, that incorporate a Superconducting Link [20, 21]. A Superconducting Link transfers the current from the power converters, at room temperature, to the liquid helium environment of the magnets. It consists of several cables, which are made from ex-situ $MgB_2$ superconducting wire. The $MgB_2$ wire has a diameter of 1 mm. It contains 37 $MgB_2$ filaments (50–55 $\mu$m average diameter) each surrounded by a niobium diffusion barrier, and embedded in a matrix consisting of Nickel and Monel. The Monel is copper plated ($\sim$ 15 $\mu$m) and tin plated ($\sim$ 1 $\mu$m). The $MgB_2$ wires are reacted — and therefore superconducting — before cabling. Their minimum bending diameter is about 100 mm. The $MgB_2$ cables are rated at DC currents ranging from 600 A to 18000 A [22]. They are electrically insulated and twisted together to form a multi-cable assembly that optimises electro-magnetic and mechanical behaviors. The multi-cable assembly contains up to nineteen cables and is housed inside a flexible cryostat. The cryostat is made from two concentric corrugated pipes. It provides the thermal insulation and the cryogenic environment for the cables. A Superconducting Link transfers DC currents of up to 120 kA at 25 K. The multi-cable assembly is cooled by forced flow of helium gas. It has an external diameter of about 90 mm. The protection of the $MgB_2$ cables was studied during the R&D phase of the project [23]. A resistive transition in the cables is a rare event. It can be generated by an accidental event like loss of vacuum insulation inside the cryostat. If the critical temperature of the $MgB_2$ is exceeded, a voltage is generated along the superconducting cables. The resistive zone propagation along the cables is in the range of 10 cm/s–20 cm/s. The transition is detected at a voltage threshold of about 100 mV. When this voltage is reached, a power abort is triggered and the active protection system of the respective magnet circuit is activated. The amount of copper stabilizer included in the cables is such that the hot spot temperature reached during the transient is less than 60 K. This limited increase in temperature enables a fast cool-down of the Superconducting Links after the resistive transition.

## References

[1] O. Aberle, I. Béjar Alonso, O. Brüning, P. Fessia, L. Rossi, L. Tavian, M. Zerlauth, C. Adorisio, A. Adraktas, M. Ady, J. Albertone, L. Alberty, M. Alcaide Leon, A. Alekou, D. Alesini, B. A. Ferreira, P. A. Lopez, G. Ambrosio, P. Andreu Munoz, M. Anerella, D. Angal-Kalinin, F. Antoniou, G. Apollinari, A. Apollonio, R. Appleby, G. Arduini, B. A. Alonso, K. Artoos, S. Atieh, B. Auchmann, V. Badin, T. Baer, D. Baffari, V. Baglin, M. Bajko, A. Ball, A. Ballarino, S. Bally, T. Bampton, D. Banfi, R. Barlow, M. Barnes, J. Barranco, L. Barthelemy, W. Bartmann, H. Bartosik, E. Barzi, M. Battistin, P. Baudrenghien, I. B. Alonso, S. Belomestnykh, A. Benoit, I. Ben-Zvi, A. Bertarelli, S. Bertolasi, C. Bertone, B. Bertran, P. Bestmann, N. Biancacci, A. Bignami, N. Bliss, C. Boccard, Y. Body, J. Borburgh, B. Bordini, F. Borralho, R. Bossert, L. Bottura, A. Boucherie, R. Bozzi, C. Bracco, E. Bravin, G. Bregliozzi, D. Brett, A. Broche, K. Brodzinski, F. Broggi, R. Bruce, M. Brugger, O. Brüning, X. Buffat, H. Burkhardt, J. Burnet, A. Burov, G. Burt, R. Cabezas, Y. Cai, R. Calaga, S. Calatroni, O. Capatina, T. Capelli, P. Cardon, E. Carlier, F. Carra, A. Carvalho, L. Carver, F. Caspers, G. Cattenoz, F. Cerutti, A. Chancé, M. C. Rodrigues, S. Chemli, D. Cheng, P. Chiggiato, G. Chlachidze, S. Claudet, J. Coello De Portugal, C. Collazos, J. Corso, S. Costa Machado, P. Costa Pinto, E. Coulinge, M. Crouch, P. Cruikshank, E. Cruz Alaniz, M. Czech, K. Dahlerup-Petersen, B. Dalena, G. Daniluk, S. Danzeca, H. Day, J. De Carvalho Saraiva, D. De Luca, R. De Maria, G. De Rijk, S. De Silva,

B. Dehning, J. Delayen, Q. Deliege, B. Delille, F. Delsaux, R. Denz, A. Devred, A. Dexter, B. Di Girolamo, D. Dietderich, J. Dilly, A. Doherty, N. Dos Santos, A. Drago, D. Drskovic, D. D. Ramos, L. Ducimetière, I. Efthymiopoulos, K. Einsweiler, L. Esposito, J. Esteban Muller, S. Evrard, P. Fabbricatore, S. Farinon, S. Fartoukh, A. Faus-Golfe, G. Favre, H. Felice, B. Feral, G. Ferlin, P. Ferracin, A. Ferrari, L. Ferreira, P. Fessia, L. Ficcadenti, S. Fiotakis, L. Fiscarelli, M. Fitterer, J. Fleiter, G. Foffano, E. Fol, R. Folch, K. Foraz, A. Foussat, M. Frankl, O. Frasciello, M. Fraser, P. F. Menendez, J.-F. Fuchs, S. Furuseth, A. Gaddi, M. Gallilee, A. Gallo, R. G. Alia, H. G. Gavela, J. G. Matos, H. Garcia Morales, A. G.-T. Valdivieso, C. Garino, C. Garion, J. Gascon, C. Gasnier, L. Gentini, C. Gentsos, A. Ghosh, L. Giacomel, K. G. Hernandez, S. Gibson, C. Ginburg, F. Giordano, M. Giovannozzi, B. Goddard, P. Gomes, M. Gonzalez De La Aleja Cabana, P. Goudket, E. Gousiou, P. Gradassi, A. G. Costa, L. Grand-Clément, S. Grillot, J. Guillaume, M. Guinchard, P. Hagen, T. Hakulinen, B. Hall, J. Hansen, N. Heredia Garcia, W. Herr, A. Herty, C. Hill, M. Hofer, W. Höfle, B. Holzer, S. Hopkins, J. Hrivnak, G. Iadarola, A. Infantino, S. I. Bermudez, S. Jakobsen, M. Jebramcik, B. Jenninger, E. Jensen, M. Jones, R. Jones, T. Jones, J. Jowett, M. Juchno, C. Julie, T. Junginger, V. Kain, D. Kaltchev, N. Karastathis, P. Kardasopoulos, M. Karppinen, J. Keintzel, R. Kersevan, F. Killing, G. Kirby, M. Korostelev, N. Kos, S. Kostoglou, I. Kozsar, A. Krasnov, S. Krave, L. Krzempek, N. Kuder, A. Kurtulus, R. Kwee-Hinzmann, F. Lackner, M. Lamont, A. Lamure, L. L. m, M. Lazzaroni, M. Le Garrec, A. Lechner, T. Lefevre, R. Leuxe, K. Li, Z. Li, R. Lindner, B. Lindstrom, C. Lingwood, C. Löffler, C. Lopez, L. Lopez-Hernandez, R. Losito, F. Maciariello, P. Macintosh, E. Maclean, A. Macpherson, P. Maesen, C. Magnier, H. M. Durand, L. Malina, M. Manfredi, F. Marcellini, M. Marchevsky, S. Maridor, G. Marinaro, K. Marinov, T. Markiewicz, A. Marsili, P. Martinez Urioz, M. Martino, A. Masi, T. Mastoridis, P. Mattelaer, A. May, J. Mazet, S. Mcilwraith, E. McIntosh, L. Medina Medrano, A. Mejica Rodriguez, M. Mendes, P. Menendez, M. Mensi, A. Mereghetti, D. Mergelkuhl, T. Mertens, L. Mether, E. Métral, M. Migliorati, A. Milanese, P. Minginette, D. Missiaen, T. Mitsuhashi, M. Modena, N. Mokhov, J. Molson, E. Monneret, E. Montesinos, R. Moron-Ballester, M. Morrone, A. Mostacci, N. Mounet, P. Moyret, P. Muffat, B. Muratori, Y. Muttoni, T. Nakamoto, M. Navarro-Tapia, H. Neupert, L. Nevay, T. Nicol, E. Nilsson, P. Ninin, A. Nobrega, C. Noels, E. Nolan, Y. Nosochkov, F. Nuiry, L. Oberli, T. Ogitsu, K. Ohmi, O. R., J. Oliveira, P. Orlandi, P. Ortega, J. Osborne, T. Otto, L. Palumbo, S. Papadopoulou, Y. Papaphilippou, K. Paraschou, C. Parente, S. Paret, H. Park, V. Parma, C. Pasquino, A. Patapenka, L. Patnaik, S. Pattalwar, J. Payet, G. Pechaud, D. Pellegrini, P. Pepinster, J. Perez, J. P. Espinos, A. P. Marcone, A. Perin, P. Perini, T. Persson, T. Peterson, T. Pieloni, G. Pigny, J. Pinheiro de Sousa, O. Pirotte, F. Plassard, M. Pojer, L. Pontercorvo, A. Poyet, D. Prelipcean, H. Prin, R. Principe, T. Pugnat, J. Qiang, E. Quaranta, H. Rafique, I. Rakhno, D. R. Duarte, A. Ratti, E. Ravaioli, M. Raymond, S. Redaelli, T. Renaglia, D. Ricci,

G. Riddone, J. Rifflet, E. Rigutto, T. Rijoff, R. Rinaldesi, O. Riu Martinez, L. Rivkin, F. Rodriguez Mateos, S. Roesler, I. Romera Ramirez, A. Rossi, L. Rossi, V. Rude, G. Rumolo, J. Rutkovksi, M. Sabate Gilarte, G. Sabbi, T. Sahner, R. Salemme, B. Salvant, F. S. Galan, A. Santamaria Garcia, I. Santillana, C. Santini, O. Santos, P. S. Diaz, K. Sasaki, F. Savary, A. Sbrizzi, M. Schaumann, C. Scheuerlein, J. Schmalzle, H. Schmickler, R. Schmidt, D. Schoerling, M. Segreti, M. Serluca, J. Serrano, J. Sestak, E. Shaposhnikova, D. Shatilov, A. Siemko, M. Sisti, M. Sitko, J. Skarita, E. Skordis, K. Skoufaris, G. Skripka, D. Smekens, Z. Sobiech, M. Sosin, M. Sorbio, F. Soubelet, B. Spataro, G. Spiezia, G. Stancari, M. Staterao, J. Steckert, G. Steele, G. Sterbini, M. Struik, M. Sugano, A. Szeberenyi, M. Taborelli, C. Tambasco, R. T. Rego, L. Tavian, B. Teissandier, N. Templeton, M. Therasse, H. Thiesen, E. Thomas, A. Toader, E. Todesco, R. Tomás, F. Toral, R. Torres-Sanchez, G. Trad, N. Triantafyllou, I. Tropin, A. Tsinganis, J. Tuckamantel, J. Uythoven, A. Valishev, F. Van Der Veken, R. Van Weelderen, A. Vande Craen, B. Vazquez De Prada, F. Velotti, S. Verdu Andres, A. Verweij, N. V. Shetty, V. Vlachoudis, G. Volpini, U. Wagner, P. Wanderer, M. Wang, X. Wang, R. Wanzenberg, A. Wegscheider, S. Weisz, C. Welsch, M. Wendt, J. Wenninger, W. Weterings, S. White, K. Widuch, A. Will, G. Willering, D. Wollmann, A. Wolski, J. Wozniak, Q. Wu, B. Xiao, L. Xiao, Q. Xu, Y. Yakovlev, S. Yammine, Y. Yang, M. Yu, I. Zacharov, O. Zagorodnova, C. Zannini, C. Zanoni, M. Zerlauth, F. Zimmermann, A. Zlobin, M. Zobov, and I. Zurbano Fernandez, *High-Luminosity Large Hadron Collider (HL-LHC): Technical design report.* CERN Yellow Reports: Monographs, CERN, Geneva (2020). doi: 10.23731/CYRM-2020-0010. URL `https://cds.cern.ch/record/2749422`.

[2] F. Rodriguez-Mateos, D. Carrillo, S. Balampekou, K. Dahlerup-Petersen, M. Favre, J. Mourao, and B. Panev, Design and Manufacturing of the First Industrial-Grade CLIQ Units for the Protection of Superconducting Magnets for the High-Luminosity LHC Project at CERN. Design and Manufacturing of the First Industrial-Grade CLIQ Units for the Protection of Superconducting Magnets for the High-Luminosity LHC Project at CERN, *IEEE Trans. Appl. Supercond.* **28** (3), 4702504. 4 p (2018). doi: 10.1109/TASC.2018.2794473. URL `https://cds.cern.ch/record/2676704`.

[3] M. Valette, L. Bortot, A. F. Navarro, B. Lindstrom, M. Mentink, E. Ravaioli, R. Schmidt, E. Stubberud, A. Verweij, and D. Wollmann, Impact of superconducting magnet protection equipment on the circulating beam in hl-lhc, *Proceedings of the 9th Int. Particle Accelerator Conf.* **IPAC2018** (2018). SeriesInformation Proceedings of the 9th Int. Particle Accelerator Conf., IPAC2018, Vancouver, BC, Canada.

[4] B. Lindstrom, P. Bélanger, L. Bortot, R. Denz, M. Mentink, E. Ravaioli, F. R. Mateos, R. Schmidt, J. Uythoven, M. Valette, A. Verweij, C. Wiesner, D. Wollmann, and M. Zerlauth, Fast failures in the lhc and the future high luminosity lhc, *Phys. Rev. Accel. Beams.* **23**, 81001 (8, 2020). doi: 10.1103/PhysRevAccelBeams.23.081001. URL `https://link.aps.org/doi/10.1103/PhysRevAccelBeams.23.081001`.

[5] B. H. F. Lindstrom, Criticality of fast failures in the high luminosity large hadron collider (2021).

[6] B. Todd, A Beam Interlock System for CERN High Energy Accelerators. URL `https://cds.cern.ch/record/1019495` (2006). Presented on 20 Nov 2006.

[7] C. Hernalsteens, M. V. Basco, O. K. Tuormaa, B. H. Lindstrom, E. Ravaioli, C. Wiesner, and D. Wollmann, Effect of a spurious cliq firing on the circulating beam in hl-lhc, *Proceedings of the 13th Int. Particle Accelerator Conf.* **IPAC2022** (2022). SeriesInformation Proceedings of the 13th Int. Particle Accelerator Conf., IPAC2022, Bangkok, Thailand.

[8] S. Redaelli, R. Appleby, R. Bruce, O. Brüning, A. Kolehmainen, G. Ferlin, A. Foussat, M. Giovannozzi, P. Hermes, D. Mirarchi, D. Perini, A. Rossi, and G. Stancari, Hollow electron lenses for beam collimation at the high-luminosity large hadron collider (HL-LHC). **16** (3), P03042 ISSN 1748-0221. doi: 10.1088/1748-0221/16/03/P03042. URL `https://iopscience.iop.org/article/10.1088/1748-0221/16/03/P03042`. tex.ids= redaelliHollowElectronLenses2021a publisher: IOP Publishing.

[9] C. Hernalsteens, M. V. Basco, O. K. Tuormaa, B. H. Lindstrom, E. Ravaioli, C. Wiesner, and D. Wollmann, Effect of a spurious cliq firing on the circulating beam in hl-lhc, *Proceedings of the 13th Int. Particle Accelerator Conf.* **IPAC2022** (2022). SeriesInformation Proceedings of the 13th Int. Particle Accelerator Conf., IPAC2022, Bangkok, Thailand.

[10] R. Denz, E. de Matteis, A. Siemko, and J. Steckert, Next Generation of Quench Detection Systems for the High-Luminosity Upgrade of the LHC, *IEEE Trans. Appl. Supercond.* **27** (4), 4700204 (2017). doi: 10.1109/TASC.2016.2628031.

[11] J. Steckert, J. Kopal, H. Bajas, R. Denz, S. Georgakakis, E. De Matteis, J. Spasic, and A. Siemko, Development of a digital quench detection system for nb3sn magnets and first measurements on prototype magnets, *IEEE Transactions on Applied Superconductivity.* **28** (4), 1–4 (2018). doi: 10.1109/TASC.2018.2792682.

[12] J. Steckert, R. Denz, S. Mundra, T. Podzorny, J. Spasic, and D.-G. Vancea, Application of the universal quench detection system to the protection of the high-luminosity lhc magnets at cern, *IEEE Transactions on Applied Superconductivity.* **32** (6), 1–5 (2022). doi: 10.1109/TASC.2022.3152125.

[13] A. Will, G. D'Angelo, R. Denz, D. Hagedorn, A. Monteuuis, E. Ravaioli, F. Rodriguez Mateos, A. Siemko, K. Stachon, A. Verweij, D. Wollmann, A.-S. Mueller, and A. Bernhard, Characterization of the radiation tolerance of cryogenic diodes for the high luminosity lhc inner triplet circuit, *Phys. Rev. Accel. Beams.* **23**, 053502 (May, 2020). doi: 10.1103/PhysRevAccelBeams.23.053502. URL `https://link.aps.org/doi/10.1103/PhysRevAccelBeams.23.053502`.

[14] A. Will, Damage mechanisms in superconductors due to the impact of high energy proton beams and radiation tolerance of cryogenic diodes used in particle accelerator magnet systems (2021).

[15] M. Blaszkiewicz, A. Apollonio, T. Cartier-Michaud, B. Panev, M. Pojer, and

D. Wollmann, Reliability analysis of the hl-lhc energy extraction systems, *Proceedings of the 13th Int. Particle Accelerator Conf.* **IPAC2022** (2022). SeriesInformation Proceedings of the 13th Int. Particle Accelerator Conf., IPAC2022, Bangkok, Thailand.

[16] V. Raginel, M. Bonura, D. Kleiven, K. Kulesz, M. Mentink, C. Senatore, R. Schmidt, A. Siemko, A. Verweij, A. Will, and D. Wollmann, First experimental results on damage limits of superconducting accelerator magnet components due to instantaneous beam impact, *IEEE Transactions on Applied Superconductivity.* **28** (4), 1–10 (2018). doi: 10.1109/TASC.2018.2817346.

[17] V. Raginel, Study of the Damage Mechanisms and Limits of Superconducting Magnet Components due to Beam Impact. URL `https://cds.cern.ch/record/2628622` (Jun, 2018). Presented 04 Jul 2018.

[18] A. Will, Y. Bastian, A. Bernhard, M. Bonura, B. Bordini, L. Bortot, M. Favre, B. Lindstrom, M. Mentink, A. Monteuuis, A.-S. Müller, A. Oslandsbotn, R. Schmidt, C. Senatore, A. Siemko, K. Stachon, A. Usoskin, M. Vaananen, A. Verweij, and D. Wollmann, Beam impact experiment of 440 GeV/p protons on superconducting wires and tapes in a cryogenic environment. p. THPTS066. 4 p (2019). doi: 10.18429/JACoW-IPAC2019-THPTS066. URL `https://cds.cern.ch/record/2690322`.

[19] J. Schubert, Damage Study on Single Strand Nb3Sn Ultra-Fast Beam Impact in Cryogenic Environment Simulation with Finite Element Method. URL `https://cds.cern.ch/record/2724326` (2020). Presented 2020.

[20] A. Ballarino, Development of superconducting links for the large hadron collider machine, *Superconductor Science and Technology.* **27** (4), 044024 (Mar, 2014). doi: 10.1088/0953-2048/27/4/044024. URL `https://dx.doi.org/10.1088/0953-2048/27/4/044024`.

[21] A. Ballarino, C. Cruikshank, J. Fleiter, Y. Leclercq, V. Parma, and Y. Yang, *Cold Powering of the superconducting circuits.* CERN Yellow Reports: Monographs, CERN, Geneva (2020). doi: 10.23731/CYRM-2020-0010. URL `https://cds.cern.ch/record/2749422`.

[22] K. Konstantopoulou, A. Ballarino, A. Gharib, A. Stimac, M. G. Gonzalez, A. T. P. Fontenla, and M. Sugano, Electro-mechanical characterization of mgb2 wires for the superconducting link project at cern, *Superconductor Science and Technology.* **29** (8), 084005 (Jun, 2016). doi: 10.1088/0953-2048/29/8/084005. URL `https://dx.doi.org/10.1088/0953-2048/29/8/084005`.

[23] S. Giannelli, G. Montenero, and A. Ballarino, Quench Propagation in Helium-Gas-Cooled MgB 2 Cables, *IEEE Trans. Appl. Supercond.* **26** (3), 5400705 (2016). doi: 10.1109/TASC.2016.2524449.

© 2024 The Author(s)
https://doi.org/10.1142/9789811280184_0011

# Chapter 11

# Overview of the ATLAS HL-LHC upgrade programme

Francesco Lanni
*European Organization for Nuclear Research (CERN),*
*Esplanade des Particules 1, 1211 Geneva 23, Switzerland*

ATLAS is planning to upgrade most of its detector systems to prepare for the operations during the HL-LHC era. A tenfold increase of the luminosity delivered to the experiment by the accelerator complex, compared to the integrated luminosity of the first three LHC runs, will impose significant challenges in terms of rates and radiation resistance. An overview of the ATLAS upgrade programme is hereinafter summarised.

## 1. Introduction

The high-luminosity upgrade of the LHC (HL-LHC)[1] will enable ATLAS to enter a new era, extending its potential for new discoveries through precision measurements of several Standard Model (SM) processes (in the Higgs sector in particular), which may unravel deviations from the theoretical expectations, and searches of new signatures and phenomena irreconcilable with the SM predictions. The HL-LHC operations will set an unprecedented challenging environment to the ATLAS detector systems as to particle rates and radiation levels:

- An ultimate, leveled instantaneous luminosity of $\mathcal{L} = 7.5 \times 10^{34}$ cm$^{-2}$s$^{-1}$ at $\sqrt{s} = 14$ TeV center-of-mass will produce, on average, up to 200 minimum bias interactions (in-time pileup) at each bunch crossing, i.e. every 25 ns.
- To cope with the increased rate of collisions, and to maintain the performance of the experiment at the electroweak scale, the first level of trigger and the detector readout shall sustain rates up to

This is an open access article published by World Scientific Publishing Company. It is distributed under the terms of the Creative Commons Attribution 4.0 (CC BY) License.

1 MHz, compared to the 100 kHz nominal value of the original system.

- The trigger architecture shall be redesigned to implement more sophisticated algorithms in order to maximize selection capabilities by maintaining high efficiency and improving background rejection, which is particularly challenging in very high pileup conditions for the hadronic and missing transverse energy signatures.
- Accordingly, the granularity of the detector systems shall be re-optimized (in particular in the inner tracker) to reduce the occupancy and the data throughput of the front-end electronics to acceptable levels. In the endcap region, which in ATLAS is more sensitive to performance deterioration due to pile-up, a significant mitigation can be achieved by tagging tracks and calorimeter clusters with precise timing information.
- Integrated over the expected lifetime of the HL-LHC, the luminosity delivered to the experiment will be up to 4 $ab^{-1}$, which is approximately a tenfold increase compared to the total luminosity integrated by the end of the first three runs of the LHC.
- Consequently, the ATLAS detectors will be exposed to unprecedented radiation levels. In the innermost pixel layers, a hadron fluence up to $2.3 \times 10^{16}$ $n_{eq}/cm^2$ and total ionizing doses in excess of 5 MGy are expected. Special precautions and refined design techniques for both sensors and on-detector electronics have to be taken to guarantee immunity to single-event effects. This also holds for detectors relatively distant from the interaction point.

The HL-LHC upgrades of the ATLAS detector systems are designed to cope with these conditions and requirements. The scope of the ATLAS upgrade programme has been defined in the years 2012–2015.[2,3] It is illustrated concisely in Fig. 1. The tracking system is replaced by a new all Silicon detector covering pseudo-rapidity values up to $|\eta| \simeq 4.0$. Both the on-detector front-end electronics and the back-end modular electronics of the Liquid Argon (LAr) and the Tile Calorimeters (TileCal) are replaced. The Muon spectrometer upgrades include (i) the installation of new inner barrel (BI) detectors, i.e. Resistive Plate Chambers (RPCs) and small diamater Monitored Drift Tubes (MDTs), (ii) the replacement of the barrel-endcap transition Thin Gap Chambers (TGCs), (iii) the upgrade of the trigger, and readout electronics of the existing RPC, TGC and MDT detectors. A High Granularity Timing Detector (HGTD), based on a novel

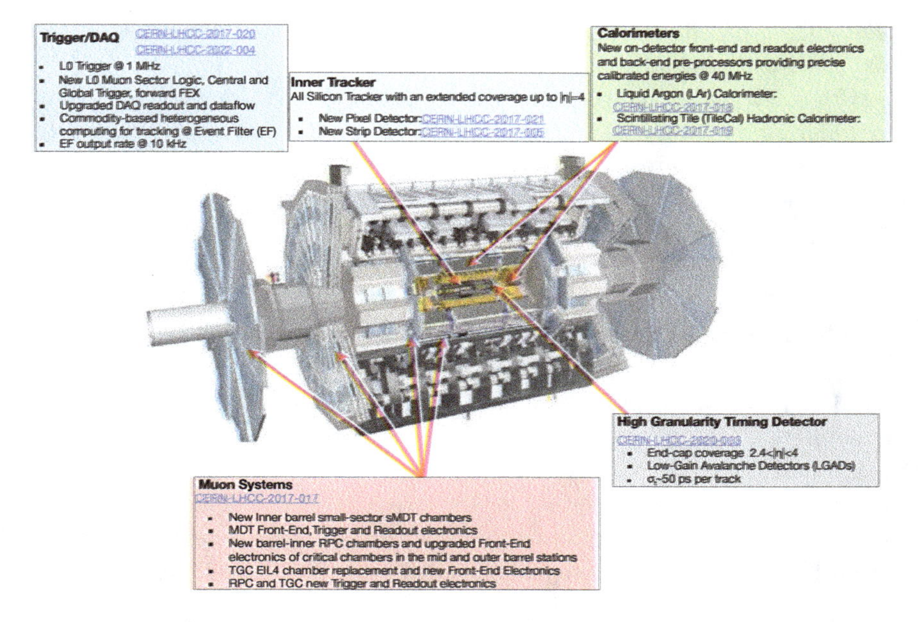


Fig. 1. ATLAS detector and the HL-LHC upgrade programme of its detector systems.

Silicon technology, the Low Gain Avalanche Detectors (LGADs), is installed in the end-cap region in the gap between the barrel and endcap calorimeters. Finally, the Trigger and Data Acquisition (TDAQ) system features an improved Level-0 (L0) Trigger with maximum rates of 1 MHz and 10 $\mu$s latency. The output rate to permanent storage system is 10 kHz.

The collaboration consolidated the detector upgrade programme by defining six upgrade projects, described through a set of Technical Design Reports[4–9] (TDRs), eventually reviewed by the CERN LHCC committee and approved by CERN Research Board in the years 2017–2019, and with the late addition of the HGTD upgrade project in 2020.[10]

Between 2021–2022 the TDAQ architecture was internally reviewed by the Collaboration. The baseline architecture described in the TDR[4] assumed the possibility of evolving the first level of hardware trigger into a system that could use tracking in the earliest stage of the trigger chain. The evolution would have been possible by implementing track reconstruction in the Event Filter sub-system in electronics modules housing Associative Memory application specific integrated circuits (ASICs). Eventually, the evolution option has been dropped, and the opportunity to avoid tight latency constraints favoured the implementation of the Event Filter by a purely CPU-based commodity solution that can possibly benefit accelerators as co-processors.[11]

Presently, i.e. at the time of writing, all the ATLAS upgrades projects are finalising their designs and entering the pre-production cycle, while early productions have already started for components (e.g. front-end ASICs) required for their assembly and integration on detector.

## 2. The Inner Tracker

The Inner Tracker (ITk) layout is shown in Fig. 2. Five layers of pixel detector modules are installed at the inner radii around the beam pipe in the barrel region. Several pixel rings in the forward region extend the pseudo-rapidity coverage of the tracker system to $|\eta| < 4$. The outer tracker is made of four barrel layers and six end-cap disks of strip detectors modules on both sides of the layers, covering up to $|\eta| < 2.7$. The Pixel and Strip Detector volumes are separated by a Pixel Support Tube (PST). The ITk layout has been optimised to reach, at the HL-LHC conditions, similar or better performance than the current tracker: to cope with the higher pileup, the granularity of the sensors is increased, resulting in an average occupancy of 0.16% in the Pixel and 1.2% in the Strip detectors. The ITk design targets to have about half as much material as compared to the current Inner Detector, minimising the effects of losses due to hadronic interactions and bremsstrahlung.

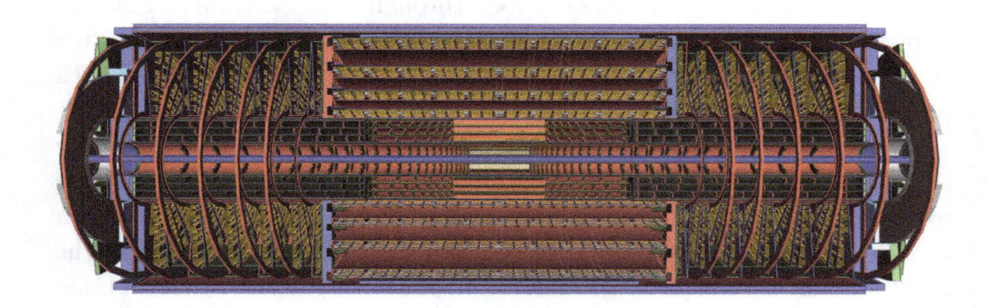

Fig. 2. Layout of the HL-LHC Inner Tracker (ITk) of ATLAS.

### 2.1. *ITk Pixel Detector*

The design of the ITk-Pixel system[5] features a short central barrel region, inclined modules that cover the intermediate $\eta$-region, and rings perpendicular to the beam direction in the very forward region. The active sensor area is about 14 $m^2$ of silicon sensors for a total of 5 billions pixels. The innermost two Pixel layers are designed to be replaceable, as the maximum expected neutron fluence exceeds $2 \times 10^{16}$ 1 MeV $n_{eq}cm^{-2}$. Different sensor technologies are used in the different regions of the detector: 3D sensors and thin (100 $\mu$m) n-in-p planar sensors in the inner layers, 150 $\mu$m thick n-in-p planar sensors in the outer three layers and in the end-cap rings. The pixel size is $25 \times 100$ $\mu m^2$ in the central barrel region of the first layer and $50 \times 50$ $\mu m^2$ elsewhere.

Pixel modules are made by bump-bonding the silicon sensors on the front-end read-out ASICs, fabricated in 65 nm CMOS technology. The modules are glued to a flex circuit interfacing to the readout data transmission, and power distribution systems. 9,000 modules will be used to assemble the ITk Pixel detector.

### 2.2. *ITk Strip Detector*

The ITk outer system is made of approximately 18000 modules of silicon Strip sensors.[6] A module is built by gluing kapton flexible hybrids to the sensors. Also, the hybrids host the readout ASICs in 130 nm CMOS technology. The modules are assembled onto $CO_2$ cooled carbon fibre structures forming rectangular "staves" and trapezoidal "petals", respectively, in the barrel and end-cap regions.

Staves and petals are assembled to form cylindrical layers and disks using large mechanical support structures. 392 staves are installed in the four barrel layers. Each barrel stave is populated with 28 Strip modules on both the top and bottom side. The Strip sensors are 24.1 mm long (short-strips) in the two innermost layers, and 48.2 mm (long-strips) in the outer two. In the end-caps 32 identical petals, each housing 9 modules on each side, are assembled on every disk. Six different sensor geometries allow the wedge-shaped petal surface to be covered, i.e. pointing to the beam axis.

The entire detector system consists of 18,000 modules for approximately 6 millions readout channels and for an active area of 165 $m^2$ of Silicon sensors.

## 3. The High Granularity Timing Detector

The High Granularity Timing Detector (HGTD) is a precision timing system based on Low Gain Avalanche Detectors (LGAD), installed in the region $2.4 < |\eta| < 4.0$.[10] It improves the rejection of pileup jets with $30 < p_T < 50$ GeV in the forward region by up to 40%, and reduces the inefficiencies of forward lepton isolated tracks by a factor two. HGTD comprises two layers of pixelated sensors ($1.3 \times 1.3$ mm$^2$) installed on the end-cap cryostats for an active area of 6.3 m$^2$ (see Fig. 3(a)) and approximately $3.5 \times 10^6$ readout channels. Full-sized ($15 \times 15$) LGAD arrays, shown in Fig. 3(b), have achieved in testbeams a time resolution of 30 ps.

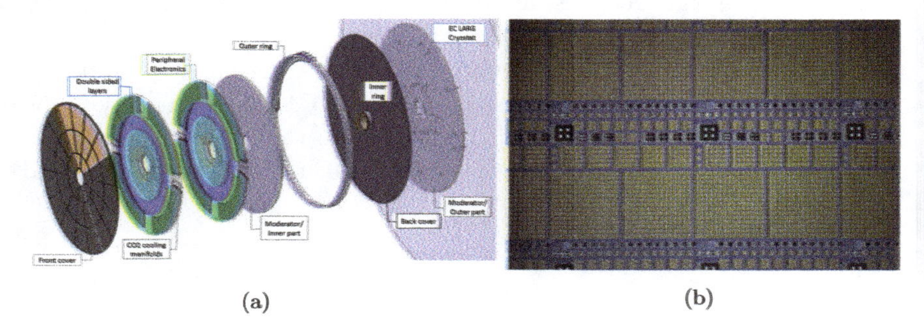


Fig. 3. (a) HGTD detector layout and mechanical structure. (b) Microscope picture of an HGTD LGAD prototype with full size sensors tested on beam.

## 4. Liquid Argon calorimeter electronics

The on-detector and off-detector readout electronics of the LAr calorimeter[7] are entirely replaced during the Phase-II upgrades. Figure 4(a) shows a high-level diagram of the readout electronics.

In the LAr system 1524 new Front-End Boards (FEB2s) are installed on detector, each processing the signals from 128 calorimeter cells in custom front-end crates. Each FEB2 processes the signals from 128 calorimeter cells: after a first amplification stage the signals are splitted into two overlapping linear gain scales and filtered by a bipolar (RC)-(CR)$^2$ shaper. Both gain scales are digitized at 40 MHz, multiplexed and transmitted off detector optically. The front-end amplifiers and the ADCs are deployed as two ASICs in TSMC CMOS 65nm and 130nm technology respectively.

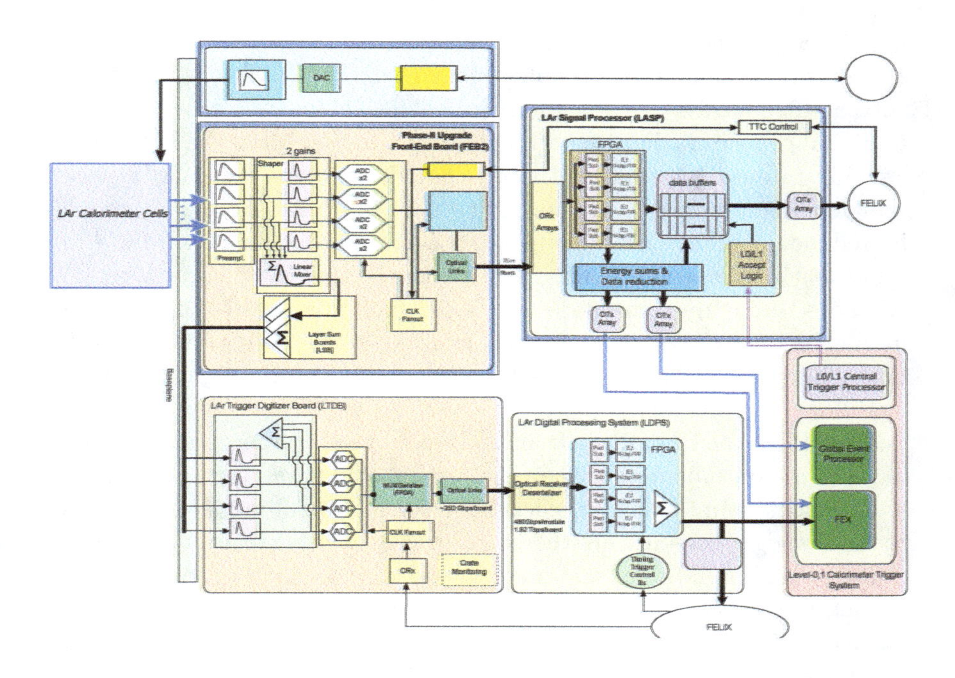


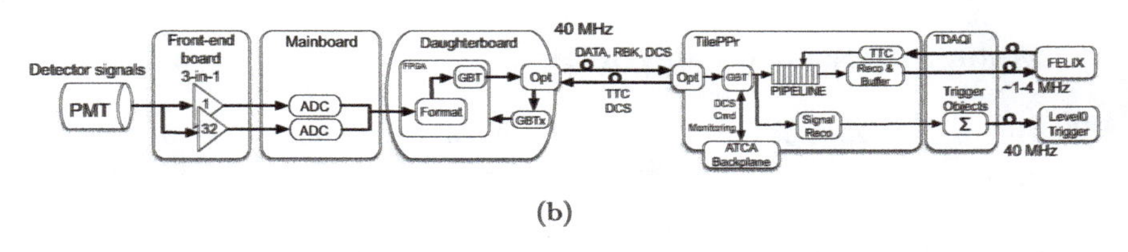


(b)

Fig. 4. High-level diagrams of the LAr (a) and TileCal (b) calorimeter readout.

372 LAr Signal Processor (LASP) modules receive off-detector the front-end data through approximately 31900 fibers, for a throughput in excess of 280 Tb/s. In addition, an upgraded calibration system allows the calibration signals to be injected directly into the LAr cells with a precision better than 0.1% over the full 16-bit dynamic range.

## 5. Tile Calorimeter

The Tile Calorimeter (TileCal) also will replace the on-detector front-end and the processing modules in USA-15.[8] Figure 4(b) represents a high-level schematics of the TileCal readout: the Front-End cards amplify and shape the signals of each photomultiplier (PMT). Two gain settings, with a $\times 32$ gain ratio, cover the required 17-bit dynamic range. Analog-to-digital converters digitize both gains signals at 40 MHz, the LHC bunch crossing rate on the front-end motherboards, and the resulting digitised samples are transferred off-detector to the Pre-Processor (PPr) modules by redundant optical links installed on dedicated daughter-boards. A total of 4096 fibres, each running at 9.6 Gb/s bandwidth, i.e. for a total of 1.3 Tb/s throughput, interface the on-detector front-end to 32 PPr modules. The off-detector PPr modules interface to the DAQ system and to both calorimeter and muon level-0 trigger processors through the TDAQi modules.

In TileCal the PMTs, the front-end electronics, the power and the cooling services are supported by new mechanical structures, segmented in four independent "Mini-drawers", each servicing a TileCal module, i.e. one of the 256 wedges constituting the barrel and end-cap detectors. The Mini-drawers will ease the installation and the maintenance operations, minimising the impact of component failures on the overall ATLAS data, in particular if compared to today's super-drawers whose failures would result in large dead regions ($\Delta\eta \times \Delta\phi = 0.7 \times 0.1$ radians).

## 6. Muon Spectrometer upgrades

The main challenge for the ATLAS Muon Spectrometer[9] is to maintain excellent selection and tracking capability at the HL-LHC conditions in terms of background rates, pile-up and integrated radiation doses. Figure 5 shows the views of the spectrometer upgraded detectors (in red) in the two orthogonal planes.

### 6.1. *RPC and MDT upgrades in the barrel spectrometer*

New RPCs with increased rate capability are installed in the inner barrel layer (BI) to improve the acceptance and robustness of the trigger selection. The upgrade addresses a fundamental limitation of the existing detectors: stable operations at the HL-LHC could be guaranteed only configuring the chambers at a reduced gas gain, i.e. at lower voltage bias, with loss of efficiencies up to 35% in the regions of highest rates. The installation of the BI chambers is challenging in terms of available space, in particular in the small sectors of the muon spectrometer, where it is possible only

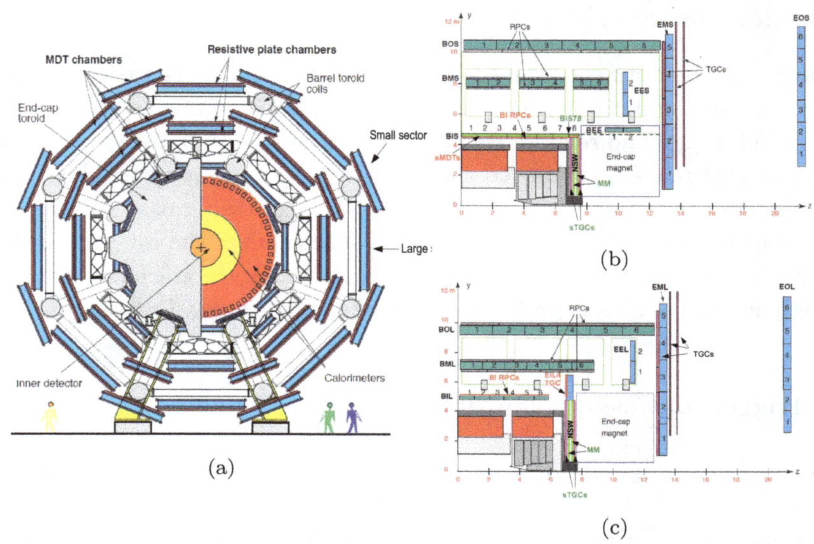


Fig. 5. Muon spectrometer layout in the r-$\phi$ plane (a), and in the r-z plane for the small (b) and large (c) sectors. In red the upgraded detectors systems contributing to the muon level-0 trigger and offline reconstruction.

if the existing Monitored Drift Tubes (MDT) chambers are replaced with sMDT detectors. Also, selected RPC chambers in the middle (BM) and outer barrel (BO) layers in the areas of highest rate, i.e. at $|\eta| > 0.8$, will be refurbished during the winter shutdowns after LS3, with new electronics and readout planes, to operate the chambers at reduced high voltage without efficiency loss.

## 6.2. *TGC upgrade in the barrel-endcap transition*

In the barrel-endcap transition region new Thin Gap Chambers (TGC) triplets replace the current TGC doublets (EIL4 in Fig. 5(c)). The triplets, with finer readout granularity, allow to implement a more robust majority logic, i.e. requiring hits in two out of three planes, and to use a smaller coincidence window, suppressing the rate of random coincidences generated by low-$p_T$ charged particles (typically slow protons) produced inside the endcap toroid cryostats.

### 6.3. *Electronics upgrades*

A large fraction of the on- and off-detector readout and trigger electronics is upgraded for compatibility with the Level-0 trigger requirements. The RPC and TGC trigger and readout chains are redesigned with data streamed off-detector and made available to the Level-0 trigger processors. The front-end electronics of the MDT detectors is also upgraded: raw data are sent to dedicated processors where precise measurements of the hit coordinates allow the Level-0 trigger processor to sharpen trigger efficiency turn-on curves at high $p_T$, and significantly reduce the background rate.

## 7. Trigger and Data Acquisition

The baseline configuration[4] features a Level-0 (L0) hardware trigger with a readout rate of 1 MHz and a latency of 10 $\mu$s, followed by the Event Filter (EF) system performing the final event selection, and outputting data at 10 kHz, see Fig. 6.

### 7.1. *Level-0 Trigger System*

The Level-0 trigger decision is made with information from the calorimeters and the muon detectors. The existing Muon Trigger processors are entirely replaced with upgraded modules that process the information from the RPC, TGC and sTGC and MicroMegas detectors in the forward region (New Small Wheel). They are complemented by additional processors that improve the precision of the muon $p_T$ measurement using the information of the MDT detectors.

The Calorimeter Feature Extractors (FEXes) modules, installed for the ATLAS Phase-I upgrades during the long shutdown 2, are maintained during the HL-LHC operations, and their firmware optimized for the expected pile-up conditions and the extra-latency available in Run-4. Additional FEXes units are installed to process data from the forward calorimeter.

The Global Trigger sub-system performs offline-like algorithms, and executes topological algorithms extending the functionality of the Run-2/Run-3 Topological Processor. The sub-system will be implemented in an architecture with multiplexing modules routing at each bunch crossing all the trigger primitives generated by the detectors to a set of modules each implementing all the selection and signature algorithms.

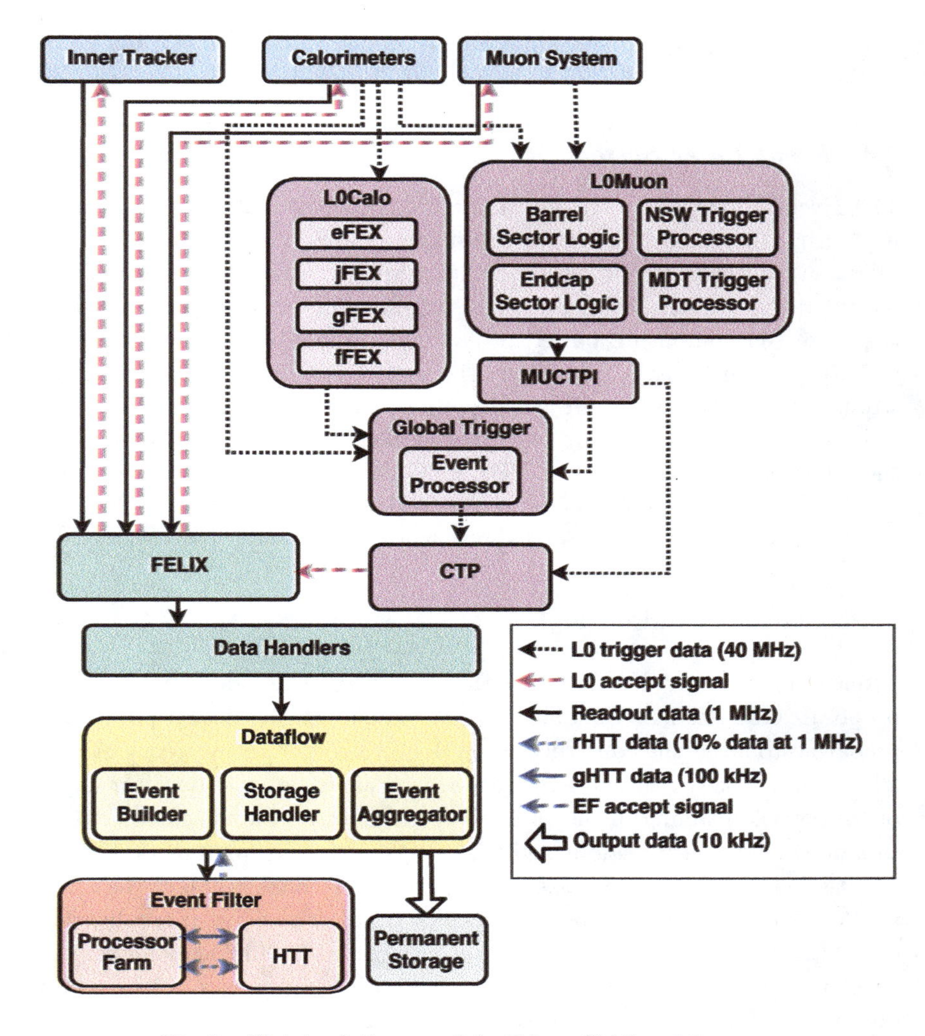


Fig. 6. High-level diagram of the Trigger/DAQ architecture.

## 7.2. *Data Acquisition System*

Detector data are transmitted to the Front-End Link eXchange (FELIX) nodes, a common interface between the detector-specific links and the commodity network downstream. Along the network, data are received by the Data Handlers, where detector-specific processing is implemented, and transferred to the Dataflow sub-system, which transports, aggregates, buffers and compresses event data for utilization in the Event Filter. The system is designed to sustain a throughput of 5.2 TB/s.

### 7.3. *Event Filter System*

The Event Filter (EF) system, a large commodity CPU-based processing farm, is upgraded to match the detector upgrades and maintain the rejection by employing offline-type particle reconstruction, and achieve an output rate of 10 kHz. New selection software will be developed to exploit the increased computing power available through advances in computing technology. The output rate is 10 kHz, and the total data throughput to storage is 60 GB/s. The project will continuously evaluate the evolution of heterogeneous architecture and the advancement of the technologies for the different accelerators, e.g. GPUs, FPGAs, Neural engines, before committing to a decision on the final configuration of the processing farm that will be procured during the long shutdown 3.

## 8. Conclusions and outlook

After more than a decade since the initial HL-LHC detector upgrade proposal,[2] the ATLAS Collaboration is completing the design phase and preparing to enter the construction of all the detector sub-systems. A new all-Silicon tracker with extended forward coverage, a novel detector capable of precision timing in the endcap region, improved muon identification capabilities, and new readout and trigger systems. The upgrade of ATLAS is a vast endeavour, comparable in cost and size to almost half of the original detector and engaging more than 1,000 collaborators for approximately two decades. Innovative detector and electronics technologies have been developed, resulting in an experiment that will be capable of coping with the upgraded accelerator complex. The HL-LHC detector upgrade will lay the foundations for the exploration of the energy frontier, searching for new phenomena and performing precision measurements in the Higgs and Standard Model sectors, in an unprecedented challenging environment.

## References

1. I. Béjar Alonso, O. Brüning, P. Fessia, M. Lamont, L. Rossi, L. Tavian and M. Zerlauth, *High-Luminosity Large Hadron Collider (HL-LHC): Technical Design Report*, CERN-2020-010.

2. The ATLAS Collaboration, *Letter of Intent for the Phase-II Upgrade of the ATLAS Experiment*, CERN-LHCC-2012-022.
3. The ATLAS Collaboration, *ATLAS Phase-II Upgrade Scoping Document*, CERN-LHCC-2015-020.
4. The ATLAS Collaboration, *Technical Design Report for the Phase-II Upgrade of the ATLAS Trigger and Data Acquisition System*, CERN-LHCC-2017-020.
5. The ATLAS Collaboration, *Technical Design Report for the ATLAS Inner Tracker Pixel Detector*, CERN-LHCC-2017-021.
6. The ATLAS Collaboration, *Technical Design Report for the ATLAS Inner Tracker Strip Detector*, CERN-LHCC-2017-005.
7. The ATLAS Collaboration, *ATLAS Liquid Argon Calorimeter Phase-II Upgrade Technical Design Report*, CERN-LHCC-2017-018.
8. The ATLAS Collaboration, *Technical Design Report for the Phase-II Upgrade of the ATLAS Tile Calorimeter*, CERN-LHCC-2017-019.
9. The ATLAS Collaboration, *Technical Design Report for the Phase-II Upgrade of the ATLAS Muon Spectrometer*, CERN-LHCC-2017-017.
10. The ATLAS Collaboration, *Technical Design Report: A High-Granularity Timing Detector for the ATLAS Phase-II Upgrade*, CERN-LHCC-2020-007.
11. The ATLAS Collaboration, *Technical Design Report for the Phase-II Upgrade of the ATLAS Trigger and Data Acquisition System - Event Filter Tracking Amendment*, CERN-LHCC-2022-004.

© 2024 The Author(s)
https://doi.org/10.1142/9789811280184_0012

# Chapter 12

# The CMS HL-LHC Phase II upgrade program: Overview and selected highlights

Marcello Mannelli
*CERN*

## 1. Introduction

With the unprecedented instantaneous luminosity of the HL-LHC, the Phase II program aims to provide about one order of magnitude more integrated luminosity compared to Phase I (~3'000–4'000 fb$^{-1}$ vs ~300–450 fb$^{-1}$) and offers the potential for an extensive program of precision measurements of Standard Model processes and possible deviations from the predictions. This includes a detailed characterization of the Higgs sector, together with direct searches for rare processes and subtle and/or exotic signatures in the search for physics beyond the Standard Model. Realizing that potential in the HL-LHC environment, with up to 200 almost-simultaneous interactions (in-time pile up) per 40 MHz bunch crossing rate, together with unprecedented levels of radiation exposure to the detector, presents a series of formidable experimental challenges, which define the requirements and drive the scope of the CMS Phase II upgrade program.

Briefly, these can be summarized as follows:

- The High Radiation Environment necessitates:
  - The complete replacement of the Tracker and Endcap Calorimeter systems, which cannot continue to operate at integrated luminosities much higher than the 300~450 fb$^{-1}$ foreseen for the Phase I program. By then, the performance will have already been substantially degraded due to radiation effects;

This is an open access article published by World Scientific Publishing Company. It is distributed under the terms of the Creative Commons Attribution 4.0 (CC BY) License.

  - Cold operation (around 9°C compared to 18°C during Phase I) of the Barrel ECAL to mitigate radiation induced dark currents in the Avalanche Photo Diodes (APDs), which would otherwise result in unacceptable performance degradation through increased noise.

- The High Pile Up motivates:
  - Improved granularity wherever possible, and extended Tracking capability to cover higher pseudorapidity;
  - Novel approaches to in-time pile up mitigation, and in particular the use of Precision Timing (~30 ps) to discriminate between particles, both charged and neutral, originating from the collision of interest and those resulting from pileup collisions within the same single bunch crossing (~180 ps RMS).
- The High Luminosity requires:
  - Substantially improved L1 Trigger primitives for better selectiveness, despite the high pileup, which in turn necessitates a longer L1 trigger latency, and motivates the inclusion of L1 Tracking Trigger capability as an integral part of the CMS Tracker upgrade, together with high granularity calorimeter information read-out at 40 MHz, and the introduction of additional End-Cap Muon Stations for use in the L1 Trigger;
  - Adapting the Front-End read-out electronics, L1 Trigger and Data Acquisition (DAQ) systems, to accommodate longer L1 trigger latency, 12 $\mu$s compared to the current 4 $\mu$s, accommodate an increased event size and allow higher read-out bandwidth, 750 kHz compared to the present 100 kHz, to provide high efficiency across the broader set of physics signatures of interest;
  - Dimensioning the High-Level Trigger (HLT) computer farm to run online event reconstruction on the 750 kHz input stream and select up to 7.5 kHz for permanent storage and further off-line analysis.

Taken together, these considerations lead to a set of CMS Phase II upgrade projects[1] which include:

- Replacement of the existing Silicon Tracker with a new Silicon

Tracker[2] featuring, in addition to improved radiation hardness, adequate for 4'000 $fb^{-1}$;

- A fourth Inner Pixel layer and higher Outer Tracker granularity to maintain sufficiently low cell occupancy to provide reliable pattern recognition in the high pile up HL-LHC environment;
- An on-detector front-end system capable of locally identifying, selecting, and transmitting Track vectors for charged particles with $p_T > 2$ GeV/c, together with a back-end system able to process these and reconstruct tracks at 40 MHZ for use in the L1 Trigger;
- Extended tracking coverage, up to $|\eta| \sim 4$, mostly achieved through an extended Inner Pixel Tracker;
- Substantially reduced material within the Tracker volume, to improve momentum resolution for charged tracks, as well as improving the calorimeter performance and quality of electron and photon reconstruction, while also reducing tracking inefficiencies and/or reconstruction errors resulting from hadrons undergoing nuclear interactions.

• Deployment of dedicated novel precision timing detectors, in front of both Barrel and End-Cap calorimeters, capable of providing 30 ∼ 40 ps precision for Minimum Ionizing Particles (MIPs).[3,4]
• Major overhaul of the existing Barrel ECAL (EB) and HCAL (HB) detector systems.[5] The overhaul:

- Is compatible with the revised L1 latency time and rate and enables independent read-out of every EB crystal at 40 MHz for the L1 trigger;
- Implements a lower APD operating temperature, 9°C compared to the current 18°C, which mitigates the effect of radiation induced dark current and the corresponding noise contribution to preserve the intrinsic Barrel ECAL energy resolution;
- Leads to 30 ps timing precision for electromagnetic (EM) showers $E > 30$ GeV, such as photons from $H \to \gamma\gamma$ decays, to help mitigate the effects on in-time pile up;
- Replaces HCAL Hybrid Photodiodes (HPDs) with higher efficiency and more radiation tolerant SiPMs — already done as part of the CMS Phase I upgrade program.

- Replacement of the Endcap calorimeters (Electromagnetic and Hadronic) with a radiation tolerant, integrated high granularity sampling calorimeter,[6] known as HGCAL. The salient features include:
  - Radiation tolerance, which drives the choice of silicon sensors in the regions of the calorimeter with the highest radiation exposure — front sections and highest pseudorapidity regions of the rear section — with SiPM on plastic scintillator tiles in the region less exposed to radiation;
  - High granularity, with the longitudinal segmentation being driven by energy resolution requirements, which the need to maintain the ability to track the effects of radiation damage and calibrate the calorimeter with MIPs translates to small read-out cell transverse sizes to reduce cell capacitance and electronic noise, resulting in a calorimeter with good imaging capability and the ability to resolve nearby showers in a dense environment, and in the presence of extreme pile up;
  - A novel front-end read out ASIC design which, in addition to satisfying the very large dynamic range requirements, also exploits the intrinsic timing characteristics of the silicon signal response to provide timing information with a precision of 20 ps or better for EM showers with $p_T > 2$ GeV, and 30–40 ps precision with an efficiency better than 90% for hadrons of $p_T > 5$ GeV.
- Major overhaul of the Barrel and End-Cap muon read-out systems, to comply with the increased L1 Trigger latency and read-out bandwidth and ensure adequate radiation tolerance for the HL-LHC operation.[7]
- Installation of an additional GEM based End-Cap Muon station (ME0) in the space liberated downstream of the HGCAL, which is denser and thus shorter than the existing Endcap calorimeter system:[7] together with the extended Tracker coverage, this extends the CMS muon acceptance from the present $\eta \approx 2.4$ up to $\eta \approx 2.8$, covering most of the area shadowed by the HGCAL (up to $\eta \approx 3$), and improves the muon trigger performance in the high $\eta$ region.
- Replacement of the L1 Trigger,[8] and of the DAQ and HLT systems[9] to meet the more demanding requirements of the CMS Phase II operation.

A full overview and in-depth discussion of the CMS upgrade program is well beyond the scope of this article: instead in what follows we briefly summarize some of the salient features of the upgrades, with a focus on the novel techniques they introduce, and their enabling technologies.

## 2. Tracker upgrade, and L1 Tracking Trigger

The upgraded Outer Tacker coverage will remain limited to $\eta < 2.4$, while the Inner (Pixel) Tracker extends the tracking acceptance out to $\eta < 4$. In addition to substantially improving (VBF/VBS) forward jet reconstruction and MET resolution, the extended tracking allows lepton identification and reconstruction over the full acceptance of the new CMS HGCAL endcap calorimeter, which covers the region up to $\eta < 3$.

Material within the tracking volume adversely affects performance in several ways: multiple scattering may dominate charged particle momentum resolution up to very high $p_T$, a fraction (proportional to the radiation length of the material present) of photons will convert, and electrons will shower, nuclear interactions may confuse track reconstruction and create displaced vertices which will show up in the background when searching for long lived neutral particles, etc.

The Phase II CMS Silicon Tracker employs thousands of modules, which generate large data flows and dissipate in the order of 100 kW, which in turn requires substantial optical and electrical cable plants and necessitate distributed cooling, all of which must be supported by rigid mechanical support structures. Minimizing the material within the Tracking volume is thus a key challenge for such a detector. Following the experience from the Phase I CMS Tracker, the first all-silicon Tracker ever deployed at a collider experiment, the CMS Phase II Tracker achieves a very large relative decrease in the material both between the Inner (Pixel) Tracker (IT) and Outer Tracker (OT) and in the Outer Tracker volume itself, reducing the total amount of material in the tracking volume by about a factor of two over the range of $0.5 \sim \eta \sim 1.5$. Notably, this is achieved despite the extension of the tracking acceptance, and the inclusion of L1 Track Trigger capability, discussed below.

A key novel feature of the CMS Phase II Outer Tracker is that it fully integrates a L1 Track Trigger capability. This is achieved through the introduction of so called "$p_T$" modules. As sketched out in Fig. 1(Left), $p_T$ modules integrate a pair of closely spaced silicon sensors, which can be used to determine not only the location but also the incident angle, in

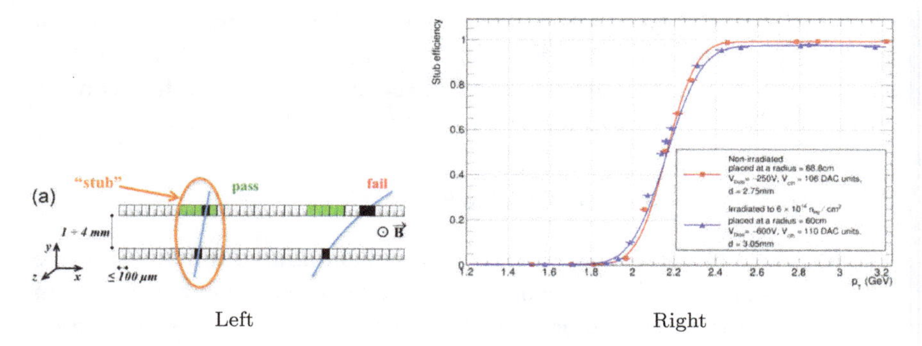


Fig. 1. Left: Illustration of the $p_T$ module concept. (a) Correlation of signals in closely-spaced sensors enables rejection of low-$p_T$ particles; the channels shown in green represent the selection window to define an accepted "stub". (c) For the endcap discs, a larger spacing between the sensors is needed to achieve the same discriminating power as in the barrel. Center: Stub reconstruction efficiency for a non-irradiated (red) and an irradiated (blue) 2S mini-module. Right: The 2S module (left) and PS module (right) of the Outer Tracker. Shown are views of the assembled modules (top), details of the module parts (centre) and sketches of the front-end hybrid folded assembly and connectivity (bottom).

the r-$\varphi$ plane, of charged particles traversing the module. This provides a measurement of the $p_T$ of the incident particle, with sufficient resolution to discriminate efficiently between particles with $p_T$ above or below a $\sim 2$ GeV threshold. This is shown in Fig. 1(Center), where the performance of modules built with irradiated and non-irradiated sensors are also compared.

Hits from the two sensors are brought together in a single readout ASIC, so that pairs of hits belonging to charged particles with $p_T > 2$ GeV, constituting a small fraction of the total number of hits, can be locally selected on-module, and transmitted to the off-detector L1 Track Trigger system, which in turns reconstructs the corresponding tracks and makes them available to the L1 Global Trigger.

The CMS Phase II Outer Tracker uses two different types of $p_T$ modules. At smaller radii, below about 60 cm, so-called Pixel-Strip or PS $p_T$ Modules are used, in which a strip sensor with 100 $\mu$m pitch and $2 \times 24$ mm strip length is paired with a second sensor with 100 $\mu$m pitch and 1.5 mm macro-pixel length. This granularity is well matched to the higher particle density in the inner region of the tracker, and the use of 1.5 mm long macro-pixels allows longitudinal primary vertex reconstruction with a precision adequate for the needs of the L1 Trigger. At radii above 60 cm, where the particle density is lower and the distance from the beam too large to allow a useful improvement of the vertex reconstruction, so called Strip-Strip or 2S $p_T$ modules are used, which use a pair of strip sensors with 90 $\mu$m pitch and $2 \times 48$ mm strip length. These are simpler in construction compared to the PS modules, dissipate less power, and allow both cost reduction and decreased material within the tracking volume.

## 3. MIP Timing Detectors, MTD

If left unaddressed, particles from 140 to 200 collisions within a single bunch crossing (in-time pile up) will degrade the ability to correctly reconstruct the collision of interest. In addition to having adequate detector granularity to retain accurate reconstruction of individual particle tracks and showers in the resulting crowded environment, an effective means of differentiating reconstructed particles originating from the collision of interest from those due to pile up collisions can greatly aid in maintaining the overall event reconstruction quality.

The LHC luminous region is several millimeters long, so that longitudinal track vertex reconstruction can help distinguish charged particle tracks (from the event vertex) from those originating from pile up vertices. At very high pile up, however, the vertex density is such that these overlap so much in space such that it is no longer possible to resolve them with track vertex reconstruction alone. Figure 2(Left) shows how the addition of precision timing information to charged particle tracks can help resolve such overlapping pile up vertices. It is becoming customary to refer to this use of timing information in vertex reconstruction as "4D reconstruction", as opposed to the usual 3D reconstruction.

The benefits of this are substantial for all aspects of event reconstruction. As an example, Fig. 2(Right) shows the improvement for the b-tagging performance of CMS.

Precision timing information for neutral particles can also play an important role in limiting the effect of pile up on event reconstruction. The upgrade of the Barrel Electromagnetic Calorimeter (EB) aims at providing $\approx 30$ ps time resolution for high $p_T$ photons, such as those from Higgs to $\gamma\gamma$ decays.

Two distinct technologies are adopted for the CMS MTD, as follows.

For the Barrel Timing Layer (BTL), which requires a relatively large surface to be instrumented (38 $m^2$), and where the maximum fluence remains below $\approx 2 \times 10^{14}$ (1 MeV neq)/$cm^2$, LYSO crystals doped with small

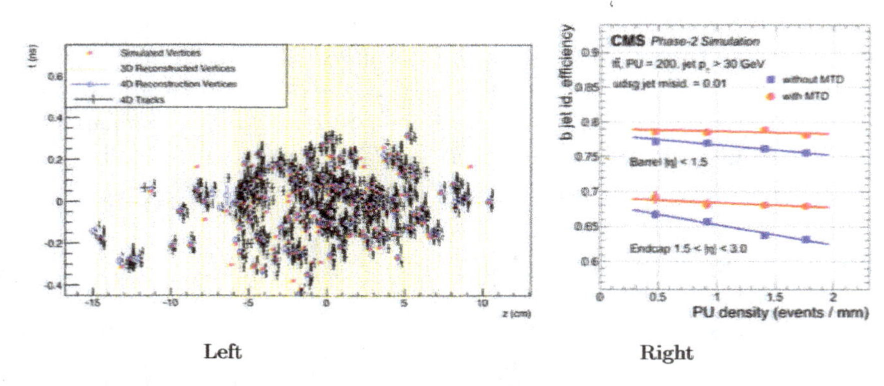

Left Right

Fig. 2. Left: Simulated and reconstructed vertices in a bunch crossing with 200 pile up collisions, assuming a MIP timing detector covering the barrel and endcaps. The vertical lines indicate 3D-reconstructed vertices, with instances of vertex merging visible throughout the event display. Center: Secondary vertex tagging ROC curves for light-quark and charm-quark jets for $|\eta| < 1.5$ (left) and for $1.5 < |\eta| < 3.0$ (centre), and b-tagging efficiency vs. average pile up density, with a constant light-jet efficiency of 0.01 (right). Results with and without precision timing information are compared to the zero pile up case. Right: Space-time diagrams illustrating the concept of hermetic timing for $H \to \gamma\gamma$ events. The reconstructed time for the photons at each vertex (green open dots), with error bars from the uncertainty on the time measurement of photons, can be cross referenced with the time information of the 4D vertices. The green straight lines are drawn to guide the eye. The pile up is reduced to an average of 20 in this case, to improve clarity. For photons with a small rapidity gap, shown here, photon timing alone is not sufficient and the coincidence with a 4D-vertex is necessary to enable accurate vertex location.

amounts of Cerium (LYSO: Ce) with SiPM read-out have been chosen. Early "proof of principle" prototype assemblies of such devices are shown in Fig. 3(Left), which also shows an arrangement used for test-beam studies of the time resolution. Results from such tests are shown in Fig. 3(Right), which demonstrate the target resolution of $\approx$ 30 ps for un-irradiated devices. It is interesting to note that the R&D for the CMS BTL benefitted from the already ongoing studies of similar devices, as part of an R&D aimed at improving the 3D-imaging capability of PET devices for medical purposes with precision timing (so called "TOF PET"). This is a notable example of a spin-off from High Energy Physics detector development coming around to contribute back into the field.

Maintaining sufficient timing resolution in the face of radiation damage is challenging and has been the focus of extensive optimization studies following the MTD TDR.[4] Among other things, these have led to the adoption of thermo-electrical coolers to reduce the effect of radiation induced

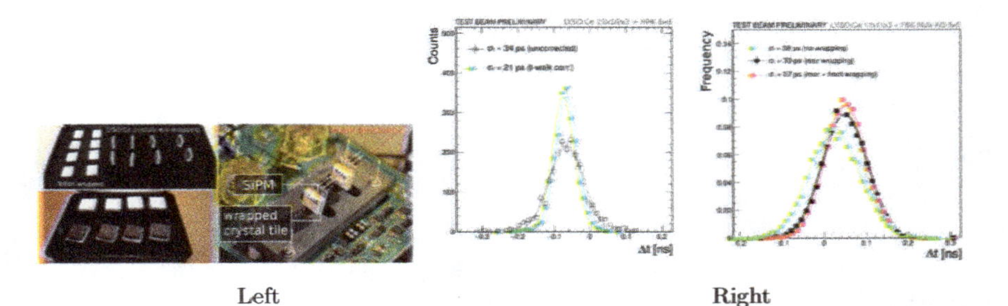


Fig. 3. Left: Top left: Set of $11 \times 11 \times 3$ mm$^3$ LYSO:Ce crystals with depolished lateral faces, before and after Teflon wrapping. Bottom left: $6 \times 6$ mm$^2$ HPK SiPMs glued on LYSO crystals. Right: Crystal+SiPM sensors plugged on the NINO board used for test beam studies. Right: Distribution of the time difference in a pair of LYSO:Ce tiles exposed to a 3 mm wide beam of MIPs hitting the centre of the tiles. Left: Results before and after time walk correction for $10 \times 10 \times 3$ mm$^3$ crystals read out with $6 \times 6$ mm$^2$ HPK SiPMs. Right: Results for $11 \times 11 \times 3$ mm$^3$ crystals read out with $5 \times 5$ mm$^2$ FBK SiPMs under different wrapping configurations.

SiPM noise. By lowering their operating temperature several degrees below the $-35$°C minimum temperature that the $CO_2$ cooling system is able to provide, as well as careful tuning of the LYSO crystal geometry and SiPM characteristics and size.

Even so, the substantially higher fluence (up to $\approx 2 \times 10^{15}$ (1 MeV neq)/cm$^2$) precludes the use of this technology in the End-Cap region, and the End-Cap Timing Layer (ETL) is based instead on Low Gain Avalanche Diodes (LGADs). This is a relatively recent technology, an extension of the more mature planar silicon technology widely used for strip, pixel, and pad silicon detectors, and is still in a phase of rapid development. The basic novel feature, as illustrated in Fig. 4(Left), is the introduction of an additional implant immediately below the usual surface implant, which generates a sufficiently high local gradient to induce avalanche multiplication of charges, effectively amplifying the original signal.

Barring the constant term, the time resolution is inversely proportional to the signal over noise (S/N) ratio. Achieving time resolutions of order 30 ps for a single MIP signal in traditional silicon sensors would require prohibitively high levels of power dissipation in the front-end read-out electronics for it to be practical over the $\approx 14$ m$^2$ of the CMS endcaps. The signal amplification provided by the LGADs allows the necessary S/N to be achieved with relatively modest electronics power dissipation and thus provides a viable solution. In the context of the ETL design, it has been found that gain of 10 allows a 30–40 ps timing resolution to be achieved.

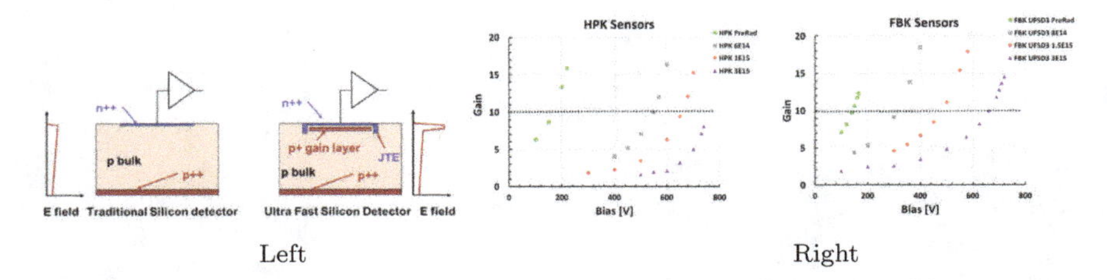


Left Right

Fig. 4. Left: Cross-sectional diagrams comparing a standard silicon detector and a UFSD, with an additional p implant providing the electric field for charge multiplication. Right: Gain as a function of bias voltage for different neutron fluences for LGADs manufactured by HPK, and FBK.

Radiation damage affects both the collection of primary charges (in a similar way as for standard planar silicon sensors) as well as the dopant concentration of the implants and the resulting gain; maintaining the necessary level of performance has been a key goal of the R&D program. Figure 4(Right) shows the gain as function of bias voltage for prototype LGADs from two different producers, after exposure to fluences ranging up to $3 \times 10^{15}$ (1 MeV neq)/cm$^2$; these devices can reach a gain of 10, at a bias voltage below 800 V, and have been shown to achieve a timing resolution in the range of 30–50 ps up to the highest fluence.

## 4. Endcap High Granularity Calorimeter, HGCAL

The existing CMS calorimeter systems were designed for an integrated luminosity up to 500 fb$^{-1}$. As stated in the Introduction, it is possible to refurbish the Barrel ECAL and HCAL in such a way as to enable their continued exploitation for Phase II operation at the HL-LHC. In the endcap region, however, radiation levels are expected to reach up to $1 \times 10^{16}$ (1 MeV neq)/cm$^2$ at 3'000 fb$^{-1}$, some fifty times higher than in the Barrel. The resulting performance degradation of the existing End Cap calorimeters much beyond $\approx 450$ fb$^{-1}$ would be such that these cannot continue to be used, and their replacement is an essential part of the CMS Phase II upgrade.

As the result of several years of dedicated R&D, and having examined several potential alternative designs and technologies, the CMS collaboration settled on HGCAL as the replacement for the existing endcap calorimeters.

The CMS HGCAL uses silicon sensors for the electromagnetic section of the calorimeter, as well as for parts of the hadronic section that are exposed to the highest radiation levels. Plastic scintillator tiles with direct (on-tile) SiPM readout are used for those sections of the hadronic calorimeter that will be exposed to less than $\approx 5 \times 10^{13}$ (1 MeV neq)/cm$^2$ after 3'000 fb$^{-1}$. The CMS HGCAL employs almost 600 m$^2$ of silicon sensors with $\approx$ 6 M readout channels, and close to 370 m$^2$ of plastic scintillator with about 240'000 readout channels.

The choice of HGCAL was motivated by the demonstration that simple planar silicon sensors (single-sided, DC-coupled, n-on-p) can tolerate the required levels of radiation, while retaining adequate signal charge collection efficiency even after exposure to fluences of $1.5 \times 10^{16}$ (1 MeV neq)/cm$^2$. On the other hand, the commitment of key industrial partners towards silicon sensor production on 8" lines made possible cost effective sensor production on the very large scale required for the CMS HGCAL, without undue interference with the silicon sensor production for the Phase II ATLAS and CMS Trackers on well-established 6" production lines.

The silicon sensors are of hexagonal shape, the largest tile-able polygon, which allows the most efficient use of the sensor wafer. In combination with the use of 8" wafers this minimizes the number of modules to be assembled and integrated into the system, reducing it by well over a factor two compared to the more typical square sensors produced on 6" wafer sensors.

Even so, with approximately 26'000 silicon modules, there is strong emphasis on a simple, mechanically robust module design well adapted to automated robotic assembly and ease of handling. The HGCAL silicon modules, shown schematically in Fig. 5(Left (a)), include a base plate onto which the sensor is glued. A front-end read out PCB "Hexaboard" is then glued on top of the silicon, which covers the full area of the sensor. The Hexaboard is connected to the sensor by wire bonds via through holes in the PCB. The base plate is made of Cu/W for the Electromagnetic section of the calorimeter, where it forms part of the absorber, and a carbon fiber plate for the hadronic part. A Kapton$^{\text{TM}}$ foil is laminated onto the baseplate to provide both bias (high) voltage DC protection as well as AC de-coupling of the silicon sensor backside.

For the scintillator part of the HGCAL, building on a design developed by the CALICE Collaboration,[10] modules consist of large area PCB boards with surface mounted SiPMs onto which wrapped scintillator tiles, with a central dimple to house the SiPM, are glued (SiPM-on-tile) (Fig. 5(Left (b))).

The requirement to calibrate the detector through the MIP Landau

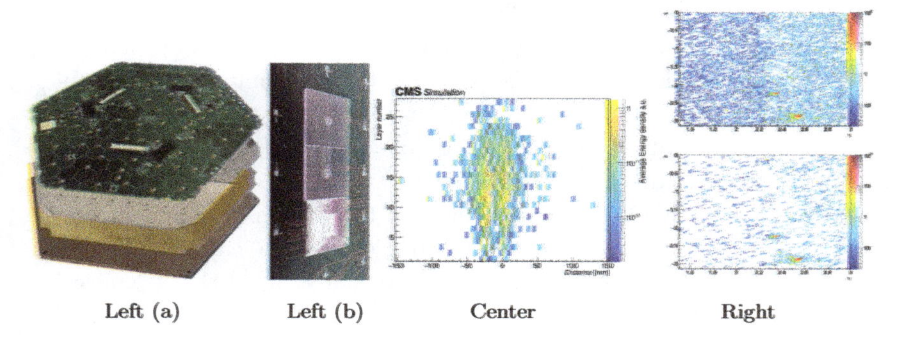


Fig. 5. Left: (a) An HGCAL Silicon Module, from Electromagnetic section of the calorimeter, showing the stacked layers, next to (b) an example of three CALICE 3 × 3 cm$^2$ scintillator tiles mounted on a PCB that holds one SiPM per tile. The left two scintillators are unwrapped to show the SiPM within the small dome at the centre of the tile, while right-most tile is wrapped with reflective foil. Center: Energy deposited in HGCAL cells by pairs of unconverted photons. The photons have an energy of E = 80 GeV ($p_T$ = 14.4 GeV) at $\eta = 2.4$ in the HGCAL, and are separated by $\Delta R = 0.05$ (in a random orientation), corresponding to a separation distance of about 30 mm. Reconstructed hits are projected onto the plane defined by the axes of the two showers. The colour code represents energy density. Right: Hits with a charge $>$ 12 fC, from a VBF Higgs $\rightarrow \gamma\gamma$ event, projected to the front face of the calorimeter: (upper plot) without a timing requirement, and (lower plot) after removal of hits with $|\Delta t| > 90$ ps.

peak drives several of the basic design HGCAL parameters. For the silicon sensors, it limits the input capacitance, thereby setting the cell size, which is $\sim 1.2$ cm$^2$ over most of the detector, and 0.5 cm$^2$ in the regions where the highest radiation level is expected and thinner (120 $\mu$m) sensors are used, and sets demanding requirements for the front-end readout chip of a dynamic range from 0.2 fC to 10 pC with a noise of less than 2'500 e$^-$ for a 60 pF input capacitance. In turn, the corresponding S/N level enables a timing resolution of better than 100 (20 ps) for cells with 3 (10) MIPs in 300 $\mu$m silicon. Similarly, for the SiPM-on-tile the MIP calibration requirement drives the choice of combination of tile sizes (which range from 4 cm$^2$ at the inner radii to 30 cm$^2$ at outer radii), plastic scintillator material (both cast and molded are used), and SiPM sizes (which range from 9 mm$^2$ at inner radii to 4 mm$^2$ towards the outer radii).

Figure 5(Center) illustrates the shower imaging capability and resolving power in the HGCAL Electromagnetic section which results from such a granularity, and an example of the impact of precision timing on pile up removal and mitigation is shown in Fig. 5(Right).

## 5. Summary and Outlook

Following a long period of reflection and focused R&D effort, the CMS Collaboration has engaged in a coherent program of refurbishing and overhauling where possible, and replacing and upgrading where necessary, the infrastructure and detector systems of the CMS experiment to make effective use of the physics potential afforded by HL-LHC. In response to the demands of precision physics at very high luminosity the CMS upgrade will deploy, in addition to many incremental improvements, several novel experimental features which include a full L1 Tracking Trigger, High Granularity Endcap Calorimeters, as well as charged and neutral particle precision timing for pile up mitigation. These will be complemented by the development of novel analysis techniques and the deployment of advanced computing architectures such as GPU's, to make best use of the upgraded detector while maintaining a cost-efficient computing model.

This is a challenging program, for which the R&D phase is by now mostly reaching completion and which is entering into the final qualification phase to prepare for the start of construction. Its success continues to depend crucially on the substantial commitment by the Collaboration with the strong backing of the Funding Agencies involved and of the CERN Laboratory. It will set the ground for at least another decade of scientific exploitation of the LHC complex, building on the discovery of the Higgs Boson in 2012 and the precision measurements made since then, in view of further exploring physics beyond the Standard Model.

## References

1. The Compact Muon Solenoid Phase II Upgrade Technical Proposal / CERN-LHCC-2015-010 / CMS-TDR-15-02 / 01-06-2015.
2. The Phase-2 Upgrade of the CMS Tracker Technical Design Report / CERN-LHCC-2017-009 / CMS-TDR-014 / 01-07-2017.
3. Technical Proposal for a MIP Timing Detector in the CMS Experiment Phase 2 Upgrade / LHCC-P-009 / 27-11-2017.
4. A MIP Timing Detector for the CMS Phase-2 Upgrade Technical Design Report / CERN-LHCC-2019-003 / CMS-TDR-020 / 29-03-2019.
5. The Phase-2 Upgrade of the CMS Barrel Calorimeters Technical Design Report / CERN-LHCC-2017-011 / CMS-TDR-015 / 12-09-2017.
6. The Phase-2 Upgrade of the CMS Endcap Calorimeter Technical Design Report / CERN-LHCC-2017-023 / CMS-TDR-019 / 09-05-2018.
7. The Phase-2 Upgrade of the CMS Muon Detectors Technical Design Report / CERN-LHCC-2017-012 / CMS-TDR-016 / 12-09-2017.

8. The Phase-2 Upgrade of the CMS Level-1 Trigger Technical Design Report / CERN-LHCC-2020-004 / CMS-TDR-021 / 10-03-2020.
9. The Phase-2 Upgrade of the CMS Data Acquisition and High-Level Trigger Technical Design Report / CERN-LHCC-2021-007 / CMS-TDR-022 / 17-06-2021.
10. Design, Construction and Commissioning of a Technological Prototype of a Highly Granular SiPM-on-tile Scintillator-Steel Hadronic Calorimeter / e-Print: 2209.15327 [physics.ins-det] / 30-09-2022 / CALICE-PUB-2022-003 Note: to be submitted to JINST.

© 2024 The Author(s)
https://doi.org/10.1142/9789811280184_0013

# Chapter 13

# LHCb Upgrades for the high-luminosity heavy-flavour programme

Matteo Palutan

*INFN Laboratori Nazionali di Frascati, Frascati, Italy*

## 1. LHCb Upgrades

The very successful operation of LHCb during Run 1 and Run 2 of LHC vindicated the concept and design of a dedicated heavy flavour physics experiment at a hadron collider. The detector was able to run at an instantaneous luminosity of $4 \times 10^{32}$ cm$^{-2}$s$^{-1}$, twice the design value, and to collect 9 fb$^{-1}$ of data by the end of Run 2.

In order to be able to continue the LHCb physics programme, a first upgrade of the detector (Upgrade I) was approved in 2012,[1] which is now in its first year of operation with colliding beams. The key concept of this upgrade is that the bottleneck of the level 0 hardware trigger is removed. By reading out the full detector at the LHC bunch crossing rate of 40 MHz and implementing all trigger decisions in software, it is possible to increase the luminosity without suffering a compensating loss in efficiency. By increasing the instantaneous luminosity by a factor of five, to $2\times10^{33}$ cm$^{-2}$s$^{-1}$, and improving the trigger efficiency for hadronic modes by a factor of two, the annual yields in most channels will be an order of magnitude larger than during Run 2. The target integrated luminosity for the Upgrade I phase is around 35 and 65 fb$^{-1}$ by the end of Run 3 and Run 4, respectively (Fig. 1). The upgraded detector has been designed to meet these specifications, and to withstand the higher occupancies foreseen at Run 3, while keeping performance comparable to Run 2.

Further data collection with the Upgrade I detector beyond Run 4 will not be attractive, on account of the excessive "data-doubling" time, and

This is an open access article published by World Scientific Publishing Company. It is distributed under the terms of the Creative Commons Attribution 4.0 (CC BY) License.

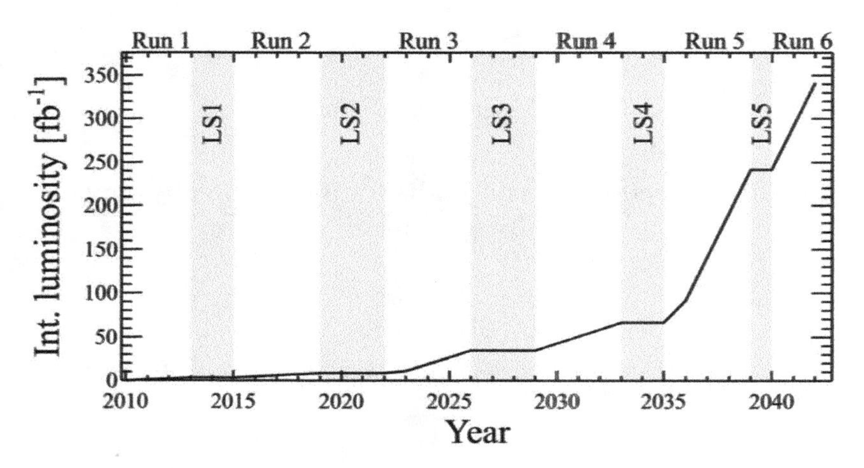


Fig. 1. Integrated luminosity profile for the original LHCb (Runs 1 and 2), and the one expected for Upgrade I (Runs 3 and 4) and Upgrade II (Runs 5 and 6).

also due to the fact that many of its components will have reached the end of their natural lifespan in terms of radiation exposure. There is therefore a strong motivation to perform a second upgrade (Upgrade II) of the detector,[2] in order to fully realise the flavour physics potential of the HL-LHC. Upgrade II is proposed for installation during LS4, and it is expected to take data at a maximum instantaneous luminosity of $1.5 \times 10^{34}$ cm$^{-2}$s$^{-1}$, with the target of accumulating $\sim$ 300 fb$^{-1}$ during Run 5 and Run 6 of LHC (Fig. 1). The flavour physics data sample will be at this point significantly larger than that of any other planned experiment, and will lead to improvements in the precision of a large number of key observables without being limited by systematic uncertainties.[3]

In the following sections more details are given on the Upgrade I detector and on the proposed design for the Upgrade II detector, respectively.

## 1.1. *The Upgrade I detector*

A new detector has been installed to cope with the increase in luminosity and pile-up by a factor of five, reaching values of $2 \times 10^{33}$ cm$^{-2}$s$^{-1}$ and 6, respectively. Particular focus has been put on increasing the detector granularity and radiation tolerance. An exception is represented by the calorimeter modules and the muon chambers, which have been recycled from the previous run. The readout scheme has also been upgraded for all subdetectors, in order to be able to readout the events at 40 MHz and

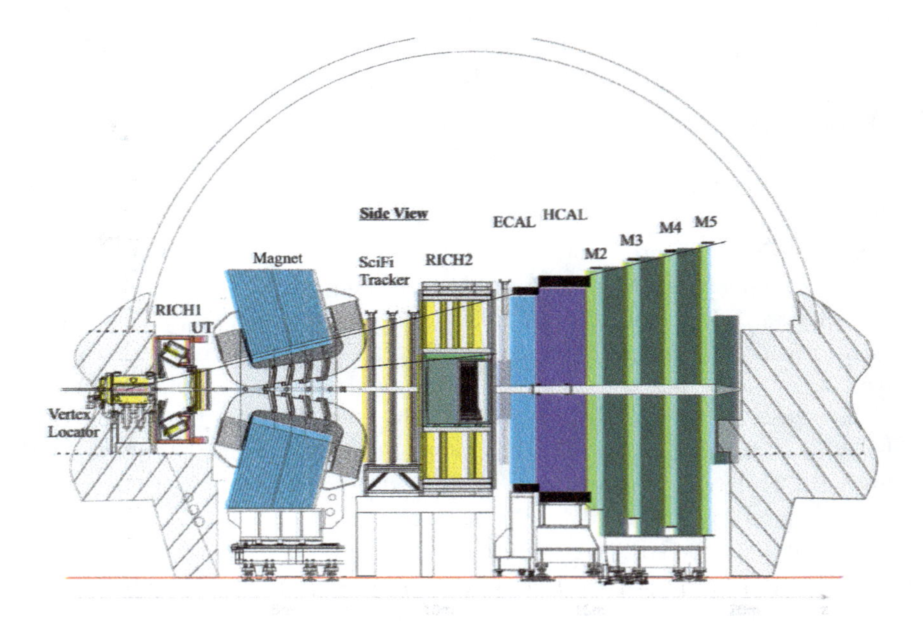


Fig. 2. Layout of the LHCb Upgrade I detector.

implement the new software trigger concept. Figure 2 shows a sketch of the upgraded detector.

The Vertex Locator (VELO) is the tracking detector devoted to the precise reconstruction of primary vertices and displaced vertices of short-lived particles. The previous version of the VELO, made of silicon microstrips, has been replaced by 26 tracking layers based on $50 \times 50\ \mu\text{m}^2$ pixel technology, that will ensure a better hit resolution and simpler track reconstruction. The upgraded VELO is closer to the beam axis, from the 8.4 mm of the previous detector down to 5.1 mm of the present one, and the particles will see substantially less material before the first measured point, from 4.6% to 1.7% radiation length. This design is expected to improve the impact parameter resolution by $\sim$40%, to increase the tracking efficiency, especially for low momentum tracks, and to provide a better decay time resolution.

The Upstream Tracker (UT) will be used for reconstruction of long-lived particles decaying after the VELO, and consists of 4 layers based on silicon strip technology. The tracking system is completed by the Scintillating

Fibre tracker (SciFi), placed downstream the magnet, and consisting of 12 detector planes with transversal dimension of about $6 \times 5$ m$^2$. Each plane is made of 6 layers of plastic scintillating fibres, 2.4 m length and 250 $\mu$m diameter, arranged along the vertical direction. The fibres are readout by SiPMs placed on the top and the bottom of the detector planes, which are cooled to a temperature of $-40^\circ$C in order to decrease the radiation damage and the dark noise. The usage of UT hits in the track extrapolation from VELO to SciFi detector will allow the number of fake tracks reconstructed by the tracking algorithms to be reduced, thus improving the trigger timing.

The two RICH detectors of LHCb are used for $p$, $\pi$ and $K$ particle identification. The optical layout of the RICH1, which is closer to the interaction point, has been modified to handle the higher particle occupancy of the upgrade conditions. In particular, the focal length of the mirrors has been increased by a factor $\sqrt{2}$, thus halving the occupancy. The readout of both RICH detectors, previously performed by HPDs at 1 MHz rate, has been replaced by multi-anode PMTs, working at 40 MHz.

Finally, for the software trigger a fast reconstruction entirely running on GPUs, that aims at selecting inclusive signatures of beauty and charm decays as well as high $p_T$ muons has been implemented. This is followed by a full reconstruction on CPUs, that indicate the signals of interest.

### 1.2. *The Upgrade II detector*

Performing flavour physics in the forward region at a peak luminosity of $1.5 \times 10^{34}$ cm$^{-2}$s$^{-1}$ presents significant experimental challenges. The expected number of interactions per crossing is around 40, producing $\sim$ 2000 charged particles within the LHCb acceptance. Radiation damage also becomes a greater concern for most detectors, *e.g.* with neutron fluences reaching $6 \times 10^{16}$ 1 MeV $n_{\rm eq}$/cm$^2$ in the innermost region of the VELO. The design proposed in Ref. 2 is based on the present spectrometer footprint, with all the detector components being upgraded in order to meet the desired specifications.

Among the distinctive features of the new design is the capability of providing fast-timing information with resolution of few tens of ps per particle, which at very high pile-up becomes an essential attribute for suppressing combinatorial background. As an example, a new VELO detector will be designed to provide a similar spatial resolution as in Upgrade I, but with a 50 ps resolution time-stamp per hit, thus becoming the first 4D-tracking device of this type. To meet the above challenges, the VELO ASIC will be designed in 28 nm technology, and new radiation-hard silicon sensors will be developed, with R&D results identifying already 3D sensors as a promising candidate for this purpose.

For the tracking system, high granularity pixel sensors appear as a solution to cope with high particle density in the UT and in the central region of downstream tracker, and to minimise the incorrect matching of upstream and downstream track segments. The emerging radiation-hard MAPS technology is a strong candidate for the above detectors. The outer region of the downstream tracker will be still covered by scintillating fibres, as in Run 3. However significant developments are required to cope with the increased radiation damage.

The RICH system will be a natural evolution of the current detector, with SiPMs replacing the multi-anode PMTs due to their higher granularity and excellent timing performances. In particular, this will allow a significant suppression of the combinatorial background. The new ECAL will implement a SpaCal design for the innermost highly irradiated region, while keeping Shashlik modules for the outer part. For SpaCal, a combination of tungsten absorber coupled with novel very radiation-hard crystals fibres or lead absorber coupled with polysterene fibres will be used for expected doses above and below 200 kGy, respectively. In order to achieve a timing resolution of few tens of ps, a double readout with longitudinal segmentation of the modules is foreseen on the whole calorimeter, which will guarantee the needed background reduction. Finally, for the muon system, new detectors will be needed to replace MWPCs in the innermost region of all stations, with a design possessing both high granularity and high rate capability. A promising candidate for this purpose is the $\mu$-RWELL, a new type of Micro-Pattern Gaseous Detector based on the same principle as the GEM, and exploiting a very similar manufacturing process.

The examples discussed above give some ideas of the technological developments needed to face the very challenging experimental conditions of HL-LHC moving forward. That will certainly represent a bridge towards projects based at future accelerators.

## References

1. LHCb collaboration. Framework TDR for the LHCb Upgrade. Technical Report CERN-LHCC-2012-007, CERN, Geneva (2012).
2. LHCb collaboration. Framework TDR for the LHCb Upgrade II. Technical Report CERN-LHCC-2021-012, CERN, Geneva (2022).

3. LHCb collaboration. Physics case for an LHCb Upgrade II. Technical Report CERN-LHCC-2018-027, CERN, Geneva (2018).

© 2024 The Author(s)
https://doi.org/10.1142/9789811280184_0014

# Chapter 14

# ALICE upgrades for the high-luminosity heavy-ion programme

Jochen Klein

*CERN, Esplanade des Particules 1, 1211 Geneva 23, Switzerland*

## 1. High-luminosity heavy-ion programme

The successful heavy-ion campaign with different collision systems during LHC Runs 1 and 2 has produced many new results improving our understanding of the quark–gluon plasma, see Ch. 5. Despite the significant progress, many fundamental questions about its nature remain open and require new and/or refined measurements. The areas we lack knowledge in include: the properties of the initial stage (e.g. temperature and chiral symmetry restoration), the nature of the interaction of high-energy partons with the plasma, the mechanisms of equilibration in the plasma, and the transition of partons to hadrons (hadronisation). The unique potential of the LHC to study the hottest and longest-lived quark–gluon plasma available in the laboratory, with large heavy-flavour abundances at vanishing baryo-chemical potential, calls for an extensive heavy-ion programme in future runs. Further experimental progress relies on improved detector performance in combination with the accumulation of larger data samples. This has motivated upgrade programmes for both the accelerator and the experiments.

Until Run 2, the instantaneous Pb–Pb luminosities were limited to $\sim$8 kHz by the acceptable rate of bound-free pair production in the LHC. With the upgrade of the collimation system during LS2, Pb–Pb interaction rates of $\sim$50 kHz become possible, now limited by the bunch intensities available from the injector chain. This will allow the accumulation of an integrated Pb–Pb luminosity of $\sim$13 nb$^{-1}$ during the course of Runs 3 and

This is an open access article published by World Scientific Publishing Company. It is distributed under the terms of the Creative Commons Attribution 4.0 (CC BY) License.

4. To arrive at even larger data samples, possible measures to increase the available bunch intensities are studied for Runs 5 and 6, including the option of lighter ions to mitigate space charge effects in the SPS and LEIR.

Significant upgrades of the experiment are required to benefit from the increased instantaneous luminosities and to meet the requirements on the detector performance, e.g. the pointing resolution needed for secondary vertexing. Many measurements rely on the extraction of a signal on top of combinatorial background. The resulting need to record all collisions without trigger or event selection has driven the major upgrade to ALICE 2, which has been installed during LS2, see Sec. 2. Some further extensions (ITS3, FoCal) are planned for installation during LS3, see Sec. 3. Recently, ALICE 3 has been conceived and proposed as the next-generation heavy-ion experiment for Run 5 and 6, see Sec. 4.

## 2. ALICE 2 (Run 3)

The needs for the measurements of dielectrons and heavy-flavour hadrons in Runs 3 and 4 have driven the conception of the LS2 upgrades with the recording of all collisions with improved vertex reconstruction as a core requirement.[1] The upgrades were completed within budget and on schedule for the start of Run 3,[2] leading to the experimental setup shown in Fig. 2. The Time Projection Chamber (TPC) constitutes the main tracking detector. At interaction rates of 50 kHz, every drift time interval of ~100 µs contains an average of 5 collision. As such, the TPC has to be read out continuously. This was made possible by replacing the MWPC-based read-out chambers with new chambers based on GEM foils, which limits the ion backflow to about 1 %, hence no gating is required to suppress the accumulation of space charge in the drift volume.[3,4]

In order to improve the pointing resolution, a thinner and lighter Inner Tracking System (ITS2) was constructed.[5] With an active area of $\sim 10\,\mathrm{m}^2$ equipped with ALPIDE sensors,[6] it is the largest tracker based on CMOS monolithic pixel sensors. The three innermost layers, with radii of 23 mm, 31 mm and 39 mm, each consisting of material corresponding to 0.35 % of a radiation length, form the inner barrel, which improves the pointing resolution by a factor of 3. The outer barrel is composed of two times two layers and covers the radii up to the inner radius of the TPC.

The muon system has seen a complete overhaul of the readout electronics and some consolidation on the detectors. It has been further extended by the Muon Forward Tracker (MFT), an assembly of 5 tracking disks installed on the muon side close to the interaction point. The detector is based on the same ALPIDE sensor developed for the ITS2 and allows the propagation of muon tracks to the primary vertex.

In addition, a new Fast Interaction Trigger (FIT) has been installed, which serves as interaction trigger, online luminometer, indicator of the vertex position, and forward multiplicity counter. It comprises two arrays of quartz Cherenkov radiators (FT0-A/C), a large, segmented scintillator disk (FV0), and two arrays of scintillator pads (FDD).

In addition to consolidation work, all detectors have upgraded their readout to either implement continuous readout or increase the trigger rates. The change of the readout paradigm from triggered to continuous operation has also required a completely new approach to the data processing. An integrated online-offline software framework has been developed to receive and process the incoming data. The detector data arrive at the so-called First-Level Processors (FLP) at a rate of $\sim$3.5 TB s$^{-1}$, where they are pre-processed (e.g. zero-suppressed) and then sent to Event Processing Nodes (EPN) at a rate of about $\sim$600 GB s$^{-1}$. The EPN farm comprises 2000 GPUs in 250 nodes to run the synchronuous reconstruction, whose output is stored on grid storage at a rate of $\sim$100 GB s$^{-1}$. The output of a subsequent aynchronuous reconstruction pass, running on the LHC computing grid with improved calibration input, then forms the basis for the physics analyses.

The upgraded detector will enable new measurements during the course of Runs 3 and 4, which will also add pO and OO to the already established collision systems (pp, p–Pb, Pb–Pb). The prospects and expected physics performance have been discussed extensively in the report on the physics prospects at the HL-LHC.[7]

## 3. ALICE 2.1 (Run 4)

A further upgrade of the inner barrel of the ITS (ITS3) and the installation of a Forward Calorimeter (FoCal) are under preparation for installation during LS3 and operation in Run 4.

The ITS3 project aims at enhancing the physics capabilities for the measurements of dielectrons and heavy-flavour probes by further improving the pointing resolution. The notion that the active silicon constitutes only about 15 % of the material in the three innermost layers of the ITS2 motivates their replacement with wafer-scale, cylindrical sensors without external services and almost no support structures.[8] As part of the R&D

activities, it has already been shown that thinned silicon wafers ($\leq$50 µm) can be bent and stabilised in the form of half-cylinders with very little support material (carbon foam). Furthermore, it has been demonstrated that the performance of ALPIDE sensors is unaffected by the bending.[9] The ongoing R&D studies aim at establishing the production of wafer-scale sensors through the stitching of repeated sensor units in the Tower 65 nm CIS process. Together with the reduced power density and cooling by a forced air flow, this allows the construction of an ultra-lightweight detector consisting only of the silicon cylinders and carbon foam wedges in the active area resulting in ~0.05 % of a radiation length per layer. In combination with the reduction of the beam pipe radius and thickness as well as the inner radius of the first ITS3 layer, this will further improve the pointing resolution as well as reduce the conversion probability in the first layer, see Fig. 1.

A central objective of the FoCal project is to constrain the gluon PDFs down to very low $x \approx 10^{-5}$. This can be achieved by measuring isolated (non-decay) photons in the pseudo-rapidity region $3.4 < \eta < 5.8$. To this end, a highly granular electromagnetic calorimeter will be combined with a hadronic calorimeter. The former is based on a stack of tungsten absorber plates with silicon pixel and silicon pad layers for read-out. This provides excellent resolution for the shower profile and two-photon separation. The hadronic calorimeter will be based on copper tubes with integrated scintillating fibres. The impact of the detector on the measurement of the gluon PDFs is shown in Fig. 1.

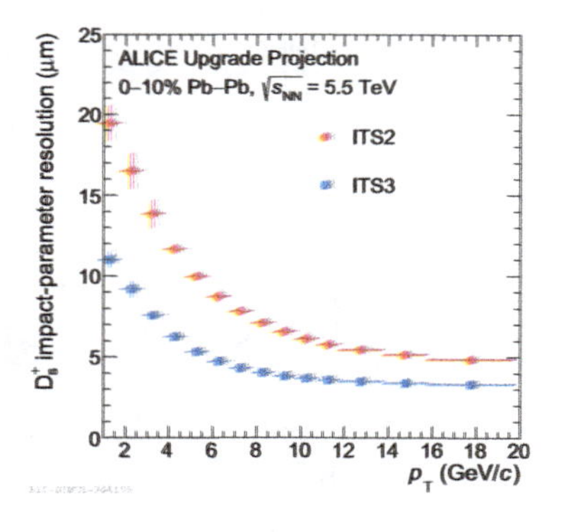


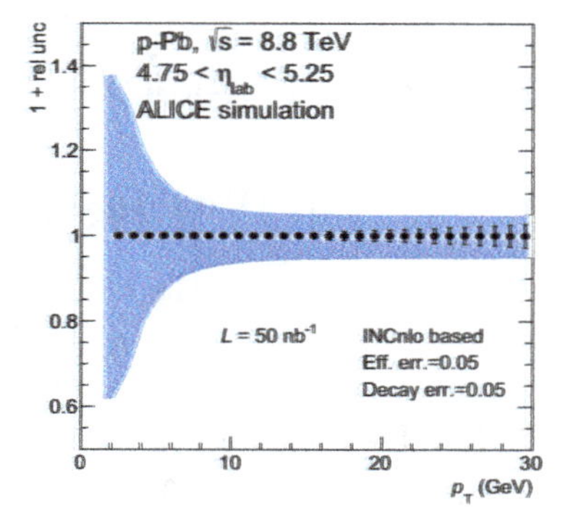


Fig. 1. ALICE 2.1 performance. Left: Improvement of pointing resolution with ITS3. Right: Relative uncertainties for the measurement of the ratio of isolated photons produced in p–Pb and pp collisions.[10]

## 4. ALICE 3 (Runs 5 and 6)

Two key challenges of the heavy-ion programme of LHC Runs 5 and 6 are the multi-differential measurement of dielectrons and the systematic measurement of (multi-)heavy-flavoured states.[11] The former is crucial for understanding the time dependence of thermal electromagnetic emission from the plasma and the mechanisms responsible for chiral symmetry restoration. This requires the measurement of a very clean electron sample down to low transverse momenta and the effective rejection of electrons from background processes, such as weak decays of heavy-flavoured hadrons and photon conversion before the first detection layer. The goal of the heavy-flavour programme is to understand the transport properties of the quark-gluon plasma, the mechanisms of equilibration in the plasma, and the effects relevant for the formation of hadrons. For the latter, multi-charmed hadrons are of particular interest since they are expected to show enhancements by orders of magnitude if the charm quarks (produced independently in early hard scatterings) can combine. The measurement of heavy-flavoured probes demands the best possible pointing resolution for the reconstruction of the decay vertices in combination with the highest possible statistics, which in turn requires a large acceptance of the detector. The latter is also required for the measurement of correlated charm production and the measurement of the pseudo-rapidity dependence. A further goal is the systematic measurement of hadron-hadron correlations in the charm sector to extract the interaction potentials of the strong interaction and the nature of (exotic) bound states. In addition to these pillars of the physics programme, there is a wide area of additional topics, including the measurements of baryon fluctuations, jet substructure, ultra-soft photons as well as searches for beyond the standard model phenomena.

The physics programme cannot be carried out at any other existing or planned experiment and defines the main experimental requirements. Foremost, the detector must provide good tracking and particle identification over a large acceptance and down to very low transverse momenta ($<100\,\mathrm{MeV}/c$). Excellent pointing resolution is required for background rejection and the reconstruction of the decay chains of heavy-flavoured probes. The detector is further optimised for the required high interaction rates and the resulting hit densities as well as radiation load.

These requirements have led to the detector concept shown in Fig. 2. Charged particle tracking in a magnetic field is realised by an all-silicon tracker arranged in barrel layers and forward disks in a cylindrical volume

Fig. 2. Left: Overview of the ALICE 2 detector for Run 3, the insert shows the inner tracking system and the muon forward tracker (refer to the text for further details).[2] Right: Overview of the ALICE 3 concept with the planned detector systems.[11]

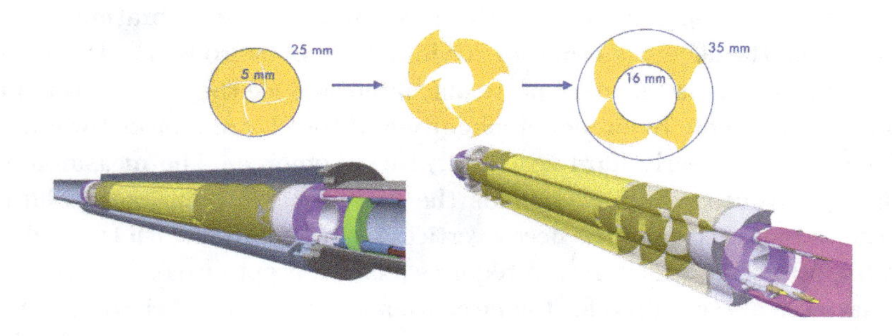


Fig. 3. Conceptual study of an iris-like mechanics for the retractable vertex detector.

of $R \approx 80\,\mathrm{cm}$ and $L = \pm 4\,\mathrm{m}$. This poses the challenge to minimise the material over a large surface in order to achieve the best possible momentum resolution. The magnetic field is provided by a superconducting magnet system, for which the coil configuration can be optimised starting from the baseline of a solenoidal coil providing $B = 2\,\mathrm{T}$. To achieve the required pointing resolution, a vertex detector is proposed for installation inside the beam pipe to measure the first hit as close as possible to the interaction point, i.e. at a radius of $\sim$5 mm. While such a small aperture is possible at the LHC's top energy, a larger aperture of $\sim$16 mm is required at injection energy. This implies the need for a retractable detector, for which the concept of an iris-like mechanism is studied, see Fig. 3. The tracker is complemented by a particle identification system comprising time-of-flight and RICH detectors to cover the low-to-intermediate transverse momentum range. The former consists of an inner and outer barrel layer as well as a forward disk on both sides of the experiment, all of which are equipped with silicon timing sensors providing a TOF resolution of 20 ps. The RICH detectors, based on aerogel radiators and silicon photon sensors, are installed behind the outer TOF layers. The system is further extended by an electromagnetic calorimeter covering the area of the central barrel and one endcap. In addition to the instrumentation with Pb-scintillator technology, a high-resolution segment is equipped with PbWO4 crystals for measurements of photons at very low energies. Outside of the magnet system, a hadron absorber is installed and followed by two layers of muon chambers for the identification of muons. A dedicated forward conversion tracker allows the measurement of photons at very low transverse momenta, making use of the Lorentz boost in the forward direction.

Two flavours of pixel sensors will be optimised for use in the vertex detector and the outer tracker. While the former requires high spatial resolution (~2.5 µm) on wafer-scale sensors, the latter requires a sensor with moderate spatial resolution (~10 µm) that is optimised for high yield and power consumption. In both cases a time resolution on the order of 100 ns is required. As baseline the technology node of the ITS3, i.e. the Tower 65 nm CIS process, is pursued but alternatives could be considered. For the silicon timing sensors, the primary goal is to implement a gain layer in monolithic CMOS sensors. Alternatively, LGADs or SPADs could be considered as sensors. For the photon detection in the RICH detector, a monolithic implementation of SPADs is targeted. The proposed detector is based on technologies of general interest for particle detectors and the R&D programme is relevant for the field at large.

The performance of ALICE 3 has been studied for the running scenario of six years of Pb–Pb collisions over the course of Runs 5 and 6.[11] Figure 4

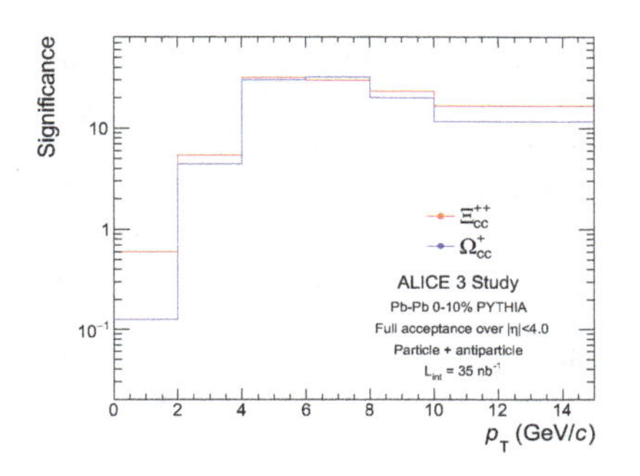


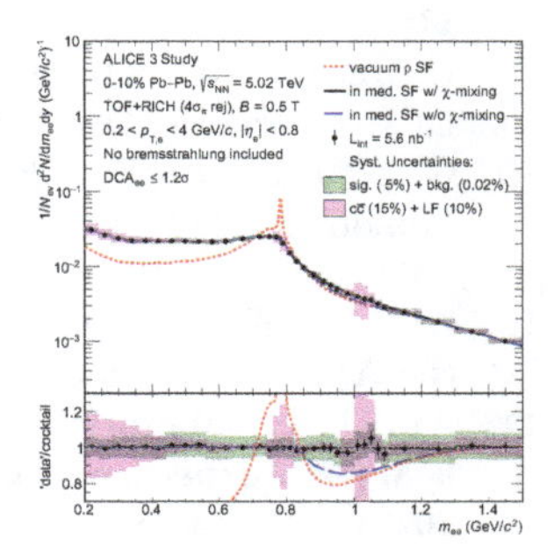


Fig. 4. ALICE 3 physics performance.[11] Left: Significance of multi-charm reconstruction using strangeness tracking. Right: Dielectron invariant mass spectrum.

shows the expected performance for the measurement of multi-charm states and the invariant mass spectrum of dielectrons.

## 5. Conclusions

The continued heavy-ion programme is an important aspect of the full exploitation of the LHC physics potential. With the upgrades for Runs 3 and 4, there are excellent prospects for new results over this decade. Beyond that, the ALICE 3 programme provides a roadmap for exciting heavy-ion physics in the 2030s with a novel and innovative detector concept.

## References

1. B. Abelev et al., Upgrade of the ALICE Experiment: Letter Of Intent, *J. Phys. G.* **41**, 087001 (2014). doi: 10.1088/0954-3899/41/8/087001.
2. ALICE collaboration, ALICE upgrades during the LHC Long Shutdown 2, *accepted by JINST* (2, 2023).
3. C. Lippmann, Upgrade of the ALICE Time Projection Chamber (3, 2014).
4. J. Adolfsson et al., The upgrade of the ALICE TPC with GEMs and continuous readout, *JINST.* **16** (03), P03022 (2021). doi: 10.1088/1748-0221/16/03/P03022.
5. B. Abelev et al., Technical Design Report for the Upgrade of the ALICE Inner Tracking System, *J. Phys. G.* **41**, 087002 (2014). doi: 10.1088/0954-3899/41/8/087002.
6. G. Aglieri Rinella, The ALPIDE pixel sensor chip for the upgrade of the ALICE Inner Tracking System, *Nucl. Instrum. Meth. A.* **845**, 583–587 (2017). doi: 10.1016/j.nima.2016.05.016.
7. Z. Citron et al., Report from Working Group 5: Future physics opportunities for high-density QCD at the LHC with heavy-ion and proton beams, *CERN Yellow Rep. Monogr.* **7**, 1159–1410 (2019). doi: 10.23731/CYRM-2019-007.1159.
8. L. Musa. Letter of Intent for an ALICE ITS Upgrade in LS3. Technical report, CERN, Geneva (Dec, 2019). URL `https://cds.cern.ch/record/2703140`.
9. G. A. Rinella et al., First demonstration of in-beam performance of bent Monolithic Active Pixel Sensors, *Nucl. Instrum. Meth. A.* **1028**, 166280 (2022). doi: 10.1016/j.nima.2021.166280.
10. ALICE collaboration, Letter of Intent: A Forward Calorimeter (FoCal) in the ALICE experiment (6, 2020).
11. ALICE Collaboration. Letter of intent for ALICE 3: A next generation heavy-ion experiment at the LHC. Technical report, CERN, Geneva (Mar, 2022). URL `https://cds.cern.ch/record/2803563`.

© 2024 The Author(s)
https://doi.org/10.1142/9789811280184_0015

# Chapter 15

# Higgs Physics at HL-LHC

Aleandro Nisati
*Istituto Nazionale di Fisica Nucleare, Sezione di Roma, P.le A. Moro 2, Rome, 00185, Italy*

Vivek A. Sharma
*Department of Physics, University of California San Diego, 9500 Gilman Drive, La Jolla, CA 92093-0319, USA*

In this chapter we review the projected reach of the ATLAS and the CMS upgraded detectors with full HL-LHC dataset of about 3000 fb$^{-1}$ at $\sqrt{s} = 14$ TeV (per experiment)in the measurement of key properties of the Higgs boson such as its mass and natural width, coupling to SM fermions and bosons, and its self-coupling.

## 1. Introduction

The discovery[1,2] of the Higgs boson ($H$) represents a major milestone in the understanding of the Electroweak Symmetry Breaking mechanism (EWSB) in nature. Since its discovery, a major push at the LHC has been to determine whether this object is the elementary boson predicted by the Standard Model (SM) of particle physics, or whether it represents the first observation of a beyond SM (BSM) particle. With the first $\approx$ 140 fb$^{-1}$ of p-p collisions at the LHC taken mainly at $\sqrt{s} = 13$ TeV, an impressive set of Higgs boson properties, including its spin-parity have been measured. The Higgs boson mass, a free parameter in the theory, has been measured with better than per mille accuracy. The Higgs mass of $\approx$ 125 GeV allows for measurements of its coupling to a variety of fermions. Its coupling to gauge bosons and the heavy fermions of the third generation has been measured

This is an open access article published by World Scientific Publishing Company. It is distributed under the terms of the Creative Commons Attribution 4.0 (CC BY) License.

with a precision of $\approx 5\%$ and 10% respectively. All the major Higgs boson production modes have been observed and the first evidence of its coupling to muons has been established.

While a general portrait[3–5] of the Higgs boson has emerged, the experimental exploration of the Higgs sector is in its infancy. With about 20× more data that HL-LHC is expected to deliver at $\sqrt{s} = 14$ TeV, the two major thrusts in Higgs physics are on (i) measurement of its mass, natural width and precise measurements of its couplings to fermions and bosons (ii) the first measurements of Higgs self-interaction.

The motivation for precision coupling measurement stems from the fact that rates for Higgs-related processes could be impacted by the contributions of BSM particles which may be too heavy to be directly produced at the LHC but still contribute to its properties via quantum loops. A suggested rule of thumb[6] is that increasing the coupling precision by a factor of four doubles the BSM mass scale that can be indirectly probed. The second important line of HL-LHC probe is on the shape of the Higgs potential. After EWSB, the Higgs potential gives rise to cubic and quartic terms in the Higgs boson field, resulting in a self-coupling term. Given a precise Higgs boson mass, this self-coupling is precisely predicted in the SM and can be measured in Higgs boson pair production (HH) processes. Any significant deviation from SM predictions signifies BSM physics and has major consequences for our understanding of this universe.

This chapter is organised as follows: The ATLAS and CMS detector upgrade for HL-LHC is summarised in Sec. 2, followed by a brief description of the procedures used for the projections in Sec. 3. Section 4 projects the precision on measurements of Higgs mass and its natural width, Sec. 5 summarizes the Higgs boson couplings measurements prospects. Section 6 is dedicated to the HL-LHC potential for the measurements of Higgs self-coupling, followed by a summary in Sec. 7.

## 2. Upgraded ATLAS and CMS detectors for the HL-LHC

In order to operate in the high intensity environment of the HL-LHC, ATLAS[7] and CMS[8] experiments are planning significant modification of their detectors, see Chapters 11 and 12 for a summary.[9] These upgrades, with increased granularity detectors with larger acceptances, targets efficient data taking and event reconstruction at increased luminosity and pileup (PU), up to ~200 additional inelastic interactions per bunch crossing and up to an order of magnitude larger radiation doses. For both detectors, in order to maintain or even lower the trigger thresholds with respect to Run 2, several trigger subsystems will be replaced or upgraded as well. New charged particle tracking systems will be installed, extending the tracking coverage up to pseudo-rapidities of $|\eta| = 4$. The addition of new timing detectors with a precision of about 35 ns, covering up to $|\eta| = 3$ for CMS and $2.4 \leq |\eta| \leq 4.0$ for ATLAS, will introduce new PU rejection capabilities in the HL-LHC environment. The existing ATLAS Liquid Argon and Tile calorimeters as well as the CMS barrel electromagnetic and hadron calorimeters will be upgraded with new electronics. The CMS endcap electromagnetic and hadron calorimeters will be replaced with a new high-granularity sampling calorimeter. Finally, the muon systems of ATLAS and CMS will be upgraded with fast electronics to deal with the extreme rates of secondary particles produced in HL-LHC collisions and additional muon chambers will be added to increase acceptance and redundancy.

## 3. Prognostication on Higgs studies in the HL-LHC era

As Niels Bohr once said, "Prediction is very difficult, especially if it's about the future!". We need to keep these sage words in mind while prognosticating the future. The ATLAS and CMS Collaborations have followed a few conservative strategies to project the precision of the Higgs measurements with the data collected at the end of HL-LHC era. The majority of the Higgs boson studies presented in this chapter generally follow the techniques used in Run 2 analyses. In some cases, the Run 2 analyses have been extrapolated to HL-LHC, taking into account the superior performance expected from the upgraded ATLAS and CMS apparatuses, and the large event pile-up expected at the HL-LHC. In other cases, such as for Higgs self-coupling measurements, Monte Carlo simulation studies were performed to assess the physics prospects at HL-LHC.

The expected performance at HL-LHC depends on both the signal event statistics, as well as on the systematic uncertainties that affect the event reconstruction in the unprecedented pile-up produced in the p-p collisions. Several scenarios have been identified to describe the impact of these uncertainties on the measurements. In the baseline scenario, the systematic uncertainties are set according to the technical recommendations[10] for HL-LHC projections. The expected uncertainty on the integrated luminosity of the full HL-LHC dataset is assumed to be 1%, lower than the Run 2 uncertainty of 1.7%. Theoretical systematic uncertainties are reduced by a factor of two with respect to those used in the Run 2 analyses, under the assumption of continued progresses on theoretical computations of perturbative corrections, PDFs, $\alpha_s$ etc. The statistical components of the experimental uncertainties are scaled according to $1/\sqrt{L}$.

In terms of analysis tools, the rapid deployment of Deep Machine Learning methods in Higgs measurements is expected to bring substantial gains in the HL-LHC era. Finally, it should be noted that in general, at particle colliders, due to continued innovations in analysis techniques, the precision in the final measurements of an electroweak observable have by far exceeded their prior projections.

## 4. Mass and natural width

The Higgs boson mass, $m_H$, is an unknown parameter of the Standard Model which should be measured as precisely as possible. The Higgs boson mass, together with the top quark mass, impacts the electroweak vacuum stability, and it has important consequences in cosmology.

With the Run 2 data, the Higgs boson mass, $m_H$, has been measured to be $m_H = 124.97 \pm 0.24$ GeV (ATLAS[11]) and $m_H = 125.38 \pm 0.14$ GeV (CMS[12]). These are the result of the combination of the independent $H \to ZZ^{(*)} \to 4l$ and $H \to \gamma\gamma$ mass measurements in each experiment.

At HL-LHC, the mass measurement accuracy will be limited by the systematic uncertainties dominated by the uncertainty on the electron, photon and muon energy scales. The projection studies assumed (conservatively) that the energy scale achieved in Run 2, namely 0.01% (0.15%) for the muon (electron), holds for HL-LHC as well. Consequently, the accuracy on the Higgs mass measurement will be dominated by the 4-lepton final states, particularly the $H \to ZZ^{(*)} \to \mu\mu\mu\mu$ final state, which are statistically limited in the current measurements. In the $H \to ZZ^{(*)} \to \mu\mu\mu\mu$ channel, an overall systematic uncertainty of 15 MeV is projected for the $m_H$ measurement. In the $H \to ZZ^{(*)} \to \mu\mu ee$ and $H \to ZZ^{(*)} \to eeee$ channels, the $m_H$ measurements are less precise, with a total uncertainty of about 100 MeV. A statistical accuracy of about 28 MeV is expected with 3000 fb$^{-1}$ from the $H \to ZZ^{(*)} \to \mu\mu\mu\mu$ channel, that reduces to 22 MeV when combined with the other 4-lepton final states. Finally, by combining all $H \to ZZ^{(*)} \to 4l$ final states, a total uncertainty of 30 MeV on $m_H$ is projected for the HL-LHC data set.

The total width ($\Gamma_H$) is another very important Higgs boson observable for probing new physics contributions. In the SM, the Higgs boson width is predicted to be 4.1 MeV for $m_H = 125$ GeV. The invariant mass resolution of the two Higgs final states that can be fully reconstructed ($H \to \gamma\gamma$ and $H \to ZZ^{(*)} \to 4l$) is much larger ($O$(1 GeV)) than the SM prediction, therefore only a model-independent experimental upper limit can be set. The projected upper limits at 95% C.L. of the Higgs boson width are 94 MeV (statistical uncertainty only) and 177 MeV (statistical and systematic uncertainties combined).

More precise (but model-dependent) estimates of $\Gamma_H$ can be obtained by relating its off-shell production to its on-shell production.[13–16] With Run 2 data, CMS measurement of the Higgs boson off-shell production processes $H \to ZZ^{(*)} \to 4l$ and $H \to ZZ^{(*)} \to 2l2\nu$ yield $\Gamma_H = 3.2^{+2.4}_{-1.7}$ MeV.[17] A measurement[18] by ATLAS yields $\Gamma_H = 4.6^{+2.6}_{-2.5}$ MeV. The observed (expected) upper limit on the total width is 9.7(10.2) MeV at 95% confidence level in the asymptotic approximation. With this method and by combining the CMS and ATLAS results with the HL-LHC data, a precision on the Higgs boson width of about 0.8 MeV, dominated by theoretical uncertainties, can be obtained.

## 5. The Higgs boson production, decay and couplings

The leading Feynman diagrams for Higgs boson production in pp collisions at the LHC, decay and pair production are shown in Fig. 1.

### 5.1. *Production and Decay*

The projected precision in the measurements of various production modes are shown in Fig. 2(left). The expected precision ranges from 1.6% (ggH) to 5.7% ($VH$) and begins to be dominated by the theoretical uncertainties within the phase space of the experimental measurement. The major Higgs decay rate measurements should reach precision of about 3% for $H \to \gamma\gamma$, $H \to ZZ^{(*)}$, $H \to WW^{(*)}$ and $H \to \tau\tau$, and 4% for ($H \to b\bar{b}$). They are all expected to be dominated by theoretical uncertainties. The measurements of rare decays $H \to \mu^+\mu^-$ and $Z\gamma$ will be statistically limited with a branching ratio uncertainty of about 8% and 19% respectively.

### 5.2. *Higgs Boson Couplings to bosons and fermions*

BSM phenomena are expected to affect the Higgs production modes and its decay channels in a correlated way if they are governed by similar interactions. Any modification in the interaction between the Higgs boson and, e.g. the W bosons and top quarks would not only affect the $H \to WW^{(*)}$

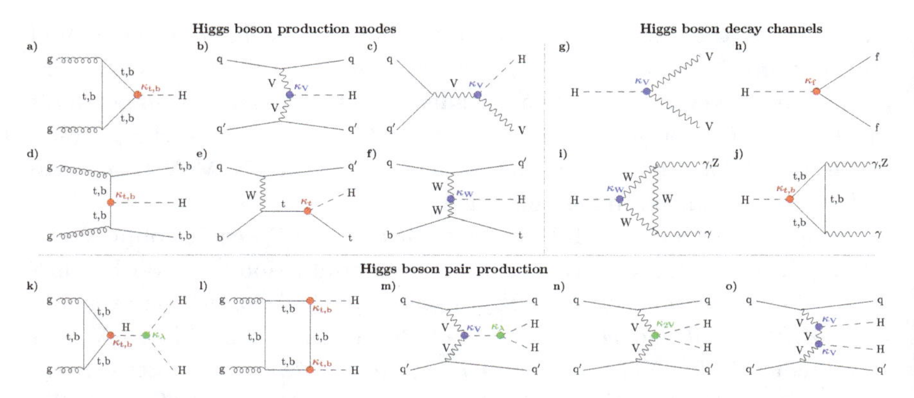


Fig. 1. Feynman diagrams for the leading order Higgs boson interactions: Higgs boson production in (a) gluon-gluon fusion ($gg \to H$), (b) vector boson fusion (VBF), (c) associated production with a W or Z (V) boson ($VH$), (d) associated production with a top or bottom quark pair ($ttH$ or $bbH$), (e, f) associated production with a single top quark ($tH$); with Higgs boson decays into (g) heavy vector boson pairs, (h) fermion-antifermion pairs, and (i, j) photon pairs or $Z\gamma$; Higgs boson pair production: (k, l) via gluon-gluon fusion, and (m, n, o) via vector boson fusion. The different Higgs boson interactions are labelled with the coupling modifiers $\kappa$, and highlighted in different colours for Higgs-fermion interactions (red), Higgs-gauge-boson interactions (blue), and multiple Higgs boson interactions (green).[3]

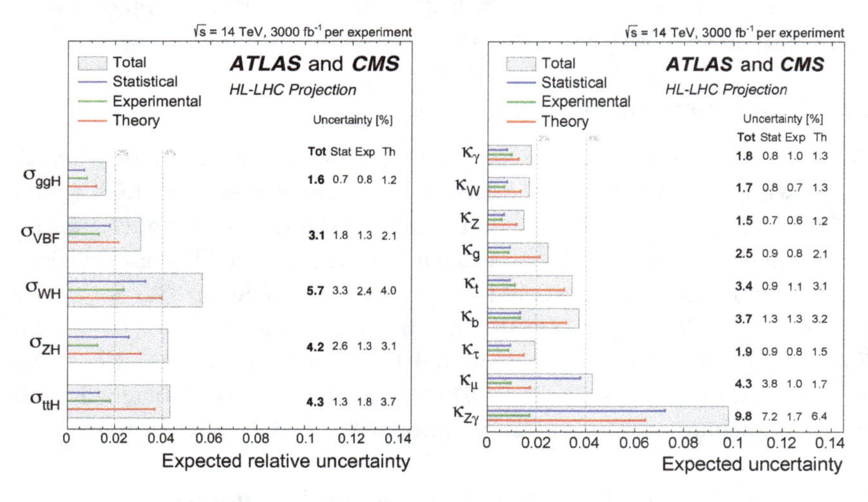


Fig. 2. (Left) projected precision in Higgs production cross section measurement, (right) projected precision in the modifiers of Higgs coupling to bosons and fermions.

and $H \to \gamma\gamma$ decay rates but also the production cross-section for gluon fusion production process ggH, $VBF$ and $VH$. In order to probe small contribution of BSM particles to the SM predictions, the $\kappa$-framework[19] developed to analyse Run 2 data is used. For a given production process or decay mode $j$, a coupling modifier $\kappa_j$ is defined such that;

$$\kappa_j^2 = \sigma_j/\sigma_j^{\mathrm{SM}} \quad \text{or} \quad \kappa_j^2 = \Gamma^j/\Gamma_{\mathrm{SM}}^j. \tag{1}$$

In the SM, all $\kappa_j$ values are positive and equal to unity. Six coupling modifiers $\kappa_W$, $\kappa_Z$, $\kappa_t$, $\kappa_b$, $\kappa_\tau$ and $\kappa_\mu$ corresponding to tree-level Higgs couplings are introduced along with effective coupling modifiers $\kappa_g$, $\kappa_\gamma$ and $\kappa_{Z\gamma}$ addressing the loop-process in $ggH$, $H \to \gamma\gamma$ and $H \to Z\gamma$. The total width of the Higgs boson, relative to the SM prediction, varies with the coupling modifiers as $\Gamma_H/\Gamma_H^{\mathrm{SM}} = \sum_j \mathrm{B}_{\mathrm{SM}}^j \kappa_j^2/(1 - \mathrm{B}_{\mathrm{BSM}})$, where $\mathrm{B}_{\mathrm{SM}}^j$ is the SM branching fraction for the $H \to jj$ channel and $\mathrm{B}_{\mathrm{BSM}}$ is the Higgs boson branching fraction to BSM final states. In the results for the $\kappa_j$ parameters presented here $\mathrm{B}_{\mathrm{BSM}}$ is fixed to zero and only decays to SM particles are allowed. The projected precision on Higgs boson coupling modifiers are shown in Fig. 2(right).

It should be noted that the $\kappa$ framework merely compares the experimental measurement to their best values computed within the SM and does not require any BSM calculation. It is based on assumptions and has limitations in its ability to describe general deformations of the SM. A systematic and powerful way to capture the deviations in Higgs coupling due to BSM phenomena comes from SM Effective Field Theory[5] and are being studied in the context of HL-LHC.[20]

### 5.3. *Rare Higgs Decays*

A dataset of 3000 $\mathrm{fb}^{-1}$ will allow probe of several rare or hard-to-detect Higgs boson decays. Having measured Higgs boson coupling to the vector bosons and fermions of the third generation to $\approx 10\%$ precision, the attention will focus on measurements of Higgs boson coupling to the second and first generation fermions. So far, due to tiny rates and large backgrounds, there are no model-independent and sensitive strategies at LHC to directly measure the Higgs coupling to the first generation fermions. For example, the best 95% CL limit[21] on $H \to e^+e^-$ decay rate with Run 2 data is about $3 \times 10^{-4}$ to be compared with the expected SM branching ratio of $\approx 5 \times 10^{-9}$. But by searching for and not finding such decays in the Run 2 data, we have learnt already that Higgs couplings to fermions are not universal.

### 5.3.1. $H \to \mu^+\mu^-$

At the HL-LHC, the relatively most accessible channel to probe Higgs coupling to the second generation is $H \to \mu^+\mu^-$. With a branching rate of $2 \times 10^{-4}$ and an irreducible $Z/\gamma^* \to \mu^+\mu^-$ background with a rate several orders of magnitude higher, it is also one of the most difficult Higgs decay channels to probe. The $H \to \mu^+\mu^-$ signal appears as a narrow peak over a smoothly falling Drell-Yan background. The CMS detector, with an excellent charged particle momentum resolution of $\approx$ 1–2.5% and precise background shape modeling, reported[22] the first evidence-level measurement of this decay with the Run 2 data. After the phase II upgrade, CMS and ATLAS will benefit from the improved charged particle tracking acceptance and momentum resolution leading to improvement in $m_{\mu\mu}$ resolution (for example, the expected momentum resolution improvement in CMS is about 30%). These improvements will enable a direct measurement of $\kappa_\mu$ with an uncertainty of about 4%.[23]

### 5.3.2. $H \to c\bar{c}$

Due to small charm hadron masses and low charged particle multiplicity in their decay, the identification of charm quark jets in the multi-jet environment of LHC is difficult. This limitation, along with the contamination from b quark jets and the small decay rate of $H \to c\bar{c} \approx 3\%$ makes the direct measurement of Higgs coupling to charm very challenging. In the recent years, ML techniques have been employed to better isolate $H \to c\bar{c}$ signal from multitudes of backgrounds. At the HL-LHC, a 95% upper limit on the signal strength of VH($H \to c\bar{c}$) production mode of 6.4 times the SM prediction and a constraint of $|\kappa_c| < 3.0$ is projected.[24]

### 5.3.3. $H \to Z\gamma$

The SM Higgs boson can decay into $H \to Z\gamma$ through loop diagrams and the branching ratio is predicted to be $BR(H \to Z\gamma) \approx 1.5 \times 10^{-3}$. It is an interesting mode because the measured rate can differ from the SM prediction in many BSM scenarios.[25–27] Due to the small branching fraction in the SM, the $H \to Z\gamma$ decay has not yet been observed at the LHC. An upper limit[28] at 95% confidence level on the production cross-section times the branching ratio for $pp \to H \to Z\gamma$ has been set at 3.6 times the SM prediction. The extrapolation to HL-LHC uses simple scaling approach on the Run 2 analysis. The measurement is expected to be statistically limited.

For example, with the 3000 fb$^{-1}$ dataset, ATLAS expects only a 5 standard deviations observation of this mode and a signal strength measurement of 23% precision.[20] The effective coupling $\kappa_{Z\gamma}$ is expected to be measured with an uncertainty of about 10%.[24]

#### 5.3.4. $H \to$ invisible ($H \to$ inv)

While major decay modes of the Higgs boson have been measured with some accuracy, there is enough room for the Higgs boson to decay into[29] a pair of BSM particles such as Dark Matter (DM) candidates. If kinematically accessible, non-interacting DM particles, such as neutralinos in supersymmetry models or graviscalars in models with extra dimensions could manifest themselves as invisible decays of the Higgs boson. Current[5] 95% CL limits on the branching fraction to invisible decays, dominated by the sensitivity in the VBF topology, is about 10%. The irreducible background in this search occurs at 0.1% and arises from $H \to ZZ^{(*)}$ decays where both Z bosons decay into neutrinos. At the HL-LHC, the major challenge in the search for $H \to$ inv stems from the impact of high pileup conditions on the reconstruction of the $E_{\mathrm{T}}^{\mathrm{miss}}$ and its resolution. Feasibility studies[23] in the VH and VBF modes using a variety of $E_{\mathrm{T}}^{\mathrm{miss}}$ threshold project a 95% CL upper limit on $BR(H \to \mathrm{inv}) < 2.5\%$.

## 6. The Higgs boson self-coupling

The study of the Higgs boson self-coupling represents an important test of the Standard Model, and hence is one of the primary goals of the HL-LHC. Deviations from SM predictions would indicate the presence of new physics beyond this theory. Furthermore, the Higgs boson self-interactions are of primary importance for cosmological theories involving, for example, the vacuum stability and inflation.[5]

Figure 3 shows the theoretical predictions for the total rates at proton-proton colliders with up to $\sqrt{s} = 100$ TeV, see Ref. 30 and references therein. The production cross section at $\sqrt{s} = 14$ TeV is about 39 fb. Figure 4 shows the dependence of the total HH production cross section as a function of the self-coupling $\lambda$ in units of the SM predicted value. The HH production rate is particularly low for a coupling strength around the SM value, $\lambda/\lambda_{SM} = 1$.

At the HL-LHC, the Higgs boson self-interactions are probed by measuring the $HH$ production rate. The main physics $HH$ final states studied are $HH \to b\bar{b}b\bar{b}$, $HH \to b\bar{b}\gamma\gamma$ and $HH \to b\bar{b}\tau^+\tau^-$. Other channels, such as

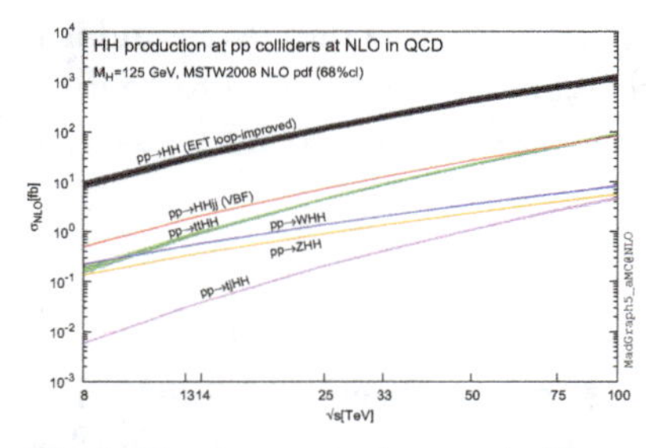


Fig. 3. Total cross sections at the Next-to-the-Leading-Order (NLO) in QCD for the six largest HH production channels at pp colliders. The thickness of the lines corresponds to the scale and Parton Distribution Functions (PDF) uncertainties added linearly.[30]

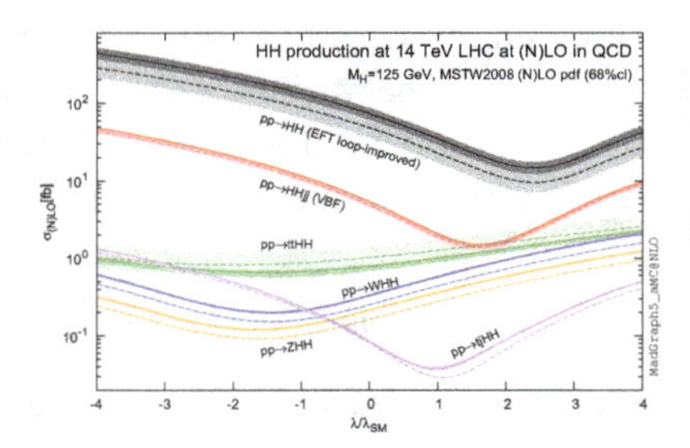


Fig. 4. Total cross section at the Leading Order (LO) and NLO in QCD for HH production channels, at $\sqrt{s}$ = 14 TeV LHC as a function of the self-interaction coupling $\lambda$. The dashed (solid) lines and light-(dark-)colour bands correspond to the LO (NLO) results and to the scale and PDF uncertainties added linearly.[30]

$HH \to b\bar{b}W^+W^-$, $HH \to b\bar{b}ZZ$, $HH \to W^+W^-\gamma\gamma$ and $HH \to \tau^+\tau^-\gamma\gamma$ have been also explored. Detailed studies have been reported in the HL-LHC CERN Yellow Report[20] (from now on called "Yellow Report"). In response to the Snowmass 2021 process,[31] new studies and several updates have been produced regarding the HL-LHC projections for non-resonant $HH$ processes, by both ATLAS[24] and CMS[23] Collaborations.

The combined minimum negative-log-likelihoods of ATLAS and CMS projections are shown in Fig. 5. As seen, two minima are found. The first[a] is at $\kappa_\lambda = 1$, as expected for Standard Model. The 68% Confidence Intervals (CIs) are $0.52 \leq \kappa_\lambda \leq 1.5$ and $0.57 \leq \kappa_\lambda \leq 1.5$ with and without systematic uncertainties respectively. The overview of the 68% CI for each channel in each experiment, as well as the combination, is shown in Fig. 6. The 68% CI. for $\kappa_\lambda$ are $0.52 \leq \kappa_\lambda \leq 1.5$ and $0.57 \leq \kappa_\lambda \leq 1.5$ with and without systematic uncertainties respectively.

ATLAS has updated the HL-LHC projections for non-resonant $HH$ production in the $HH \to b\bar{b}\gamma\gamma$[32] and $HH \to b\bar{b}\tau^+\tau^-$[33] final states, taking

[a]The presence of a second minimum, located at $\kappa_\lambda{\sim}7$, can be mostly explained by the result of the $HH \to b\bar{b}\tau^+\tau^-$ analysis. At $\kappa_\lambda$ larger than 1, the $HH$ production cross section increases with increasing $\kappa_\lambda$, and at the same time the signal acceptance decreases, leading to a similar shape to the $\kappa_\lambda = 1$ signal. Together, the degeneracy of the second minimum, originating mainly from the $HH \to b\bar{b}\tau^+\tau^-$ channel, is largely removed by the result of the $HH \to b\bar{b}\gamma\gamma$ analysis. This second minimum of the likelihood can be excluded at 99.4% CL.

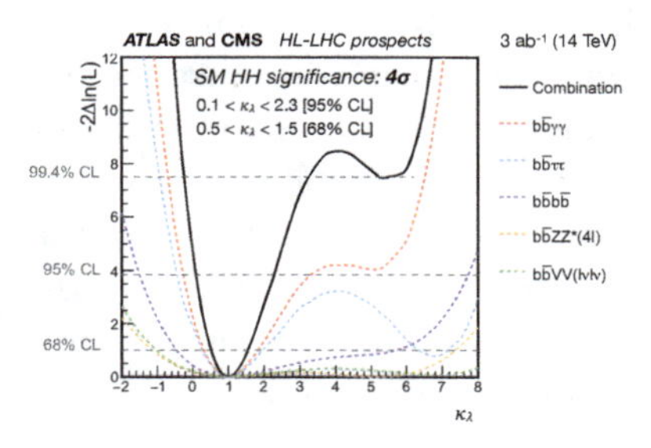


Fig. 5. Minimum negative log-likelihood as a function of $\kappa_\lambda$, calculated by performing a conditional signal+background fit to the background and SM signal. The coloured dashed lines correspond to the combined ATLAS and CMS results by channel, and the black line to their combination. The likelihoods for the channels are normalised to 6000 fb$^{-1}$.

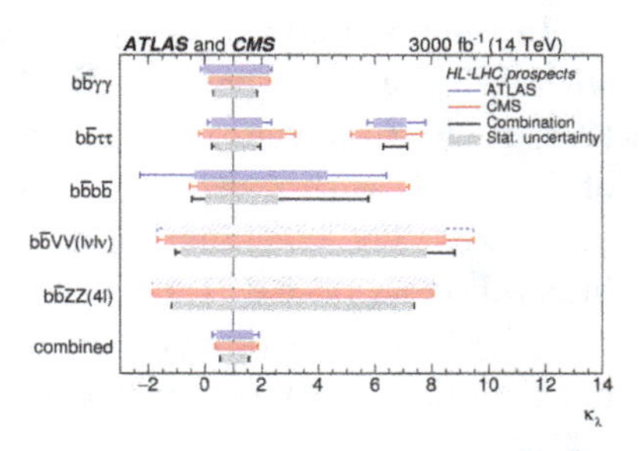


Fig. 6. Expected measured values of $\kappa_\lambda$ for the different channels for the ATLAS in blue and the CMS experiment in red, as well as the combined measurement. The lines with error bars show the total uncertainty on each measurement while the boxes correspond to the statistical uncertainties.

advantage of the analysis methodologies which were updated to the latest Run 2 analyses.[34,35]

### 6.1. *The $bb\gamma\gamma$ final state*

The $HH \to b\bar{b}\gamma\gamma$ analysis exploits the large $H \to b\bar{b}$ branching ratio in combination with the excellent ATLAS photon resolution, allowing the search for a narrow $H \to \gamma\gamma$ signal in the smoothly falling distribution of the $\mathrm{m}_{\gamma\gamma}$ di-photon mass spectrum.

The ATLAS Run 2 analyses studied a categorization based on the output of a BDT discriminant and the modified four-body mass $m^*_{bb\gamma\gamma} = m_{bb\gamma\gamma} - (m_{bb} - 125) - m_{\gamma\gamma} - 125)$ (units in GeV) allowed to increase significantly the sensitivity of this search with respect to previous analyses.

In the Yellow Report, the projection results are obtained using the profile likelihood ratio.[36] Signal and background distributions in the Run 2 categories are first scaled by a uniform scaling factor defined as the ratio between the target integrated luminosity of 3000 fb$^{-1}$ to the Run 2 integrated luminosity (139 fb$^{-1}$). The change in the center-of-mass energy from $\sqrt{s} = 13$ TeV to $\sqrt{s} = 14$ TeV is accounted for with an additional scaling factor, which depends on the physics process considered. Finally, the projected results in individual analyses are obtained by considering different scenarios of systematic uncertainties.

In the updated projection,[32] the significance of the SM signal ($\kappa_\lambda = 1$) with (without) the baseline HL-LHC systematic scenario increases to 2.2 (2.3) standard deviations (s.d.), an improvement with respect to the values of 2.0 (2.1), with (without) systematic uncertainties, found in the previous study. The combination of all categories results in a 1 s.d. confidence interval on $\kappa_\lambda$ of [0.3, 1.9] ([0.4, 1.8]) with (without) systematic uncertainties. This represents again an improvement with respect to the previous projection,[37] which established the 1 s.d. CI for $\kappa_\lambda$ to be [−0.2, 2.5] ([−0.1, 2.4]).

The CMS Collaboration has also updated its projection using the latest studies of the detector layout in the high-luminosity upgrade and corresponding reconstruction algorithms. In addition to the gluon-gluon fusion process, the new study includes the VBF production mode of $HH$, which provides a unique access to the HHVV (V = $W^\pm$ or $Z$ bosons) coupling. In the early study this process was neglected. The Vector Boson Fusion process is the second largest production mode of the non-resonant $HH$ production. The study[38] was performed using Monte Carlo samples emulated in the CMS upgrade detector with the DELPHES fast and parametric simulation package tuned for $\sqrt{s} = 14$ TeV and assuming an average pile-up of 200 events per bunch crossing.

The invariant mass distributions derived from the selected photon and $b$-jets pairs are studied to categorise events according to their signal sensitivities. Figure 7 shows the distribution of the $\gamma\gamma$ invariant mass, $m_{\gamma\gamma}$, for the selected pseudo-data events by this new CMS analysis. The curves correspond to continuum background only (green dashed), total background (continuum + single Higgs boson) (solid blue), and the signal + background (solid red). The signal contribution is shown in solid magenta line at the bottom of the plot.

The expected significance for the inclusive di-Higgs signal is 2.16 s.d. including systematic uncertainties. This is an improvement over the value of 1.83 s.d. reported in the previous study.[39]

### 6.2. *The $bb\tau^+\tau^-$ final state update*

New $b$-tagging performance studies expected with the ATLAS detector at HL-LHC were performed by ATLAS since the publication of the Yellow Report, taking into account, in particular, the latest developments in the ITk simulation. The $b$-tagging performance is not expected to be significantly worse than in Run 2, in spite of the significantly larger pile-up expected for HL-LHC. The expected $b$-tagging performance will be very beneficial to the search for the di-Higgs process, in particular, for the study of the $HH \to b\bar{b}b\bar{b}$ and $HH \to b\bar{b}\tau^+\tau^-$ final states. Concerning the $HH \to b\bar{b}\tau^+\tau^-$ final state, the updated projection leads to a signal significance of 2.8 s.d., while the previous extrapolation yielded a signal significance of 2.2 s.d.[33]

Assuming baseline uncertainties, the updated $HH \to b\bar{b}\tau^+\tau^-$ projection provides 1 s.d. Confidence Interval $0.3 \le \kappa_\lambda \le 1.9$ and $5.2 \le \kappa_\lambda \le 1.9$ which, compared to the previous projection, results in an improvement of 28%. While the full Run 2 search is dominated by statistical uncertainties, the current projection clearly shows that systematic uncertainties will become a limiting factor of the $HH \to b\bar{b}\tau^+\tau^-$ analysis at the HL-LHC.

### 6.3. *Combination of $HH \to b\bar{b}\tau^+\tau^-$ and $HH \to b\bar{b}\gamma\gamma$ analyses*[40]

ATLAS has also combined the updated projections of the $HH \to b\bar{b}\tau^+\tau^-$ and $HH \to b\bar{b}\gamma\gamma$ final states. The combination of the these updated projections[40] is performed through multiplication of the single analysis likelihoods into a combined likelihood function. The different searches are then fit to the data in order to constrain simultaneously the parameters of interest and the nuisance parameters. The systematic uncertainties of various analyses are correlated following the strategy of Ref. 41.

Values of the negative log-profile-likelihood ratio as a function of $\kappa_\lambda$ for various uncertainty scenarios are shown in Fig. 8. If the baseline scenario is assumed, the combination of $HH \to b\bar{b}\tau^+\tau^-$ and $HH \to b\bar{b}\gamma\gamma$ yields a significance of 3.2 s.d. This result is an improvement on the projection presented by ATLAS in the Yellow Report and in subsequent updates, in particular with respect to the analysis where ATLAS combined the projections for the $HH \to b\bar{b}b\bar{b}$, $HH \to b\bar{b}\tau^+\tau^-$ and $HH \to b\bar{b}\gamma\gamma$ final states,[42] where a significance of 2.9 s.d. was estimated.

Combining, the 1-standard deviation confidence intervals on $\kappa_\lambda$ are found to be in the interval [0.5, 1.6] in the baseline scenario and [0.6, 1.5] without systematic uncertainties. These intervals show an improvement in sensitivity with respect to the previous projection[42] which reported a 1-standard deviation CI at [0.25, 1.9] ([0.4, 1.7]) with (without) systematic uncertainties.

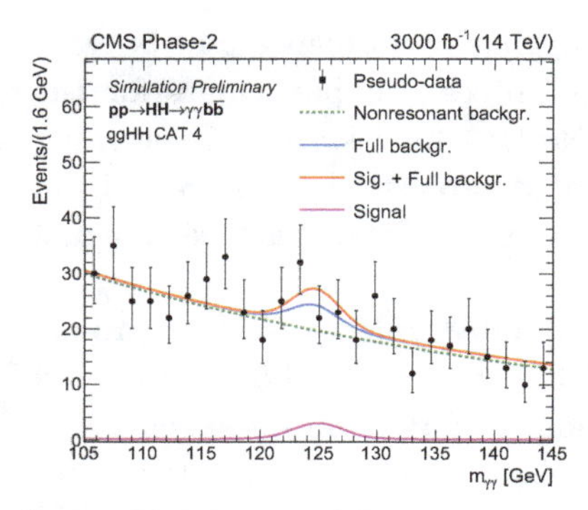


Fig. 7. Distribution of the photon pair invariant mass $m_{\gamma\gamma}$ for the selected pseudo-data events (black points) corresponding to L = 3000 fb$^{-1}$ along with the expectations as estimated from the simulation.

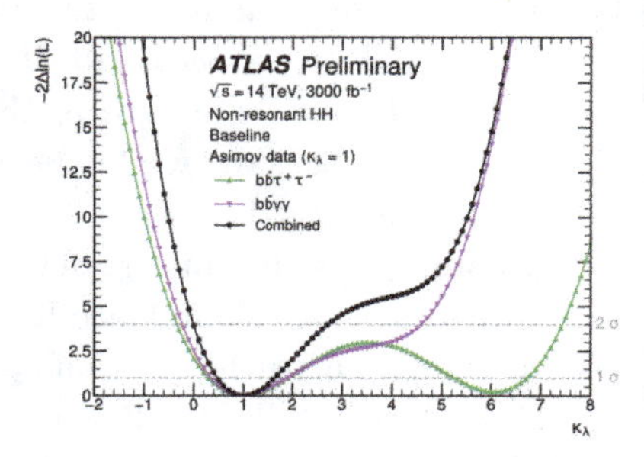


Fig. 8. Negative log-profile-likelihood ratio as a function of $\kappa_\lambda$ projected to 3000 fb$^{-1}$ and $\sqrt{s}$ = 14 TeV, assuming Standard Model Processes ($\kappa_\lambda = 1$) and assuming the four uncertainty scenarios described in the Yellow Report. Dashed horizontal lines correspond to 1 s.d. and 2 s.d. confidence intervals.

### 6.4. *The $t\bar{t}HH$ final state and other $HH$ final states*

The production of a pair of Higgs bosons in association with a $t\bar{t}$ pair offers the possibility to explore the interplay between the $HH$ and the $t\bar{t}H$ measurements. This process is also highly sensitive to potential BSM contributions. A preliminary study of this process has been carried out by CMS.[23] In this study, the analysis is based on the reconstruction of events with two Higgs boson decaying each to b-quark pairs, and semileptonic decays of the top-antitop quark pair, which leads to final states with a single lepton, multiple jets, multiple b-jets and moderate missing transverse momentum. The results of this prospect study shows that an upper limit to the $t\bar{t}HH$ production cross section of $3.14^{+1.27}_{-0.9}\times$ *SM prediction* (0.948 fb$^{-1}$). CMS performed a di-Higgs search study also in the $HH \to W^+W^-\gamma\gamma$ and $HH \to \tau^+\tau^-\gamma\gamma$ channels, which benefits from the sensitive $H \to \gamma\gamma$ process and provides a clean and distinguishable signature. Combining all these final states, the expected significance for signal is 0.22 s.d., including systematic uncertainties at integrated luminosity of 3000 fb$^{-1}$ at the HL-LHC.

### 6.5. *Summary of the projected precision on self-coupling*

Table 1 shows a summary of the signal significancy of $HH$ final states produced by prospects studies performed by the ATLAS and CMS Collaborations. As seen, the update studies made for the Snowmass process further increased the findings reported in the Yellow Report. The SM double Higgs boson production yield is expected to be measured with a significance greater than 4 s.d. Correspondingly, the Higgs self-coupling modifier $\kappa_\lambda$ should be measured with an uncertainty of about 50% by each experiment. The combination of the results from two experiments should yield an uncertainty significantly better than 50%. Further refined studies will consolidate the current projections, and most likely new Deep learning based analyses techniques could increase the possibility of observing (5 s.d.) the Standard Model $HH$ production at HL-LHC, if realised in nature. On the other hand, a statistically significant measurement of $\kappa_\lambda \neq 1$ would be very interesting.

Table 1. Prospects for the signal significance of the $HH$ final states studied by the ATLAS and CMS Collaborations, reported in the CERN Yellow Report. The numbers between [] reports the results presented in the Snowmass paper. The number between () represents a simple combination of the most recent signal significance results available.

| $HH$ final state | ATLAS significance (s.d.) | CMS significance (s.d.) |
|---|---|---|
| $HH \to b\bar{b}b\bar{b}$ | 0.61 | 0.95 |
| $HH \to b\bar{b}\gamma\gamma$ | 2.0 [2.2] | 1.8 [2.16] |
| $HH \to b\bar{b}\tau^+\tau^-$ | 2.1 [2.8] | 1.4 |
| $HH \to b\bar{b}VV(\text{ll}\nu\nu)$ | — | 0.56 |
| $HH \to b\bar{b}ZZ(4\text{l})$ | — | 0.37 |
| $HH \to W^+W^-\gamma\gamma$ + $HH \to \tau^+\tau^-\gamma\gamma$ | — | — [0.22] |
| overall combination: 4.0[20] (∼4.3) | | |

## 7. Summary

The luminosity upgrade of LHC, the HL-LHC, represents a unique opportunity for precision measurements of the Higgs boson properties by ATLAS and CMS. Any significant deviation of these measurements from the Standard Model predictions will indicate presence of new physics beyond SM at the energy scale of the LHC or colliders proposed for future high-energy physics exploration.

Higgs boson couplings to gauge bosons will be measured with an accuracy of about 2%, while the couplings to the fermions of the third generation and the muons of the second generation will be measured with an accuracy of about 4%. Because of the very small branching ratios, the model-independent measurement of the Higgs coupling to the fermions of the first generation will be very challenging.

The measurement of the Higgs boson mass, an unpredicted parameter in the Standard Model, can be performed at the HL-LHC with an uncertainty of tens of MeV. The direct measurement of the Higgs boson natural width is limited by the energy resolution of muon systems and electromagnetic calorimeters. The $\Gamma_H \simeq 4$ MeV, predicted by Standard Model, can be probed with the study of off-shell Higgs boson production. An uncertainty of 1 MeV (or better) on this parameter is expected at the HL-LHC.

The study of the Higgs self-coupling represents one of strongest physics cases for the HL-LHC programme. The Higgs self-coupling can be studied by measuring the production of Higgs boson pairs. The contribution to the Higgs boson pair production cross section from processes induced by Higgs self-coupling, is very small and thus, large data samples are needed for a precise measurement. At the HL-LHC, the Higgs boson pairs produced by Higgs self-coupling can be observed with a significance of about 5 standard deviations, and the strength of the self-coupling can be measured with an uncertainty of about 50%.

The investigations at the HL-LHC of the Higgs couplings, particularly its self-coupling, will be unique in the world for many decades to come.

## 8. Acknowledgement

We are grateful to our Funding Agencies support in writing of this manuscript. V.S. is supported by the grant DE–SC0009919 of the United States Department of Energy.

## References

1. ATLAS Collaboration, Observation of a new particle in the search for the Standard Model Higgs boson with the ATLAS detector at the LHC, *Phys. Lett. B.* **716**, 1 (2012). doi: 10.1016/j.physletb.2012.08.020.
2. CMS Collaboration, Observation of a new boson at a mass of 125 GeV with

the CMS experiment at the LHC, *Phys. Lett. B.* **716**, 30 (2012). doi: 10.1016/j.physletb.2012.08.021.
3. CMS Collaboration, A portrait of the higgs boson by the CMS experiment ten years after the discovery, *Nature.* **607** (7917) URL `https://doi.org/10.1038/s41586-022-04892-x`.
4. ATLAS Collaboration, A detailed map of higgs boson interactions by the ATLAS experiment ten years after the discovery, *Nature.* **607** (7917) (Jul, 2022). URL `https://doi.org/10.1038/s41586-022-04893-w`.
5. R. L. Workman and Others, Status of Higgs Boson Physics in the Review of Particle Physics, *PTEP.* **2022**, 083C01 (2022). doi: 10.1093/ptep/ptac097.
6. G. P. Salam, L.-T. Wang, and G. Zanderighi, The Higgs boson turns ten, *Nature.* **607** (7917) (2022). URL `https://doi.org/10.1038/s41586-022-04899-4`.
7. ATLAS Collaboration, ATLAS Phase 2 Upgrade Technical Design Reports, URL `https://cds.cern.ch/record/2652549`.
8. CMS Collaboration, Technical proposal for the phase ii upgrade of the cms detector, URL `https://cds.cern.ch/record/2020886`. CMS-TDR-15-02.
9. F. Lanni, Overview of the ATLAS HL-LHC upgrade programm; M. Mannelli, The CMS HL-LHC Phase II upgrade program; this book.
10. Addendum to the report on the physics at the HL-LHC, and perspectives for the HE-LHC: Collection of notes from ATLAS and CMS. Technical report, CERN, Geneva (Dec, 2019). URL `http://cds.cern.ch/record/2651134`, `https://arxiv.org/abs/1902.10229`.
11. Measurement of the Higgs boson mass in the $H \to ZZ^{(*)} \to 4l$ and $H \to \gamma\gamma$ channels with s = 13 TeV pp collisions using the ATLAS detector, *Physics Letters B.* **784**, 345–366 (2018). URL `https://doi.org/10.1016/j.physletb.2018.07.050`.
12. A.M. Sirunyan *et al.*, A measurement of the Higgs boson mass in the diphoton decay channel, *Physics Letters B.* **805**, 135425 (2020). URL `https://doi.org/10.1016/j.physletb.2020.135425`.
13. N. Kauer and G. Passarino, Inadequacy of zero-width approximation for a light higgs boson signal, *Journal of High Energy Physics.* **2012** (8) (Aug, 2012). doi: 10.1007/jhep08(2012)116. URL `https://link.springer.com/article/10.1007/JHEP08(2012)116`.
14. F. Caola and K. Melnikov, Constraining the higgs boson width with zz production at the LHC, *Physical Review D.* **88** (5) (Sep, 2013). doi: 10.1103/physrevd.88.054024. URL `https://journals.aps.org/prd/abstract/10.1103/PhysRevD.88.054024`.
15. J. M. Campbell, R. K. Ellis, and C. Williams, Bounding the higgs width at the LHC using full analytic results for gg $\to e^- e^+ \mu^- \mu^+$, *Journal of High Energy Physics.* **2014** (4) (Apr, 2014). doi: 10.1007/jhep04(2014)060. URL `https://link.springer.com/article/10.1007/JHEP04(2014)060`.
16. J. M. Campbell, R. K. Ellis, and C. Williams, Bounding the higgs width at the LHC: Complementary results from $h \to ww$, *Physical Review D.* **89** (5) (Mar, 2014). doi: 10.1103/physrevd.89.053011. URL `https://journals.aps.org/prd/abstract/10.1103/PhysRevD.89.053011`.

17. CMS Collaboration, Measurement of the Higgs boson width and evidence of its off-shell contributions to ZZ production, *Nature Phys.* **18** (11), 1329–1334 (2022). doi: 10.1038/s41567-022-01682-0.
18. Evidence of off-shell Higgs boson production and constraints on the total width of the Higgs boson in the $ZZ \to 4l$ and $ZZ \to 2l2\nu$ decay channels with the ATLAS detector. Technical report, CERN, Geneva (Nov, 2022). URL `http://cds.cern.ch/record/2842520`.
19. LHC Higgs Cross Section Working Group, C. Potter et al., Handbook of LHC Higgs Cross Sections: 3. Higgs Properties: doi: 10.5170/CERN-2013-004. URL `http://cds.cern.ch/record/1559921`.
20. A. Dainese, M. Mangano, A. Meyer, A. Nisati, G. Salam, and M. A. Vesterinen. Report on the Physics at the HL-LHC, and Perspectives for the HE-LHC. Technical report, Geneva, Switzerland (2019). URL `https://cds.cern.ch/record/2703572`.
21. CMS Collaboration, Search for the Higgs boson decay to a pair of electrons in proton-proton collisions at $\sqrt{s} = 13$ TeV. URL `https://doi.org/10.48550/arxiv.2208.00265`.
22. C. Collaboration, Evidence for Higgs boson decay to a pair of muons, *JHEP.* **01**, 148 (2021). doi: 10.1007/JHEP01(2021)148.
23. Snowmass White Paper Contribution: Physics with the Phase-2 ATLAS and CMS Detectors. Technical report, CERN, Geneva (2022). URL `https://cds.cern.ch/record/2806962`.
24. Snowmass White Paper Contribution: Physics with the Phase-2 ATLAS and CMS Detectors. Technical report, CERN, Geneva (Apr, 2022). URL `https://cds.cern.ch/record/2805993`.
25. C.-W. Chiang and K. Yagyu, Higgs boson decays to $\gamma\gamma$ and $Z\gamma$ in models with Higgs extensions, *Phys. Rev. D.* **87** (3) (2013). doi: 10.1103/PhysRevD.87.033003.
26. C.-S. Chen, C.-Q. Geng, D. Huang, and L.-H. Tsai, New Scalar Contributions to $h \to Z\gamma$, *Phys. Rev. D.* **87**, 075019 (2013). doi: 10.1103/PhysRevD.87.075019.
27. P. Archer-Smith, D. Stolarski, and R. Vega-Morales, On new physics contributions to the Higgs decay to Z$\gamma$, *JHEP.* **10** (2021). doi: 10.1007/JHEP10(2021)247.
28. A. Collaboration, A search for the $Z\gamma$ decay mode of the Higgs boson in $pp$ collisions at $\sqrt{s} = 13$ TeV with the ATLAS detector, *Phys. Lett. B.* **809** (2020). doi: 10.1016/j.physletb.2020.135754.
29. A. Djouadi, A. Falkowski, Y. Mambrini, and J. Quevillon, Direct Detection of Higgs-Portal Dark Matter at the LHC, *Eur. Phys. J. C.* **73** (6) (2013). doi: 10.1140/epjc/s10052-013-2455-1.
30. R. Frederix, S. Frixione, V. Hirschi, F. Maltoni, O. Mattelaer, P. Torrielli, E. Vryonidou, and M. Zaro, Higgs pair production at the LHC with NLO and parton-shower effects, *Physics Letters B.* **732**, 142–149 (May, 2014). doi: 10.1016/j.physletb.2014.03.026. URL `https://www.sciencedirect.com/science/article/pii/S0370269314001828`.

31. Snowmass21: DPF Community Planning Exercise. URL `https://snowmass21.org/energy/start`.
32. Measurement prospects of Higgs boson pair production in the $b\bar{b}\gamma\gamma$ final state with the ATLAS experiment at the HL-LHC. Technical report, CERN, Geneva (Jan, 2022). URL `https://cds.cern.ch/record/2799146`.
33. Projected sensitivity of Higgs boson pair production in the $bb\tau\tau$ final state using proton-proton collisions at HL-LHC with the ATLAS detector. Technical report, CERN, Geneva (Dec, 2021). URL `https://cds.cern.ch/record/2798448`.
34. Search for resonant and non-resonant Higgs boson pair production in the $b\bar{b}\tau^+\tau^-$ decay channel using 13 TeV $pp$ collision data from the ATLAS detector. Technical report, CERN, Geneva (Jul, 2021). URL `https://cds.cern.ch/record/2777236`.
35. ATLAS Collaboration. Search for higgs boson pair production in the two bottom quarks plus two photons final state in $pp$ collisions at $\sqrt{s} = 13$ Tev with the atlas detector. URL `https://arxiv.org/abs/2112.11876` (2021).
36. G. Cowan, K. Cranmer, E. Gross, and O. Vitells, Asymptotic formulae for likelihood-based tests of new physics, *The European Physical Journal C*. **71** (2) (Feb, 2011). doi: 10.1140/epjc/s10052-011-1554-0. URL `https://doi.org/10.1140%2Fepjc%2Fs10052-011-1554-0`.
37. Measurement prospects of the pair production and self-coupling of the Higgs boson with the ATLAS experiment at the HL-LHC. Technical report, CERN, Geneva (Dec, 2018). URL `https://cds.cern.ch/record/2652727`.
38. Prospects for non-resonant Higgs boson pair production measurement in bb$\gamma\gamma$ final states in proton-proton collisions at $\sqrt{s} = 14$ TeV at the High-Luminosity LHC. Technical report, CERN, Geneva (2022). URL `https://cds.cern.ch/record/2803918`.
39. Prospects for HH measurements at the HL-LHC. Technical report, CERN, Geneva (2018). URL `https://cds.cern.ch/record/2652549`.
40. Projected sensitivity of Higgs boson pair production combining the $b\bar{b}\gamma\gamma$ and $b\bar{b}\tau^+\tau^-$ final states with the ATLAS detector at the HL-LHC. Technical report, CERN, Geneva (Feb, 2022). URL `http://cds.cern.ch/record/2802127`.
41. Combination of searches for non-resonant and resonant Higgs boson pair production in the $b\bar{b}\gamma\gamma$, $b\bar{b}\tau^+\tau^-$ and $b\bar{b}b\bar{b}$ decay channels using $pp$ collisions at $\sqrt{s} = 13$ TeV with the ATLAS detector. Technical report, CERN, Geneva (Oct, 2021). URL `https://cds.cern.ch/record/2786865`.
42. Expected $b$-tagging performance with the upgraded ATLAS Inner Tracker detector at the High-Luminosity LHC. Technical report, CERN, Geneva (Mar, 2020). URL `https://cds.cern.ch/record/2713377`.

© 2024 The Author(s)
https://doi.org/10.1142/9789811280184_0016

# Chapter 16

# High Luminosity LHC: Prospects for New Physics

Marie-Hélène Genest*, Greg Landsberg† and Marie-Hélène Schune‡

**Univ. Grenoble Alpes, CNRS, Grenoble INP, LPSC-IN2P3, 38000 Grenoble, France*

*†Brown University, Dept. of Physics, 182 Hope St., Providence, RI 02912, USA*

*‡Université Paris-Saclay, CNRS/IN2P3, IJCLab, Orsay, France*

## 1. Introduction

While there has been no clear deviation from the Standard Model (SM) seen in the LHC data analysed so far, the data set available at the end of Run 2 corresponds to only about 5% of the full integrated luminosity that will become available by the end of the HL-LHC. Consequently, there are multiple ways in which the searches could benefit from the high-luminosity phase. Considering only the increase in the integrated luminosity ($L$) of the data samples, the sensitivity will improve as $\sqrt{L}$ and, for processes with no known SM backgrounds, as $L$, making these searches particularly interesting in the early years of the HL-LHC's running. However, this will not be the only source of improvement; indeed, increased data sets will also benefit performance studies, reducing systematic uncertainties associated with the searches and further increasing their sensitivity. This will allow the analyses to probe for new physics (NP) at even higher masses, either directly, by, e.g. looking for a resonance in a given mass spectrum, or indirectly, by carefully looking for deviations in the tails of specific kinematic distributions. There could also be new phenomena at low masses, which would have escaped detection due to very weak couplings, and that still could be observed with a larger data set if innovative methods are employed to reduce

This is an open access article published by World Scientific Publishing Company. It is distributed under the terms of the Creative Commons Attribution 4.0 (CC BY) License.

the background and to trigger on these events. Production mechanisms or decays, which were not covered in previous analyses due to their rarity, could also become accessible. Finally, going to the high-luminosity phase of the LHC will not imply merely repeating what has already been done before: the upgraded detectors and innovative analysis approaches, notably exploiting fast-developing machine-learning techniques, will most certainly allow the LHC physicists to investigate a much broader range of signatures than originally foreseen, e.g. new types of long-lived particles. Examples of these expected improvements will be discussed in the subsequent sections.

## 2. Where to Look for New Physics at the HL-LHC

### 2.1. *Probing Higher Masses*

Multiple NP theories predict the existence of new bosons, and the heavier version of the $Z$ boson, the spin-1 $Z'$, is a standard benchmark when it comes to study the potential of colliders. A common benchmark is the $Z'_{SSM}$, for which the fermion couplings of the $Z'$ are taken to be that of the $Z$ boson. Multiple final states can be used to search for this new boson ($e^+e^-$, $\mu^+\mu^-$, $q\bar{q}$, ...) by looking for a resonance in the invariant mass spectrum of the decay products over a smoothly falling background. Sensitivity studies for 3 ab$^{-1}$ of data at $\sqrt{s} = 14$ TeV have shown[1,2] that in the most sensitive channel combining the $e^+e^-$ and the $\mu^+\mu^-$ final states, a discovery of the $Z'_{SSM}$ up to a mass of 6.4 (6.8) TeV would be possible with the ATLAS[1] (CMS[2]) detector, given that the current limits from these experiments are at 5.1 (5.2) TeV with the full Run 2 data set.[3,4]

As in the SM case, in many NP models a $U(1)$ $Z'$ boson would be accompanied by the $SU(2)$ $W^{\pm'}$ bosons, which could be sought in the leptonic or dijet decay channels. For the case of the SM-like couplings, $W'_{SSM}$ bosons can be discovered (excluded) in CMS with masses up to 6.4 (7.0) TeV based on decays into a $\tau$ lepton and a neutrino, followed by a hadronic $\tau$ lepton decay.[1] For a right-handed $W'$ boson decaying into a top and bottom (anti)quarks, ATLAS can discover (exclude) it up to masses of 4.3 (4.9) TeV.[1]

A new vector boson could also act as a portal, allowing the production of weakly interacting massive particles (WIMPs, $\chi$), which can be dark matter candidates, through an $s$-channel interaction $q\bar{q} \to Z' \to \chi\chi$. The production rate in a simplified model, which only considers the $Z'$ and $\chi$ as new particles, will depend on the couplings of this boson to quarks, $g_q$,

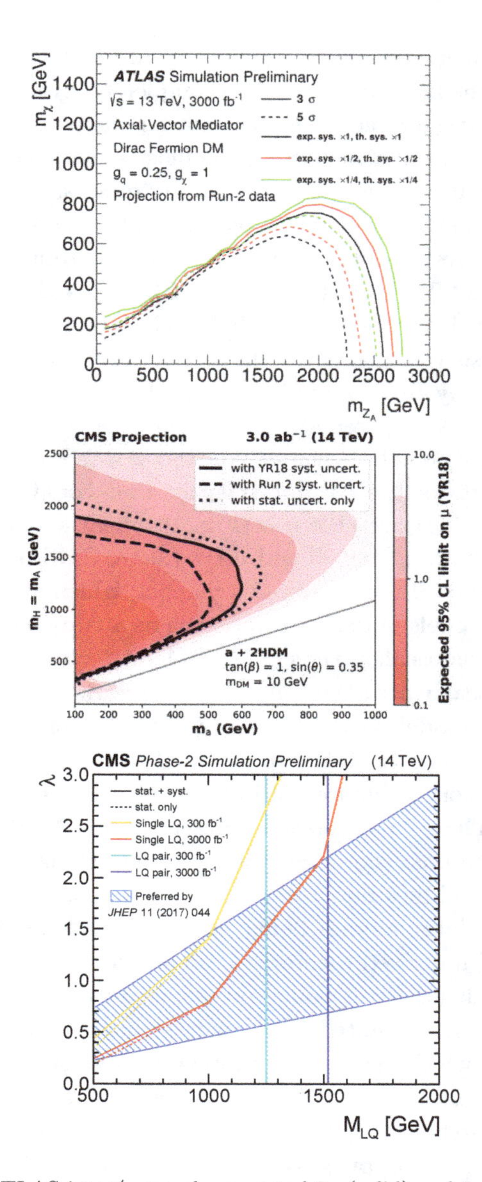


Fig. 1. (Upper) ATLAS jet+$\not{p}_T$ search: expected $3\sigma$ (solid) and $5\sigma$ (dashed) discovery contours in the $m_\chi$ versus $m_{Z'_A}$ mass plane as described in the text. Three systematic uncertainty scenarios are considered. The regions below the curves would be within reach. (Middle) Projected sensitivity of the CMS mono-$Z$ search in the 2HDM+a model with an extra pseudoscalar, as a function of its mass. (Lower) Projected sensitivity of the CMS LQ search in the plane of the LQ-lepton-quark coupling $\lambda$ versus the leptoquark mass. The areas on the left of the curves are within the reach. Figures taken from Ref. 1.

and to the dark matter particles, $g_\chi$. As $\chi$ interacts only weakly and escapes detection, the final state would be completely invisible to the detector (and therefore untriggerable), if it were not for the emission of initial-state QCD or electroweak radiation (ISR), the most sensitive channel being the emission of a gluon by one of the interacting quarks, leading to a jet+$\not{p}_{\mathrm{T}}$ final state. A prospective study was performed by the ATLAS experiment, extrapolating the result obtained with 36 fb$^{-1}$ of Run 2 data to 3 ab$^{-1}$ of data at $\sqrt{s} = 14$ TeV. Contrary to the $Z'_{SSM}$ search mentioned above, for which the sensitivity is mainly limited by the integrated luminosity, the jet+$\not{p}_{\mathrm{T}}$ analysis is impacted by the systematic uncertainties, which will likely be reduced by further studies at the HL-LHC. The impact of different scenarios for the systematic uncertainty (same as before or reduced by factors of 2 or 4) on the sensitivity of the search was thus investigated. The resulting parts of the parameter space that could lead to a 3 (5) $\sigma$ evidence (observation) are shown in Fig. 1(upper) for an axial-vector $Z'$ (here called $Z_A$) with $g_q = 0.25$ and $g_\chi = 1.0$. A CMS projection is based on a complementary $Z$+$\not{p}_{\mathrm{T}}$ (mono-$Z$) channel, which also offers access to other simplified models of dark matter, such as a two-Higgs doublet model with an extra pseudoscalar particle (2HDM+a),[5] via the $H \to Za$ decay, with the pseudoscalar $a$ further decaying into dark-matter particles. The reach within this model, as a function of the $a$ boson mass and the mass of a heavy Higgs boson $H$ (taken to be equal to the mass of a heavy pseudoscalar Higgs boson $A$), expressed in terms of 95% CL limits, is shown in Fig. 1(middle). These conclusions are based on the Run 2 CMS analysis,[6] under two assumptions: the same systematic uncertainties as in the Run 2 analysis and reduced by a factor of two, which is expected to be achievable, thanks to larger data samples being used to constrain the experimental uncertainties and improved theoretical calculations.

Mounting evidence for flavour non-universality in $b \to c\ell\bar{\nu}$ and $b \to s\ell^+\ell^-$ transitions has resulted in a renewed interest in searches for leptoquarks (LQs), which have been proposed as a theoretically preferred solution to these anomalies. Direct searches for LQs are therefore complementary to searches for flavour anomalies in $b$ hadron decays discussed in Sec. 2.3. An HL-LHC projection by the CMS experiment for searches for a scalar LQ decaying to a $b$ quark and a $\tau$ lepton, based on the Run 2 analyses,[7,8] shown in Fig. 1(lower) indicates that the combination of pair and single production would allow to probe these LQs up to a mass of 1.5 TeV,[1] which covers a significant range of the parameter space preferred by the theoretical models explaining flavour anomalies[9] and corresponds to about 500 GeV improvement on the sensitivity achieved so far.

### 2.2. *Probing Lower Masses and Couplings*

While HL-LHC offers a sizable improvement to the sensitivity at high masses, it will also allow to significantly extend the reach of searches for NP at lower masses, which could still have evaded the detection at the LHC and earlier machines because of small couplings and/or small production cross section and overwhelming backgrounds.

A significant development during the LHC Run 1 and (especially) Run 2 was the implementation of the trigger-level, or *scouting* analyses by ATLAS and CMS, which effectively increase the bandwidth of specific triggers dramatically by writing out only very compressed, trigger-level information about specific objects, instead of full events. These techniques have been used for searches for low-mass dijet resonances[10–13] and low-mass dimuon resonances.[14] While no projections for HL-LHC for analyses using these techniques have been done so far, the upgraded trigger systems of ATLAS and CMS will allow this program to continue at the HL-LHC and likely include additional final states, thus significantly expanding the program of low-mass NP searches. The "triggerless" approach taken by the LHCb experiment for the HL-LHC upgrade would allow for similar analyses in the forward rapidity region to be conducted as well.

Another important new technique pioneered at the LHC was the extensive use of ISR not just for dark matter particle searches (as discussed in Sec. 2.1), but also to look for low-mass dijet resonances (including the Higgs boson decaying to a $b$ quark-antiquark pair) recoiling against an ISR jet or photon. The presence of a resonance is inferred from its merged decay products reconstructed as a single large-radius jet with a characteristic two-prong substructure. The jet substructure techniques help to reduce the large background from the QCD production of dijets. Using this approach allowed ATLAS and CMS to probe dijet resonances with masses as low as 10 GeV, and with couplings as low as 10% of the electromagnetic coupling,[15–17] exploring new territory which had not been accessible to $Sp\bar{p}S$, LEP, or the Tevatron experiments, as those techniques were either not available or not used.

The search for supersymmetry (SUSY), a theory which predicts a new particle (a sparticle) differing by half a unit of spin for every SM particle, has been one of the major axes of NP searches at the LHC. While the focus of the Run 1 LHC analyses was mainly to look for the pair production of strongly produced sparticles; the gluinos and squarks — the superpartners of the gluons and the quarks — the searches soon diversified to a broader spectrum of signatures. With the limits pushing these states to higher masses, it becomes more and more interesting to probe for the elusive direct production, through electroweak interactions, of charginos ($\tilde{\chi}_i^{\pm}$) and neutralinos ($\tilde{\chi}_j^0$), the mixed charged and neutral states formed by the superpartners of the electroweak and Higgs bosons, ordered in increasing mass for $i = 1, 2$ and $j = 1, 2, 3, 4$. (In many models, $\tilde{\chi}_1^0$ is stable and can be a dark matter candidate). Indeed, if the masses of the gluinos and squarks are beyond the 3–4 TeV range, charginos and neutralinos may dominate the SUSY production at the HL-LHC energy and, if they exist, could be the first sparticles to be discovered. Prospective studies were performed in various channels.

One such example is the search for the production of $\tilde{\chi}_1^{\pm}\tilde{\chi}_2^0$, where $\tilde{\chi}_1^{\pm} \to W^{\pm}\tilde{\chi}_1^0$ and $\tilde{\chi}_2^0 \to h\tilde{\chi}_1^0$. This analysis channel underlines the speed at which analyses can evolve at the LHC: the once elusive Higgs boson is used here, via its main decay channel ($h \to b\bar{b}$), not as the focus for Higgs-boson studies, but as a tool to search for new particles. As the lightest neutralino is stable and escapes detection, the final state is hence composed of $\not{p}_{\mathrm{T}}$, two b-tagged jets whose invariant mass is compatible with the SM-like Higgs boson, and the decay products of the $W$ boson. In ATLAS prospective studies,[1] the leptonic decay of the $W$ is considered, although the fully hadronic decay was later found, in the full Run 2 analysis, to yield even greater sensitivity at higher masses.[18] In the leptonic analysis, in order to better separate this signal from the SM backgrounds, which are dominated by $t\bar{t}$ events, a boosted-decision-tree (BDT) approach is used, taking in input seven kinematic variables such as the transverse mass of lepton and the $\not{p}_{\mathrm{T}}$ or the angular separation between the two b-jets. Three different BDTs are trained, each one targeting a different range of mass *compression*, that is of the mass difference between the produced $\tilde{\chi}_1^{\pm}/\tilde{\chi}_2^0$ and the final-state $\tilde{\chi}_1^0$. The expected reach at the HL-LHC is shown in Fig. 2: for a massless $\tilde{\chi}_1^0$, $\tilde{\chi}_1^{\pm}/\tilde{\chi}_2^0$ up to 1.08 TeV could be discovered. However, as mentioned above, this channel would not be expected to be the most sensitive, and this should hence be considered as a conservative reach, but it is still interesting to note that masses at the TeV scale will be within the reach of the HL-LHC even in some scenarios of electroweak production.

### 2.3. *Probing Rare b- and c-Hadron Decays*

The absence of tree-level Flavour Changing Neutral Current (FCNC) transitions is a feature that is highly specific to the SM. Consequently decays

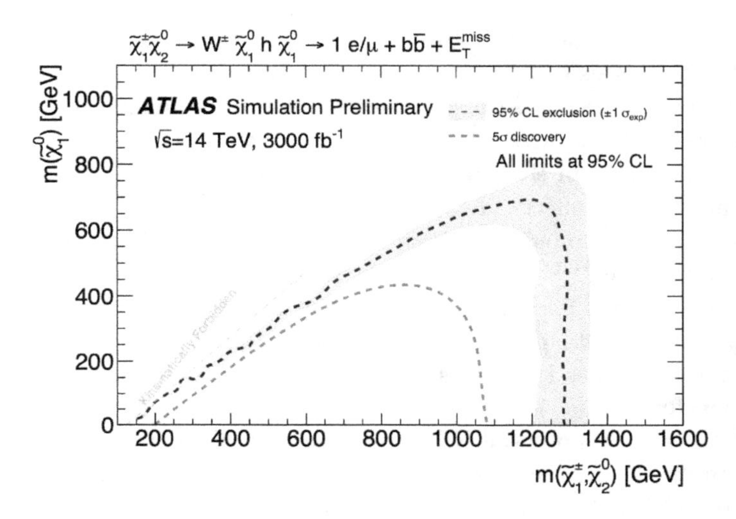


Fig. 2. Expected 95% CL exclusion and $5\sigma$ discovery potential of the $W(\rightarrow \ell\nu)h(\rightarrow b\bar{b})$ analysis in the $\tilde{\chi}_1^0$ versus $\tilde{\chi}_1^{\pm}/\tilde{\chi}_2^0$ mass plane. Figure taken from Ref. 1.

that can only proceed through FCNCs are sensitive NP probes. Since they are only allowed at the one-loop level in the SM, their studies are very well suited as a discovery mode, as any NP effect should be more pronounced.

The leptonic decays $B_s \rightarrow \mu^+\mu^-$ and $B_0 \rightarrow \mu^+\mu^-$ are helicity suppressed in the SM and, for the latter, further suppressed by the CKM matrix structure. In terms of the SM predictions, this results in tiny branching fractions with a very small relative uncertainty. The ratio between these two branching fractions is a powerful test for SUSY and for NP scenarios with minimal flavour violation. With the Upgrade II data set, LHCb will furthermore have the unique ability to measure the parameters $A^{\mu\mu}_{\Delta\Gamma}$ and $S_{\mu\mu}$ of the $B_s \rightarrow \mu^+\mu^-$ decay-time distribution. These are considered as "smoking gun" observables that, if different from their SM expectations of unity and zero respectively, would provide unambiguous evidence for NP.[19] The ratio of branching fractions is expected to be measured with an 11% precision and the $S_{\mu\mu}$ parameter with an uncertainty of $\sim 20\%$.[20]

An observation of the $B^0 \rightarrow \mu^+\mu^-$ decay and precision measurement of its branching fraction is one of the key goals of the HL-LHC in this sector. The 30% improvement in the dimuon mass resolution offered by the CMS Phase II tracker Upgrade would allow CMS to establish this rare process at a significance of $\sim 7$ standard deviations and measure its branching fraction to a precision of 16%.[1] Moreover, the effective lifetime of $B_s$ meson in this decay channel will be measured with a precision of 0.05 ps.[1]

Beyond these modes, the study of $b \to s\ell^+\ell^-$ transitions offers a unique opportunity to seek out signs of NP due to their rich phenomenology. The most powerful tests consist of the study of analyses of angular distributions and theoretically clean Lepton Flavour Universality (LFU) violation tests. Indeed, since LFU in the gauge couplings is exact in the SM, and corrections due to mass effects are calculable to good precision, it provides an excellent testing ground for NP theories where the LFU can be violated. The quantity $R_X$,[21,22] the ratio of $B \to X\mu^+\mu^-$ and $B \to Xe^+e^-$ decay rates for a specific hadronic system $X$ and in a defined range of $q^2$, has proven to be a particularly interesting test of LFU.[23–26] With a 300 $\text{fb}^{-1}$ data sample, further LFU tests will be accessible; comparing angular distributions for $b \to s\mu^+\mu^-$ and $b \to se^+e^-$ transitions will allow the axial or vector nature of a potential NP particle to be probed at.[27] The expected precision on the $R_K$ and $R_{K^*}$ ratios are 0.007 and 0.009, respectively, in the 1 to 6 $\text{GeV}^2/c^4$ dilepton invariant mass squared region.[20] The expected constraints obtained on the difference between the muon and electron contributions to the Wilson coefficients $C_9$ and $C_{10}$ from such prospective measurements, including angular observables sensitive to the differences between the $b \to se^+e^-$ and $b \to s\mu^+\mu^-$ couplings, are shown in Fig. 3. The precision achievable with the LHCb Upgrade II data set will not only allow NP contributions to be established with an overwhelming significance, but will also provide a characterisation of the NP that will be essential to distinguish between theoretical models.

Due to the respective size of the CKM matrix elements involved, $b \to d\ell^+\ell^-$ transitions are suppressed by a factor of about 20 compared to the corresponding $b \to s\ell^+\ell^-$ processes. The large data sample will therefore provide a unique possibility to study $b \to d\ell^+\ell^-$ transitions with more precision than the one currently obtained for $b \to s\ell^+\ell^-$. It will allow comparisons between transitions of $b$ quarks to the second ($s$) and first ($d$) quark generations and thus to discriminate between different models.

## 2.4. *Probing CP Violation in b and c Decays*

Searches for CP-violating NP require a precise determination of the SM benchmarks which are presented in Sec. 14. The results of the SM benchmark measurements allow prediction of the SM value for the CP-violating weak phase $\phi_s$. This parameter is a particularly sensitive probe for NP models as it is both extremely small and very precisely predicted in the SM. It is accessible though the study of a range of channels mediated by

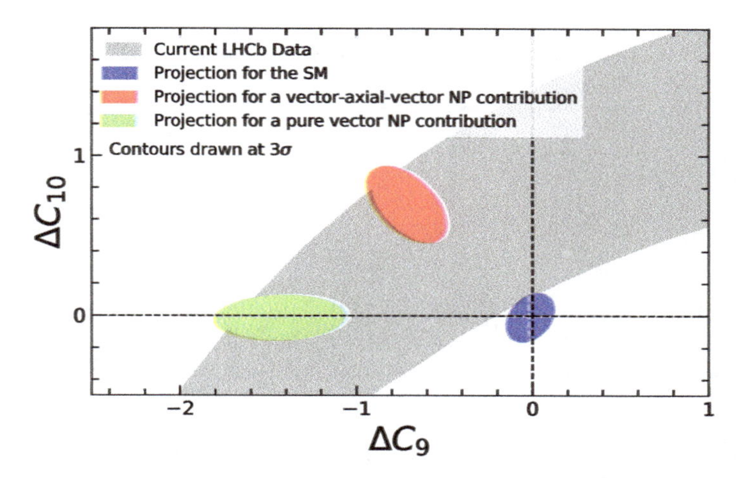


Fig. 3. Projected sensitivity to the difference between the muon and electron mode contributions to the vector, $C_9$, and axial-vector, $C_{10}$, Wilson coefficients for different scenarios: the SM (blue), NP with a pure vector contribution (green), or NP with both vector and axial-vector contributions (red). The blue, red and green filled regions show the $3\sigma$ uncertainty contours for each scenario with the LHCb Upgrade II data set. The grey region shows the current $3\sigma$ uncertainty. Figure taken from Ref. 20.

the $b \to c\bar{c}s$ transition but also from decays to final states that proceed through pure loop processes. Their comparison will allow to probe for the potential existence of NP contributing to the loop processes.

A further theoretically clean approach to search for CP-violating NP is through the parameters of $B^0_{(s)} - \overline{B}^0_{(s)}$ mixing. These are denoted $a^{d,s}_{\rm sl}$ since they are typically determined using semileptonic decays, and their tiny values are very precisely predicted in the SM[28] while being highly sensitive to NP that could enter the mixing loops. Copious signal yields will be available, and the main challenge will be to control potential systematic uncertainties due to production, detection and background asymmetries. These can, however, be determined from control samples. The expected precision compared with the SM predictions is shown in Table 1.

Charm hadrons provide a unique opportunity to measure CP violation associated with processes mediated by up-type quark transitions. These can be affected by NP contributions in fundamentally different ways to the down-type quarks that make up the kaon and beauty systems. Since the amount of CP violation is expected to be very small, $\mathcal{O}(10^{-4})$, significant deviations would be clear signs of NP. However, contrary to the $b$-hadron sector, SM predictions are more challenging and a full understanding

Table 1. Results of the current theoretical and expected experimental determination of the semileptonic asymmetries $a_{\mathrm{sl}}^{d}$ and $a_{\mathrm{sl}}^{s}$.[20,29]

| | Current theory | LHCb (300 $\mathrm{fb}^{-1}$) |
|---|---|---|
| $\delta a_{\mathrm{sl}}^{d}$ $[10^{-4}]$ | 0.6 | 2 |
| $\delta a_{\mathrm{sl}}^{s}$ $[10^{-4}]$ | 0.03 | 3 |

requires the study of a significant number of decay modes. LHCb Upgrade II will have the best sensitivity on the fundamental parameters of CP violation in charm mixing, which has not been observed yet.

### 2.5. *Benefiting from Detector Upgrades and Improved Analysis Techniques*

The detector upgrades (see Sec. 12) will also improve searches for NP. One sector which may particularly benefit from these upgrades are searches for long-lived particles (LLPs). The LLPs naturally occur in many NP theories, their detector-size lifetimes being due to heavy mediators, small couplings, or compressed mass spectra reducing the available decay phase space. The signatures for such particles involve decays in various parts of the detectors, away from the interaction point. This can lead to, e.g. displaced vertices (DVs) reconstructed in the inner detector. For ATLAS, this part of the detector will be fully replaced in view of the HL-LHC by a full-silicon tracker (ITk) which will significantly impact the DV-based LLP searches. Not only will the ITk cover a larger range in pseudo-rapidity, extending the tracking capabilities from $|\eta| < 2.5$ to $|\eta| < 4.0$, but the tracking reconstruction performances for decays happening at large radii will also be improved, as shown in Fig. 4(upper). In this figure, a SUSY model with long-lived gluinos is considered; these gluinos hadronise into so-called $R$-hadrons, which eventually decay into a pair of SM quarks and a stable $\tilde{\chi}_1^0$ (masses of 2 TeV and 100 GeV are considered here for the $R$-hadron and the neutralino respectively). In the current detector, the tracking efficiency for charged particles with $p_T > 1$ GeV worsens drastically beyond a radius of 300 mm, which corresponds to the first layer of the SCT, as a certain number of hits are required in this subdetector to insure the track quality. With the ITk, the larger spacing between the layers provides a high efficiency up to 400 mm instead, with some tracks fulfilling the requirements even at the $R$-hadron decay radii of 500–600 mm.

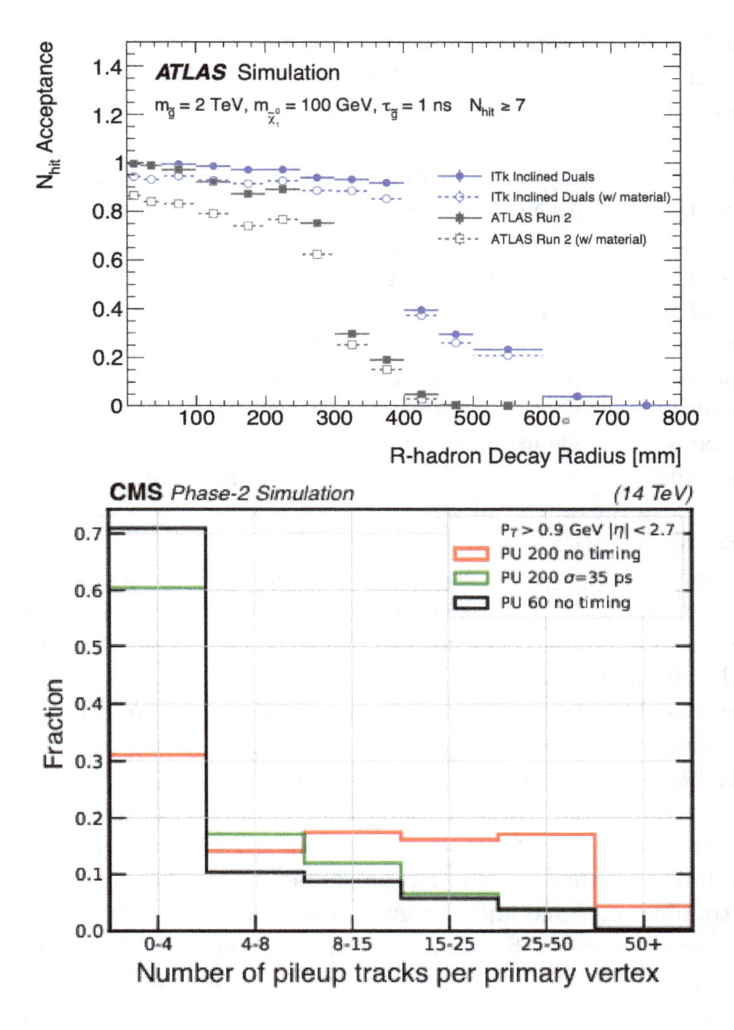


Fig. 4. (Upper) Probability of a charged particle to be reconstructed as a good-quality track by ATLAS as a function of its radius of origin for the Run 2 and HL-LHC detector configurations (considering or not the effect of material producing hadronic interactions). See the text for details. Figure taken from Ref. 1. (Lower) Distribution of the number of incorrectly associated tracks with or without the use of a $3\sigma$ (where $\sigma = 35$ ps) selection on timing information. The vertical axis is the fraction of primary vertices, which have the number of pile-up tracks shown on the horizontal axis associated with them. Figure taken from Ref. 30.

Another important tool to improve the sensitivity of searches is to use precision timing to help separating particles coming from pile-up interactions from a signal. In the case of CMS, this will be achieved via a minimum ionizing particle timing detector (MTD) with the time resolution of $\sim 30$ ps per track. The precision timing allows separation of the pile-up vertices from the signal hard-scattering vertex not only in space, but also in time, thus allowing for a four-dimensional vertex reconstruction and separation. Figure 4(lower) shows the fraction of primary vertices which have a certain number of tracks from pile-up interactions incorrectly associated with them, with and without the MTD. As can be seen from this plot, at a pile-up of 200, on average there are about 15 tracks incorrectly associated with the primary vertex if no timing information is used. Utilizing this information, the average number of such tracks drops to about 5, which is effectively equivalent to the reduction of pile-up from 200 to 60, similar to the number of pile-up interactions at the beginning of the LHC fills in Run 2. Additionally, the MTD can be used to improve the sensitivity to LLPs by using the time-of-flight information to detect tracks left by massive particles moving with velocities $\beta < 1$.

On the LHCb side, the upgraded detector will be also capable of providing fast-timing information with resolution of few tens of ps per particle. This is an essential attribute for dealing with the very challenging experimental conditions of HL-LHC in the forward direction at very high pile-up.

Novel analyses techniques are also expected to improve the coverage of the possible NP parameter space. While machine-learning techniques have already been used for quite some time in particle physics, from BDTs to deep neural networks, recently there has been a rise in the frequency and breadth of their use.[31] Some machine-learning techniques can target specific well-defined theories better than simple "cut-and-count" analyses by using training on signal and background samples, especially when the signal extraction from the background relies on minute details in a variety of kinematic distributions. However, the results of these analyses can highly depend on the model under study. An exciting recent area is the search for anomalies in the data themselves, focusing on a search for objects, which would differ significantly from the data-based training sample, without requiring a prior knowledge of the signal characteristics, as, for example, has been done recently in an ATLAS analysis.[32] These developments are gaining momentum and they will very likely become one of the standard NP search tools by the time of the HL-LHC.

## 3. Conclusions

In this chapter, we summarised just a handful of projections made in the most active areas of the LHC NP search programme to date, which are inspired either by well-motivated theoretical models or by the existing hints of various anomalies that survived more than a decade of the LHC data. There are many other avenues of NP searches which will be pursued at the HL-LHC, including low-mass NP searches and searches for long-lived particles, where only a few precise projections have been made so far.

One should also keep in mind that the assumptions made in deriving these projections are typically very conservative, as it is very hard to quantify the rapid progress the LHC physics analyses have been continuously achieving. Some of the earlier HL-LHC projections have already been realised or nearly achieved with the existing data sets, thanks to much more elaborated analysis methods, powerful machine-learning techniques used for both particle identification and for refining the analysis strategy, more precise theoretical calculations, and the ingenuity of the LHC physicists tirelessly working on pushing these analyses to the best sensitivity possible. We believe that the enormous progress seen in the last decade will continue in the next one, making the HL-LHC an ultimate place to witness another major discovery, which may change the very way we think about modern physics, bringing us another step closer to solving the remaining mysteries of our universe.

## References

1. X. Cid Vidal et al., Report from Working Group 3: Beyond the Standard Model physics at the HL-LHC and HE-LHC, *CERN Yellow Rep. Monogr.* **7**, 585–865 (2019). doi: 10.23731/CYRM-2019-007.585.
2. ATLAS and CMS Collaborations. Snowmass White Paper Contribution: Physics with the Phase-2 ATLAS and CMS Detectors. URL `https://cds.cern.ch/record/2805993/files/ATL-PHYS-PUB-2022-018.pdf`. ATL-PHYS-PUB-2022-018, CMS PAS FTR-22-001 (2022).
3. ATLAS Collaboration, Search for high-mass dilepton resonances using 139 $\text{fb}^{-1}$ of $pp$ collision data collected at $\sqrt{s} = 13$ TeV with the ATLAS detector, *Phys. Lett. B.* **796**, 68–87 (2019). doi: 10.1016/j.physletb.2019.07.016.
4. CMS Collaboration, Search for resonant and nonresonant new phenomena in high-mass dilepton final states at $\sqrt{s} = 13$ TeV, *JHEP.* **07**, 208 (2021). doi: 10.1007/JHEP07(2021)208.
5. M. Bauer, U. Haisch, and F. Kahlhoefer, Simplified dark matter models with

two Higgs doublets: I. Pseudoscalar mediators, *JHEP.* **05**, 138 (2017). doi: 10.1007/JHEP05(2017)138.

6. CMS Collaboration, Search for new physics in events with a leptonically decaying Z boson and a large transverse momentum imbalance in proton–proton collisions at $\sqrt{s} = 13$ TeV, *Eur. Phys. J. C.* **78** (4), 291 (2018). doi: 10.1140/epjc/s10052-018-5740-1.
7. CMS Collaboration, Search for third-generation scalar leptoquarks decaying to a top quark and a $\tau$ lepton at $\sqrt{s} = 13$ TeV, *Eur. Phys. J. C.* **78**, 707 (2018). doi: 10.1140/epjc/s10052-018-6143-z.
8. CMS Collaboration, Search for heavy neutrinos and third-generation leptoquarks in hadronic states of two $\tau$ leptons and two jets in proton-proton collisions at $\sqrt{s} = 13$ TeV, *JHEP.* **03**, 170 (2019). doi: 10.1007/JHEP03(2019) 170.
9. D. Buttazzo, A. Greljo, G. Isidori, and D. Marzocca, B-physics anomalies: a guide to combined explanations, *JHEP.* **11**, 044 (2017). doi: 10.1007/ JHEP11(2017)044.
10. CMS Collaboration, Search for narrow resonances in dijet final states at $\sqrt{s} = 8$ TeV with the novel CMS technique of data scouting, *Phys. Rev. Lett.* **117** (3), 031802 (2016). doi: 10.1103/PhysRevLett.117.031802.
11. ATLAS Collaboration, Search for low-mass dijet resonances using trigger-level jets with the ATLAS detector in *pp* collisions at $\sqrt{s} = 13$ TeV, *Phys. Rev. Lett.* **121** (8), 081801 (2018). doi: 10.1103/PhysRevLett.121.081801.
12. CMS Collaboration, Search for narrow and broad dijet resonances in proton-proton collisions at $\sqrt{s} = 13$ TeV and constraints on dark matter mediators and other new particles, *JHEP.* **08**, 130 (2018). doi: 10.1007/JHEP08(2018) 130.
13. CMS Collaboration, Search for dijet resonances using events with three jets in proton-proton collisions at s = 13 TeV, *Phys. Lett. B.* **805**, 135448 (2020). doi: 10.1016/j.physletb.2020.135448.
14. CMS Collaboration, Search for a Narrow Resonance Lighter than 200 GeV Decaying to a Pair of Muons in Proton-Proton Collisions at $\sqrt{s} =$ TeV, *Phys. Rev. Lett.* **124** (13), 131802 (2020). doi: 10.1103/PhysRevLett.124.131802.
15. ATLAS Collaboration, Search for low-mass resonances decaying into two jets and produced in association with a photon using *pp* collisions at $\sqrt{s} =$ 13 TeV with the ATLAS detector, *Phys. Lett. B.* **795**, 56–75 (2019). doi: 10.1016/j.physletb.2019.03.067.
16. CMS Collaboration, Search for Low-Mass Quark-Antiquark Resonances Produced in Association with a Photon at $\sqrt{s} = 13$ TeV, *Phys. Rev. Lett.* **123** (23), 231803 (2019). doi: 10.1103/PhysRevLett.123.231803.
17. CMS Collaboration, Search for low mass vector resonances decaying into quark-antiquark pairs in proton-proton collisions at $\sqrt{s} = 13$ TeV, *Phys. Rev. D.* **100** (11), 112007 (2019). doi: 10.1103/PhysRevD.100.112007.
18. ATLAS Collaboration, Search for charginos and neutralinos in final states with two boosted hadronically decaying bosons and missing transverse momentum in *pp* collisions at $\sqrt{s} = 13$ TeV with the ATLAS detector, *Phys. Rev. D.* **104** (11), 112010 (2021). doi: 10.1103/PhysRevD.104.112010.

19. K. De Bruyn, R. Fleischer, R. Knegjens, P. Koppenburg, M. Merk, A. Pellegrino, and N. Tuning, Probing new physics via the $B_s^0 \to \mu^+\mu^-$ effective lifetime, *Phys. Rev. Lett.* **109**, 041801 (2012). doi: 10.1103/PhysRevLett.109.041801.
20. LHCb Collaboration. Framework TDR for the LHCb Upgrade II Opportunities in flavour physics, and beyond, in the HL-LHC era. Technical report, CERN, Geneva (Jul, 2021). URL `http://cds.cern.ch/record/2776420`.
21. G. Hiller and F. Krüger, More model-independent analysis of $b \to s$ processes, *Phys. Rev.* **D69**, 074020 (2004). doi: 10.1103/PhysRevD.69.074020.
22. Y. Wang and D. Atwood, Rate difference between $b \to s\mu^+\mu^-$ and $b \to se^+e^-$ in supersymmetry with large tan $\beta$, *Phys. Rev.* **D68**, 094016 (2003). doi: 10.1103/PhysRevD.68.094016.
23. LHCb Collaboration, Test of lepton universaility in $b \to s\ell^+\ell^-$ decays, `https://arxiv.org/abs/2212.09152`. Accepted for publication in *Phys. Rev. Lett.*
24. LHCb Collaboration, Test of lepton universality using $\Lambda_b \to pK^-\ell^+\ell^-$ decays, *JHEP.* **05**, 040 (2020). doi: 10.1007/JHEP05(2020)040.
25. LHCb Collaboration, Test of lepton universality in beauty-quark decays. *Nat. Phys.* **18**, 277–282 (2022).
26. LHCb Collaboration, Tests of lepton universality using $B^0 \to K_S^0\ell^+\ell^-$ and $B^+ \to K^{*+}\ell^+\ell^-$ decays (2021). *Phys. Rev. Lett.* **128** (2022) 191802.
27. B. Capdevila, Assessing lepton-flavour non-universality from $b \to k^*\ell^+\ell^-$ angular analyses, *Journal of Physics: Conference Series.* **873**, 012039 (Jul, 2017). ISSN 1742-6596. doi: 10.1088/1742-6596/873/1/012039. URL `http://dx.doi.org/10.1088/1742-6596/873/1/012039`.
28. M. Artuso, G. Borissov, and A. Lenz, CP violation in the $B_s^0$ system, *Rev. Mod. Phys.* **88**, 045002 (2016). doi: 10.1103/RevModPhys.88.045002.
29. LHCb Collaboration, Physics case for an LHCb Upgrade II — Opportunities in flavour physics, and beyond, in the HL-LHC era. (CERN-LHCC-2018-027 LHCb-PUB-2018-009) (2018).
30. CMS Collaboration. A MIP Timing Detector for the CMS Phase-2 Upgrade. URL `https://cds.cern.ch/record/2667167/files/CMS-TDR-020.pdf`. CERN-LHCC-2019-003 (2019).
31. G. Karagiorgi, G. Kasieczka, S. Kravitz et al., Machine learning in the search for new fundamental physics. *Nat. Rev. Phys.* **4**, 399–412 (2022).
32. ATLAS Collaboration, Dijet resonance search with weak supervision using $\sqrt{s} = 13$ TeV *pp* collisions in the ATLAS detector, *Phys. Rev. Lett.* **125** (13), 131801 (2020). doi: 10.1103/PhysRevLett.125.131801.

© 2024 The Author(s)
https://doi.org/10.1142/9789811280184_0017

# Chapter 17

# Precision SM Physics

Jan Kretzschmar*, Alexander Savin† and Mika Vesterinen‡
*University of Liverpool, U.K.
†University of Wisconsin-Madison, U.S.A.
‡University of Warwick, U.K.

## 1. Introduction

As the successor of the LEP $e^+e^-$ collider at CERN, the LHC was originally conceived mainly as a discovery machine, not as much to perform *precision Standard Model measurements.* However, this notion has been challenged throughout Run 1 and Run 2, not just by the *forward precision* experiment LHCb, but also by a variety of percent- and permille-level measurements by ATLAS and CMS. Even though large data sets will be available at the end of Run 3 in 2025, the High-Luminosity phase of the LHC (HL-LHC) from 2029 is expected to bring further significant improvements to these measurements. Prospects for this program are discussed in detail in the "Yellow Report on the Physics at the HL-LHC, and Perspectives for the HE-LHC"[1] and in the more recent "Snowmass White Paper Contribution: Physics with the Phase-2 ATLAS and CMS Detectors"[2] and "Future physics potential of LHCb"[3] and we will give a synthesis of this in the following sections.

The obvious implication of the HL-LHC operation will be the larger data set available. The total integrated luminosity is expected to exceed what will be available by the end of Run 3 by a factor 10, reaching a total of 3–4 $\text{ab}^{-1}$ for the ATLAS and CMS experiments and at least 300 $\text{fb}^{-1}$ for LHCb. This will directly lead to improvements for measurements that are statistically limited.

This is an open access article published by World Scientific Publishing Company. It is distributed under the terms of the Creative Commons Attribution 4.0 (CC BY) License.

A huge program of upgrades to the detectors and their data acquisition systems is required to fully exploit the increased luminosity delivered by the HL-LHC machine as is explained in this book. The high-intensity operation will come with a significant increase of the number of simultaneous *pp* interactions per bunch crossing (generally referred to as pile-up) up to 200. All detector systems will have to be more granular and precise, additional features such as timing measurements will be implemented, and the trigger systems need to be more capable to select the interesting collision events. Key examples of these upgrades are the inner tracking systems of ATLAS and CMS that will have higher performance and extend into the forward region $|\eta| < 4$ (as opposed to $|\eta| < 2.5$ now) and a more efficient full-software trigger for the first upgrade of LHCb. After further developments in the reconstruction algorithms, the significant increase in data volumes will also allow for a better understanding of experimental systematic uncertainties. To avoid theoretical uncertainties becoming dominant effects in measurements, the tools, calculations and supporting measurements will need continuous effort.

In the following sections we will first describe our expectations for the measurements of global SM parameters, then discuss the electroweak production of multiboson final states and precise differential cross section measurements with top quarks, jets and photons, before closing with some examples from quark flavour physics.

## 2. The SM parameters and global Electroweak fit

Even though the number of free parameters is often brought up as a limitation, the SM is a highly predictive theory and relations between parameters such as the masses and couplings are well known. The discovery of the Higgs boson and its precise mass measurement delivered the final input needed to overconstrain the global Electroweak fit. Exploiting the SM relations, parameters can be determined indirectly and compared to direct measurements. Any deviations observed in these comparisons will indicate the presence of new physics.

While not (yet) providing a similar diversity of EW precision observables as LEP, the masses of the $W$ boson, $m_W$, and the top quark, $m_t$, as well as the effective mixing parameter $\sin^2\theta_{\mathrm{eff}}^{\mathrm{lept}}$ are those where the LHC already surpasses $e^+e^-$ results or can be expected to do so. An update of the global fit of the EW precision observables was performed in Ref. 1 with the HEPFit package. The experimental constraints using the current and the expected

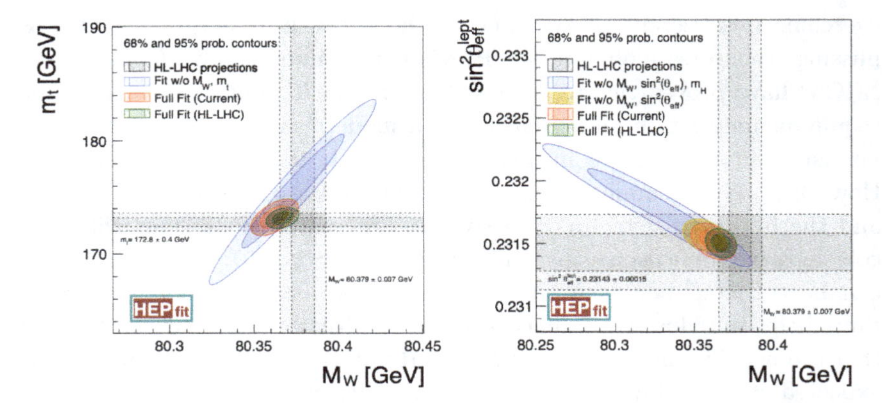


Fig. 1. Comparison of the indirect constraints on $M_W$ and $m_t$ with the current experimental measurements and the expected improvements at the HL-LHC (left). The same in the $M_W$ and $\sin^2 \theta_{\text{eff}}^{\text{lept}}$ plane (right).[1]

future HL-LHC data are shown in Fig. 1 for the relation between $m_W$ and $m_t$ (left) as well as $m_W$ and $\sin^2 \theta_{\text{eff}}^{\text{lept}}$ (right). The expected precision of the HL-LHC measurements will improve the constraints significantly and would potentially increase existing tensions between indirect and direct measurements. As also highlighted by the recent $m_W$ measurement by CDF,[4] these measurements are able to challenge the SM and are among the main goals of the HL-LHC scientific program.

The measurement of the $W$-boson mass at the LHC is performed via leptonic decays to electrons or muons and the major challenge is to

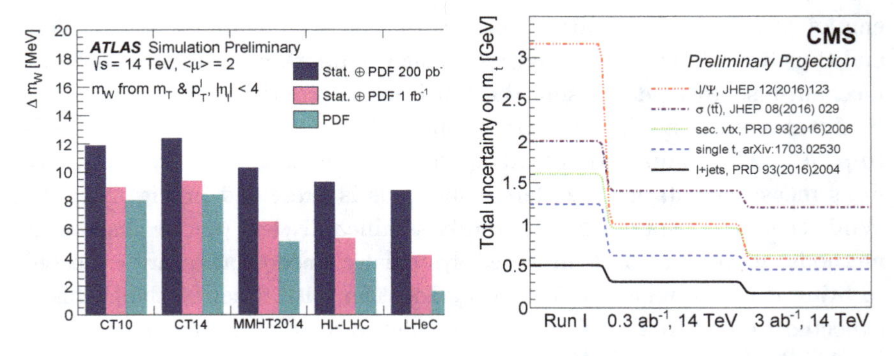


Fig. 2. Left. PDF uncertainty on the $W$-boson mass $m_W$ measurement using different PDF sets.[5] Right. Total uncertainty on top quark mass ($m_t$) obtained with different measurement methods and their projections to the HL-LHC.[6]

overcome missing information due to the decay neutrino only visible as missing transverse momentum.[5] Previous measurements by ATLAS[7] and LHCb[8] have proved this to be possible. In fact it would appear that the requirements for this measurement — a modest sample collected at low-intensity running — are in disagreement with the HL-LHC conditions. However, the measurement can exploit the increased detector acceptance and the improved detectors. In special, low pile-up conditions with two collisions per bunch crossing on average, about $2 \times 10^6$ $W$ boson events can be collected per week of operation. The extended detector acceptance helps to reduce the impact from parton distribution functions (PDFs) that are usually among the largest uncertainties in the measurement. The expected statistical and PDF uncertainties are summarised in Fig. 2(left) for different PDF sets. While the CT10 and CT14 PDF sets are found to give about 8 MeV uncertainty, the MMHT2014 set gives about 5 MeV. All three projected HL-LHC PDF sets with anticipated additional LHC data constraints[9] give uncertainties that are slightly lower than 4 MeV. A qualitatively different level would be reached by incorporating additional DIS data from the LHeC[10] that would halve the PDF uncertainty to about 2 MeV. Depending on the available PDF knowledge, one can expect to reach 7 MeV uncertainty or lower at the HL-LHC.

The most precise top-quark mass measurements at the LHC stem from the so-called "direct measurements" that reconstruct information from the top-quark decay products. The typical uncertainty with these methods is of the order of 500 MeV, with a recent CMS measurement reaching 380 MeV.[11] These uncertainties are dominated by the theoretical modelling and, specifically, non-perturbative QCD effects. It is thus of interest to employ a variety of alternative methods that do not rely on jet observables[6] and thus have different sensitivities to the top quark production and decay mechanisms. Instead of using the full b-jet information, one may choose final states where a b-hadron has fragmented into a $J/\psi$ meson that decays to $\mu^+\mu^-$. The comparison of extrapolated uncertainties on the top quark mass measurements using different methods is presented in Fig. 2(right). While the "$J/\psi$" approach is currently significantly less precise than other methods, more data and improved systematic uncertainties are expected to bring it to the region of 500 MeV. Together with other methods this is expected to lead to a measurement of the top mass with an accuracy of a few hundred MeV at the HL-LHC.

The presence of both vector and axial-vector couplings of electroweak bosons to fermions leads to a forward-backward asymmetry $A_{FB}$ in the production of Drell–Yan lepton pairs that can be used to extract the effective weak mixing angle that is (at tree level) directly related to the ratio of the $W$ and $Z$ boson masses. First competitive measurements by ATLAS, CMS and LHCb[12–14] have been performed with Run 1 data. As this measurement is purely based on selecting a di-electron or di-muon pair, it will directly benefit from the large HL-LHC dataset, as well as the extended rapidity coverage of the central detectors.[15,16] Figure 3(left) shows an example of the $A_{FB}$ distributions for CMS in bins of dimuon mass and rapidity for different energies and pseudorapidity acceptances. In symmetric $pp$ collisions, $A_{FB}$ is generated from the valence quark contributions. Extending the pseudorapidity acceptance increases the coverage for larger parton $x$-values in the production and reduces both the statistical and PDF uncertainties. As shown in Fig. 3(right), the PDF uncertainty restricts the precision of the measurement already for less than 100 fb$^{-1}$. Exploiting the PDF-dependence of the $A_{FB}$ with dilepton rapidity and mass, one may constrain this PDF uncertainty, which hence decreases depending on the amount of data. One can expect a total uncertainty of about $12 \cdot 10^{-5}$, about a factor of two better than the current hadron collider combination. Similar conclusions have been drawn for a future measurement with LHCb.[17] Additional constraints from LHeC data have the potential to reduce this dominating PDF uncertainty by up to a factor of five.

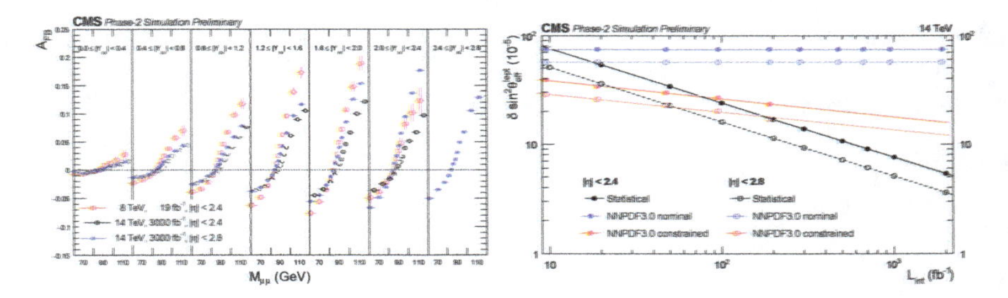


Fig. 3. Left. Forward-backward asymmetry distribution, $A_{FB}$, in dimuon events. Right. Projected statistical, nominal PDF and constrained PDF uncertainties for $|\eta| < 2.4$ and 2.8 acceptance selections for the muons.[15]

## 3. Multiboson production

The production of pairs of heavy bosons $V = W$ or $Z$ has been an important topic of study to establish the presence of triple-gauge interactions as predicted by the SM. The frontier has now moved to the topic of purely electroweak diboson production $VVjj$. This process class was proposed a long time ago[18,19] as being sensitive to the nature of mass generation as well as quartic-gauge interactions. It is characterised by a very low cross section and will only be conclusively investigated with data samples available at the HL-LHC. The total EW $VVjj$ production may proceed in different polarization states where each boson can be longitudinally (L) or transversely (T) polarized, leading to a total of three possibilities: LL, TL, TT. The LL component directly probes the unitarization mechanism of the vector boson scattering amplitude production through the Higgs boson.

Establishing the EW $VVjj$ production at a hadron collider over the "strong production" mechanism is already non-trivial as it requires the presence of two vector bosons in the central part of the detector and two jets separated by a large rapidity gap with reduced hadronic activity. The observation was already reported using LHC Run 2 data, starting with the golden same-charge $W^{\pm}W^{\pm}$ channel and later, also for the channels with more background, $WZ$ and $ZZ$. With the increased luminosity of the HL-LHC one can expect to measure the EW $VVjj$ cross sections with a few percent uncertainty as shown in Fig. 4(left).[20] For the $W^{\pm}W^{\pm}$ channel the uncertainty at 3000 fb$^{-1}$ is expected to reach 5%, while for $WZ$ and $ZZ$ 10% may be reached.

Figure 4(right) demonstrates the expected significance of the LL measurement as a function of integrated luminosity.[20] When using the $WW$ rest frame, the sensitivity of the measurement is expected to reach $> 5\sigma$ beyond 4000 fb$^{-1}$ such that a solid observation is expected when CMS and ATLAS will combine their results. The sensitivity of the $ZZ$ LL

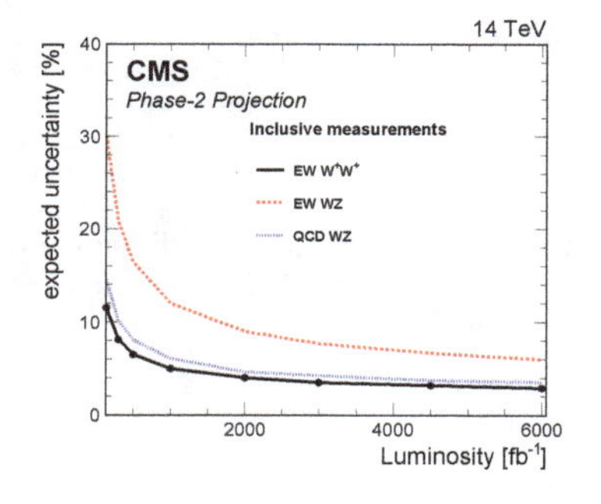


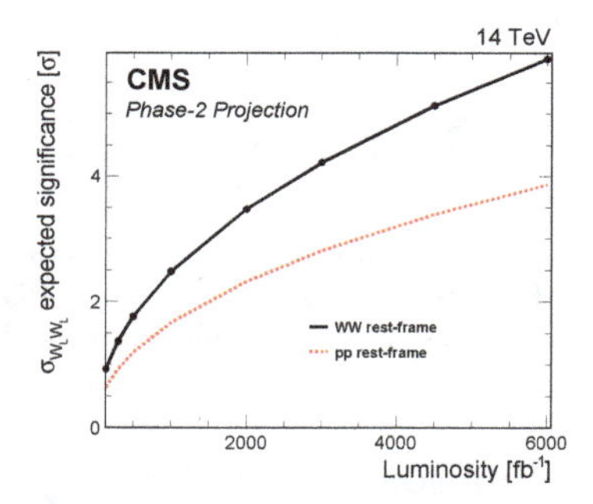


Fig. 4. Left. Projected estimated uncertainty in the EW WW, EW WZ, and QCD WZ cross section measurements as a function of the integrated luminosity. Right. Projected estimated significance for the EW $W_L W_L$ process as a function of the integrated luminosity for the WW and parton-parton center-of-mass reference frames.[20]

measurement will be much lower. The study of triboson final states such as $WWW$, $WWZ$, $WZZ$ has also just barely passed the observation threshold with Run 2 data and will be only fully explored at the HL-LHC.[21]

## 4. Differential cross sections measurements

In this section, we discuss a few examples of differential measurements that will be significantly improved in the HL-LHC phase because the data will allow an increase in precision, a reduction of the bin sizes in the *bulk phase space*, and the exploration of new phase space at higher energy or momentum.

### 4.1. *Measurements with top-quark pairs*

The differential $t\bar{t}$ cross-section measurements will improve at the HL-LHC because the enormous amount of data is expected to reduce dominant uncertainties related to jets. The extended $\eta$-coverage will allow fine-binned measurements at high rapidity.[22] The HL-LHC data recorded by LHCb will permit high precision measurements of asymmetries in $t\bar{t}$ production at large rapidities.

Double-differential cross-sections will be used to constrain PDF. As shown in Fig. 5(left), the uncertainties of the medium and high $x$ gluon distribution are expected to reduce drastically when new $t\bar{t}$ data are added to the current NNPDF3.1 fit. The improvement reaches up to a factor of 10 at $x = 0.5$, the edge of the kinematic reach. Nevertheless, improving PDF uncertainties with $pp$ data will remain difficult even at the HL-LHC. As hinted before, this can be resolved through high-luminosity $ep$ data delivered concurrently by the LHeC as discussed in this book in Chapter 20, "Resolving the Dynamics of Partons in Protons and Nuclei".

Another example can be found by studying the kinematic properties of a top-quark pair production in association with a photon ($t\bar{t}\gamma$) that probes the electroweak $t\gamma$ coupling, provides important constrains for effective field theory and information about $t\bar{t}$ spin correlation and production charge asymmetry.[23] Deviations in the transverse momentum spectrum of the photon from the SM prediction could point to new physics through anomalous dipole moments of the top quark. Figure 5(right) illustrates how such analyses will evolve at the HL-LHC: the statistical uncertainties in all bins of the differential distribution decrease significantly and additional bins at high $p_T$ can be added, compared to published analyses with partial Run 2 data. Also, the precise study of rare top-quark processes, such as four-top-quark production, is expected to become feasible at the HL-LHC.[24,25]

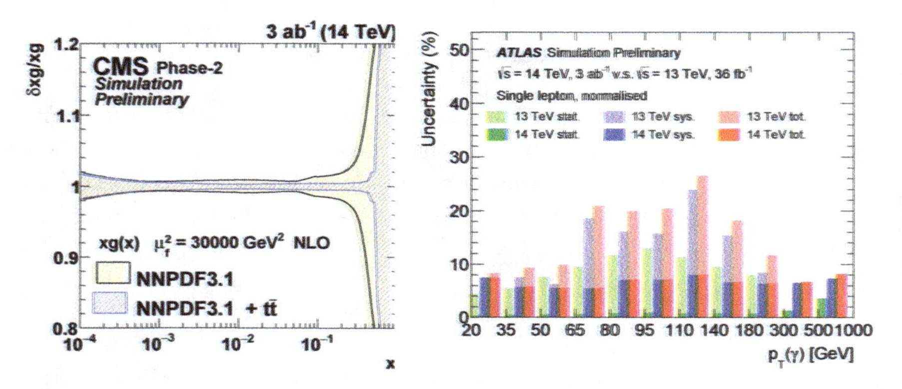


Fig. 5. Left. The relative gluon uncertainties of the original and profiled NNPDF3.1 PDF set after adding the $t\bar{t}$ information.[22] Right. Comparison of statistical/systematic/total uncertainties for the normalised differential cross-sections as a function of the photon $p_T$ in the single-lepton channel of the $t\bar{t}\gamma$ analysis.[23]

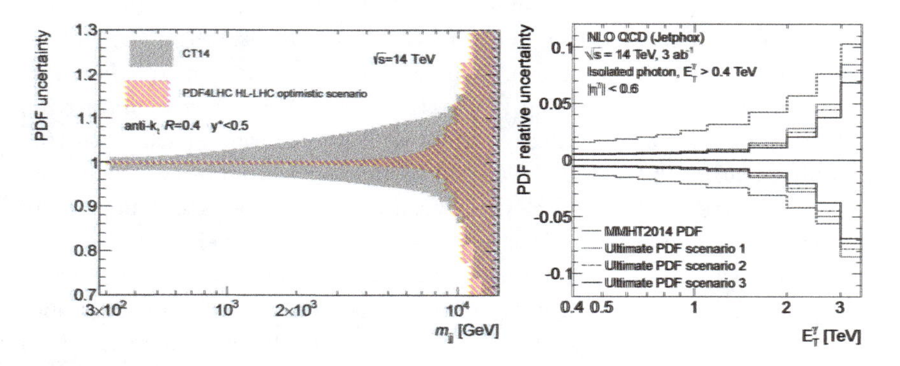


Fig. 6. Left. Comparison of the PDF uncertainty in the dijet cross sections calculated using the CT14 PDF and PDF4LHC HL-LHC sets. Right. Relative uncertainty in the predicted number of inclusive isolated photon events due to the uncertainties in the PDFs as a function of $E_T^{\gamma}$. The relative uncertainty due to the PDFs is shown for different PDF sets: the MMHT2014 PDF set (dashed lines) as well as the Ultimate PDF set in different HL-LHC scenarios.[26]

### 4.2. *Inclusive jet and photon measurements*

Inclusive jet production at a hadron collider is the QCD process that probes the highest accessible scales and has been used to constrain PDFs at highest $x$ and extract the running of the strong coupling at high scales. At the HL-LHC, the accessible dijet mass range will reach up to 10 TeV, close to the kinematic limit and the highest ever reached value at colliders.[26,27] A similar case can be made for inclusive photon production where small measurement uncertainties will be reached up to many TeV. Figure 6 demonstrates how the PDF uncertainties may improve after including dijet and inclusive photon measurements into PDF fits when assuming further improvements in theory uncertainties.

## 5. Quark flavour physics

In the SM there are three families of fermions, distinguished by their flavour. The fermion mass hierarchy is governed by Yukawa interactions with the Higgs field. The mis-alignment between the mass and weak-interaction eigenstates of the quarks is characterised by the Cabibbo-Kobayashi-Maskawa (CKM) matrix, with its four degrees of freedom corresponding to three angles and one phase, which is the only source of $CP$-violation in the SM. The six quark masses and four mixing parameters are free parameters of the SM and an open question is: Whether there is a deeper explanation for the hierarchical pattern in their values. The study of quark flavour changing transitions may reveal amplitudes involving new fields with a different flavour patterns.

While previously the field of quark flavour physics was determined almost entirely by $e^+e^-$ collision data collected at the $\Upsilon(4S)$ resonance, the first decade of the LHC has changed the landscape. The cross-section for beauty hadron production at the LHC is roughly five orders of magnitude larger at about 1 mb. Furthermore, beauty hadrons are produced inclusively, meaning all conceivable meson, baryon, or exotic bound states are available. Charmed hadrons are produced at an even higher rate. The LHCb experiment has demonstrated emphatically that high precision studies of beauty and charm hadrons is possible at the LHC with its unique instrumentation in the forward pseudorapidity range of $2 < \eta < 5$. The ATLAS and CMS experiments have contributed studies of beauty hadron decays particularly with decays into dimuon final states that are more easily triggered.

### 5.1. *The unitarity triangle*

The search for BSM physics in quark flavour requires the precise determination of the free CKM parameters of the SM using processes unlikely to be influenced by BSM physics, because they are dominated by tree-level amplitudes in the SM. Deviations are then searched for in processes that occur at loop-level in the SM. Six of the unitarity conditions of the CKM matrix can be represented by triangles. The triangle representing the condition $V_{ud}V_{ub}^* + V_{cd}V_{cb}^* + V_{td}V_{tb}^* = 0$ is usually referred to as *the* unitary triangle (UT). Figure 7 shows the projected constraints on the UT with data from the LHCb Upgrade II. The length of the left hand side of the triangle is determined from tree-level semileptonic decays of beauty hadrons via $b \to c$ and $b \to u$ transitions. LHCb can contribute, in particular, with decays of a wide range of beauty hadrons, having already demonstrated first observations of $V_{ub}$ decays with $B_s$ mesons[28] and $\Lambda_b$ baryons.[29] LHCb Upgrade II will allow unprecedented precision on these rare decays and further permit studies of similar decays of $B_c$ mesons, which are currently beyond reach.

The slope of the left hand side of the triangle corresponds to the phase $\gamma = \arg\left[\frac{V_{ud}V_{ub}^*}{V_{cd}V_{cb}^*}\right]$, which can be determined via the family of $B \to DK$ decays. These decays are mediated via tree-level $b \to u$ and $b \to c$ transitions and the interference between the corresponding amplitudes causes $CP$ asymmetries that depend on $\gamma$. The $CP$ asymmetries also depend on the relative magnitudes and phases of the amplitudes but these can be simultaneously determined from the data. A recent combination of measurements of $CP$ observables in such decays from LHCb resulted in a determination of $\gamma = (63.8^{+3.5}_{-3.7})^\circ$.[30] These measurements are extremely clean since most experimental systematic uncertainties cancel very effectively. However, these decays are to fully hadronic final states, for which the trigger efficiency is a limitation of the current experiment. With the full LHCb Upgrade II dataset a determination of $\gamma$ will be made possible, with a precision of around $0.3^\circ$. With the apex of the UT fixed by measurements of $|V_{ub}|/|V_{cb}|$ and $\gamma$ via tree-level decays, the slope and length of the side proportional to $V_{td}V_{tb}^*$ can be determined with loop processes that are extremely sensitive to BSM physics.

### 5.2. *Exotic hadrons*

The first decade of the LHC has already rewritten the textbooks on bound quark states, following a proliferation of states that are indicative of $Q\bar{Q}q\bar{q}$

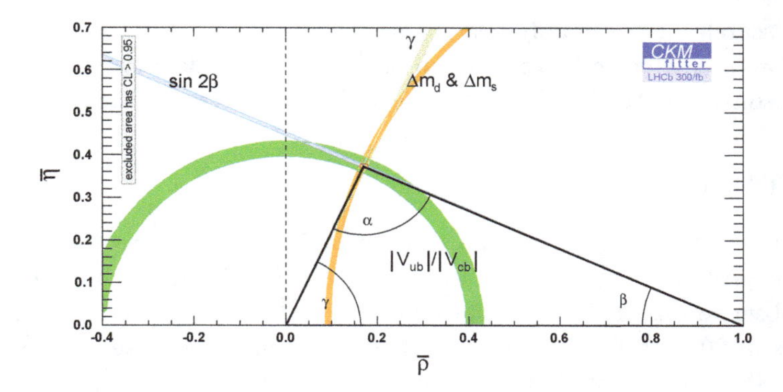


Fig. 7. A projection of the constraints on the Unitarity Triangle with 300 $\mathrm{fb}^{-1}$ of data from the LHCb Upgrade II.[31]

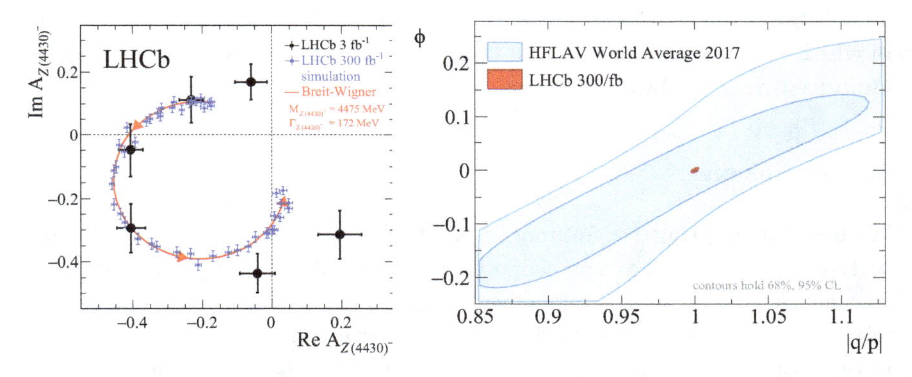


Fig. 8. Left: Argand diagram of the $Z(4430)^-$ amplitude ($\mathrm{A}_{Z(4430)^-}$) in bins of $m^2_{\psi(2S)\pi^-}$ from a fit to the $B^0 \to \psi(2S)K^+\pi^-$ decays. The black and blue points correspond to Run I data and LHCb Upgrade II projections, respectively.[31] Right: Current and projected constraints on charm indirect $CP$-violation parameters are delineated by the blue and red contours, respectively.[31]

quark content; the discovery of $J/\psi p$ structures with $c\bar{c}uud$ quark content by LHCb in 2015;[32] the discovery of a doubly charmed tetraquark by LHCb in 2021.[33] The existence of such states has been anticipated since the birth of the quark model in the 1960s but a complete understanding of their dynamics requires far more experimental studies. Figure 8(left) illustrates how the characteristic resonance pattern in the Argand diagram of the $Z(4430)^-$ state could be resolved with LHCb Upgrade II dataset. A particularly interesting area that is beyond the reach of current experiments is the search for the doubly heavy baryons $\Xi_{bc}$ and $\Omega_{bc}$, whose production rates are predicted to be roughly an order of magnitude lower than that of $B_c$ mesons. The HL-LHC data offer exciting prospects for their discovery.

### 5.3. *Charm*

A rich set of measurements and discoveries in the charm sector has been an unexpected legacy of the first decade of the LHC. This includes LHCb's first observations of $D-\overline{D}$ oscillations in 2012[34] and $CP$-violation in charm decays in 2019.[35] $CP$-violation in charm systems is expected to be small in the SM at a rate of $\mathcal{O}(10^{-4})$ or less, but can easily be enhanced by BSM physics. The study of *indirect* $CP$-violation, characterised by the parameters $\phi$ and $|q/p|$, remains currently out of reach. However, a dramatic improvement from the HL-LHC data is expected, as shown in Fig. 8(right). This is because the measurements proceed through asymmetry observables in which uncertainties cancel to a large degree and the sensitivity is limited almost entirely by data statistics.

## 6. Conclusions

Without attempting to summarise all topics discussed in the preceding sections, it is clear that the scope to test the SM through precision measurements with HL-LHC data is very significant. The questions addressed are of fundamental nature, such as the structure and symmetries of the SM interactions, the mechanism for mass generation, the role of flavour, and the study of the strong interaction from lowest to highest scales.

## References

1. A. Dainese et al., Report on the Physics at the HL-LHC, and Perspectives for the HE-LHC. (CERN-2019-007) (2019). URL `https://cds.cern.ch/record/2703572`.
2. ATLAS and CMS Collaborations, Snowmass White Paper Contribution: Physics with the Phase-2 ATLAS and CMS Detectors (Apr, 2022). URL `http://cds.cern.ch/record/2805993`.
3. LHCb Collaboration. Future physics potential of LHCb. Technical report, CERN, Geneva (Apr, 2022). URL `https://cds.cern.ch/record/2806113`.
4. CDF collaboration, High-precision measurement of the W boson mass with the CDF II detector, *Science*. **376** (6589), 170 (2022).
5. ATLAS Collaboration. Prospects for the measurement of the $W$-boson

mass at the HL- and HE-LHC. Technical Report ATL-PHYS-PUB-2018-026 (2018). URL `https://cds.cern.ch/record/2645431`.
6. CMS Collaboration, ECFA 2016: Prospects for selected standard model measurements with the CMS experiment at the High-Luminosity LHC. (CMS-PAS-FTR-16-006) (2017). URL `https://cds.cern.ch/record/2262606`.
7. ATLAS collaboration, Measurement of the $W$-boson mass in pp collisions at $\sqrt{s} = 7$ TeV with the ATLAS detector, *Eur. Phys. J. C.* **78** (2), 110 (2018). doi: 10.1140/epjc/s10052-017-5475-4. [Erratum: Eur. Phys. J. C 78, 898 (2018)].
8. LHCb collaboration, Measurement of the W boson mass, *JHEP.* **01**, 036 (2022). doi: 10.1007/JHEP01(2022)036.
9. R. Abdul Khalek, S. Bailey, J. Gao, L. Harland-Lang, and J. Rojo, Towards Ultimate Parton Distributions at the High-Luminosity LHC, *Eur. Phys. J. C.* **78** (11), 962 (2018). doi: 10.1140/epjc/s10052-018-6448-y.
10. P. Agostini et al., The Large Hadron-Electron Collider at the HL-LHC, *J. Phys. G.* **48** (11), 110501 (2021). doi: 10.1088/1361-6471/abf3ba.
11. CMS Collaboration, A profile likelihood approach to measure the top quark mass in the lepton+jets channel at $\sqrt{s} = 13$ TeV (2022). URL `https://cds.cern.ch/record/2806509`.
12. LHCb collaboration, Measurement of the forward-backward asymmetry in $Z/\gamma^* \to \mu^+\mu^-$ decays and determination of the effective weak mixing angle, *JHEP.* **11**, 190 (2015). doi: 10.1007/JHEP11(2015)190.
13. CMS collaboration, Measurement of the weak mixing angle using the forward-backward asymmetry of Drell-Yan events in pp collisions at 8 TeV, *Eur. Phys. J. C.* **78** (9), 701 (2018). doi: 10.1140/epjc/s10052-018-6148-7.
14. ATLAS Collaboration, Measurement of the effective leptonic weak mixing angle using electron and muon pairs from $Z$-boson decay in the ATLAS experiment at $\sqrt{s} = 8$ TeV (Jul, 2018). URL `https://cds.cern.ch/record/2630340`.
15. CMS Collaboration, A proposal for the measurement of the weak mixing angle at the HL-LHC. (CMS-PAS-FTR-17-001) (2017). URL `http://cds.cern.ch/record/2294888`.
16. ATLAS Collaboration, Prospect for a measurement of the Weak Mixing Angle in $pp \to Z/\gamma^* \to e^+e^-$ events with the ATLAS detector at the High Luminosity Large Hadron Collider. (ATL-PHYS-PUB-2018-037) (2018). URL `https://cds.cern.ch/record/2649330`.
17. W. J. Barter. Prospects for measurement of the weak mixing angle at LHCb. Technical report, CERN, Geneva (Nov, 2018). URL `https://cds.cern.ch/record/2647836`.
18. M. J. G. Veltman, Second Threshold in Weak Interactions, *Acta Phys. Polon. B.* **8**, 475–492 (1977).
19. B. W. Lee, C. Quigg, and H. B. Thacker, The Strength of Weak Interactions at Very High-Energies and the Higgs Boson Mass, *Phys. Rev. Lett.* **38**, 883–885 (1977). doi: 10.1103/PhysRevLett.38.883.
20. CMS Collaboration. Prospects for the measurement of vector boson scattering production in leptonic $W^\pm W^\pm$ and $WZ$ diboson events at $\sqrt{s} =$

14 TeV at the High-Luminosity LHC. Technical Report CMS-PAS-FTR-21-001 (2021). URL https://cds.cern.ch/record/2776773.
21. ATLAS Collaboration. Prospect studies for the production of three massive vector bosons with the ATLAS detector at the High-Luminosity LHC. Technical Report ATL-PHYS-PUB-2018-030 (2018). URL https://cds.cern.ch/record/2647220.
22. CMS Collaboration. Projection of measurements of differential $t\bar{t}$ production cross sections in the $e/\mu$+jets channels in $pp$ collisions at the HL-LHC. Technical Report CMS-PAS-FTR-18-015 (2018). URL http://cds.cern.ch/record/2651195.
23. ATLAS Collaboration. Prospects for the measurement of $t\bar{t}\gamma$ with the upgraded ATLAS detector at the High-Luminosity LHC. Technical Report ATL-PHYS-PUB-2018-049 (2018). URL https://cds.cern.ch/record/2652168.
24. ATLAS Collaboration. HL-LHC prospects for the measurement of the Standard Model four-top-quark production cross-section. Technical Report ATL-PHYS-PUB-2018-047 (2018). URL https://cds.cern.ch/record/2651870.
25. CMS Collaboration. Expected sensitivities for $t\bar{t}t\bar{t}$ production at HL-LHC and HE-LHC. Technical Report CMS-PAS-FTR-18-031 (1900). URL http://cds.cern.ch/record/2650211.
26. ATLAS Collaboration. Prospects for jet and photon physics at the HL-LHC and HE-LHC. Technical Report ATL-PHYS-PUB-2018-051 (2018). URL https://cds.cern.ch/record/2652285.
27. CMS Collaboration. High-$p_T$ jet measurements at the HL-LHC. CMS Physics Analysis Summary CMS-PAS-FTR-18-032 (2018). URL http://cds.cern.ch/record/2651219.
28. LHCb collaboration, First observation of the decay $B_s^0 \to K^-\mu^+\nu_\mu$ and Measurement of $|V_{ub}|/|V_{cb}|$, *Phys. Rev. Lett.* **126** (8), 081804 (2021). doi: 10.1103/PhysRevLett.126.081804.
29. LHCb collaboration, Determination of the quark coupling strength $|V_{ub}|$ using baryonic decays, *Nature Phys.* **11**, 743–747 (2015). doi: 10.1038/nphys3415.
30. Simultaneous determination of the CKM angle $\gamma$ and parameters related to mixing and CP violation in the charm sector. Technical report, CERN, Geneva (2022). URL https://cds.cern.ch/record/2838029.
31. R. Aaij et al., Physics case for an LHCb Upgrade II - Opportunities in flavour physics, and beyond, in the HL-LHC era (8, 2018).
32. LHCb collaboration, Observation of $J/\psi p$ Resonances Consistent with Pentaquark States in $\Lambda_b^0 \to J/\psi K^- p$ Decays, *Phys. Rev. Lett.* **115**, 072001 (2015). doi: 10.1103/PhysRevLett.115.072001.
33. LHCb collaboration, Observation of an exotic narrow doubly charmed tetraquark (9, 2021).
34. LHCb collaboration, Observation of $D^0 - \overline{D}^0$ oscillations, *Phys. Rev. Lett.* **110** (10), 101802 (2013). doi: 10.1103/PhysRevLett.110.101802.
35. LHCb collaboration, Observation of direct $CP$ violation in $D^0$ meson decays at LHCb. pp. 263–270 (2019). doi: 10.1393/ncc/i2020-20041-4.

© 2024 The Author(s)
https://doi.org/10.1142/9789811280184_0018

# Chapter 18

# High Luminosity Forward Physics

M. Deile* and M. Taševský†

*CERN, 1211 Genève 23, Switzerland

†FZU - Institute of Physics of the Czech Academy of Sciences, Na Slovance 2, 18221 Prague, Czech Republic

## 1. Introduction

Forward physics and experimentation at the LHC are innovative areas that test the Standard Model more extensively and may unravel new physics, including new Higgs physics. The HL-LHC, expected to provide about ten times more integrated luminosity than collected in Runs 1–3, offers a unique opportunity for detailed studies of central production — a part of forward physics promising from the view of discovery potential — consisting of exclusive and semi-exclusive processes. In these rare events, signals of new physics can emerge through advantageous signal-to-background ratios thanks to the additional kinematic information (compared to inclusive processes) in conjunction with keeping combinatorial backgrounds from high pile-up under control. For the latter, timing detectors with excellent resolutions are crucial.

The central (semi-) exclusive production processes in proton-proton collisions, $pp \to p \oplus X \oplus p$, are defined by one intact proton on each side of the interaction point (IP), a state X produced at central rapidities and (almost) no other activity in the central detector, manifesting itself in large rapidity gaps (LRG, denoted by $\oplus$). The intact protons are characterized by very small scattering angles relative to the beam and by small momentum losses, $\Delta p_{1/2}$, of proton 1 on one side and proton 2 on the other side of the IP with respect to the momentum $p$ of the incoming proton,

This is an open access article published by World Scientific Publishing Company. It is distributed under the terms of the Creative Commons Attribution 4.0 (CC BY) License.

permitting the creation of the central state X. The fractional momentum loss, $\xi_{1/2} := \Delta p_{1/2}/p$, typically amounts to only a few percent. Thus these protons stay very close to the beam and escape the central detector. They are measured by dedicated forward proton detectors (FPD) that need to be placed far from the IP and very close to the circulating beam (a few millimeters away).

Already in the first two LHC runs — and to be continued in Run 3 — both interaction points IP1 and IP5 have been equipped with FPDs, first pioneered by TOTEM [1] and ATLAS-ALFA [2] for measurements of elastic and soft diffractive scattering in special runs. In Long Shutdown LS1, these systems were complemented and upgraded for high luminosity operation in all regular LHC runs, yielding the ATLAS-AFP [3] and CMS-PPS [4] (initially CMS-TOTEM PPS) subdetectors dedicated to measurements of processes with much lower cross sections than the ones of elastic and soft diffractive scattering. Both spectrometers are equipped with trackers to measure kinematic quantities and time-of-flight (ToF) detectors to measure proton arrival times, the latter enabling the reconstruction of the longitudinal vertex position, which is essential at the high pile-up levels of LHC, even more so at the HL-LHC. Both are also an integral part of the data acquisition system and central trigger of the main detectors.

During the Long Shutdown LS3, the Long Straight Sections LSS1 and LSS5 will be redesigned and the present AFP and PPS systems uninstalled. This offers an opportunity to develop improved detector systems and to place them in optimised locations, building on the experience gained in the first LHC runs (see Refs. [5] and [6] for more details).

## 2. Physics Objectives

The experimental signature of (semi-) exclusive processes are usually explained by an exchange of a colourless object with vacuum quantum numbers between the colliding particles. This object can be a gluonic state (Pomeron) or a photon. Pomeron (photon) exchanges dominate at central system masses below (above) roughly 150 GeV.

In the presence of high pile-up, the LRGs are filled by particles from unrelated interactions, and one has to rely on other exclusivity criteria. These include matching at two levels: kinematic quantities and the primary vertex measured by the central detector should equal those measured by the FPD. The requirement on LRGs is then replaced by a track veto: zero or very few tracks are allowed around the primary vertex.

Although double-proton tagging is necessary for the ToF method to work for the FPDs alone, time information can also be provided by the central detector (see HGTD [7] and MTD [8] — fast time detector upgrades for HL-LHC in ATLAS and CMS, respectively), in which case single-tag events are a viable alternative that extends the physics reach to smaller masses and adds single-proton dissociation to the signal processes [9, 10]. This increases the event yield but also leads to a more serious background contamination. Since the background mostly comes from Standard Model (SM) processes, their precise measurements are indispensable in any search for physics beyond the SM (BSM). Published measurements with tagged protons (exclusive and semi-exclusive production of pairs of photons [11], top quarks [14], W and Z bosons [15, 16] and leptons [17, 18]) demonstrated the viability of using FPDs in high pile-up environments and paved the way to measuring even rarer (semi-) exclusive processes with similar final states but in harsher HL-LHC conditions. Phenomenological studies suggest promising prospects for extracting (semi-) exclusive signals of SM processes (e.g. $t\bar{t}$ production [19, 20]) and various BSM processes, even with HL-LHC pile-up levels. Photon-induced processes, for example, give access to anomalous magnetic moments [21], searches for Dark Matter [22, 23], axion-like particles [24], and anomalous gauge couplings [26, 27] studied via Effective Field Theory. On one hand, measuring rates or properties of Higgs bosons (e.g. quantum numbers or the $Hb\bar{b}$ Yukawa coupling [28–30]), born exclusively via Pomeron fusion, can help inclusive analyses to determine the SM or BSM nature of the Higgs boson discovered at a mass of 125.5 GeV. Alternatively, on the other hand, this can potentially shed light on the excess at masses around 95 GeV [31]. Such low masses will be reachable only if detector stations are placed about 420 m from the IP. In general, if deviations from the SM are observed experimentally, a categorised analysis (no-tag, single-tag, double-tag) would be advisable for disentangling different sources of new physics effects.

## 3. AFP and PPS Proton Spectrometer Layouts at HL-LHC

The search for suitable detector locations is driven by the physics programme striving for the coverage of the widest possible range of masses in diffractive processes. The mass of the central system in (semi-) exclusive processes is obtained via the relation:

$$M^2 = \xi_1 \, \xi_2 \, s \,, \tag{1}$$

where $\sqrt{s} = 14$ TeV is the centre-of-mass energy.

The protons are deflected from the beam centre in a direction determined by the horizontal and vertical dispersion of the LHC bending magnets. They are then detected by sensors approaching the beam in movable beam-pipe insertions, e.g. so-called Roman Pots (RPs). In the case of horizontal beam crossing (to be implemented in IP1), the dispersion is almost purely horizontal, whereas for vertical crossing (to be implemented in IP5) there is also a substantial vertical component. In both cases the best detector acceptance for leading protons with $\xi > 0$ is obtained with sensors approaching the beam horizontally, i.e. along $x$.

The minimum accessible $|\xi|$ is given by the ratio of the closest detector approach to the beam, defined as a multiple of the horizontal beam width, $\sigma_x$, and the horizontal dispersion $D_x$ at that location: $|\xi|_{\min} \propto \sigma_x/|D_x|$. In LSS1 and LSS5, locations with small $|\xi|_{\min}$ lie around the quadrupole Q6 and at distances greater than 300 m from the IP.

Events with large $M$ or $\xi$ have protons moving far away from the beam centre, hence their acceptance is determined by the tightest aperture limitations upstream of the detection point. Since the dominant aperture bottlenecks are the debris collimators (TCLX4, TCL5 and TCL6), it is advantageous to place detector stations immediately upstream of them.

A layout solution taking into account these acceptance considerations and the available space in the beam line is shown in Fig. 1. In both AFP and PPS, detector stations are proposed at ~196 m (RP1), ~220 m (RP2), ~234 m (RP3) and, at a later stage, at ~420 m (RP4) from the IP, on both sides. Each of these stations will consist of two horizontal detector units with a few meters of lever arm to allow the measurement of track angles.

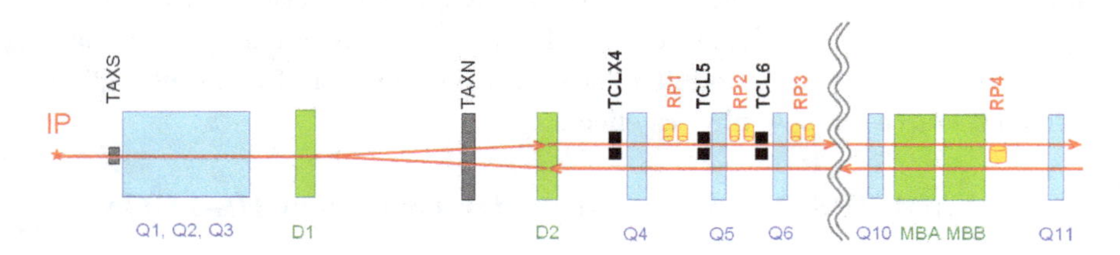


Fig. 1. Schematic overview of the planned AFP/PPS stations in Sector 1-2 / 5-6 (outgoing Beam 1) (adapted from Ref. 32). The dipoles (D1, D2, MBA, MBB), quadrupoles (Q$n$) and higher order correctors (not shown) define the beam optics. TAXS and TAXN are shower absorbers. The instrumentation in Sector 8-1 / 4-5 (outgoing Beam 2) is mirror-symmetric.

## 4. Kinematic Acceptance

The central mass acceptance for a generic central diffractive process, where the mass in double-tag events is evaluated using Eq. (1), varies over a fill because the luminosity levelling procedure [33] gradually changes the beam optics determining the proton trajectories from the IP to the detectors. During most of the fill, for optics characterized by $\beta^{*a}$ values between 50 and 15 cm, the variations of the mass limits amount to only a few tens of GeV. On average, PPS will reach the mass intervals (1150, 2750), (500, 900) and (200, 350) GeV for the stations RP1, RP2 and RP3, respectively. For AFP, due to a lower horizontal dispersion, the horizontal crossing scheme in IP1 leads to substantially higher masses compared to the vertical crossing in IP5, the corresponding intervals are: (2000, 5000), (1000, 4000), (600, 1600) GeV. The stations RP4, envisaged for a later stage, will add acceptance at low masses in the region (50, 160) GeV. The single-tag configuration in general leads to the mass minimum of a few tens of GeV. The exact value is given by the kinematical limit and central detector acceptance for a given process.

## 5. Instrumentation

### 5.1. *Movable Detector Vessels*

For the locations up to 245 m from the IPs, the well-proven Roman Pot technology is an adequate solution for housing the detectors and moving them towards the beam. Only details (size and shape) of the detector chambers need to be adapted to the expected hit distributions at HL-LHC in each location. Each pot will contain a stack of tracking and timing detectors. The strongly peaked radiation near the beam will lead to local fluences up to $10^{16}$ p/cm$^2$ after an integrated luminosity of 300 fb$^{-1}$, which can be reached within a year of running. Such fluence peaks are currently beyond the radiation hardness of most detector and readout electronics technologies. The local radiation damage is mitigated by periodic vertical displacements of the detector package, which distributes the irradiation over the detectors. This principle has been successfully tested in Run 2 by raising or lowering entire RP units relative to the beam centre in short technical stops during the running seasons [34]. At HL-LHC, however, it is expected that the harsh radiation will limit tunnel accesses to only

[a] $\beta^*$ is the value of the betatron function at the IP.

regular End-of-Year technical stops. Therefore, remote-controlled vertical movement systems have been developed and will be tested in Run 3.

The RP location RP4 at 420 m, considered for a later stage, would be substantially different from the others: since the horizontal dispersion has the opposite sign, the signal proton tracks will end up in the narrow space between the outgoing and incoming beam pipes, which excludes the conventional RP technology and requires new developments. In addition, this location lies in the cryogenic region and is presently occupied by the empty cryostat LEGR, which would thus have to be replaced with a cryogenic bypass. Possible solutions are explored in cooperation with the collimation team, which has already developed such a bypass for the installation of new TCLD collimators in other interaction regions. Alternatively, one can also make use of solutions developed for the FP420 R&D project and described in great detail in its TDR [35].

## 5.2. *Detector Technologies*

### 5.2.1. *Tracking Detectors*

Both AFP and PPS spectrometers base their tracking on a 3D Silicon pixel technology which excels in: i) a very narrow dead area (around 50 $\mu$m), ii) very good spatial resolution (sub-10 $\mu$m), and iii) radiation hardness. This provides: i) detector efficiency as close as possible to the beam (maximizing the mass acceptance), ii) precise measurement of the central system mass and iii) tolerance of high radiation levels. Both projects plan to maintain this technology for HL-LHC.

The 3D Silicon sensors to be used in Run 3 by AFP [36, 37] and PPS [34, 38, 39] are segmented in pixels of $(50 \times 250)$ $\mu\text{m}^2$ and $(100 \times 150)$ $\mu\text{m}^2$, respectively, and are expected to withstand radiation doses of up to $5 \times 10^{15}$ p/cm$^2$. Smaller pixel cells with higher radiation tolerance have been successfully tested for HL-LHC conditions [40].

### 5.2.2. *Timing Detectors*

ToF detectors were first installed in Run 2 by both AFP and PPS. While AFP used a matrix of 16 L-shaped *quartz bars* with the smallest transverse size of 2 mm, detecting Cherenkov light [41], PPS instead relied on single crystal Chemical Vapour Deposit (scCVD) *diamond detectors* in "double diamond" readout configuration [34, 42] with a crystal size of $(4.5 \times 4.5)$ mm$^2$. Low pile-up runs and testbeam measurements demonstrated time resolutions of 20–30 ps and 50 ps for the quartz bars and the double-diamond plane, respectively. From the point of view of the long-term LHC operation, both systems suffered radiation damage to their readout chains.

A pile-up multiplicity of 200 interactions per bunch crossing, as expected ultimately at HL-LHC, leads to a vertex density of about 1.5/mm at $z = 0$ (where this density is highest) [7] and up to 3 protons ending up in the FPD acceptance. To reduce this combinatorial background, equipping FPD with a ToF detector with a resolution of the order of 10 ps and a granularity of 1 mm seems to be imperative [43]. An additional improvement will come from the combination of ToF information from the FPD and the central detector (see HGTD [7] and MTD [8]).

There are presently three viable ToF detector options in R&D phases and on good track to attain the required resolution and granularity: Quartz bars and diamonds as described above, along with radiation-hard read-out electronics, and Low Gain Avalanche Diodes (LGAD). The latter were developed for the CMS Endcap Timing Layer [8, 44] and for HGTD [7], making them a conceivable option for AFP and PPS with maximum synergy inside the respective collaborations. Similar to the tracker, the non-uniform irradiation would be tackled by moving the whole detector package according to the fluence.

## References

[1] The TOTEM Collaboration, The TOTEM Experiment at the CERN Large Hadron Collider, JINST. **3**, S08007 (2008).
[2] S. Abdel Khalek et al., The ALFA Roman Pot Detectors of ATLAS, JINST. **11**(11), P11013 (2016).
[3] L. Adamczyk et al. Technical Design Report for the ATLAS Forward Proton Detector. CERN-LHCC-2015-009; ATLAS-TDR-024 (2015).
[4] The CMS and TOTEM Collaborations. CMS-TOTEM Precision Proton Spectrometer. CERN-LHCC-2014-021; TOTEM-TDR-003; CMS-TDR-13 (2014).
[5] The CMS Collaboration. The CMS Precision Proton Spectrometer at the HL-LHC – Expression of Interest. arXiv:2103.02752.
[6] L. Adamczyk et al. Initial Design Report for the ATLAS Forward Proton Detector, in preparation. (2022).
[7] G. Aad et al. Technical Design Report: A High-Granularity Timing Detector for the ATLAS Phase-II Upgrade. CERN-LHCC-2020-007 (2020).
[8] The CMS Collaboration. A MIP Timing Detector for the CMS Phase-2 Upgrade. CERN-LHCC-2019-003; CMS-TDR-020 (2019).

[9] L. A. Harland-Lang et al., A new approach to modelling elastic and inelastic photon-initiated production at the LHC: SuperChic 4, Eur. Phys. J. C. **80**(10), 925 (2020).
[10] S. Bailey et al. Modelling $W^+W^-$ production with rapidity gaps at the LHC. arXiv:2201.08403 (2022).
[11] A. Tumasyan et al., First Search for Exclusive Diphoton Production at High Mass with Tagged Protons in Proton-Proton Collisions at $\sqrt{s} = 13$ TeV, Phys. Rev. Lett. **129**(1), 011801 (2022). doi: 10.1103/PhysRevLett.129.011801.
[12] The CMS and TOTEM Collaborations, Search for exclusive diphoton production with intact protons in PPS, CMS-PAS-EXO-21-007; TOTEM-NOTE-2022-005 (2022) `https://cds.cern.ch/record/2810862`.
[13] Aad, Georges and others, Search for an axion-like particle with forward proton scattering in association with photon pairs at ATLAS, arXiv 2304.10953, CERN-EP-2023-049 (2023).
[14] The CMS and TOTEM Collaborations. Search for central exclusive production of top quark pairs in proton-proton collisions at $\sqrt{s} = 13$ TeV with tagged protons. CMS-PAS-TOP-21-007; TOTEM-NOTE-2022-002 (2022). URL `https://cds.cern.ch/record/2803843`.
[15] Search for high-mass exclusive $\gamma\gamma \to$ WW and $\gamma\gamma \to$ ZZ production in proton-proton collisions at $\sqrt{s} = 13$ TeV (11, 2022).
[16] A search for new physics in central exclusive production using the missing mass technique with the CMS detector and the CMS-TOTEM precision proton spectrometer (3, 2023).
[17] G. Aad et al., Observation and Measurement of Forward Proton Scattering in Association with Lepton Pairs Produced via the Photon Fusion Mechanism at ATLAS, Phys. Rev. Lett. **125**(26), 261801 (2020).
[18] The CMS and TOTEM Collaborations, Observation of proton-tagged, central (semi)exclusive production of high-mass lepton pairs in pp collisions at 13 TeV with the CMS-TOTEM precision proton spectrometer, JHEP. **07**, 153 (2018).
[19] V. P. Gonçalves et al., Top quark pair production in the exclusive processes at the LHC, Phys. Rev. D. **102**(7), 074014 (2020).
[20] D. E. Martins, M. Tasevsky, and V. P. Goncalves, Challenging exclusive top quark pair production at low and high luminosity LHC, Phys. Rev. D. **105**(11), 114002 (2022). doi: 10.1103/PhysRevD.105.114002.
[21] L. Beresford and J. Liu, New physics and tau $g-2$ using LHC heavy ion collisions, Phys. Rev. D. **102**(11), 113008 (2020).
[22] L. A. Harland-Lang et al., LHC Searches for Dark Matter in Compressed Mass Scenarios: Challenges in the Forward Proton Mode, JHEP. **04**, 010 (2019).
[23] L. Beresford and J. Liu, Search Strategy for Sleptons and Dark Matter Using the LHC as a Photon Collider, Phys. Rev. Lett. **123**(14), 141801 (2019).
[24] C. Baldenegro et al., Searching for axion-like particles with proton tagging at the LHC, JHEP. **06**, 131 (2018).

[25] L. A. Harland-Lang and M. Tasevsky, New calculation of semiexclusive axionlike particle production at the LHC, arXiv 2208.10526, doi: 10.1103/PhysRevD.107.033001, *Phys. Rev. D* vol. 107(3), p. 033001 (2023).
[26] S. Fichet et al., Light-by-light scattering with intact protons at the LHC: from Standard Model to New Physics, JHEP. **02**, 165 (2015).
[27] E. Chapon et al., Anomalous quartic W W gamma gamma, Z Z gamma gamma, and trilinear WW gamma couplings in two-photon processes at high luminosity at the LHC, Phys. Rev. D. **81**, 074003 (2010).
[28] S. Heinemeyer et al., Studying the MSSM Higgs sector by forward proton tagging at the LHC, Eur. Phys. J. C. **53**, 231–256 (2008).
[29] B. E. Cox et al., Detecting Higgs bosons in the $b\bar{b}$ decay channel using forward proton tagging at the LHC, JHEP. **10**, 090 (2007).
[30] M. Tasevsky, Review of Central Exclusive Production of the Higgs Boson Beyond the Standard Model, Int. J. Mod. Phys. A. **29**, 1446012 (2014).
[31] T. Biekötter et al. Mounting evidence for a 95 GeV Higgs boson. arXiv:2203.13180 (3, 2022).
[32] J. Oliveira et al. Layouts of HL-LHC Insertions - IR5, IR6, IR7, IR8, drawing LHCLSXGH0002 v.AB, EDMS 1557086 (2020).
[33] G. Arduini et al. HL-LHC Run 4 proton operational scenario. CERN-ACC-2022-0001 (2022). URL `https://cds.cern.ch/record/2803840`.
[34] The CMS Collaboration. The evolution and performance of the CMS detector at the CERN LHC, Chapter 6. PRF-21-001 (2022).
[35] M. G. Albrow et al., The FP420 R&D Project: Higgs and New Physics with forward protons at the LHC, JINST. **4**, T10001 (2009).
[36] J. Lange et al., 3D silicon pixel detectors for the ATLAS Forward Physics experiment, JINST. **10**(03), C03031 (2015).
[37] J. Lange et al., 3D silicon pixel detectors for the High-Luminosity LHC, JINST. **11**(11), C11024 (2016).
[38] F. Ravera, The CT-PPS tracking system with 3D pixel detectors, JINST. **11**(11), C11027 (2016).
[39] M. M. Obertino, The PPS tracking system: performance in LHC Run2 and prospects for LHC Run3, JINST. **15**(05), C05049 (2020).
[40] S. Terzo et al., A new generation of radiation hard 3D pixel sensors for the ATLAS upgrade, Nucl. Instrum. Meth. A. **982**, 164587 (2020).
[41] L. Chytka et al., Timing resolution studies of the optical part of the AFP Time-of-flight detector, Optics Express. **26(7)**, 8028–8039 (2018).
[42] E. Bossini, The CMS Precision Proton Spectrometer timing system: performance in Run 2, future upgrades and sensor radiation hardness studies, JINST. **15**(05), C05054 (2020).
[43] K. Černý et al., Performance studies of Time-of-Flight detectors at LHC, JINST. **16**(01), P01030 (2021).
[44] M. Ferrero, The CMS MTD Endcap Timing Layer: Precision timing with Low Gain Avalanche Diodes, Nucl. Instr. Meth. A. **1032**, 166627 (2022).

© 2024 The Author(s)
https://doi.org/10.1142/9789811280184_0019

# Chapter 19

# The FASER Experiment

J. Boyd
*Experimental Physics Department, CERN*
*1211 Geneva 23, Switzerland*

FASER is a new, small experiment that was installed into the LHC complex during Long Shutdown 2. It is designed to search for light, long-lived, new particles that could be produced in the LHC collisions and decay inside the detector. In addition, with the FASER$\nu$ sub-detector FASER will be able to detect and study neutrino's produced in the LHC collisions.

## 1. Introduction

The ForwArd Search ExpeRiment, or FASER, is a small experiment that was installed into the LHC tunnel during the LHC Long Shutdown 2 (between 2019 and 2021). It is designed to search for new long-lived and weakly interacting particles which could be produced in the LHC collisions at the ATLAS collision point (IP1), and with the addition of the FASER$\nu$ sub-detector, study high-energy neutrinos produced in the LHC collisions as well. The detector is placed on the collision-axis line-of-sight (LOS), 480 m away from the collision point, after the LHC has bent away from the LOS, as shown in Fig. 1. As can be seen in the figure, the detector is placed in an unused service tunnel TI12, formerly used as an injection beamline for the old LEP collider. The TI12 tunnel is an excellent location for such an experiment, since backgrounds from particles produced in the IP1 collisions are largely suppressed by the location; such particles would need to travel through strong LHC magnets (sweeping away charged particles) as well as 100 m of rocks before they reach FASER.

---

This is an open access article published by World Scientific Publishing Company. It is distributed under the terms of the Creative Commons Attribution 4.0 (CC BY) License.

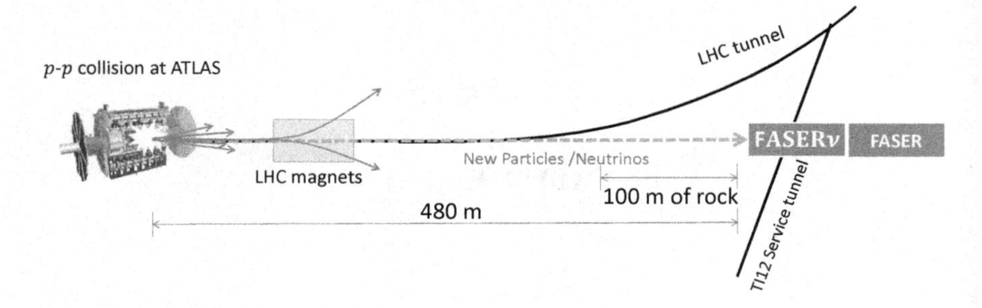


Fig. 1. A sketch of the FASER location, situated on the collision axis LOS in the TI12 tunnel in the LHC complex.

The FASER active detector is only 20 cm across in the transverse plane, thus it covers an extremely small angular region of 0.1 mrad around the LOS (and only $10^{-8}$ of the full solid angle). However, given that light particles are predominately produced in the LHC collisions at very small angles to the beam collision axis, even such a small detector has excellent prospects for searching for new particles in interesting regions of parameter space.

Studying neutrinos produced at particle colliders was initially proposed nearly 30 years ago,[1] but has yet to be realised. The FASER$\nu$ sub-detector, placed in front of the main FASER detector will enable such studies for the first time.[a] This will allow all three neutrino flavours, electron neutrinos ($\nu_e$), muon neutrinos ($\nu_\mu$) and tau neutrinos ($\nu_\tau$), to be detected and studied in an energy regime in which no measurements currently exist.

## 2. Brief History

The FASER experiment was first proposed in 2017,[2] and after discussions with the LHC Experiments Committee (LHCC) in 2018, and securing funding, it was approved by CERN in March 2019. The detector design, prototyping and individual component commissioning happened during 2019, and the production and testing of the final components in 2020. The full experiment was installed into the TI12 tunnel in March 2021, followed by about nine months of in situ commissioning.

The additional FASER$\nu$ sub-detector was proposed as an extension of FASER in 2019[3] and approved by CERN in December of that year.

[a] In addition to FASER$\nu$ the SND@LHC experiment discussed in Chapter 20 will also start data taking in 2022, with the aim to study neutrinos produced in the LHC collisions.

As an important step in the preparation of FASER, in situ measurements were taken during 2018 LHC running. The measured radiation levels and background particle rates in TI12[4] confirmed the feasibility of the location for an experiment and helped define the detector requirements. As part of these measurements, a small pilot neutrino detector (with about 1% of the target mass of the final FASER$\nu$ sub-detector) was installed for 1 month of LHC's running. Analysis of the pilot detector data led to, for the first time, the observation of neutrino interaction candidates at a collider.[5]

## 3. Physics Prospects

FASER will search for light, extremely weakly-interacting particles which may be produced in the LHC collisions, travel long distances through rock without interacting and then decay to visible particles in the detector. Such new particles are motivated by dark-sector models, which can explain dark-matter as long as the mass ($m$) and couplings ($\epsilon$) of the new-particles are in certain ranges. An interesting region of the theory parameter space is for low masses and very weak couplings, where FASER would have excellent prospects to discover these particles. As an example, Fig. 2 shows the region of parameter space in which FASER will be sensitive to dark-photons for different amounts of LHC luminosity. The sensitivity for other dark-sector models is shown in Ref. 6.

The FASER$\nu$ sub-detector will allow measurements of the interaction

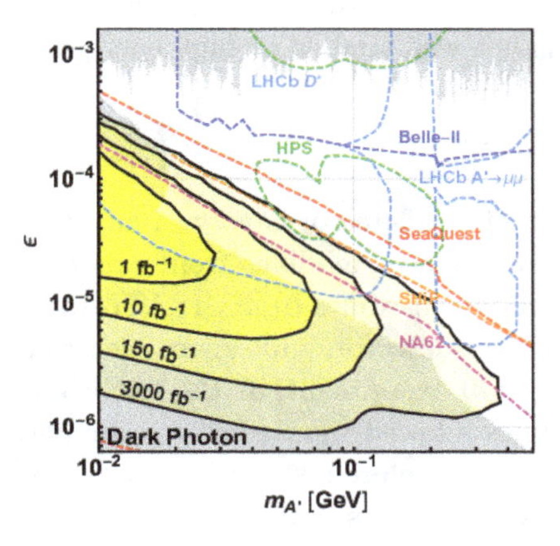


Fig. 2. The expected sensitivity of FASER for dark photons as a function of the mass ($m_{A'}$) and coupling ($\epsilon$), for different values of integrated luminosity. The regions of parameter space already excluded by experiment are shown in grey.

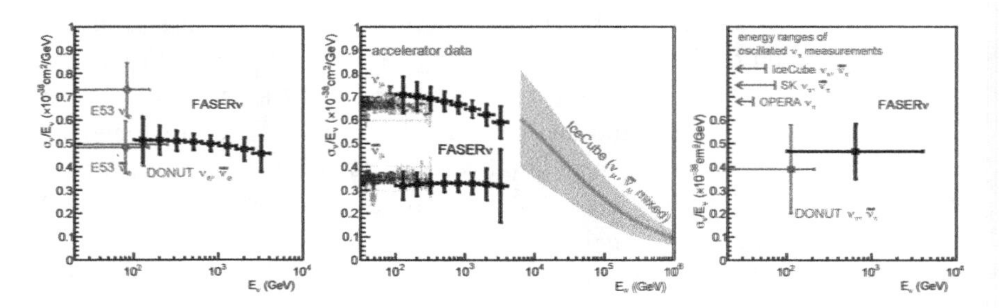


Fig. 3. FASERν's estimated cross section sensitivity for $\nu_e$ (left), $\nu_\mu$ (center), and $\nu_\tau$ (right) for Run 3 of the LHC. Existing constraints are shown in gray.[7] The error bars correspond to the expected statistical uncertainties, while systematic uncertainties have not been fully assessed and so are not shown.

probability (or cross-section) of neutrinos to be made in an un-probed energy regime (for $\nu_e$ and $\nu_\tau$ this will be the highest energy measurements ever made, whereas for $\nu_\mu$ the measurements will be at energies between those from accelerator based experiments, and those from neutrino telescopes). Figure 3 shows the expected statistical precision for the FASER neutrino cross-section measurements. Currently there have only been around 20 $\nu_\tau$ interactions directly detected by experiment. FASERν is expected to detect a similar number of such interactions, although with large uncertainties on this estimate. Since the neutrinos originate from the decay of hadrons, where the hadrons are produced in the very forward direction of the LHC collisions, these measurements provide important insight on forward hadron production which is not well studied at high energies.

## 4. The FASER Detector

Since the active area of the detector is very small, the experiment was able to make use of spare detector modules from existing experiments, where silicon microstrip detector modules from the ATLAS experiment were used in the FASER tracking detector, and spare electromagnetic calorimeter modules from the LHCb experiment, as the FASER calorimeter. The use of these spare modules allowed the detector to be constructed and tested in the short time available before installation, and significantly reduced the overall cost of the experiment as well.

A sketch showing the different components of the FASER detector is shown in Fig. 4. The detector, which is described in detail in Ref. 8, is made up of a number of sub-detectors as detailed below:

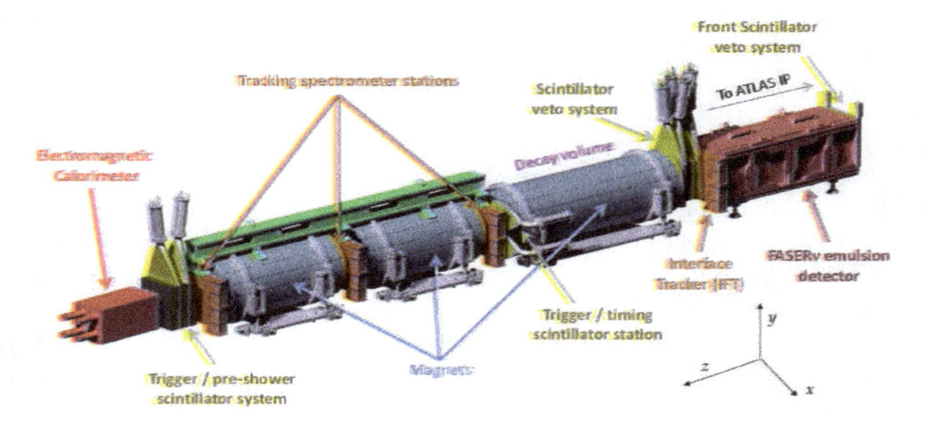


**Fig. 4. A sketch of the FASER detector, showing the different sub-detector systems.**

- Scintillator detector stations. These are used to be able to veto the presence of charged particles entering the detector in the physics analysis. They are also used to select events of interest to be recorded for analysis (so-called triggering), and for precise time measurements of particles in the detector.
- A decay volume, immersed in a 0.6 T dipole magnetic field. This is 1.5 m long, and is the region in which the decay of new physics particles would be able to be detected by the experiment.
- A tracking spectrometer, made up of three tracking stations placed before, in the middle and after two 1-m long dipole magnets (each with a magnetic field strength of 0.6 T). The tracking spectrometer will measure the position and momentum of charged particles arising from new particles decaying in the decay volume. The tracking detector is described in detail in Ref. 9.
- An electromagnetic calorimeter, placed at the back of the detector to measure the energy of particles traversing the detector.
- The FASER$\nu$ neutrino detector. Situated in front of the main FASER detector, this is made up of interleaved tungsten plates and nuclear-emulsion films, with 770 of each forming a detector that is 1.1 m long and weighs 1.1 tonnes. The detector acts as both the target for the neutrinos to interact with, as well as a detector to observe the particles produced from the interaction. The emulsion films must be removed from the detector, developed in a dark-room, and scanned with powerful optical microscopes before the trajectories of the particles can be

reconstructed. Therefore, the full FASER$\nu$ detector will be removed from the TI12 location, and replaced with a new detector, three times a year during Technical Stops of the LHC.

In order to be able to place the detector on the collision axis LOS within a few cm, a small trench was excavated from the floor of the TI12 tunnel, which the detector is installed in. All the detector components except for the emulsion based-detector, have their signals read-out following a trigger which fires when a charged particle traverses one of the scintillators, or when a particle leaves significant energy in the calorimeter. The trigger signal is distributed to all of the detector components which sends the associated data to computers on the surface that will then assemble the data into a complete detector event. The expected trigger rate in FASER for Run 3 of the LHC is about 500 Hz, very low compared to other experiments at the LHC, due to the excellent shielding provided by the detector location. More details on the FASER trigger and data acquisition system are given in Ref. 10.

The detector was installed into the TI12 tunnel in March 2021, and has been under commissioning for the rest of 2021, first using cosmic-rays to exercise the detector. In October 2021 the LHC circulated and collided a few proton bunches in the accelerator as part of a pilot beam test. During this test, FASER observed its first particles in the detector from beam operations, and was able to adjust the timing of the detector components to be ready for high energy, high luminosity running in 2022. The long period of in situ commissioning means the FASER experiment is in excellent shape for physics running from the start of LHC Run 3.

A strong physics case is building up for future larger-versions of the FASER and FASER$\nu$ detectors to operate at the HL-LHC, possibly as part of the proposed Forward Physics Facility.[11]

## References

1. A. De Rujula and R. Ruckl, Neutrino and muon physics in the collider mode of future accelerators. In *SSC Workshop: Superconducting Super Collider Fixed Target Physics*, pp. 571–596 (5, 1984). doi: 10.5170/CERN-1984-010-V-2.571.
2. J. L. Feng, I. Galon, F. Kling, and S. Trojanowski, ForwArd Search ExpeRiment at the LHC, *Phys. Rev. D.* **97** (3), 035001 (2018). doi: 10.1103/PhysRevD.97.035001.

3. FASER Collaboration, Detecting and Studying High-Energy Collider Neutrinos with FASER at the LHC, *Eur. Phys. J. C.* **80** (1), 61 (2020). doi: 10.1140/epjc/s10052-020-7631-5.
4. FASER Collaboration, Technical Proposal for FASER: ForwArd Search ExpeRiment at the LHC (12, 2018). `https://arxiv.org/abs/1812.09139`.
5. FASER Collaboration, First neutrino interaction candidates at the LHC, *Phys. Rev. D.* **104** (9), L091101 (2021). doi: 10.1103/PhysRevD.104.L091101.
6. FASER Collaboration, FASER's physics reach for long-lived particles, *Phys. Rev. D.* **99** (9), 095011 (2019). doi: 10.1103/PhysRevD.99.095011.
7. M. Tanabashi et al., Review of Particle Physics, *Phys. Rev. D.* **98** (3), 030001 (2018). doi: 10.1103/PhysRevD.98.030001.
8. H. Abreu et al., The FASER Detector (Jul, 2022). CERN-FASER-2022-001. `https://arxiv.org/abs/2207.11427`.
9. H. Abreu et al., The tracking detector of the FASER experiment, *Nucl. Instrum. Meth. A* Vol. 1034 p. 166825 (2022). `https://arxiv.org/abs/2112.01116`. doi: 10.1016/j.nima.2022.166825.
10. FASER Collaboration, The trigger and data acquisition system of the FASER experiment, *JINST.* **16** (12), P12028 (2021). doi: 10.1088/1748-0221/16/12/P12028.
11. J. L. Feng et al., The Forward Physics Facility at the High-Luminosity LHC, *J. Phys. G* Vol. 50, p. 030501 (2023). `https://arxiv.org/abs/2203.05090`. doi: 10.1088/1361-6471/ac865e.

© 2024 The Author(s)
https://doi.org/10.1142/9789811280184_0020

# Chapter 20

# The SND@LHC experiment

Giovanni De Lellis
*University of Naples "Federico II", Physics Department, via Cintia 19, 80126, Naples, Italy*

## 1. Introduction

As the accelerator with the highest beam energy, the LHC is also the source of the most energetic human-made neutrinos. Indeed, the LHC produces an intense and strongly collimated beam of TeV-energy neutrinos along the direction of the proton beams. Notably, this neutrino beam includes a sizable fraction of tau neutrinos, mainly produced via the $D_s \to \tau\nu_\tau$ decay and subsequent $\tau$ decays, and hence provides a novel opportunity to study their properties.

Already in 1984, De Rujula and Rückl proposed to use the LHC neutrino beam by placing a neutrino experiment in the far forward direction.[1] This idea of detecting LHC neutrinos was revisited several times in the following decades.[2,3] More recently, a feasibility study was carried out, resulting in the estimate of the physics potential and in the identification of a proper location underground in the LHC tunnel for such an experiment to operate during the Run 3 of the LHC.[4,5] In 2018, the FASER collaboration installed a suitcase size pilot detector employing emulsion films and recently reported the first neutrino interaction candidates at the LHC.[6]

SND@LHC, Scattering and Neutrino Detector @ LHC, is a compact experiment designed to perform measurements with neutrinos produced at the LHC in the unexplored pseudo-rapidity region of $7.2 < \eta < 8.4$, complementary to all the other experiments at the LHC, including FASER.[7] The Collaboration submitted an LoI in August 2020[8] and a Technical Proposal in January 2021.[9] The experiment was approved in March 2021.

This is an open access article published by World Scientific Publishing Company. It is distributed under the terms of the Creative Commons Attribution 4.0 (CC BY) License.

## 2. Experiment concept

The experiment is located 480 m downstream of IP1 in the TI18 tunnel, an injection tunnel during LEP operation. The detector consists of a hybrid system based on an 830 kg target mass of tungsten plates, interleaved with emulsion and electronic trackers, followed downstream by an hadronic calorimeter and a muon identification system, as shown in Fig. 1. The emulsion films with their micrometric accuracy[10] constitute the vertex detector while trackers in the target region, based on the Scintillating Fibre technology,[11] provide the time stamp to the events and complement emulsion for the electromagnetic energy reconstruction. Nuclear emulsion films are readout by state-of-the-art, fully automated, optical scanning systems.[12–15] The hadronic calorimeter and muon system comprises of eight layers of scintillating bar planes interleaved with 20 cm-thick iron slabs. The three most downstream stations are made of fine grained bars with both horizontal and vertical orientation (to trace the penetrating muons). Every scintillating bar as well as every fibre module is viewed by SiPMs.

The detector configuration allows efficient differentiation between all three neutrino flavours, as well as searching for Feebly Interacting Particles via signatures of scattering in the detector target.[16] The first phase aims at operating the detector throughout Run 3 to collect about 290 $\text{fb}^{-1}$.

The detector takes full advantage of the space available in the TI18 tunnel to cover the desired range in pseudo-rapidity. Figure 2 shows the top and side views of the detector positioned inside the tunnel. It is worth noting that the tunnel floor is sloped, as can be seen from the side view, with the floor sloping down along the longitudinal axis of the detector. As shown in the top view, the nominal collision axis from IP1 comes out of the

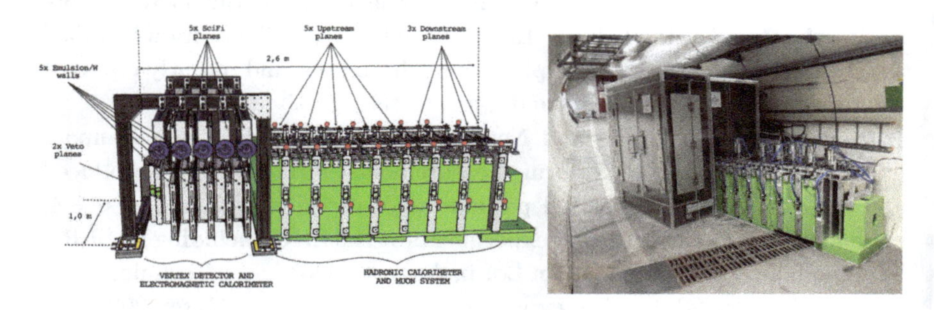


Fig. 1. Detector layout (left) and picture of the detector installed in TI18 (right).

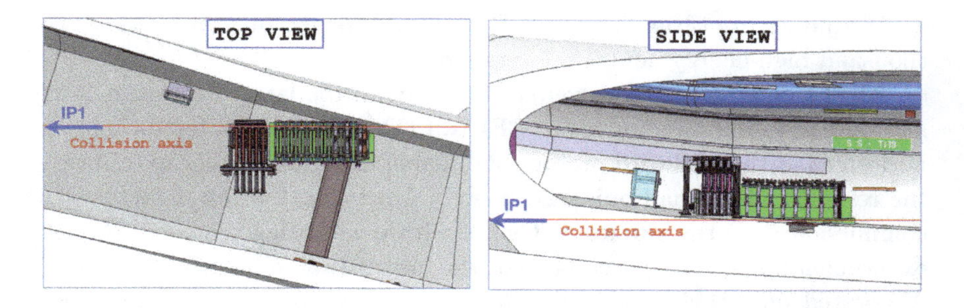


Fig. 2. Side and top views of the SND@LHC detector in the TI18 tunnel.[9]

floor very close to the wall of the tunnel. The location is ideal to explore the off-axis region.[17] Since no civil engineering work could have been done in time for the operation in Run 3, the tunnel geometry imposed several constraints. The following guidelines were adopted for the optimisation of the detector design: a good calorimetric measurement of the energy requires about 10 $\lambda_{\text{int}}$; a good muon identification efficiency requires enough material to absorb hadrons; for a given transverse size of the target region, the azimuthal angular acceptance decreases with the distance from the beam axis. The energy measurement and the muon identification set a constraint on the minimum length of the detector. With the constraints from the tunnel, this requirement competes with the azimuthal angular acceptance that determines the overall flux intercepted and therefore the total number of observed interactions. The combination of position and size of the proposed detector is an optimal compromise between these competing requirements. The geometrical constraints also restrict the detector to the first quadrant only around the nominal collision axis, as shown in the top view of the detector in Fig. 2.

The result is a compact detector, 2.6 m in length. The energy measurement and the muon identification limit the target region to a length of about 80 cm. The transverse size downstream of about 80(H) $\times$ 60(V) cm$^2$ is limited by the constraint of the tunnel side wall. The transverse size of the target region is proportionally smaller in order to match the acceptance of the energy measurement and the muon identification for the vertices identified in the target volume. In order to maximise the number of neutrino interactions, tungsten has been selected as the passive material. The emulsion target will be replaced a few times each year, during technical stops of the LHC.

With data from Run 3, SND@LHC will be able to study more than two thousand high-energy neutrino interactions.

All the detector systems were constructed in the labs by Summer 2021 and were assembled and tested at CERN. In October 2021, a test-beam was performed at the SPS with protons of different energies in order to calibrate the response of the hadronic calorimeter. Moreover, the full detector was commissioned on the surface at CERN with penetrating muons in the H6 experimental hall. On November 1st, the installation underground started. A borated polyethylene shielding box was added to surround the target and absorb low-energy neutrons originated from beam-gas interactions: its installation was completed by March 15$^{\text{th}}$ 2022 as can be seen in the right picture of Fig. 1. The detector installation was completed on April 7$^{\text{th}}$ 2022 by adding the target walls with emulsion films, and it is now taking data with the Run 3 of the LHC. The apparatus and its performance during the commissioning phase are extensively described in a dedicated paper.[18]

In the following sections we review the physics case of the experiment and the future prospects.

### 2.1. *QCD measurements*

Electron neutrinos in $7.2 < \eta < 8.4$ range are mostly produced by charm decays. Therefore, $\nu_e$s can be used as a probe of charm production in an angular range where the charm yield has a large uncertainty, to a large extent coming from the gluon parton distribution function (PDF). Electron neutrino measurements can thus constrain the uncertainty on the gluon PDF in the very small (below $10^{-5}$) $x$ region. The interest therein is two-fold: gluon PDF in this $x$ domain will be relevant for Future Circular Collider (FCC) detectors; secondly, the measurement will reduce the uncertainty on the flux of very-high-energy atmospheric neutrinos produced in charm decays, essential for the evidence of neutrinos from astrophysical sources.[19,20] The charm measurement in Run 3 will be affected by a systematic uncertainty at the level of 30% and by a statistical uncertainty of 5%.

The left plot of Fig. 3 shows the ratio between charm measurements in different $\eta$ regions normalised to the LHCb measurement:[21] gluon PDF uncertainty provides the largest contribution. SND@LHC will measure charm in the $7.2 < \eta < 8.4$ region where the PDF uncertainty is dominant.

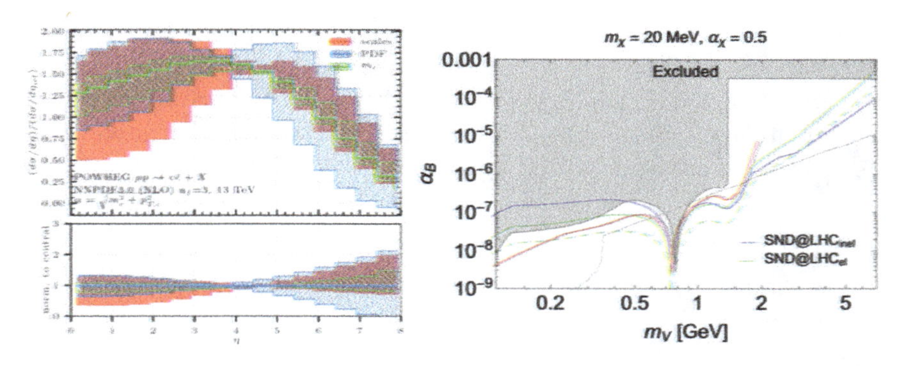


Fig. 3. Left: Ratio between the differential cross-section at 13 TeV and the differential cross-section at 7 TeV, with the latter evaluated in the pseudo-rapidity range $4 < \eta < 4.5$.[9] Right: Sensitivity of the SND@LHC experiment to the leptophobic portal[16]

## 2.2. *Lepton flavour universality with neutrino interactions*

In the pseudo-rapidity range of interest, tau neutrinos are essentially only produced in $D_s \to \tau\nu_\tau$ and the subsequent $\tau$ decays. One can thus assume that the source of both $\nu_e$ and $\nu_\tau$ is essentially provided by semi-leptonic and fully leptonic decays of charmed hadrons. Unlike $\nu_\tau$s produced only in $D_s$ decays, $\nu_e$s are produced in the decay of all charmed hadrons, essentially: $D^0$, $D$, $D_s$, and $\Lambda_c$. Therefore, the $\nu_e/\nu_\tau$ ratio depends only on the charm hadronisation fractions and decay branching ratios. The systematic uncertainties due to the charm-quark production mechanism cancel out, and the ratio becomes sensitive to the $\nu$-nucleon interaction cross-section ratio of the two neutrino species. The measurement of this ratio can thus be considered a lepton flavour universality test in neutrino interactions.

Charmed hadron fractions and $\nu$ branching ratios in the experiment acceptance produce a systematic uncertainty on this ratio of about 22% while the statistical uncertainty is dominated by the low statistics of the $\nu_\tau$ sample, which corresponds to a 30% uncertainty.[9] The systematic uncertainty was evaluated by studying the fluctuations of the ratio using different event generators, after having equalised the branching ratio $D_s \to \tau\nu_\tau$ to the PDG value.[22]

Lepton flavour universality can also be tested with the electron to muon neutrino ratio. The $\nu_\mu$s are much more abundant but heavily contaminated by $\pi$ and $K$ decays, and therefore the production mechanism cannot be considered the same as in the case of $\nu_e$. However, this contamination is mostly concentrated at low energies. Above 600 GeV, the contamination is predicted to be reduced to about 35%, and stable with the energy. Moreover, charmed hadron decays have practically equal branching ratios into electron and muon neutrinos. As a result, the $\nu_e/\nu_\mu$ ratio provides a test of the lepton flavour universality with an uncertainty of 15%, with an equal 10% statistical and systematic contribution.[9]

### 2.3. *Feebly Interacting Particles*

The experiment is also capable of performing model-independent direct searches for FIPs. They may be produced in the *pp* scattering at the LHC interaction point, propagate to the detector and decay or scatter inside it. The background from neutrino interactions can be rejected by a time-of-flight measurement.

A recent work[16] summarises the experiment's sensitivity to physics beyond the Standard Model, by considering the scatterings of light dark matter particles $\chi$ via leptophobic $U(1)_B$ mediator, as well as decays of Heavy Neutral Leptons, dark scalars and dark photons. The excellent spatial resolution of nuclear emulsions makes SND@LHC suited to search for neutral mediators decaying into two charged particles.

SND@LHC is unique in its capability to perform a direct dark matter search at accelerators. The right plot of Fig. 3 shows the sensitivity of the experiment to the leptophobic portal under the assumption that $m_\chi = 20$ MeV and the coupling of the mediator to $\chi$ particles is $\alpha_\chi = 0.5$. The considered signatures are the elastic scattering off protons (green line, 10 signal events) and the deep-inelastic scattering (blue line, 100 signal events). The dashed line corresponds to the upgraded setup that may operate during Run 4. The red line shows the 100 event contour for the DUNE experiment.[23]

### 2.4. *Future upgrade*

An advanced version of the SND@LHC detector is envisaged for the HL-LHC. It will consist of two detectors: the FAR detector placed in the same $\eta$ region as SND@LHC and the NEAR detector in the region $4 < \eta < 5$. The FAR detector will perform the charm production measurement and lepton flavour universality tests with neutrinos at the percent level, and the NEAR detector will benefit from the overlap with LHCb to reduce systematic uncertainties and will perform neutrino cross-section measurements. In order to increase the azimuth angle coverage of the NEAR detector, a location in existing caverns, closer to the interaction point, will be searched for.

Each detector will be made of three elements. The upstream one is the target region for the vertex reconstruction and the electromagnetic energy measurement with a calorimetric approach. It will be followed downstream by a muon identification and hadronic calorimeter system. The third and most downstream element will be a magnet for the muon charge and momentum measurement, thus allowing for neutrino/anti-neutrino separation for $\nu_\mu$ and for $\nu_\tau$ in the muonic decay channel of the $\tau$ lepton.

The target will be made of thin sensitive layers interleaved with tungsten plates, for a total mass of a few tons. Given that the use of nuclear emulsion at the HL-LHC may be incompatible with technical stops, the Collaboration is investigating the use of compact electronic trackers with high spatial resolution fulfilling both tasks of vertex reconstruction with micrometric accuracy and electromagnetic energy measurement. The hadronic calorimeter and the muon identification system will also be optimised.

## References

1. A. De Rujula and R. Ruckl, Neutrino and muon physics in the collider mode of future accelerators. In *SSC Workshop: Superconducting Super Collider Fixed Target Physics*, pp. 571–596 (5, 1984). doi: 10.5170/CERN-1984-010-V-2.571.
2. A. De Rujula, E. Fernandez, and J. Gomez-Cadenas, Neutrino fluxes at future hadron colliders, *Nucl. Phys. B.* **405**, 80–108 (1993). doi: 10.1016/0550-3213(93)90427-Q.
3. H. Park, The estimation of neutrino fluxes produced by proton-proton collisions at $\sqrt{s} = 14$ TeV of the LHC, *JHEP.* **10**, 092 (2011). doi: 10.1007/JHEP10(2011)092.
4. S. Buontempo, G. M. Dallavalle, G. De Lellis, D. Lazic, and F. L. Navarria, CMS-XSEN: LHC Neutrinos at CMS. Experiment Feasibility Study (4, 2018).
5. N. Beni et al., Physics Potential of an Experiment using LHC Neutrinos, *J. Phys. G.* **46** (11), 115008 (2019). doi: 10.1088/1361-6471/ab3f7c.
6. H. Abreu et al., First neutrino interaction candidates at the lhc, *Phys. Rev. D.* **104**, L091101 (Nov, 2021). doi: 10.1103/PhysRevD.104.L091101. URL `https://link.aps.org/doi/10.1103/PhysRevD.104.L091101`.
7. H. Abreu et al., Detecting and Studying High-Energy Collider Neutrinos with FASER at the LHC, *Eur. Phys. J. C.* **80** (1), 61 (2020). doi: 10.1140/epjc/s10052-020-7631-5.
8. SND@LHC Collaboration. Scattering and Neutrino Detector at the LHC. Technical report, CERN, Geneva (Aug, 2020). URL `http://cds.cern.ch/record/2729015`.
9. C. Ahdida et al. SND@LHC - Scattering and Neutrino Detector at the LHC.

Technical report, CERN, Geneva (Jan, 2021). URL https://cds.cern.ch/record/2750060.

10. C. W. Fabjan and H. Schopper, eds., *Particle Physics Reference Library: Volume 2: Detectors for Particles and Radiation*. Springer Nature, Cham (2020). ISBN 978-3-030-35317-9, 978-3-030-35318-6. doi: 10.1007/978-3-030-35318-6.
11. LHCb Collaboration. LHCb Tracker Upgrade Technical Design Report. Technical report (Feb, 2014). URL http://cds.cern.ch/record/1647400.
12. A. Alexandrov et al., A new fast scanning system for the measurement of large angle tracks in nuclear emulsions, *JINST*. **10** (11), P11006 (2015). doi: 10.1088/1748-0221/10/11/P11006.
13. A. Alexandrov et al., A new generation scanning system for the high-speed analysis of nuclear emulsions, *JINST*. **11** (06), P06002 (2016). doi: 10.1088/1748-0221/11/06/P06002.
14. A. Alexandrov et al., The Continuous Motion Technique for a New Generation of Scanning Systems, *Sci. Rep.* **7** (1), 7310 (2017). doi: 10.1038/s41598-017-07869-3.
15. A. Alexandrov, G. De Lellis, and V. Tioukov, A Novel Optical Scanning Technique with an Inclined Focusing Plane, *Sci. Rep.* **9** (1), 2870 (2019). doi: 10.1038/s41598-019-39415-8.
16. A. Boyarsky, O. Mikulenko, M. Ovchynnikov, and L. Shchutska, Searches for new physics at SND@LHC, *JHEP*. **03**, 006 (2022). doi: 10.1007/JHEP03(2022)006.
17. N. Beni et al., Further studies on the physics potential of an experiment using LHC neutrinos, *J. Phys. G*. **47** (12), 125004 (2020). doi: 10.1088/1361-6471/aba7ad.
18. G. Acampora et al., SND@LHC: The Scattering and Neutrino Detector at the LHC, https://arxiv.org/pdf/2210.02784.pdf, to appear on JINST.
19. A. Bhattacharya et al., Prompt atmospheric neutrino fluxes: perturbative QCD models and nuclear effects, *JHEP*. **11**, 167 (2016). doi: 10.1007/JHEP11(2016)167.
20. Y. S. Jeong et al., Neutrinos from charm: forward production at the LHC and in the atmosphere, *PoS*. **ICRC2021**, 1218 (2021). doi: 10.22323/1.395.1218.
21. R. Aaij et al., Measurements of prompt charm production cross-sections in $pp$ collisions at $\sqrt{s} = 13$ TeV, *JHEP*. **03**, 159 (2016). doi: 10.1007/JHEP03(2016)159. [Erratum: JHEP 09, 013 (2016), Erratum: JHEP 05, 074 (2017)].
22. PDG, P. A. Zyla et al., Review of Particle Physics, *PTEP*. **2020** (8) (08, 2020). ISSN 2050-3911. doi: 10.1093/ptep/ptaa104. URL https://doi.org/10.1093/ptep/ptaa104. 083C01.
23. S. Naaz, J. Singh, and R. B. Singh, DUNE prospect for leptophobic dark matter, *Advances in High Energy Physics*. **2020**, 1–9 (Nov, 2020). doi: 10.1155/2020/9047818. URL https://doi.org/10.1155%2F2020%2F9047818.

© 2024 The Author(s)
https://doi.org/10.1142/9789811280184_0021

# Chapter 21

# Gamma Factory

Mieczyslaw Witold Krasny
*LPNHE, CNRS-IN2P3, University Paris Sorbonne and CERN, BE-ABP*

**At the core of the Gamma Factory (GF) proposal[1] was the identification of an enormous, but thus far hidden, potential of the CERN accelerator infrastructure in conducting a new research programme at the crossroads of fundamental, particle, nuclear, atomic, and applied physics. All these disciplines could jointly profit from the novel research tools which can be created by the Gamma Factory facility. The key GF idea is to produce, accelerate, and store highly relativistic atomic beams in the LHC rings — acting as effective atomic traps — and to resonantly excite the atomic degrees of freedom of the beam particles by laser photons to: (1) cool atomic beams, and (2) produce high-energy, polarised photon beams. Their intensity can be, in the particularly interesting gamma-ray energy domain, higher than those of the presently operating light sources by at least seven orders of magnitude. Photons in this energy range are proposed to be used to produce unprecedented-intensity tertiary beams of polarised electrons, polarised positrons, polarised muons, neutrinos, neutrons, and radioactive ions. The LHC atomic traps, the laser-cooled nuclear beams, the high-intensity, polarised photon beams, and the tertiary beams constitute the principal research tools of the proposed GF facility.**

## 1. Scientific context

It is highly unlikely that the next CERN high-energy frontier project will be approved, built, and become operational before the 2050's. The present LHC research programme will certainly reach its discovery saturation earlier, perhaps in the late 2030's. By then, there will be a strong need for a novel multidisciplinary research programme which could re-use (co-use)

**This is an open access article published by World Scientific Publishing Company. It is distributed under the terms of the Creative Commons Attribution 4.0 (CC BY) License.**

the existing CERN accelerator infrastructure (including LHC) in ways and at levels that were not conceived of when the machines were first designed.

The Gamma Factory (GF) facility and its research programme[1] can fulfill such a role. It can exploit the existing opportunities offered by the CERN accelerator complex (not available elsewhere) to conduct unique research in particle, nuclear, atomic, fundamental and applied physics.

The GF project's primary goal is to create *novel research tools and novel research methods*, rather than to execute predefined measurements. At the present moment, creating new research tools, or increasing the precision of the existing ones by several orders of magnitude is of particular importance for the accelerator-based research, since we neither have any hints for a new, high-energy frontier physics that is attainable with current existing technologies at a reasonable cost, nor a certainty that our discipline remains scientifically attractive — if it remains solely on the inertial and incremental-progress path.

Historically, new directions in science are launched by new tools much more often than by new theory concepts. The effect of a concept-driven revolution is to explain known phenomena in new ways. The effect of a tool-driven revolution is to discover new phenomena that will have to be explained.

## 2. Key principles

The primary goal of the GF project is to create and store new types of beams — the ultra-relativistic atomic beams of partially stripped ions (PSI) — and to exploit the atomic degrees of freedom of the beam particles. In the LHC rings, atomic beams can be stored at very high energies, over a large range of the Lorentz factor: $200 < \gamma_L < 3000$, at high bunch intensities: $10^8 < N_{\text{bunch}} < 5 \times 10^9$, and a bunch repetition rate of up to 20 MHz.

Lasers tuned to the atomic transition frequencies can be used to manipulate such beams. The resonant excitation of atomic levels is possible due to the high energies of the ions. For the first time, utilising the large relativistic Lorentz factor $\gamma_L$ of the LHC PSI beams, all the atomic degrees of freedom — including those of high-$Z$ atoms — can be resonantly excited by the infrared, visible, UV or EUV laser photons, thanks to the Doppler upshift of the laser-photon energies by a factor of $2\gamma_L$. Besides, spontaneously emitted photons produced in the direction of the ion beam, when seen in the LAB frame, have their energies boosted by a further factor of $2\gamma_L$, so that the process of absorption and emission results in a frequency boost of the initial laser photon of up to $4\gamma_L^2$ (by a factor of $\sim 10^8$ for the LHC beams).

The atomic beams play, in the proposed scheme, the role of highly-efficient photon frequency converters. With the present circumferential voltage of the LHC cavities and the available state-of-the-art lasers, megawatt-class photon beams, in the energy range of 10 keV–400 MeV, can be efficiently produced by the Gamma Factory. The selective photon absorption and random emission naturally opens a unique path to a very efficient manipulation and collimation of high-energy hadronic beams. In particular, it provides new methods of longitudinal and transverse beam cooling, allowing significant boost to the luminosities of the present and future hadronic colliders.

## 3. Principal Gamma Factory research tools made out of light

### 3.1. *Traps for highly-charged, small size atoms*

Highly charged high-$Z$ atoms, such as hydrogen-like or helium-like lead, are proposed to be used to probe the QCD vacuum and EW processes in compact atomic systems. They are of particular interest for the Atomic, Molecular and Optical (AMO) physics community because of their simplicity.[2] GF traps of highly-charged atoms allow the observation of the sample of $\sim 10^{10}$ atoms with 200 kHz observation/laser-manipulation frequency. Manipulation of these atoms with optical lasers becomes possible in GF owing to the large $\gamma_L$-factor of the trapped atoms. The first LHC operation with trapped hydrogen-like lead (208Pb81+) atoms demonstrated the trapping lifetime to be 20 hours at the LHC injection energy and 50 hours at top LHC energy.[3]

### 3.2. *High intensity polarised photon beams*

The concept of the GF photon source is illustrated in Fig. 1, and its characteristics can be summarised as follows:

- *point-like* — for high-$Z$, hydrogen- and helium-like atoms the distance between the laser photon absorption and fluorescence photon emission, $c\tau\gamma_L \ll 1$ cm;
- *very high intensity* — a leap in the intensity by at least seven orders of magnitude w.r.t. the electron-beam-based Inverse Compton Sources (ICS) (at the fixed $\gamma_L$ and laser pulse power);

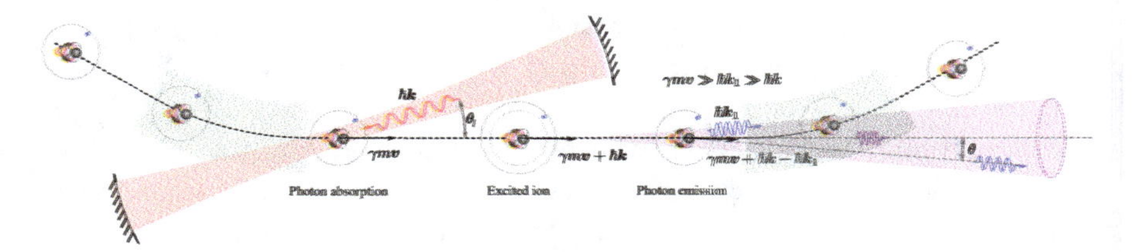


Fig. 1. The Gamma Factory concept. Laser photons with momentum $\hbar k$ (the primary photon beam) impinge onto ultra-relativistic ions (relativistic factor $\gamma$, mass $m$, velocity $v$) circulating in a storage ring. Resonantly scattered photons, as seen in the laboratory frame, are emitted in a narrow cone with an opening angle $\approx 1/\gamma$ in the direction of the motion of the ions. The energy of these secondary photons is boosted by a factor of up to $4\gamma^2$ with respect to the energy of the initial photons.

- *tuneable energy and polarisation* — the choice of: (1) the PSI beam energy (at the SPS or the LHC), (2) the ion $Z$ and $A$, (3) the number of unstripped electrons, (4) the laser type, and (5) the laser light polarisation, allows to tune the secondary photon beam energy at CERN in the energy range of 10 keV–400 MeV (extending by a factor of $\sim$ 1000 the energy range of the present and future FEL X-ray sources and providing for the first time fully polarised gamma-beams);
- *high plug-power efficiency* — PSIs lose a tiny fraction of their energy in the process of the photon emission. There is thus no need to refill the stored PSI beam. The available LHC RF power can be fully converted to the power of the GF photon beam.

### 3.3. *Laser-cooled isoscalar ion beams for precision electroweak physics at LHC*

High-luminosity collisions of isoscalar nuclei at the LHC are optimal for the LHC electroweak (EW) precision measurement programme.[4] The use of the Ca nucleus — the isoscalar nucleus with the highest atomic mass $A$ and the charge number, $Z$ — opens, in addition, the possibility of observing direct production of the Higgs boson and its decays in photon-photon collisions. The GF proposal for HL-LHC with nuclear beams[5] is to: (1) cool transversely beams of lithium-like Ca ions at the flat-top SPS energy, (2) strip the electrons in the SPS-to-LHC transfer line, and (3) collide the small transverse emittance Ca beams in the LHC.

### 3.4. *High intensity GF sources of tertiary beams*

The high-intensity photon beams with tuneable energy and polarisation open new and highly-efficient ways of creating tertiary beams of polarised electrons and positrons, polarised muons, pions, neutrinos, neutrons, and radioactive ions of unprecedented intensity — exceeding, by up to 4 orders of magnitude, the intensity of the present sources.[6,7] Polarised leptons can be produced by conversions of the GF photons in the EM field of stationary atoms.[8] Pions, produced abundantly by the photo-excitation of $\Delta$-resonances, are proposed to drive the high-intensity, low-emittance, GF muon source.[8] Muons, following their quick PWFA acceleration,[a] can be used to produce high-purity neutrino and antineutrino beams. Thanks to the muon polarisation, muon-neutrino (muon-antineutrino) beams could be separated from the electron-antineutrino (electron-neutrino) ones on the bases of their respective angular distributions.[10] Neutrons and radioactive ions can be produced by photo-excitation of the giant dipole resonances[11] and by exciting nuclear fission resonances.[12] The GF leap in the neutron source intensity, and in the rate of the fission processes, can open new avenues: (1) for the development of the photon-beam-driven advanced energy source based on subcritical reactor, and (2) for the highly efficient transmutation of nuclear waste products.

### 3.5. *Electron beam for ep collisions in the LHC interaction points*

The hydrogen-like or helium-like lead beams can be considered as the carriers of the effective electron beams circulating in the LHC rings. Collisions of such a beam with the counter propagating beam of protons provides a unique, costless option to study the electron-proton collisions in the existing LHC detectors.[13]

### 3.6. *Drive beams for plasma wakefield acceleration*

The GF high-intensity, laser-cooled atomic beams can play the role of efficient driver beams for hadron-beam-driven plasma wakefield acceleration.[14] Electrons exploited initially in the cooling and micro-bunching process of the driver beam can be subsequently used — following their stripping — to form a precisely synchronised electron witness bunches.[15]

---

[a]They can reach energy of 10 GeV over the distance of $\approx 3$ m.[9]

## 4. GF project milestones and status

The path towards full feasibility proof of the GF concepts is landmarked by the following six milestones: (1) demonstration of efficient production, acceleration, and storage of atomic beams in the CERN accelerator complex; (2) development of the requisite GF simulation tools; (3) building up the physics cases for the GF research program and attracting wide scientific communities to evaluate the merits of the GF tools in their respective research; (4) successful execution of the GF Proof-of-Principle (PoP) experiment at SPS; (5) verification of the GF performance parameters on the basis of the PoP experiment results and simulations; (6) elaboration of the GF Technical Design Report (TDR).

The first three of the six milestones have been already achieved.[3,16–21] The present status of the software development was summarised at the recent GF software workshop at CERN.[22] So far, about one hundred physicists contributed to the GF project's development. The recent studies of the physics highlights of the GF research programme has been published in a special issue of the "Annalen der Physik" journal.[23] The GF PoP experiment has already been designed[24] and its proposal, presented to the SPSC. The GF R&D on the low phase-noise laser system has achieved its goals.[25] Dedicated SPS beam tests with hydrogen-like and helium-like lead ions were performed — demonstrating sufficient stability of the atomic lead beams in the SPS and LHC rings. The GF-PoP experiment is waiting for the "go-ahead' decision by CERN.

## 5. Outlook

If realised, the proposed GF facility, shown in Fig. 2, can inject a new scientific life to the LHC storage rings, following the completion of its HL-LHC programme, by addressing new research domains in atomic, nuclear, particle, fundamental and applied physics with novel research tools of unprecedented quality. The GF research tools can significantly increase the LHC discovery potential. Such an unexpected bonus should not be missed.

## References

1. M. W. Krasny, *The Gamma Factory proposal for CERN*, arXiv:1511.07794.
2. M. S. Safronova *et al.*, Reviews of Modern Physics, **90** (2018) 025008.
3. M. Schaumann *et al.*, J. Phys. Conf. Ser. **1350** (2019) no. 1, 012071.
4. M. W. Krasny *et al.*, Eur. Phys. J. C **69** (2010) 379.

Fig. 2. The schematic view of the proposed LHC-based Gamma Factory facility. Two counter propagating, but non-interacting PSI beams, collide with the laser photons in eleven, application-specific, collision points. The clock-wise moving beam drives the low energy GF applications (highlighted with the yellow colour), while the anti-clock-wise beam drives the high energy GF applications (highlighted with the blue colour).

5. M. W. Krasny, A. Petrenko and W. Płaczek, Prog. Part. Nucl. Phys. **114** (2020), 103792, arXiv:2003.11407 [physics.acc-ph].
6. I. Chaikovska *et al.*, arXiv:2202.04939 [physics.acc-ph].
7. M. Aiba *et al.*, arXiv:2111.05788 [hep-ex].
8. A. Apyan, M. W. Krasny, W. Placzek, in preparation.
9. V. D. Shiltsev, FERMILAB-PUB-22-137-AD.
10. A. Blondel, Nucl. Instrum. Meth. A **451** (2000), 131–137.
11. D. Budker *et al.*, Annalen Phys. (2022) 2100284, arXiv:2106.06584 [nucl-ex].
12. D. Nichita *et al.*, Annalen Phys. (2021) 2100207, arXiv:2105.13058 [nucl-ex].
13. M. W. Krasny, Nucl. Instrum. Meth. A **540** (2005), 222–234.
14. A. Caldwell *et al.*, Nature Physics **5** (2009) 363.
15. D. A. Cooke *et al.*, arXiv:2006.16160 [physics.acc-ph].
16. *Gamma Factory for CERN*, CERN Yellow Report, in preparation.
17. S. Hirlaender *et al.*, doi:10.18429/JACoW-IPAC2018-THPMF015.
18. M. Krasny *et al.*, doi:10.18429/JACoW-IPAC2018-WEYGBD3.
19. A. Gorzawski *et al.*, Phys. Rev. Accel. Beams **23** (2020) no. 10, 101002.
20. W. Placzek *et al.*, Acta Phys. Polon. B **50** (2019) no. 6, 1191–1203.
21. Y. Dutheil *et al.*, doi:10.18429/JACoW-IPAC2019-MOPRB052
22. *BE-ABP Gamma Factory Software Workshop*, CERN, October, 2021.
23. *Physics Opportunities with the Gamma Factory*, Annalen Phys., special edition, March 2022, Volume 534, Issue 3,
24. M.W. Krasny *et al.*, Letter of Intent, CERN-SPSC-2019-031/SPSC-I-253.
25. A. Martens *et al.*, *Design of the optical system for the Gamma Factory Proof of Principle experiment at the CERN SPS*, to be published in Phys. Rev.

© 2024 The Author(s)
https://doi.org/10.1142/9789811280184_0022

# Chapter 22

# An Energy Recovery Linac for the LHC

S. Alex Bogacz, Bernhard J. Holzer and John A. Osborne

*European Organization for Nuclear Research (CERN), Genève, Switzerland*

*Center for Advanced Studies of Accelerators, Jefferson Lab (TJNAF), Newport News, USA*

The LHeC provides an intense, high energy electron beam to collide with the LHC as the sole opportunity for a next energy frontier electron-hadron collider. It represents the highest energy application of energy recovery linac (ERL) technology — which is increasingly recognised as one of the major pilot technologies for the development of particle physics because it utilises and stimulates superconducting RF technology progress, and it increases intensity while keeping the power consumption low. The LHeC instantaneous luminosity is determined through the integrated luminosity goal. The electron beam energy is chosen to achieve TeV cms collision energies and enable competitive searches and precision Higgs boson measurements. The wall-plug power has been constrained to about 100 MW. Two super-conducting linacs of about 900 m length, which are placed opposite to each other, accelerate the passing electrons by 8.3 GeV each. This leads to a final electron beam energy of about 50 GeV in a 3-turn racetrack energy recovery linac configuration.

## 1. Introduction

Within half a century, particle physics has established a Standard Model (SM) for the description of the fundamental constituents of matter and their electroweak and strong interactions. Besides confirming the SM in many (previously unexplored) areas, the largest contribution of the LHC to the development of particle physics, so far, has been the discovery of the

This is an open access article published by World Scientific Publishing Company. It is distributed under the terms of the Creative Commons Attribution 4.0 (CC BY) License.

Higgs boson and the study of its properties. The Standard Model, however, despite its phenomenological success, has severe deficiencies. For example, it lacks the "grand unification" of the particle interactions, has more than twenty free parameters, does not explain the existence of 3 quark and lepton families nor the difference between the nature of leptons and quarks. The strong interaction in the SM is described by Quantum Chromodynamics (QCD), which is still far from being completely developed, e.g., neither has it provided an explanation of parton confinement — a prediction about which substructure layers may exist, nor have its assumptions on non-linear dynamics in high parton density regimes been verified.

In the past decades of research on elementary particle physics, progress has been made with a threefold strategy by exploring each level of high energy with hadron-hadron, electron-positron and lepton-hadron experiments mostly based on colliders. This holds for the beginning and later for the time of the exploration of nature with the Sp$\bar{\text{p}}$S, PETRA/PEP and the fixed target electron, neutrino and muon-hadron experiments. It was repeated when the Tevatron, LEP/SLC and HERA accessed the Fermi energy scale corresponding to the masses of the weakly interacting bosons, $Z$ and $W^{\pm}$, and the top quark.

With the LHC a new era began, that of exploring the SM at even higher energies and searching for its possible extensions in the TeV energy range. A new electron-positron collider has been proposed to be built, with several candidate technologies for a new enlarged ring machine (FCC-ee at CERN and CEPC in China) or based on linear collider techniques (ILC, CLIC, or more recent concepts using high gradient or energy recovery variations). For a next generation lepton-hadron collider extending the energy frontier into the TeV region, the hadron (proton and ion) beams of the LHC provide the only realistic foundation for the coming decades. This has been recognised with the accelerator, physics and detector developments of the Large Hadron Electron Collider (LHeC), as has been documented in an initial, detailed, refereed Conceptual Design Report which was published in 2012,[1] at the time of the discovery of the Higgs Boson. A similarly detailed report[2] on the LHeC appeared in 2020, accounting for a decade of further developments of physics as well as accelerator and detector technologies.

This work, undertaken by hundreds of contributing physicists and engineers, mandated by CERN and ECFA, has provided a novel electron-hadron collider design with a few salient characteristics: i) the combination of the LHC proton beam with an about 50 GeV, 3-turn, intense energy recovery racetrack-linac accelerator resulted in a realistic, affordable design exceeding the parameters of HERA by a factor of twenty in kinematics and nearly 1000 in luminosity; ii) the LHeC design is for concurrent LHC and LHeC operation which, conceptually, does not reduce the LHC running time; iii) owing to the $10^{34}$ $cm^{-2}s^{-1}$ achievable instantaneous luminosity, the LHeC has a competitive Higgs physics and BSM search potential while providing ample luminosity for unprecedented research in the strong and electroweak interaction area[2] as is summarised in subsequent sections of this book; iv) the introduction of energy recovery technology stands out as an example for the required low power requests to the next generation of colliders for particle physics.

Each significant step of beam energy increase has provided deeper, often unexpected insights to the characteristics of elementary particle physics. With lepton-hadron experiments, a series of discoveries were made: of the finite proton size; of the existence of parton substructure and the confirmation of the quark-parton model; with the proof for the electron to weakly couple as a right-handed singlet, the decisive verification of the Glashow-Weinberg-Salam theory in 1978; later with the first measurements on parton momentum distributions and the discovery of high gluon and sea-quark densities and deep inelastic diffraction. There is hardly a quantitative understanding of the LHC pp interactions without the HERA input. The LHeC can be expected to unravel further surprises, while leading to a much deeper understanding of parton dynamics and the Higgs mechanism, and through that of LHC physics.

In order for a combined hh and eh (h = p, A) collider configuration to become a realistic, attractive scenario all essential aspects of the novel electron ERL, the main component of the LHeC, had to be carefully studied.[1,2] The following sections present an overview on the LHeC (Sec. 2) and three particularly challenging developments: the design of an optimum lattice for a 3-turn electron beam using two linacs opposite to each other in the straight sections (Sec. 2.1), the layout of an interaction region able to deal with the colliding electron and hadron beams while letting the non-colliding hadron beam pass (Sec. 2.2) and, finally, a study of the civil engineering aspects of placing a racetrack tunnel of about 6 km circumference in the vicinity of the LHC ring for enabling electron-hadron interactions (Sec. 2.3). Further sections provide overviews on various key aspects of physics at the LHeC and a design concept for a novel detector to deliver electron-proton and -ion physics in the new ranges of energy and luminosity.

The LHeC design has been adopted for the initial layout of the 3.5 TeV cms energy electron-hadron facility at the Future Circular Collider

(FCC-eh[3]). Should the LHeC be built, its key accelerator parts are considered to be relocatable to the FCC. If the LHC would be upgraded in energy, the investment in the LHeC would even more directly pay off as the electron accelerator could promptly be used for deep inelastic scattering (DIS) physics at even higher energy and intensity, at the HE-LHC. Furthermore, the electron accelerator of the LHeC may be used as the injector of a future FCC-ee machine allowing direct top-up injection at W production energies.

## 2. The ERL Configuration of the LHeC

The LHeC provides an intense, high energy electron beam to collide with the LHC. Furthermore, it pushes energy recovery linac (ERL) technology into an unprecedented beam energy regime, which is increasingly recognised as one of the major pilot technologies for the development of particle physics. Finally, it utilises and stimulates superconducting RF technology progress, and it increases intensity while keeping the power consumption manageable. The electron beam energy is chosen to achieve TeV cms collision energy and enable competitive searches and precision Higgs boson measurements. A cost-physics-energy evaluation, presented in Ref. 4, points to choosing $E_e \simeq 50$ GeV as a new default value, which was 60 GeV before. The wall-plug power has been constrained to 140 MW. The main parameters of the LHeC ERL are listed in Table 1.

Table 1. Parameters of LHeC Energy Recovery Linac (ERL).

| Parameter | Unit | Value |
|---|---|---|
| Injector energy | GeV | 0.5 |
| Maximum electron energy | GeV | 49.19 |
| Bunch charge | pC | 499 |
| Bunch spacing | ns | 24.95 |
| Electron current | mA | 20 |
| Total energy gain per linac | GeV | 8.114 |
| Frequency | MHz | 801.58 |
| Acceleration gradient | MV/m | 19.73 |
| Number of cells per cavity | | 5 |
| Cavities per cryomodule | | 4 |
| Cryomodule length | m | 7 |
| Total ERL length | km | 5.332 |

The ERL consists of two superconducting (SC) linacs operated in CW connected by at least three pairs of arcs to allow three accelerating and three decelerating passes (see Fig. 1). The length of the high energy return arc following the interaction point should be able to provide a half RF period wavelength shift to allow the deceleration of the beam in the linac structures in three passes down to the injection energy and its safe disposal. SC Cavities with an unloaded quality factor $Q_0$ exceeding $10^{10}$ are required to minimise the requirements on the cryogenic cooling power and to allow an efficient ERL operation. The choice of having three accelerating and three decelerating passes implies that the circulating current in the linacs is six times the current colliding at the Interaction Point (IP) with the hadron beam.

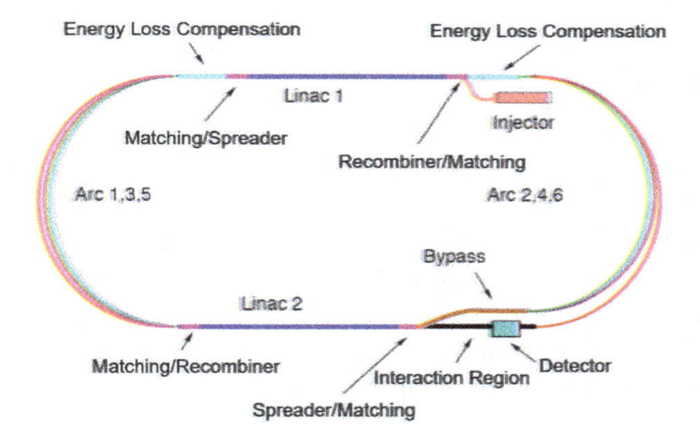


Fig. 1. Schematic layout of the LHeC design based on an Energy Recovery Linac.

## 2.1. *Linac Configuration and Multi-pass Optics*

Appropriate choice of the linac optics is of paramount importance for the transverse beam dynamics in a multi-pass ERL. The focusing profile along the linac (quadrupole gradients) needs to be set (and they stay constant), so that multiple pass beams within a vast energy range may be transported efficiently. The chosen arrangement is such that adequate transverse focusing is provided for a given linac aperture. The linac optics is configured as a strongly focusing, $130^0$ FODO. In a basic FODO cell a quadrupole is placed every four cryomodules, so that the full cell contains two groups of 16 RF cavities and a pair of quads (F, D). Energy recovery in a racetrack topology explicitly requires that both the accelerating and decelerating beams share the individual return arcs.[5] This, in turn, imposes specific requirements for

TWISS function at the linacs ends: TWISS functions have to be identical for both the accelerating and decelerating linac passes converging to the same energy and therefore entering the same arc.

As extensively discussed in Ref. 6, the corresponding accelerating and decelerating passes are joined together at the arc's entrance/exit. The optics of the two linacs are mirror-symmetric; They were optimised such that, Linac 1 is periodic for the first accelerating pass and Linac 2 has this feature for last decelerating one. In order to maximize the BBU (Beam Breakup) Instability threshold current,[7] the optics is tuned so that the integral of $\beta/E$ ($\beta$ being the betatron function and $E$, beam energy) along the linac is minimised.

#### 2.1.1. *Spreaders and Recombiners*

The spreaders are placed directly after each linac to separate beams of different energies and to route them to the corresponding arcs. The recombiners facilitate just the opposite: merging the beams of different energies into the same trajectory before entering the next linac. Each spreader starts with a vertical bending magnet, common for all three beams, that initiates the separation. The highest energy, at the bottom, is brought back to the horizontal plane with a chicane. The lower energies are captured with a two-step vertical bending adapted from the CEBAF design.[8] The vertical dispersion is suppressed by a pair of quadrupoles located in-between vertical steps. An alternative spreader design with a single vertical step has been explored as well. That option was not retained due to the superconducting technology needed for the quadrupoles that must be avoided in this highly radiative section.

#### 2.1.2. *Synchrotron Radiation — Emittance Preserving Optics*

Synchrotron radiation effects on beam dynamics, such as the energy loss, as well as the transverse and longitudinal emittance dilution induced by quantum excitations, have a paramount impact on the collider luminosity. These quantities, first introduced by M. Sands[9] are summarized below:

$$\Delta E = \frac{2\pi}{3} r_0 \, mc^2 \, \frac{\gamma^4}{\rho} \tag{1}$$

$$\Delta\epsilon_N = \frac{2\pi}{3} C_q r_0 < H > \frac{\gamma^6}{\rho^2}\,, \tag{2}$$

$$\frac{\Delta\epsilon_E^2}{E^2} = \frac{2\pi}{3} C_q r_0 \, \frac{\gamma^5}{\rho^2}\,, \tag{3}$$

where $C_q = \frac{55}{32\sqrt{3}} \frac{\hbar}{mc}$. Here, $\Delta\epsilon_E^2$ is an increment of energy square variance, $r_0$ is the classical electron radius, $\gamma$ is the Lorentz boost and $C_q \approx 3.832 \cdot 10^{-13}$ m for electrons (or positrons). Here, $H = (1+\alpha^2)/\beta \cdot D^2 + 2\alpha\, DD' + \beta \cdot D'^2$ where $D, D'$ are the bending plane dispersion and its derivative, with $< ... > = \frac{1}{\pi} \int_{\text{bends}} ... \, d\theta$.

Therefore, emittance dilution can be mitigated through appropriate choice of arc optics (values of $\alpha, \beta, D, D'$ at the bends). In the presented design, the arcs are configured with a FMC (Flexible Momentum Compaction) optics to ease individual adjustment of emittance dispersion averages, $< H >$, in various energy arcs.

Optics design of each arc takes into account the impact of synchrotron radiation at different energies. At the highest energy, it is crucial to minimise the emittance dilution due to quantum excitations; therefore, the cells are tuned to minimise the emittance dispersion, $H$, in the bending sections, as in the TME (Theoretical Minimum Emittance) lattice. The higher energy arcs (4, 5 and 6) configured with the TME cells still quasi-isochronous. All styles of FMC lattice cells, as illustrated in Fig. 2, share the same footprint for each arc. This allows us to stack magnets on top of each other or to combine them in a single design.

Cumulative dilution of the transverse, $\Delta\epsilon_N$, and longitudinal, $\Delta\sigma_{\frac{\Delta E}{E}}$, emittance due to quantum excitations calculated using analytic formulas,

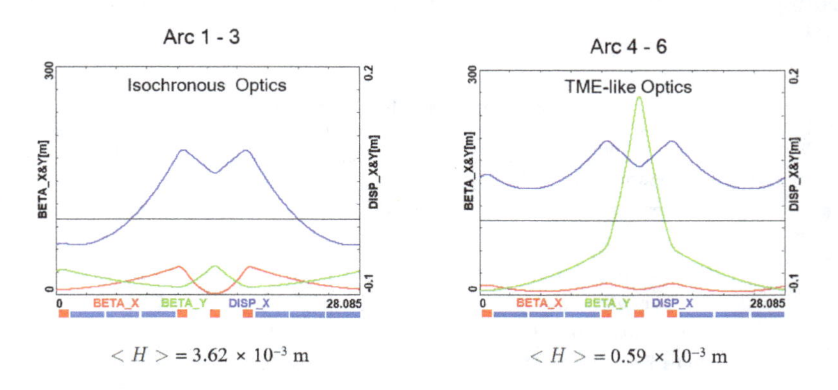


Fig. 2. Two styles of FMC cells appropriate for different energy ranges. Left: lower energy arcs (Arc 1–3) configured with *Isochronous* cells, Right: higher energy arcs configured with *TME-like* cells. Corresponding values of the emittance dispersion averages, $< H >$, are listed for both style cells.

Eqs. (1), (2) and (3), at the end of arc 6 (including spreaders, recombiners and pathelength corrcting 'doglegs') can be summarized as follows:[2] about 25 mm mrad (in both the horizontal end vertical plane) and 0.24%. Net energy loss of 836 MeV has to be replenished back to the beam, so that at the entrance of each arc the accelerated and decelerated beams have the same energy, unless separate arcs are used for the accelerated and decelerated beams. As discussed in detail in Ref. 6, the compensation makes use of a second harmonic RF at 1603.2 MHz to replenish the energy loss for both the accelerated and the decelerated beams.

### 2.2. *Interaction Region*

The Interaction Region (IR) of the ERL is one of the most challenging parts of the machine: While seeking for highest luminosity in ep-collisions, the bunches have to be separated after collision and guided to their lattice structures, to avoid parasitic bunch encounters. In addition, beam-beam effects with the second non-colliding proton beam have to be avoided. In order to meet these requirements, the design of the IR has been based on a compact magnet structure for an effective beam separation and smallest synchrotron radiation effects in the Interaction Region (IR). Following the design of the LHC upgrade project, HL-LHC, and the layout of the ERL for the electrons, the parameter list of the LHeC has been defined, Table 2, leading to a luminosity at the e-p interaction point in the order of $L = 10^{34}$ cm$^{-2}$s$^{-1}$.

Table 2. Parameter list of the LHeC.

| Parameter | Unit | Electrons | Protons |
|---|---|---|---|
| beam energy | GeV | 50 | 7000 |
| beam current | mA | 20 | 1400 |
| bunches per beam | - | 1188 | 2808 |
| bunch population | $10^{10}$ | 0.3 | 22 |
| bunch charge | nC | 0.5 | 35.24 |
| norm. emittance (at IP) | mm · mrad | 30 | 2.5 |
| beta function at IP | cm | 10.9 | 10 |
| beam-beam disruption | - | 14.3 | $1 \cdot 10^{-5}$ |
| luminosity | cm$^{-2}$s$^{-1}$ | $0.7 \cdot 10^{34}$ | |

### 2.2.1. *Electron Beam Optics and Separation Scheme*

A manifold of conditions are taken into account: Focusing of the electron beam to the required $\beta$ values in both planes, sufficient beam separation, optimisation of the beam separation for smallest critical energy and synchrotron light power, and sufficient space for the detector hardware. A separation scheme has been established[10] that combines these requirements in one lattice structure (Fig. 3). Due to the different rigidity of the beams, a separation is possible by applying a series of magnets, acting as quasi-constant deflecting field: The spectrometer dipole of the LHeC detector(*B0*) is used to establish a first separation. Right after and as close as possible to the IP, the mini-beta quadrupoles of the electron beam are located. A doublet design allows highest compactness of the IR layout and provides focusing in both planes for matched beam sizes of protons and electrons at the IP, $\beta_x(p) = \beta_x(e)$, $\beta_y(p) = \beta_y(e)$. The two quadrupoles are positioned off-center with respect to the electron beam, acting as combined function magnets to provide a continuous soft bending of the electron beam throughout the complete magnet structure.

As indicated in (Fig. 3) the co-action of an early focusing scheme of electrons, the use of off-centre quadrupoles and the minimised beam size at the separation point lead to a considerable reduction of critical energy and power of the emitted light. The presently obtained values of 250 keV and 19 kW respectively are still challenging and considered as work in progress. Especially in the context of the HL-LHC luminosity upgrade the free space available for beam separation will increase, leading to a further reduction of the synchrotron light parameters.

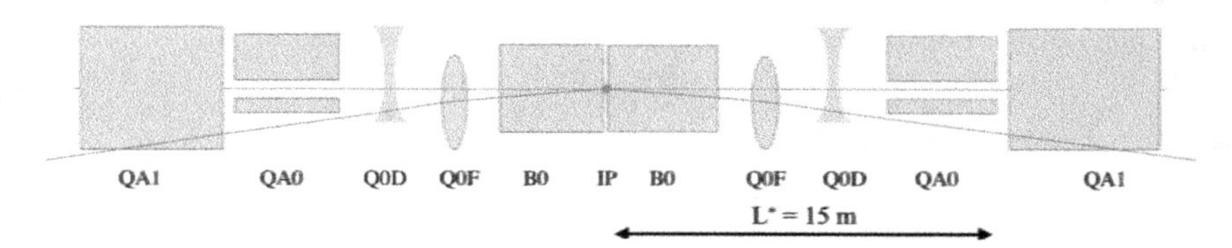


Fig. 3. Schematic view of the combined focusing - beam separation scheme.

### 2.2.2. *Proton Beam Optics*

The optics of the colliding proton beam follows the standard settings of the HL-LHC and is based on the so-called ATS scheme (achromatic telescoping squeeze). It allows smaller values of $\beta^*$ at a given collision point — and thus higher luminosity. Figure 4 shows the proton optics for values of e.g. $\beta^* = 7$ cm at the interaction point of the LHeC — embedded and well matched into the HL-LHC optics for the ATLAS and CMS interaction points. The long-ranging beta-beat, which is an essential feature of the HL-LHC optics,[11] is clearly visible on both sides of the IP.

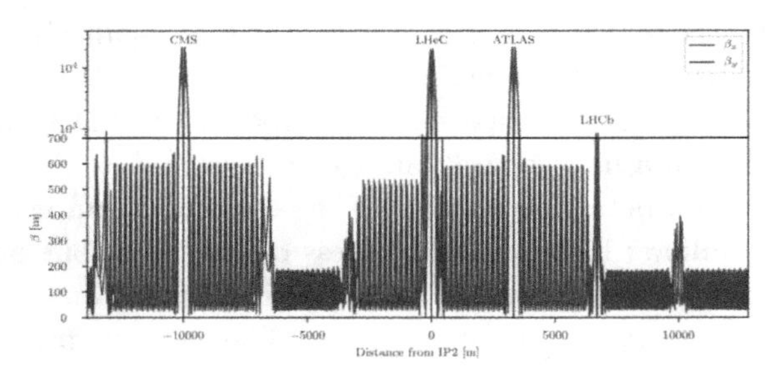


Fig. 4. LHC proton beam optics, optimised for values of $\beta = 7$ cm at the LHeC IP.

The operation of the LHeC electron-proton collisions is foreseen in parallel to the LHC standard p-p operation. As a consequence, the design orbit of the second "non-colliding" proton beam at the e-p interaction point must be included in the e-p IR layout. At the e-p interaction region, a collision of the two proton beams is avoided by selecting appropriately its location: Shifted in position and thus in time, direct collisions between the two proton beams as well as with the electron beam can be excluded. All in all the e-p interaction region, including the mini-beta structure of the electron beam, is embedded in the existing LHC lattice to allow for concurrent e-p and p-p collisions in the LHC interaction points.

### 2.2.3. *Beam-Beam Effects*

The beam-beam effect will always be a strong limitation of a particle collider and care has to be taken, to preserve the beam quality of proton and electron beam. In case of the proton beam, the beam beam effect has to be limited to preserve the proton beam emittance and allow successful parallel data taking in the p-p collision points. Due to the limited bunch population of the electron beam, this is fulfilled by design. In the case of the electron beam the beam-beam effect is determined by the proton bunch population, which is considerably higher than the electron bunch intensity and its detrimental effects on the electron emittance had to be limited to assure a successful energy recovery process in the ERL. As was comprehensively simulated in Ref. 10, the core of the beam still remains in a quasi ellipse like boundary, while tails in the transverse beam distribution are clearly visible (as consequence of the beam-beam effect). The simulation showed that these tails are still compatible with the energy recovery process.

## 2.3. *Civil Engineering*

### 2.3.1. *Introduction*

Since the beginning of the LHeC concept, various shapes and sizes of the eh collider were studied around CERN region. The conceptual study report published in 2012 focused primarily on two main options, namely the RING-RING and the LINAC-RING options. For civil engineering, these options were studied, taking into account geology, construction risks, land features as well as technical constrains and operation of the LHC. The Linac-Ring configuration was chosen as baseline for its largely decoupled CE and installation work from the nominal LHC operation.

This chapter describes the civil engineering infrastructure required for an Energy Recovery Linac (ERL) injecting into the LHC ALICE cavern at LHC Point 2. Figure 5 shows three options of different sizes proposed for the ERL, represented as fractions of the LHC circumference. This chapter focuses on the currently preferred option, specifically the 1/5 of the LHC circumference.

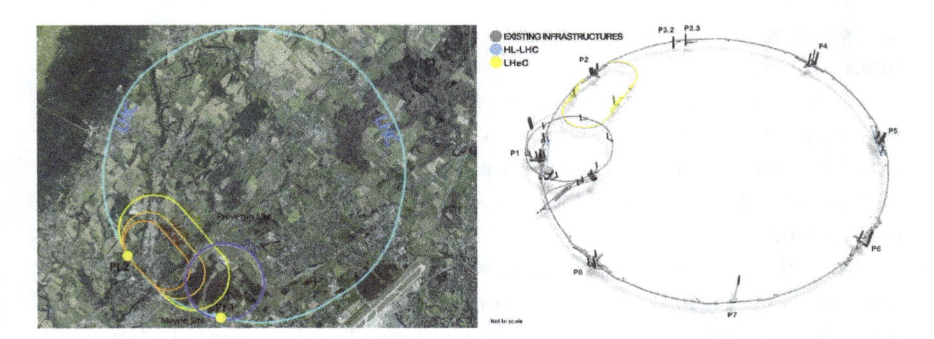


Fig. 5. Three racetrack alternatives proposed for the eh machine at LHC Point 2 (left) and 3D schematic showing the proposed racetrack of the Large Hadron electron Collider at high luminosity (right).

### 2.3.2. *Placement and Geology*

The proposed site for the LHeC is in the North-Western part of the Geneva region at the existing CERN laboratory. The proposed Interaction Region is fully located within existing CERN land at LHC Point 2, close to the village of St. Genis, in France. The CERN area is extremely well suited to housing such a large project, with well understood ground conditions having several particle accelerators in the region for over 50 years. Extensive geological records exist from previous projects such as SPS, LEP and LHC and more recently, further ground investigations have been undertaken for the High-Luminosity LHC project. Any new underground structures will be constructed in the stable molasse rock at a depth of 100–150 m in an area with very low seismic activity.

The ERL will be positioned inside the LHC Ring, in order to ensure that new surface facilities are located on existing CERN land. The proposed underground structures for a Large Hadron electron Collider (LHeC) at high luminosity aiming for an electron beam energy of 50 GeV is shown in Fig. 5. The LHeC tunnel will be tilted similarly to the LHC at a slope of 1.4% to follow a suitable layer of molasse rock.

### 2.3.3. *Underground infrastructure*

The underground structures proposed for LHeC in the proposed design with a 1/5 LHC circumference require a tunnel approximately 5.4 km long of 5.5 m diameter, including two LINACs. Parallel to the main LINAC tunnels, at 10 m distance apart, are the RF galleries, each 830 m long. Waveguides of 1 m diameter are connecting the RF galleries and LHeC main tunnel.

Two additional caverns, 25 m wide and 50 m long are required for cryogenics and technical services. These are connected to the surface via two 9 m diameter access, provided with lifts to allow access for equipment and personnel. Additional caverns are needed to house injection facilities and a beam dump.

In addition to the new structures, the existing LHC infrastructure also requires modifications. To ensure connection between LHC and LHeC tunnels, the junction caverns UJ22 and UJ27 need to be enlarged. Localised parts of the cavern and tunnel lining will be broken out to facilitate the excavation of the new spaces and the new connections, requiring temporary support.

Infrastructure works for LEP were completed in 1989, for which a design lifespan of 50 years was specified. If LHC is to be upgraded with a high energy, refurbishment, maintenance works are needed to re-use the existing infrastructure. Shaft locations were chosen such that the surface facilities are located on CERN land. The scope for surface sites is still to be defined. New facilities are envisaged for housing technical services such as cooling and ventilation, cryogenics and electrical distribution.

### 2.3.4. *Construction Methods*

A Tunnel Boring Machines (TBM) should be utilised for the excavation of the main tunnel to achieve the fastest construction. When ground conditions are good and the geology is consistent, TBMs can be two to four times faster than conventional methods. A shielded TBM could be employed, with pre-cast segmental lining, and injection grouting behind the lining.

For the excavation of the shafts, caverns and connection tunnels, conventional technique could be used. Similar construction methods as for HL-LHC, for example using roadheaders and rockbreakers, can be adopted for LHeC. Some of these machinery can be seen in Fig. 6, showing the excavation works at point 1 HL-LHC. One main constraint that dictated the equipment used for the HL-LHC excavation was the vibration limit. Considering the sensitivity of the beamline, diesel excavators have been modified and equipped with an electric motor in order to reduce vibrations that could disrupt LHC operation. A similar equipment could also be needed for LHeC if construction works are carried out during operation of the LHC.

Fig. 6. Excavator with hydraulic cutting heads being used at HL-LHC Point 1.

Existing boreholes data around IP2 shows that the moraines layer can be 25–35 m deep before reaching the molasse. Temporary support of the excavation, for example using diaphragm walls, are recommended. Once reaching a stable ground in dry conditions, common excavation methods can be adopted, for example using a roadheaders and rockbreakers. The shaft lining will consist of a primary layer of shortcrete with rockbolts and an in-situ reinforced concrete secondary lining, with a waterproofing membrane in between the two linings.

#### 2.3.5. *Cost Estimate*

A cost estimate was prepared for a 9.1 km ERL located at Point 2 of LHC, using the same measure prices as for FCC. More recently for LHeC, the cost figures were adapted to fit the smaller version, the 5.4 km racetrack at Point 2 (option 1/5 LHC).

The civil engineering costs amount to about 25% of the total project costs. In particular, for a 9.1 km ERL (1/3 LHC option) the civil engineering was estimated to 386 MCHF and for a 5.4 km configuration (1/5 LHC) the costs is 289 MCHF. These estimates include the fees for preliminary design, approvals and tender documents (12%), site investigations (2%) and contractor's profit (3%). The costs mentioned do not include surface structures. Where possible, existing surface infrastructure will be re-used.

## 3. Outlook

### 3.1. *The ERL Development Facility PERLE at Orsay*

PERLE (Powerful ERL for Experiments)[2,12] — a 'stepping stone' to the LHeC — is envisioned as a novel ERL test facility, designed to validate choices for a high energy ERL foreseen in the design of the Large Hadron electron Collider (50 GeV) and the Future Circular Collider (FCC-eh, 60 GeV). PERLE is a compact ERL, resembling the LHeC configuration, based on superconducting RF technology, expands the operational regime for ERLs to 10 MW of beam power. This is achieved with 20 mA electron current, as foreseen for the LHeC, and 500 MeV electron beam energy generated in 3 passes through two linac modules. The cryomodules equipped with four 5-cell Niobium cavities, and other elements of PERLE may directly be applied to future, more complex accelerators. One may envision PERLE being used as the injector for the LHeC. PERLE is being built at the IJC Laboratory at Orsay near Paris by an International Collaboration.

The facility was described and recognized in 2021 as a key part of the European Roadmap towards novel accelerators[13] for its unique characteristics paving the way not only for sustainable, multi-turn ERL technology, but also for pioneering industrial and low energy physics applications.

### 3.2. *The Future of the LHeC*

The Large Hadron electron Collider has been designed[2] as a novel part of the LHC facility with a far reaching physics program — both for energy frontier deep inelastic electron-hadron scattering and for empowering the exploration of proton-proton and heavy ion physics at the LHC. It builds on the complex, existing, expensive infrastructure of the LHC and represents the most economic way towards a higher precision Higgs physics program, which specifically relies on energy recovery technology at high currents. ERL is a principal means for reducing the power consumption for the next generation of lepton colliders. Operating without energy recovery, the LHeC would use GWs of power. Thanks to employing the energy recovery, the net power is reduced to 100 MW or possibly even lower. It thus is a first large scale example of an energy efficient particle physics accelerator, for which PERLE primarily provides and tests the required technology.

The book on the future of the LHC is being written, for the time beginning with and yet reaching beyond its high luminosity phase. The physics at the Fermi scale was explored about two decades ago with the hadron collider Tevatron, the $e^+e^-$ collider LEP and the first ep collider HERA. The LHeC represents the unique and timely possibility to accompany the hadron collider LHC with a partner electron-hadron collider to gain the necessary insight for particle physics to proceed. The Standard Model may then possibly be included in a fundamental theory of particles and their interactions. These developments cannot proceed without a next generation of energy frontier colliders, including one for TeV energy deep inelastic scattering. The LHeC can be realised with the HL-LHC and it may come with the HE-LHC. Its visionary prospect is the 3.5 TeV version, the FCC-eh.

## Funding Information

Work at Jefferson Lab has been supported by the U.S. Department of Energy, Office of Science, Office of Nuclear Physics under contracts DE-AC05-06OR23177 and DE-SC0012704.

## Acknowledgments

Useful discussions and conceptual input from: Kevin André, Oliver Brüning, Max Klein, Walid Kaabi, Dario Pellegrini and Alessandra Valloni at all stages of the accelerator design process are greatly acknowledged.

## References

1. J. L. Abelleira et al., A Large Hadron Electron Collider at CERN: Report on the Physics and Design Concepts for Machine and Detector, *J. Phys.* **G39**, 075001 (2012). doi: 10.1088/0954-3899/39/7/075001.
2. P. Agostini et al., The Large Hadron–Electron Collider at the HL-LHC, *J. Phys. G.* **48** (11), 110501 (2021). doi: 10.1088/1361-6471/abf3ba.
3. A. Abada et al., FCC-hh: The Hadron Collider: Future Circular Collider Conceptual Design Report Volume 3, *Eur. Phys. J. ST.* **228** (4), 755–1107 (2019). doi: 10.1140/epjst/e2019-900087-0.
4. O. Brüning, LHeC Cost Estimate, *CERN-ACC-2018-0061* (2018).
5. S. A. Bogacz et al., Novel Lattices Solutions for the LHeC, *ICFA Beam Dynamics Newsletter.* **71**, 135 (2017).
6. D. Pellegrini, A. Latina, D. Schulte, and S. A. Bogacz, Beam-dynamics driven design of the LHeC energy-recovery linac, *Phys. Rev. ST Accel. Beams.* **18** (12), 121004 (2015). doi: 10.1103/PhysRevSTAB.18.121004.
7. G. H. Hoffstaetter and I. V. Bazarov, Beam-breakup instability theory for energy recovery linacs, *Phys. Rev. ST Accel. Beams.* **7**, 054401 (2004). doi: 10.1103/PhysRevSTAB.7.054401.
8. A. Freiberger. 12 GeV CEBAF Upgrade, Reference Design: `www.jlab.org/physics/GeV/accelerator` (2012).
9. J. S. Schwinger, On radiation by electrons in a betatron, *Phys. Rev.* **70**, 798 (1946). doi: 10.2172/1195620.
10. K. D. J. André, Lattice Design and Beam Optics for the Energy Recovery Linac of the Large Hadron-Electron Collider, *Ph.D. Thesis, university Liverpool, UK* (in: prep).
11. I. Bejar Alonso et al., High Luminosity Hadron Collider, HL-LHC, technical design report, *CERN Yellow Report.* **CERN 2020-10** (2020).
12. D. Angal-Kalinin et al., PERLE - Powerful energy recovery linac for experiments. Conceptual design report, *Journal of Physics G: Nuclear and Particle Physics.* **45** (2018). ISSN 13616471. doi: 10.1088/1361-6471/aaa171.
13. C. Adolphsen et al., European Strategy for Particle Physics – Accelerator R&D Roadmap, *CERN Yellow Rep. Monogr.* **1**, 1–270 (2022). doi: 10.23731/CYRM-2022-001.

© 2024 The Author(s)
https://doi.org/10.1142/9789811280184_0023

# Chapter 23

# Electron-Hadron Scattering resolving Parton Dynamics

Néstor Armesto
*Instituto Galego de Física de Altas Enerxías IGFAE, Universidade de Santiago de Compostela, 15782 Santiago de Compostela, Galicia-Spain*

Claire Gwenlan
*Department of Physics, University of Oxford, Denys Wilkinson Building, Keble Road, Oxford. OX1 3RH. United Kingdom*

Anna Stasto
*Department of Physics, The Pennsylvania State University University Park, PA 16802, U.S.A.*

## 1. Resolving the Dynamics of Partons in Protons and Nuclei

The LHeC opens a new kinematic realm in the study of the structure of protons and nuclei through their scattering with leptons — electrons or positrons. As illustrated in Fig. 1(left), in *ep* it extends the region of the $x - Q^2$ kinematic plane studied at HERA by one order of magnitude up in $Q^2$ and down in Bjorken $x$, $Q^2 = 1 - 10^6$ GeV and $x = 10^{-6} - 0.9$.[a] The expected *ep* integrated luminosity, 1 ab$^{-1}$, exceeds that at HERA by three orders of magnitude. In electron-nucleus collisions, Fig. 1(right), the expected increase is three to four orders of magnitude down in $x$ and up in $Q^2$ compared to previous DIS experiments, with an anticipated per

This is an open access article published by World Scientific Publishing Company. It is distributed under the terms of the Creative Commons Attribution 4.0 (CC BY) License.
[a]The Electron-Ion Collider (EIC) will explore a more restricted kinematic region than HERA, but its high luminosity, the new detector techniques, and the possibility to accelerate polarised protons and vary the nuclear species, will provide valuable information on the three-dimensional structure of hadrons and nuclei and on the origin of spin.[3]

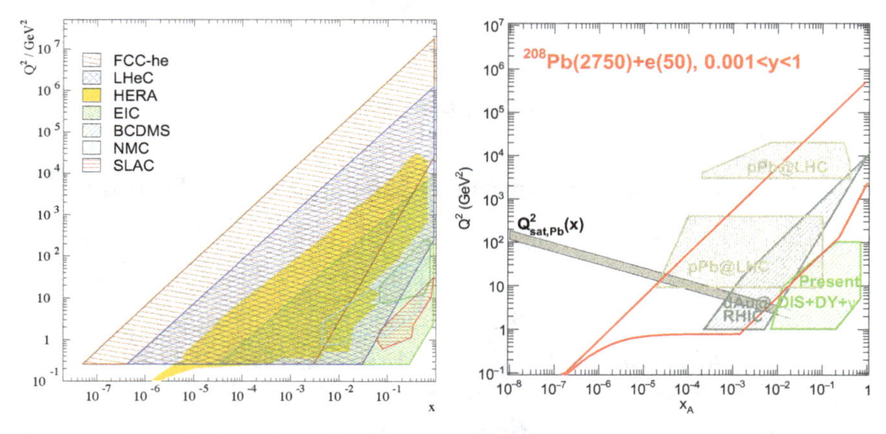


Fig. 1. Kinematic plane for $ep$ (left) and $e$Pb (right) collisions at the LHeC,[1] compared with the coverage of past and projected accelerators. Figures taken from Refs. 1,2, where further details can be found.

nucleon integrated luminosity $\sim 10$ fb$^{-1}$. Such features, together with the clean final state allowing a complete reconstruction of kinematic variables, as well as current and projected detector and theoretical developments, guarantee that the LHeC will result in a leap in our understanding of the partonic structure of protons and nuclei. Here, we briefly develop some of the possibilities studied in Ref. 1.

### 1.1. *Proton parton densities in lepton-proton collisions*

The LHeC offers the single opportunity to pin down, with high precision, the partonic structure of the proton, over an unprecedented kinematic range of DIS (see Fig. 1(left)). Through a combination of precision measurements of charged current (CC) and neutral current (NC) DIS spanning, respectively, four and six orders of magnitude in $(x, Q^2)$, supplemented by semi-inclusive measurements of strange and heavy flavour quark production, all parton distribution functions (PDFs) $xq(x, Q^2)$ and $xg(x, Q^2)$ can be separately determined in a single experiment.[b] The EIC[3] can also provide such information, but primarily in the much reduced kinematic range previously covered by HERA and fixed target experiments. The large centre-of-mass energy of the LHeC, and its higher energy version, the FCC-eh, is therefore essential to probe the full kinematic range of relevance for the HL-LHC, and later the FCC, and allows unique access to the unexplored small-$x$ regime, as discussed in Sec. 3.

[b] $q = u_v, d_v, u, \bar{u}, d, \bar{d}, s, c, b$ as well as $t$.

Maximum exploitation of the HL-LHC physics era would be made possible by an accompanying LHeC precision QCD programme. The delivery of the full complement of parton densities to unprecedented precision, together with $\alpha_s$ to per mille accuracy, derived from measurements that are experimentally and theoretically clean, and independent of the *pp* environment, would establish a new paradigm for understanding perturbative QCD and the underlying parton dynamics. Furthermore, it would enable extraordinarily precise electroweak and Higgs physics at the joint *ep* and *pp* LHC facility, and extend and facilitate the prospects for new physics discovery and interpretation.

The projected precision of the valence quark ($xu_v$, $xd_v$), anti-sea quark ($x\overline{U}$, $x\overline{D}$) and gluon ($xg$) densities, from LHeC inclusive NC and CC measurements, defined through a standard $\Delta\chi^2 = 1$ applicable for a single experiment with consistent measurements, is illustrated in Fig. 2. Ideally, LHeC data would be available at the same time as HL-LHC operation; therefore, two projections are shown: one corresponding to an initial dataset of 50 $\text{fb}^{-1}$ (yellow), and the other to the full 1 $\text{ab}^{-1}$ of inclusive DIS data (dark blue). The LHeC projections are compared to HERAPDF2.0[4] (light blue) and several other modern global fits.[5–8] Notably, in addition to large uncertainties on individual sets, especially at small and large $x$, where current data is scarce or suffers from large uncertainties, an unsatisfactory situation is evident, whereby differences between global fits can be larger than the quoted uncertainties. This arises from a combination of factors, including those related to underlying assumptions in the fits, as well as the complexity of hadron-hadron scattering data, as used in the global fits, with respect to DIS. These include issues related to initial state quark radiation, hadronisation, complex experimental uncertainty correlations, and potentially incompatible datasets, which can lead to inflation of uncertainties and/or exclusion of certain datasets. In contrast, DIS processes have clean final states and (with a precedent set by HERA) experimentally precise, compatible sets of measurements, with well understood systematic uncertainties and correlations, can be achieved, supplemented by sophisticated theoretical calculations.

#### 1.1.1. *Valence and light sea quarks*

The valence quark and light sea antiquark distributions are shown in Fig. 2 top and middle. The projection for the LHeC initial run, which corresponds to a two orders of magnitude increase in integrated luminosity compared

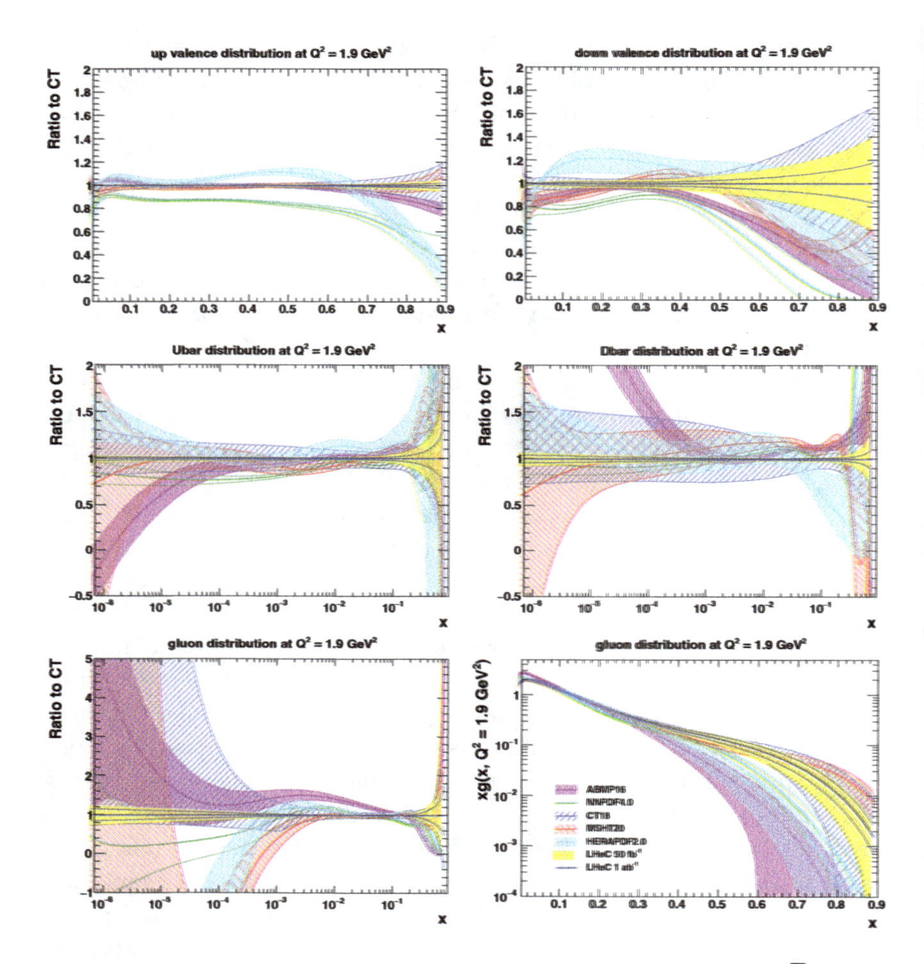


Fig. 2. Expected precision for the determination of the $u_v$ and $d_v$ (top), $\overline{U} = \bar{u}$ and $\overline{D} = \bar{d} + \bar{s}$ (middle), and gluon (bottom) PDFs from the LHeC. The gluon distribution is shown as a ratio on a log-$x$ scale (left) and as the full distribution on a linear-$x$ scale (right) to highlight both the small- and large-$x$ regions. Light blue: HERA, yellow: initial LHeC run, dark blue: full inclusive LHeC dataset, overlayed with four recent global fit results. For more information, see Ref. 1.

to that collected by the general purpose HERA experiments, shows a striking improvement in uncertainties across the full range of Bjorken-$x$, compared to today.[c] The 1 $\text{ab}^{-1}$ projection, which mainly includes $e^-p$ but

[c]Note that, following convention, while the parton distributions are shown at the starting scale of the QCD fit, $Q^2 = 1.9$ GeV$^2$, the improvements illustrated are representative, and persist from low to high scales.

also 1 $\text{fb}^{-1}$ of $e^+p$ simulated collision data, provides additional constraints, most notably at the highest $x$ values. This arises primarily from the larger integrated luminosity, which allows precise NC and CC measurements at the highest $(x, Q^2)$ values. Additionally, in the valence sector, the small amount of positron data gives access to $u_v$ and $d_v$ at small $x$ from the $e^{\pm}p$ cross section differences, as well as additional sensitivity to $d_v$ at high $x$, via the CC process. The expected precision on the valence quark distributions has strong implications for new physics searches at the HL-LHC, as well as in resolving the long-standing mystery of the unknown $d/u$ ratio at large $x$. The light (anti)quark sea distributions are currently rather poorly known, especially at small and large $x$. In the smaller $x$ region, the sea quark distributions are large and play a significant role in precision standard model measurements at hadron colliders, while at high $x$ the distributions are small, but important for searches at high mass, for which the sea and valence components must be properly distinguished. The LHeC would provide a transformation in precision (Fig. 2(middle)) as a result of the precise NC and CC measurements that probe down to the small-$x$ regime while at high $Q^2$. In particular, the combination of CC cross sections (which can be well measured for $x \gtrsim 10^{-4}$), together with NC (which has both electromagnetic and weak contributions with different dependencies on flavour composition), can distinguish between $u$- and $d$-type sea (anti)quarks. This was not possible at HERA due to the limited precision at high $Q^2$. Moreover, it is worth noting that this is a unique feature of a high energy $ep$ collider. For example, at the EIC, assuming a detection threshold of $Q^2 \approx 100$ GeV$^2$, the CC cross section will be precisely measurable only in the region above $x \approx 10^{-2}$ (see Fig. 1(left)), and so quark flavour can be disentangled only above this value.

### 1.1.2. *Strange and heavy flavour quarks*

The strange content of the proton is still poorly known, and has historically been the subject of some controversy,[9–20] yet it is highly relevant for standard model precision measurements at hadron colliders including, for instance, the $W$ mass. At the LHeC, the strange density $xs(x, Q^2)$ can be precisely mapped, for the first time, via charm tagging in the $Ws \to c$ process, in CC events. Furthermore, the LHeC provides data on charm and beauty quarks from measurements of the structure functions $F_2^c$ and $F_2^b$, extending over nearly five or six orders of magnitude, even with just a subset of the full integrated luminosity.[1] Such measurements not only serve to directly determine $xc$ and $xb$, and provide information on the correct theoretical treatment for heavy quarks, but also provide additional constraints on the gluon parton density, $xg$, and can be used to improve the determinations of the charm and beauty masses, bringing uncertainties to $\delta m_{c(b)} \simeq 3(10)$ MeV.[1] Furthermore, due to the large centre-of-mass energy and integrated luminosity, the LHeC opens up the possibility of studying top quark PDFs as a new avenue of research.

### 1.1.3. *Gluon and* $\alpha_s$

Precise knowledge of the gluon PDF across the full range of $x$ is of fundamental importance, and can be profoundly addressed at the LHeC. In principle, the projected improvements, see Fig. 2(bottom), are due to the large kinematic range and precise measurements of scaling violations,

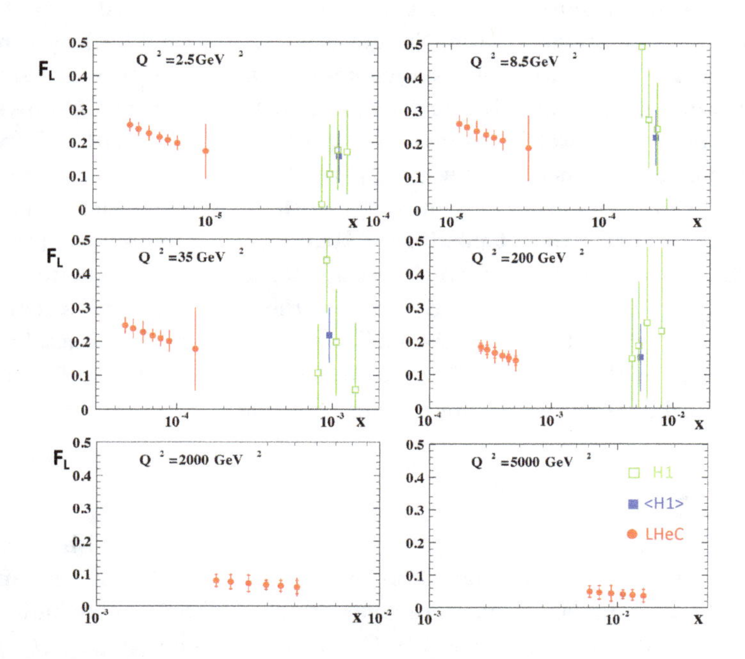


Fig. 3. H1 measurement[21] (green and blue) and LHeC projection for $F_L(x, Q^2)$ (red), derived from simulated inclusive cross section data with $E_p = 7$ TeV and $E_e = 60$, 30, 20 GeV. The LHeC inner error bars represent the statistical uncertainty, only visible for $Q^2 \geqslant 200$ GeV$^2$, and the outer error bars show the total uncertainty. The LHeC simulated data cover an $x$-range from $2 \times 10^{-6}$ to above $x = 0.01$. Full details given in Ref. 1.

$\partial F_2/\partial \ln Q^2$, as well as the fact that the inclusive NC and CC measurements, together, provide a base to fully constrain the quark distributions which, in turn, strongly constrain $xg$. The LHeC extends to smaller $x$ than HERA by more than an order of magnitude, allowing unique access to the small-$x$ region. The addition of a precise measurement of the longitudinal structure function, $F_L$, achievable at the LHeC using dedicated low energy runs, as illustrated in Fig. 3,[1] would unravel the non-linear behaviour of $xg$ at small $x$ (see also Sec. 3). Not only would this lead to a revolution in understanding the underlying parton dynamics, it also has particular significance for both signal and background in precision physics at the HL-LHC, and even more so for the FCC where, for instance, Higgs becomes small-$x$ physics, and the gluon must be accurately known, given the dominant $gg \to H$ production mechanism. The large-$x$ gluon, which has significance for searches at high masses, is also constrained, primarily via the momentum sum rule as a result of the precise determination of valence and sea quarks at high $x$. Importantly, further direct constraints on $xg$ can come from measurements of jet cross sections and $F_2^{c,b}$, as well as $F_L$ for the small-$x$ region, none of which are included in the studies represented in Fig. 1. Finally, a simultaneous QCD analysis for parton distributions and the strong coupling $\alpha_s$, using inclusive NC and CC together with DIS inclusive jet measurements, results in projected $\alpha_s$ uncertainties at the per mille level.

### 1.2. *Inclusive scattering and parton densities in electron-nucleus collisions*

Lepton-nucleus collisions at the LHeC will allow the precise extraction of the partonic nuclear structure in a completely new kinematic region, see Fig. 1(right). The EIC[3] will provide such information but only in the region covered by fixed target and $d$Au data, and limited by kinematics, e.g., by the lack of access to CC at small $x$ to determine the strangeness content of nucleons inside nuclei. Higher centre-of-mass energies are required to study the region relevant for nuclear collisions at the LHC and future hadronic colliders, with DIS providing a complete reconstruction of kinematic variables and a cleaner theoretical environment compared to proton-nucleus collisions.

While the expected integrated luminosity is considerably smaller than in $ep$, even with 1 fb$^{-1}$ a complete unfolding of the different flavours for a single nucleus will still be possible for $x \gtrsim 10^{-5}$, using the same combination of inclusive observables employed in $ep$.[1] The resulting uncertainties

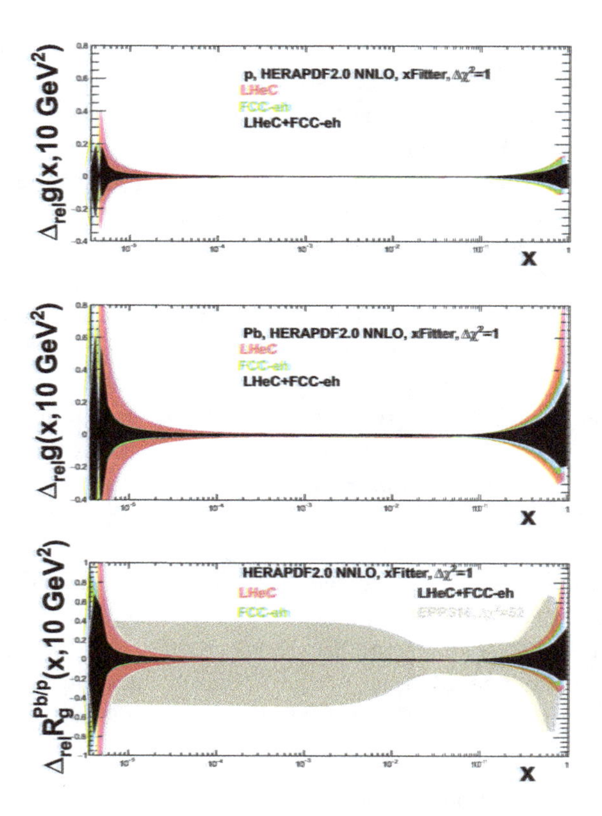


Fig. 4. Relative uncertainty of the gluon density in the proton (top), Pb (middle) and the corresponding nuclear modifications factor (bottom) in an analysis of $ep$ and $e$Pb LHeC and FCC-eh NC plus CC pseudodata using *xFitter* (both a single set of data and all combined), compared to the results of EPPS16.[22] Taken from Ref. 1.

(see Fig. 4), defined through a standard $\Delta\chi^2 = 1$ applicable for a single experiment, will be much smaller than the ones in global fits that employ a much larger tolerance and that, due to the scarcity of data for a single nuclear species, require initial conditions that depend on the nuclear size.

The determination with small uncertainties of parton densities for the different species both in $ep$ and $e$A will clarify how the partonic structure of a nucleon is affected by the nuclear medium,[1] in different kinematic regions: the origin of shadowing at small $x$ and its eventual relation to diffraction (see below), the dependence of antishadowing on the parton flavour, the $Q^2$ evolution of the EMC effect, the nuclear dependence of intrinsic charm etc.

Finally, let us note that the lack of knowledge of nuclear PDFs is among the largest sources of uncertainty in the extraction of properties of the medium produced in heavy ion collisions, the quark gluon plasma. Also note that factorisation is assumed in such studies. Therefore, they will be greatly benefited by the precise knowledge of PDFs obtained in DIS which will also allow precise tests of factorisation in proton-nucleus collisions.

## 2. Diffractive scattering and three dimensional structure

In the DIS diffractive event $e + p \to p(Y) + X$ the incoming proton $p$ is scattered elastically or dissociates into a small mass excitation $Y$, while being well separated by a ***rapidity gap*** from the diffractive system $X$, see Fig. 5. The rapidity gap is a region in the detector completely devoid of any activity. The mechanism responsible for creating rapidity gaps must involve a colour singlet exchange, so that no QCD radiation can be produced into this region.

The precise measurement of diffraction in electron-hadron collisions is of great importance for our understanding of the dynamics of the strong interaction. Since diffraction is mediated by colour neutral exchange, the exact mechanism governing this exchange can provide information about confinement. Also, an important contribution to the diffractive exchange

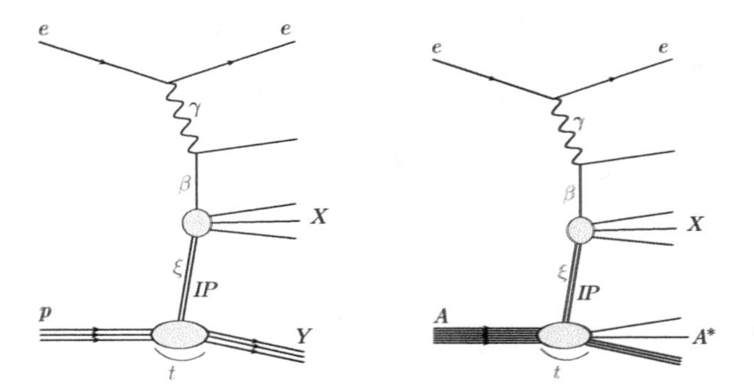


Fig. 5. Left: diffractive process in DIS. $Y$ is either the elastically scattered proton or a low mass excitation, $I\!P$ is the colour singlet exchange ('Pomeron') responsible for rapidity gap between $Y$ and $X$, and $X$ is the diffractive mass. Kinematic variables: $t$ — momentum transfer at the proton vertex, $\xi$ — longitudinal momentum fraction of the proton carried by the Pomeron, and $\beta$ — longitudinal momentum fraction of the Pomeron carried by the parton. Right: Incoherent diffraction on nuclei. The final state $A^*$ can be a nucleus in an excited state which can further disintegrate into another nucleus and any number of nucleons.

comes from gluons, and thus this process offers a unique window to study proton structure, particularly at small $x$. Diffractive structure functions can be used to pin down the details of the QCD evolution, and in particular the deviation from the linear regime. It has also been demonstrated that diffraction in $ep$ scattering is related to the mechanism of nuclear shadowing. Finally, diffractive DIS allows for the precise extraction of the diffractive PDFs and tests of the limits of collinear factorisation.

Both the LHeC and its higher energy version, the FCC-eh, offer unprecedented capabilities for studying diffraction. The extended kinematic regime in $(x, Q^2)$ of both machines translates into a wider range of available momentum fraction $\xi$ of the diffractive exchange with respect to the hadron, down to $10^{-4} - 10^{-5}$ for a wide range of the momentum fraction of the parton $\beta$ with respect to the diffractive exchange. See Fig. 5 for the definition of variables. The high luminosity, the extended lever arm both in $x$ and $Q^2$, and diffractive variables $\xi, \beta$ for the LHeC and FCC-eh would allow for much tighter constraints of the diffractive parton densities compared to HERA. Furthermore, new possibilities in diffraction open up at these machines. The higher energy allows top quark production to be studied in diffraction, which can be important, particularly at the FCC-eh. Also, charged current diffraction could be measured with much greater precision than at HERA. The dijet diffractive production can be studied in a much greater kinematic range than at HERA and thus impact on the extraction of diffractive PDFS, and the limits of diffractive factorisation could be explored.

Exclusive diffraction opens up new possibilities, particularly for exploring the spatial structure of the hadron at high energy. The diffractive exclusive vector meson production at small $x$ provides information about gluons, and is sensitive to non-linear evolution. The energy dependence of this process can provide information about changes in the dynamics from linear to non-linear. The differential cross section and the $t$-slope measurements can give insight into small-$x$ dynamics even further. Theoretical calculations show that the differential cross section will have dips, or minima, which occur when the saturated regime is reached. The position of these minima varies with the photon-hadron energy and $Q^2$ of the virtual photon, thus providing a handle to pin down saturation at small $x$. Other processes which are valuable sources for the proton structure are Deeply Virtual Compton Scattering, which can provide information about the quark distribution and its spatial extent, as well as the diffraction dissociation for protons which can be useful in the context of studying density fluctuations in the proton.

In the nuclear case, diffraction becomes a more involved process than in *ep* due to the fact that in addition to coherent $e + \mathrm{A} \to e + \mathrm{A} + X$, left plot of Fig. 5 with the replacement $p \to A$, there is also incoherent $e + \mathrm{A} \to e + \mathrm{A}^* + X$ diffraction, where $A^*$ is an excited nuclear state, see the right plot of Fig. 5. The difficulty is to distinguish the two processes or even veto one of them. Additionally, the reconstruction of the diffractive kinematic variables becomes challenging. If such difficulties, which also exist in UPCs at hadronic colliders, are resolved, a wealth of information on nuclei can be obtained using the same observables as in *ep*.[1]

By investigating coherent and incoherent diffractive scattering on nuclei, unique insight into the spatial structure of matter is obtained. On one hand, the coherent cross section, which dominates for $-t \leqslant 1/R_p^2$, is sensitive to the average spatial density distribution of gluons in transverse space. On the other hand, the incoherent cross section, dominant for $-t > 1/R_p^2$, provides information on nuclear dissociation and measures fluctuations of the gluon density inside the nucleus down to subnucleon scales. The $t$-distribution in coherent diffractive production off the nucleus gives rise to a dip-type structure for both saturation and non-saturation models. Meanwhile, in the case of incoherent production at small $|t|$, neither saturation and non-saturation models lead to dips.[23] This is in drastic contrast to the diffractive production off the proton where only saturation models lead to a dip-type structure in the $t$-distribution at values of $|t|$ that can be experimentally accessible. Therefore, diffractive production offers a unique opportunity to measure the spatial distribution of partons in the protons and nuclei. It is also an excellent tool to investigate the approach to unitarity in the high energy limit of QCD. Note that diffractive partonic densities inside nuclei are completely unknown, see the recent review[24] and[25] for studies at the LHeC and FCC-he.

Besides its intrinsic interest, spatial information on the partonic structure of nuclei is crucial for the interpretation and precise extraction of the properties of the medium created in small collision systems and in heavy ion collisions, see Ref. 1. While the EIC will produce such information, it does so for values of $x$ larger than those relevant for the LHC and future hadronic colliders. Our present knowledge of parton evolution towards smaller $x$ is largely insufficient for a reliable extrapolation of the findings at the EIC.

## 3. Small-$x$ dynamics

At the small values of $x$ accessible at very high energy electron-hadron and hadron-hadron collisions, terms containing large logarithms of $\ln 1/x$ appear that need to be included in the formalism. Thus, even at very small values of the strong coupling $\alpha_s \ll 1$, in the perturbative regime, the powers $\alpha_s \ln 1/x \sim 1$ are large and thus need to be resummed. The Balitsky-Fadin-Kuraev-Lipatov (BFKL) evolution equation[26,27] accomplishes that goal and is available at LL and NLL accuracy. It is thus predicted that the parton density evolution will be modified by the small-$x$ effects for collisions performed at sufficiently high energies. The small-$x$ evolution requires matching to the collinear DGLAP evolution, and additional constraints from kinematics are also needed to stabilize the results. Resummation procedures have been developed over the years,[28,29] and allow predictions of the growth of parton densities and structure functions at very small $x$. Studies using small-$x$ resummation in the collinear approach[30] have demonstrated that the description of the structure functions at HERA at small $x$ is improved compared to that in fixed order DGLAP evolution. In particular, detailed studies have shown that the improvement in the description is greatest for the small-$x$ and small-$Q^2$ region, exactly where the small-$x$ logarithms are expected to be significant.

This has important consequences for the predictions at higher energies accessible at LHeC. The differences between DGLAP evolution and the evolution based on the small-$x$ resummation are significant at small $x$ for inclusive quantities like $F_2$ and $F_L$. The longitudinal structure function $F_L$ is particularly sensitive, and the LHeC (even more so the FCC-eh) can easily distinguish between the different evolution scenarios, see Fig. 6.

In all its realisations, non-linear QCD dynamics leading to saturation[31,32] are density effects, i.e., parton recombination balances splitting when parton densities become large. This happens not only for small values of $x$, but also when the number of nucleons $A$ increases. Indeed, the squared saturation momentum $Q_s^2$ that provides the momentum scale below which gluon density is saturated, increases $\propto A^{1/3}$. Therefore, $e$A collisions are crucial for the discovery of saturation, as they provide an additional enhancement of the perturbative region where saturation effects should be noticeable, $\Lambda^2_{\text{QCD}} < Q^2 < Q_s^2$. They are also key in establishing the mechanism of saturation, e.g., the weak coupling one provided by the CGC.[31,32] Finally, nuclear effects should offer the possibility to distinguish between new linear QCD dynamics (resummation of small-$x$ logarithms) — which are not affected by density, and non-linear dynamics.

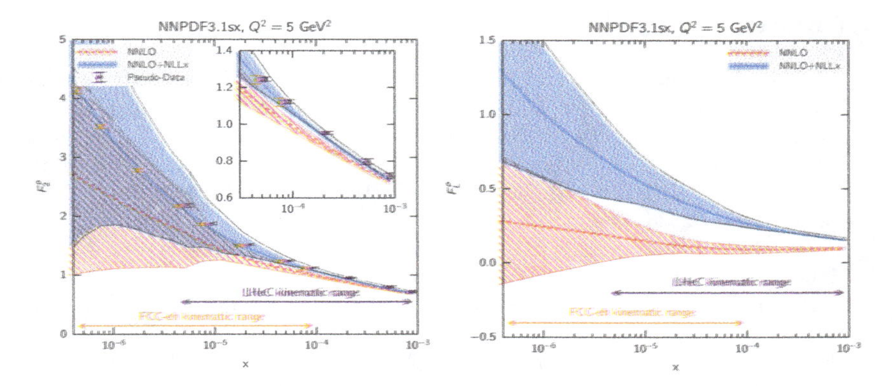


Fig. 6. Predictions for the $F_2$ and $F_L$ structure functions using the NNPDF3.1sx NNLO and NNLO+NLLx fits at $Q^2 = 5$ GeV$^2$ for the kinematics of the LHeC and FCC-eh. In the case of $F_2$, the expected total experimental uncertainties based on the simulated pseudodata are also shown, assuming the NNLO+NLLx values as the central prediction. The LHeC pseudodata have been offset by a small amount for better visibility. The inset in the left plot shows a magnified view in the kinematic region $x > 3 \times 10^{-5}$, corresponding to the reach of HERA data. Figure taken from Ref. 30.

Inclusive observables can be used to search for non-linear effects. Tension appears between the description of $F_2$ and $F_L$ saturation predictions in models based on DGLAP evolution; see Ref. 33 for a study using reweighting techniques at EIC energies. More recent studies[34] indicate that the difference in $Q^2$ evolution between linear, DGLAP based models and non-linear models may be a good candidate to observe saturation effects and that the large perturbative lever arm at small $x$ accessible at the LHeC is crucial for this effect to be quantitatively significant. On the other hand, exclusive diffraction shows significant saturation effects as commented in Sec. 2. Other observables like azimuthal correlations among particles at small $x$ are also strongly affected by saturation.[35] Nevertheless, it should be noted that conventional nuclear effects may be hardly distinguishable from weak coupling saturation.[1,36] Therefore, both *ep* and *e*A collisions will be required to establish the existence and realisation of non-linear QCD dynamics.

Finally, the dynamics of QCD at small $x$ or large energies will have strong consequences on hadronic and nuclear collisions. It will determine particle production at the initial stage of hadronic collisions. As indicated previously, our lack of knowledge on this matter limits our ability to characterise the medium produced in proton-nucleus and nucleus-nucleus collisions at LHC energies. Therefore, *e*A collisions at the LHeC become necessary for the full exploitation of the heavy-ion program at the LHC.

## References

1. P. Agostini *et al.* [LHeC and FCC-he Study Group], J. Phys. G **48**, no. 11, 110501 (2021) doi:10.1088/1361-6471/abf3ba [arXiv:2007.14491 [hep-ex]].
2. K. D. J. André, L. Aperio Bella, N. Armesto, S. A. Bogacz, D. Britzger, O. S. Brüning, M. D'Onofrio, E. G. Ferreiro, O. Fischer, C. Gwenlan *et al.*, Eur. Phys. J. C **82**, no. 1, 40 (2022) doi:10.1140/epjc/s10052-021-09967-z [arXiv:2201.02436 [hep-ex]].
3. R. Abdul Khalek, A. Accardi, J. Adam, D. Adamiak, W. Akers, M. Albaladejo, A. Al-bataineh, M. G. Alexeev, F. Ameli, P. Antonioli *et al.*, Nucl. Phys. A **1026**, 122447 (2022) doi:10.1016/j.nuclphysa.2022.122447 [arXiv:2103.05419 [physics.ins-det]].
4. H. Abramowicz *et al.* [H1 and ZEUS], Eur. Phys. J. C **75** (2015) no. 12, 580 doi:10.1140/epjc/s10052-015-3710-4 [arXiv:1506.06042 [hep-ex]].
5. R. D. Ball *et al.* [NNPDF], Eur. Phys. J. C **82** (2022) no. 5, 428 doi:10.1140/epjc/s10052-022-10328-7 [arXiv:2109.02653 [hep-ph]].
6. S. Bailey, T. Cridge, L. A. Harland-Lang, A. D. Martin and R. S. Thorne, Eur. Phys. J. C **81** (2021) no. 4, 341 doi:10.1140/epjc/s10052-021-09057-0 [arXiv:2012.04684 [hep-ph]].
7. T. J. Hou, J. Gao, T. J. Hobbs, K. Xie, S. Dulat, M. Guzzi, J. Huston, P. Nadolsky, J. Pumplin, C. Schmidt *et al.*, Phys. Rev. D **103** (2021) no. 1, 014013 doi:10.1103/PhysRevD.103.014013 [arXiv:1912.10053 [hep-ph]].
8. S. Alekhin, J. Bluemlein, S. O. Moch and R. Placakyte, PoS **DIS2016** (2016), 016 doi:10.22323/1.265.0016 [arXiv:1609.03327 [hep-ph]].
9. W. G. Seligman, C. G. Arroyo, L. de Barbaro, P. de Barbaro, A. O. Bazarko, R. H. Bernstein, A. Bodek, T. Bolton, H. S. Budd, J. Conrad *et al.*, Phys. Rev. Lett. **79** (1997), 1213–1216, doi:10.1103/PhysRevLett.79.1213 [arXiv:hep-ex/9701017 [hep-ex]].
10. M. Tzanov *et al.* [NuTeV], Phys. Rev. D **74** (2006), 012008 doi:10.1103/PhysRevD.74.012008 [arXiv:hep-ex/0509010 [hep-ex]].
11. G. Onengut *et al.* [CHORUS], Phys. Lett. B **632** (2006), 65–75 doi:10.1016/j.physletb.2005.10.062
12. J. P. Berge, H. Burkhardt, F. Dydak, R. Hagelberg, M. Krasny, H. J. Meyer, P. Palazzi, F. Ranjard, J. Rothberg, J. Steinberger *et al.*, Z. Phys. C **49** (1991), 187–224 doi:10.1007/BF01555493
13. O. Samoylov *et al.* [NOMAD], Nucl. Phys. B **876** (2013), 339–375 doi:10.1016/j.nuclphysb.2013.08.021 [arXiv:1308.4750 [hep-ex]].
14. G. Aad *et al.* [ATLAS], Phys. Rev. Lett. **109** (2012), 012001 doi:10.1103/PhysRevLett.109.012001 [arXiv:1203.4051 [hep-ex]].
15. S. Chatrchyan *et al.* [CMS], JHEP **02** (2014), 013, doi:10.1007/JHEP02(2014)013 [arXiv:1310.1138 [hep-ex]].
16. G. Aad *et al.* [ATLAS], JHEP **05** (2014), 068 doi:10.1007/JHEP05(2014)068 [arXiv:1402.6263 [hep-ex]].
17. M. Aaboud *et al.* [ATLAS], Eur. Phys. J. C **77** (2017) no. 6, 367 doi:10.1140/epjc/s10052-017-4911-9 [arXiv:1612.03016 [hep-ex]].

18. S. Alekhin, J. Blümlein and S. Moch, Phys. Lett. B **777** (2018), 134–140 doi:10.1016/j.physletb.2017.12.024 [arXiv:1708.01067 [hep-ph]].
19. A. M. Cooper-Sarkar and K. Wichmann, Phys. Rev. D **98** (2018) no. 1, 014027 doi:10.1103/PhysRevD.98.014027 [arXiv:1803.00968 [hep-ex]].
20. S. Alekhin, J. Blümlein, S. Kulagin, S. O. Moch and R. Petti, PoS **DIS2018** (2018), 008 doi:10.22323/1.316.0008 [arXiv:1808.06871 [hep-ph]].
21. V. Andreev *et al.* [H1], Eur. Phys. J. C **74** (2014) 4 doi:10.1140/epjc/s10052-014-2814-6 [arXiv:1312.4821 [hep-ex]].
22. K. J. Eskola, P. Paakkinen, H. Paukkunen and C. A. Salgado, Eur. Phys. J. C **77**, no. 3, 163 (2017) doi:10.1140/epjc/s10052-017-4725-9 [arXiv:1612.05741 [hep-ph]].
23. H. Mäntysaari and B. Schenke, Phys. Lett. B **772**, 832–838 (2017) doi:10.1016/j.physletb.2017.07.063 [arXiv:1703.09256 [hep-ph]].
24. L. Frankfurt, V. Guzey, A. Stasto and M. Strikman, Rept. Prog. Phys. **85** (2022) no. 12, 126301 doi:10.1088/1361-6633/ac8228 [arXiv:2203.12289 [hep-ph]].
25. N. Armesto, P. R. Newman, W. Słomiński and A. M. Staśto, Phys. Rev. D **100**, no. 7, 074022 (2019) doi:10.1103/PhysRevD.100.074022 [arXiv:1901.09076 [hep-ph]].
26. E. A. Kuraev, L. N. Lipatov and V. S. Fadin, Sov. Phys. JETP **45** (1977), 199–204.
27. I. I. Balitsky and L. N. Lipatov, Sov. J. Nucl. Phys. **28** (1978), 822–829.
28. M. Ciafaloni, D. Colferai, G. P. Salam and A. M. Stasto, JHEP **08** (2007), 046 doi:10.1088/1126-6708/2007/08/046 [arXiv:0707.1453 [hep-ph]].
29. G. Altarelli, R. D. Ball and S. Forte, Nucl. Phys. B **799** (2008), 199–240 doi:10.1016/j.nuclphysb.2008.03.003 [arXiv:0802.0032 [hep-ph]].
30. R. D. Ball, V. Bertone, M. Bonvini, S. Marzani, J. Rojo and L. Rottoli, Eur. Phys. J. C **78** (2018) no. 4, 321 doi:10.1140/epjc/s10052-018-5774-4 [arXiv:1710.05935 [hep-ph]].
31. Y. V. Kovchegov and E. Levin, Camb. Monogr. Part. Phys. Nucl. Phys. Cosmol. **33**, 1–350 (2012) doi:10.1017/CBO9781139022187
32. F. Gelis, E. Iancu, J. Jalilian-Marian and R. Venugopalan, Ann. Rev. Nucl. Part. Sci. **60**, 463–489 (2010) doi:10.1146/annurev.nucl.010909.083629 [arXiv:1002.0333 [hep-ph]].
33. C. Marquet, M. R. Moldes and P. Zurita, Phys. Lett. B **772**, 607–614 (2017) doi:10.1016/j.physletb.2017.07.035 [arXiv:1702.00839 [hep-ph]].
34. N. Armesto, T. Lappi, H. Mäntysaari, H. Paukkunen and M. Tevio, Phys. Rev. D **105** (2022) no. 11, 114017 doi:10.1103/PhysRevD.105.114017 [arXiv:2203.05846 [hep-ph]].
35. J. L. Abelleira Fernandez *et al.* [LHeC Study Group], J. Phys. G **39**, 075001 (2012) doi:10.1088/0954-3899/39/7/075001 [arXiv:1206.2913 [physics.acc-ph]].
36. N. Armesto, J. Phys. G **32**, R367–R394 (2006) doi:10.1088/0954-3899/32/11/R01 [arXiv:hep-ph/0604108 [hep-ph]].

© 2024 The Author(s)
https://doi.org/10.1142/9789811280184_0024

# Chapter 24

# Higgs and Beyond the Standard Model Physics

J. de Blas*, O. Fischer† and U. Klein‡

**CAFPE and Departamento de Física Teórica y del Cosmos, Universidad de Granada, Granada, Spain*

*† University of Liverpool, Department of Mathematics, Liverpool, United Kingdom*

*‡ University of Liverpool, Department of Physics, Liverpool, United Kingdom*

## 0. Introduction[a]

The discovery of the Higgs boson at the Large Hadron Collider by ATLAS[1] and CMS[2] collaborations opened a new chapter in particle physics. The Higgs Boson ($H$) is of fundamental importance for the future of particle physics and the LHC. It is related to the mechanism predicted by Refs. 3–5 in which the intermediate vector bosons of the spontaneously broken electroweak symmetry acquire masses[b] while the photon remains massless. Fermions obtain a mass via the Yukawa couplings with the Higgs field. Following the discovery of the Higgs boson, its physics has become a central theme of the experimental and theoretical programmes pursued with the LHC as well as for the high luminosity upgrade of the Large Hadron Collider, the HL-LHC, as described in this book.

This is an open access article published by World Scientific Publishing Company. It is distributed under the terms of the Creative Commons Attribution 4.0 (CC BY) License.

[a]This section has been reproduced from [P. Agostini et al., 2021 J. Phys. G: Nucl. Part. Phys. 48 110501]. ©The Author(s). Published by IOP Publishing Ltd. CC BY 4.0.

[b]The mass of the $W$ boson, $M_W$, is generated through the vacuum expectation value, $v$, of the Higgs field ($\Phi$) and given by the simple relation $M_W = gv/\sqrt{2}$ where $g$ is the weak interaction coupling. Here $v = \sqrt{-\mu^2/2\lambda}$ with the two parameters of the Higgs potential that is predicted to be $V = -\mu^2\Phi^+\Phi - \lambda(\Phi^+\Phi)^2$. The Higgs mass is given as $M_H = 2v\sqrt{\lambda}$ while the mass of the $Z$ boson is related to $M_W$ with the electroweak mixing angle, $M_Z = M_W/\cos\Theta_W$.

Any new high-energy future collider project has placed its focus on the potential to precisely study the properties of the Higgs mechanism, to understand its characteristics and hopefully open a new window into physics extending beyond the Standard Model. The LHeC, as presented in detail in Ref. 6, provides a salient opportunity for exploring the physics of the Higgs boson in *ep* collisions due to the following features: i) the production through vector boson scattering in the t-channel has a cross section of order 200 fb, comparable to that of Higgs radiation from a $Z$ boson in the so-called $e^+e^-$ Higgs factories; ii) the theoretically well controlled neutral and charged current production processes uniquely distinguish the $HZZ$ and $HWW$ vertices; iii) the clean semi-leptonic final state which is free of multiple interactions (pile-up); iv) the empowerment of the Higgs measurement programme at the hadron collider, LHC, through the resolution of uncertainties related to the strong interactions (consult the chapter on hadron structure and QCD with the LHeC), such as the unknowns related to proton structure, the value of the strong coupling and the question of possible non-linear gluon dynamics at small Bjorken-$x$.

Like in the case of the $e^+e^-$ colliders, the LHeC provides high precision measurements of the characteristics of the seven most abundant Higgs decay channels, which comprise 99.9% of all SM decays. Owing to the high gluon-gluon fusion $gg \to H$ production cross section, only a hadron collider, such as the LHC, would be able to extend these explorations to the very rare decay channels. When the *ep* results will be combined with the anticipated HL-LHC measurements, the characteristics of the SM Higgs boson could be explored at per-cent level in numerous reactions. The first significant observation of the Higgs self-coupling is in reach, with a precision to be evaluated for a combined $pp - ep$ high luminosity data analysis. This high precision programme is expected to shed light on the question of whether the Higgs mechanism potentially leads beyond the Standard Model (BSM) or not.

Despite having too many free parameters, the Standard Model provides phenomenological explanations to the observed properties and kinds of elementary particles thus far. However, there remain several fundamental questions and tasks unresolved, such as the nature of lepton-quark and baryon-antibaryon asymmetries, the quest of a grand unification of forces, possibly resolving the distinction between fermions and bosons in a supersymmetric theory, or the origin of neutrino masses. Current experimental hints as to how this may be accomplished turned out to be scarce at the LHC, while scientists are still thinking of theoretical hints. The initial

LHC data have confirmed the Higgs boson's existence to be SM-like, without much experimental hints of physics beyond the Standard Model. A special, cosmological and particle physics question regards the nature of Dark Matter. The situation resembles the time prior to the advent of the SM. It motivates the search for the widest possible options and realisation of dedicated search experiments including high energy, high intensity colliders of different types in order to lead beyond the Standard Model. This is why the HL-LHC has its special importance and is perhaps the strongest reason as to why its combination with an intense electron accelerator has been seriously brought forward.

Several anomalies in experiments at all energy scales, which could be linked to open questions as mentioned above exist.[7] However, the absence of convincing BSM signals in the currently available LHC data may indicate that "new" physics could be inaccessible at the TeV scale. And yet, it could be a very rare phenomenon possibly still hidden in the backgrounds, requiring more data and refined analysis techniques. New theoretical developments consider that as a possibility and explore the complementarity of the different collider experiments. The LHeC is projected to operate concurrent with the LHC's high luminosity phase which it may extend in time, depending on CERN's future plans. Because of its clean experimental environment and different production mechanisms, the LHeC, and future, even higher energy *ep* colliders, such as the high energy (HE) version of the LHeC (for the HE-LHC see the subsequent part of the book), or the FCC-eh,[6,8] could indeed observe BSM physics near the TeV scale, including an extended scalar particle sector, sterile neutrinos, and other exotic particles. Following the brief description of Higgs physics in *ep*, example studies for this BSM potential are presented subsequently too.

## 1. Higgs physics at high energy ep colliders

### 1.1. *Technical aspects*

As part of the updated Conceptual Design, physics, and detector of the LHeC, the technical details and the prospects of Higgs physics at the LHeC have been described in a detailed study report,[6] as well as providing further detailed references. In deep inelastic electron-proton scattering (DIS), the SM-Higgs boson is predominantly produced through $WW$ fusion in charged current DIS (CC) scattering. The next large Higgs production mode in *ep* is $ZZ \to H$ fusion in neutral current DIS (NC) scattering. The NC reaction is even cleaner than the CC process as the scattered electron fixes the kinematics more accurately than the missing energy. While in $pp$ the $WW$ and $ZZ$ Higgs boson generations are hardly distinguishable, in $ep$ they are very distinguishable, providing particularly precise constraints on the $WWH$ and $ZZH$ couplings, which can be clearly identified via the detection of large missing energy or the selection of the final state electron respectively.

Kinematics reconstruction in DIS is very precise and in neutral currents redundant, leading to important tracking and calorimeter cross calibrations and significant reduction of systematic uncertainties. Electron-proton scattering is thus an outstanding environment for most accurate measurements of particle physics processes without the pile-up complications of the LHC analyses. The energies of the electron and proton beams are quite different, causing constraints on the design of the detector to provide the necessary angular acceptance for the scattered lepton, the Higgs decay particles and the final state emerging at the virtual $W/Z$-proton interaction vertex. Studies using the MadGraph program assured, even for the most asymmetric energy configuration of the FCC-eh, that the complete final state can be very well reconstructed, with emphasis on the very forward (i.e. proton beam) direction for the hadronic final state. The Higgs decay products are well confined in the apparatus, appearing near a pseudorapidity $\eta$ value of around 2 which corresponds to a polar angle of about $\theta = 15°$, where $\eta = -\ln\tan\theta/2$.

The scattering cross sections, including the decay of the Higgs boson into a pair of particles $A_i\bar{A}_i$, can be written as:

$$\sigma^i_{CC} = \sigma_{CC} \cdot \frac{\Gamma^i}{\Gamma_H} \quad \text{and} \quad \sigma^i_{NC} = \sigma_{NC} \cdot \frac{\Gamma^i}{\Gamma_H}. \tag{1}$$

Here, the ratio of the partial to the total Higgs decay width defines the branching ratio, $br_i$, for each decay into $A_i\bar{A}_i$. The size of the $ep$ Higgs production cross section and about 1 ab$^{-1}$ luminosity prospect allows the seven most frequent SM Higgs decays to be considered, i.e. those into fermion ($b\bar{b}$, $c\bar{c}$, $\tau^+\tau^-$) and into gauge particle pairs ($WW$, $ZZ$, $gg$, $\gamma\gamma$) with high precision at the LHeC and its higher energy versions.[c]

Initially, detailed simulations and Higgs extraction studies for LHeC were made for the dominant $H \to b\bar{b}$ and the challenging $H \to c\bar{c}$ channels. These analyses were eventually updated, first using simple kinematic

[c]This paragraph has been reproduced from [P. Agostini et al., 2021 J. Phys. G: Nucl. Part. Phys. 48 110501]. ©The Author(s). Published by IOP Publishing Ltd. CC BY 4.0.

requirements and later using advanced boosted decision tree techniques (BDT). The focus on the $H \to b\bar{b}$ decay has been strongly motivated by its dominating size and difficulty reconstructing it accurately at the LHC. For *ep*, it seemed natural to extend this to the $H \to c\bar{c}$, especially because it is currently considered not to be observable at the HL-LHC, due to permutation and large background reasons. A further detailed analysis was performed for the $H \to W^+W^-$ decay based on a complete signal and background simulation and eventual BDT analysis. For this channel, CC DIS cleanly determines the $HWW$ coupling to its fourth power assuming SM production and decay. Results on other channels were obtained using an acceptance, efficiency and signal-to-background scale factor approach[6] which was successfully benchmarked with the detailed simulations for heavy quark and $W$ decays.

### 1.2. *Results on ep Higgs physics prospects*

The sum of the branching ratios for the seven Higgs decay channels accessible to *ep* at the LHeC adds up to 99.87% of the total SM range. As discussed in Ref. 6, significant constraints of the $H \to$ *invisible* decay can be set with *ep* as well, albeit not being able to exclude exotic, unnoticed Higgs decays. The accurate reconstruction of all decays considered here will present a severe constraint on the total cross section and, with that, constraint of the total decay width of the Higgs boson in the SM. Evaluation of the cross section measurement prospects for a decay channel $i$ is based on the relative signal strengths $\mu^i(NC, CC)$ with respect to the SM cross section.[d] The results for the LHeC, the HE-LHeC and the FCC-eh are displayed in Fig. 1. They are the input to joint coupling constant analyses. These can be performed in the simplest, so-called $\kappa$ framework,[9] considered subsequently, or e.g. in formalisms embedded in effective field theories.

The $\kappa$ parameters are scaling factors to the various Higgs couplings. Higgs production cross section thus scale as $\sigma_{NC/CC} \propto \kappa^2_{Z/W}$ (equal to 1 in the SM), and the channel $i$ decay width $\Gamma^i \propto \kappa_i^2$. Assuming only SM Higgs boson decays, and therefore $\Gamma_H = \Sigma_j \kappa_j \Gamma^j$, this leads to the following

[d]Reproduced from [P. Agostini et al., 2021 J. Phys. G: Nucl. Part. Phys. 48 110501]. ©The Author(s). Published by IOP Publishing Ltd. CC BY 4.0.

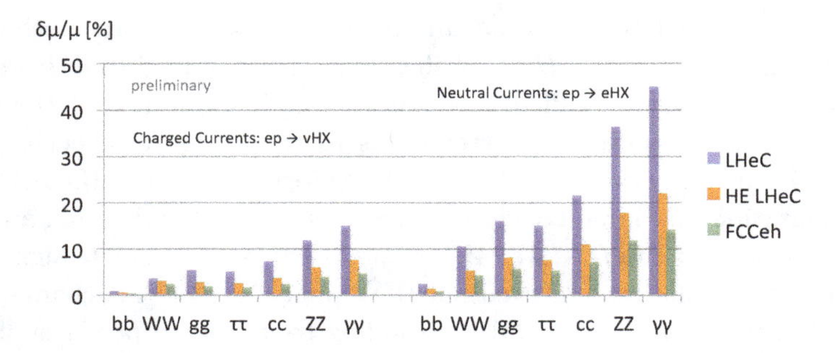


Fig. 1. Uncertainties of signal strength determinations in the seven most abundant SM Higgs decay channels for the FCC-eh (green, 2 ab$^{-1}$, $E_p = 50$ TeV), the HE LHeC (brown, 2 ab$^{-1}$, $E_p = 14$ TeV) and LHeC (blue, 1 ab$^{-1}$, $E_p = 7$ TeV), in charged and neutral current DIS production using a polarised electron beam ($P = -0.8$) of 60 GeV. From Ref. 6.

(Eq. (1)):

$$\sigma^i_{CC} = \sigma^{SM}_{CC}\ br^{SM}_i \cdot \kappa^2_W \kappa^2_i \frac{1}{\sum_j \kappa^2_j br^{SM}_j} \quad \text{and}$$
$$\sigma^i_{NC} = \sigma^{SM}_{NC}\ br^{SM}_i \cdot \kappa^2_Z \kappa^2_i \frac{1}{\sum_j \kappa^2_j br^{SM}_j}. \tag{2}$$

Here the quantities $\sigma_{NC/CC}$ and $br_j$ are understood to be the SM values. Dividing these expressions by the SM cross section predictions, one obtains the variations of the relative signal strengths, $\mu^i$, for charged and neutral currents and their $\kappa$ dependence

$$\mu^i_{CC} = \kappa^2_W \kappa^2_i \frac{1}{\sum_j \kappa^2_j br^{SM}_j} \quad \text{and} \quad \mu^i_{NC} = \kappa^2_Z \kappa^2_i \frac{1}{\sum_j \kappa^2_j br^{SM}_j}. \tag{3}$$

With seven decay channels considered in CC and NC, one finds that for each of the *ep* collider configurations there exist eight constraints on $\kappa_W$ and $\kappa_Z$ and two on the other five $\kappa$ parameters. Using the signal strength uncertainties for both NC and CC reactions, illustrated in Fig. 1, fits to all seven channels are performed in a minimisation procedure to determine the resulting uncertainties for the $\kappa$ parameters. These are done separately for each of the three *ep* collider configurations.[e] The results, listed in Table 1, exhibit an amazing precision with small systematic or theoretical uncertainties as were considered in Ref. 6.

One observes that the naive expectation of $\delta\kappa \simeq \delta\mu/2$ holds approximately for the $gg$, $\tau\tau$, $c\bar{c}$, $\gamma\gamma$ channels. However, due to the dominance

[e]Reproduced from [P. Agostini et al., 2021 J. Phys. G: Nucl. Part. Phys. 48 110501]. ©The Author(s). Published by IOP Publishing Ltd. CC BY 4.0.

Table 1. Summary of $\kappa$ uncertainty values as obtained from fits to the signal strength uncertainty estimates for the seven most abundant Higgs decay channels, in charged and neutral currents for the LHeC, the HE-LHeC and the FCC-eh.

| Setup | $\sqrt{s}$ | $L/ab^{-1}$ | $b\bar{b}$ | WW | gg | $\tau\tau$ | cc | ZZ | $\gamma\gamma$ |
|---|---|---|---|---|---|---|---|---|---|
| LHeC | 1.3 | 1 | 1.9 | 0.70 | 3.5 | 3.1 | 3.8 | 1.2 | 6.8 |
| HE-LHeC | 2.0 | 2 | 1.0 | 0.38 | 1.8 | 1.6 | 1.9 | 0.6 | 3.5 |
| FCC-eh | 3.5 | 2 | 0.60 | 0.22 | 1.1 | 0.93 | 1.2 | 0.35 | 2.1 |

of $H \to b\bar{b}$ in the total Higgs decay width and owing to the sensitivity to both the $WWH$ and $ZZH$ couplings in the initial state, there occurs a reshuffling of the precisions in the joint fit: $\kappa_b$ becomes even relatively less precise than $\mu_{bb}$ while both $\kappa_W$ and $\kappa_Z$ become more precise than naively estimated, especially when one takes into account that the $H \to WW$ decay in CC measures $\kappa_W^4$.

The potential of LHeC Higgs measurements can be compared and jointly interpreted with the anticipated results from the HL-LHC. The HL-LHC projections for the precision of the different $\kappa$ parameters from a global fit to the Higgs projected uncertainties are shown in Fig. 2[10,11] (from Ref. 6).

Numerically, the LHeC Higgs physics programme improves the precision of several HL-LHC projections. This is apparent from the same figure, where the $\kappa$ interpretation for the LHeC is displayed, both as a stand-alone prospect result and in combination with the HL-LHC. The lower panel of the same figure illustrates the improvements in these two scenarios with respect to the HL-LHC alone. Improvements are most notable for the Higgs couplings to the $W$, $Z$ bosons and to the $b$ quark, $\kappa_{W/Z}$ and $\kappa_b$, respectively, as expected from Table 1. The complementarity between the two Higgs physics programmes is indeed remarkable: the LHeC, with its clean environment, adds precision and the possibility of measuring the charm coupling, $\kappa_c$, while the HL-LHC, owing to its high luminosity and large Higgs production cross sections, leads to the rare and very rare decays, such as $H \to \mu^+\mu^-$ and $H \to Z\gamma$.

The verification of such precise results at per cent level requires very careful checks of the experimental methods and the consistency of the applied theoretical frameworks. Hence it will be of utmost importance to test the Higgs mechanism in the *pp*, *ep* and $e^+e^-$ environments with different production modes and detectors, thereby fully exploring the synergies of the diverse physics programmes. This includes: i) very rare Higgs boson decays are accessible only with *pp*; ii) a unique virtue of *ep* is the accurate resolution of strong interaction uncertainties related to the proton substructure and

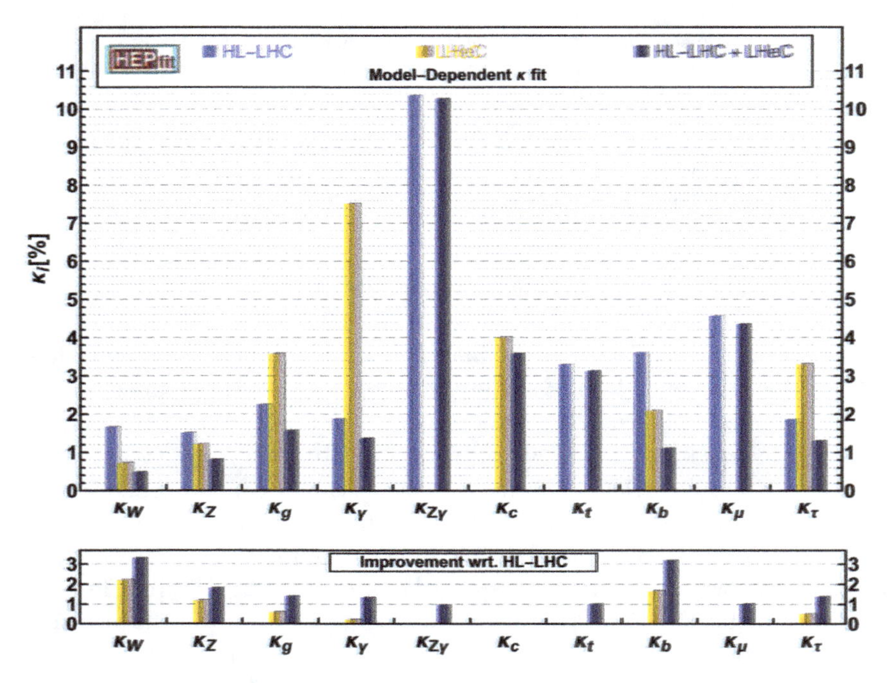


Fig. 2. Projected uncertainty (in percent) in the determination of the modified Higgs couplings in terms of the scaling parameters $\kappa_i$ at the HL-LHC (blue), LHeC (gold) and the combination HL-LHC with the LHeC (dark blue), see text. The bottom panel shows the improvement in the determination of the $\kappa$ parameters with respect to the HL-LHC prospects by adding LHeC prospected measurements. From Ref. 6.

parton dynamics as is required for precision *pp* measurements; iii) a unique result in $e^+e^-$ will be the model independent measurement of the Higgs decay width owing to the constrained initial and final state kinematics of Higgs radiation from the $Z$ boson. All these programmes require decades of preparation, operation and innovative analyses, and as such constitute a salient part of the future of particle physics and the LHC in particular. There is a host of questions on the Higgs phenomenon which extend beyond the mere verification of the Higgs boson decays. In fact, owing to the unique features of the Higgs particle, one may hope it would open a window to BSM physics. Some examples are briefly presented below, with more details in the updated CDR of the LHeC.[6]

### 1.3. *Exotic Higgs decays*

A detailed experimental characterisation of the Higgs boson includes an exhaustive study of its decay modes, which may include modes beyond the ones predicted by the SM. One of these modes is given by invisible final states, which could indicate, for example, dark matter particles. In the SM, invisible decays stem from $H \to ZZ^* \to 4\nu$ and have a tiny branching ratio of 0.1%. The latest upper bound from the ATLAS collaboration in this channel is 14.5%.[12] Invisible decays of the Higgs boson can be tested in *ep* collisions in the NC channel, where the unambiguous signal of missing energy may stem from invisible decays and allows branching ratios as small $\sim 6\%$ to be accessed at the LHeC, as shown in initial studies.

The Higgs boson may also decay into pairs of non-SM particles, for instance into a pair of light scalar bosons which in turn decay into the $4b$ final state. This signature is very difficult to test at the LHC, even if the branching ratio were to be sizeable. At the LHeC on the other hand, this process can test models with scalar masses of $\mathcal{O}(10$ GeV) and couplings to SM particles at the per-mille level. Moreover, when the scalars have macroscopic lifetimes with $\mu$m displacements in the detector, their detection prospects will improve. Scalar mixing angles as small as $\sin^2\alpha \sim 10^{-7}$ can be tested at the LHeC for scalar masses between 5 and 15 GeV.[13,14]

### 1.4. *Higgs pair production and the self-coupling*

The verification of the Higgs potential, specifically the measurement of the Higgs self-coupling via the $HHH$-vertex, is a prime target of the HL-LHC and future Higgs physics experiments. At a high-energy *ep* collider, charged current DIS di-Higgs production involves in the SM only three Higgs vertices in vector-boson fusion, i.e. $HHH$, $WWH$, and $WWHH$ (see Ref. 15), and is therefore not hampered by other di-Higgs production modes which occur at hadron colliders. It is worthwhile to point out that the $WWH$ coupling will be tested extensively at the HL-LHC and may also be very well explored at LHeC. An initial study of di-Higgs production at the FCC-eh, albeit using only the 4b decay[15] and simple kinematic requirements, indicates the important potential of *ep* colliders to disentangle the Higgs trilinear coupling, $\kappa_\lambda$, as well as to identify potential BSM contributions in the $HH$-related vertices. A recent study[16] considered modifications of the Higgs self-coupling and the $WWHH$ coupling in non-resonant $HH$ production, using also the 4b decay only and simple cuts. It suggests that the LHeC constraining power on $\kappa_\lambda$ would be weaker than the HL-LHC expectations, while being complementary in probing the $WWHH$ vertex. A more detailed analysis of the very promising, joint LHeC and HL-LHC sensitivity to the Higgs self-coupling calls for applying state-of-the-art neural network analysis techniques and incorporating all possible Higgs decay channels.

## 2. Beyond the Standard Model in ep at high energies

The physics beyond the Standard Model which can be probed with the LHeC (and at higher ep energies) has been studied for about a decade and recently been presented with the comprehensive report on the LHeC.[6] It comprises a broad spectrum of hypotheses and questions, such as supersymmetry, with prompt and long lived particle signatures, in R-parity conserving or violating models, feebly interacting particles such as heavy neutrinos or fermion triplets, dark photons or axion-like particles. It extends to anomalous gauge couplings, heavy resonances such as leptoquarks or extra $Z$ bosons, vector-like quarks, excited or colour octet leptons, quark substructure and contact interactions, etc. It so includes a very large variety of BSM physics questions which can be explored in energy frontier *ep* and *eA* collisions. It shall be noted that owing to the intense hadron beams and the energy recovery type of electron accelerator, the anticipated luminosity values exceed those of HERA by 2–3 orders of magnitude. Subsequently, a few illustrative examples are given while an interested reader is directed to Ref. 6 and the literature cited there. These studies, also as theory and LHC analyses proceed and the LHeC design progresses, will be further deepened and new ideas be incorporated. It has been recognised for quite some time, especially at the LHC, that many results may be anticipated but real data usually led to results exceeding the scope of even detailed simulations.

### 2.1. *Unique setting for BSM searches*

The unique advantages of the LHeC in the search for BSM physics are: i) the absence of towering backgrounds; ii) the absence of pile-up and complicated triggering; iii) the excellent angular acceptance and resolution of the detector to find displaced vertices for heavy flavour tagging. These properties are ideal to discover BSM that is characterised by the presence of non-prompt, long-lived particles. They are also ideal for BSM physics where the final states can be numerous and have low momenta, and which would be rejected as hadronic noise at the LHC. This renders the LHeC's discovery potential complementary to the one at the LHC and *pp* collisions in general.

Additionally, from the LHeC's very controlled asymmetric initial state, most BSM physics is created via vector-boson fusion, which suppresses the production cross section for new particles that are heavier than the weak bosons, but renders their signatures very separable from the background processes. This is particularly relevant for extensions of the SM scalar sector and neutrino mass physics, which can be well tested in this environment. An increase in the centre-of-mass energy as high as the one foreseen possibly with the HE-LHC and at the FCC would naturally boost the reach in most scenarios considerably.

### 2.2. *Extensions of the scalar sector*

The question of whether the scalar sector contains degrees of freedom beyond the Higgs boson still remains. In this regard, there are several hints in LHC data that could indicate a non-trivial scalar sector with at least two additional neutral fields.[21,22]

Several scalar extensions with interesting phenomenology have been proposed that can be studied at the LHeC, for example: i) CP violating top-Higgs interactions; ii) flavor changing neutral currents, to be tested via decays of the top quark into a charm quark and a Higgs boson; iii) doubly charged scalar bosons from the Georgi-Machacek model, produced via $W^-W^-$ fusion.

In many extensions of the SM, a number of neutral scalars are introduced, which often mix with the SM Higgs doublet. In these scenarios, additional Higgs-like resonances with reduced interaction strength are predicted, which are well testable at the LHC for masses around and above the TeV scale. Testing scalar resonances with masses at the electroweak scale is easier at the LHeC, as shown in Fig. 3.

### 2.3. *Searches for long-lived particles*

One can loosely define long-lived particles (LLP) via their lifetime, which allows them to travel measurable distances before they decay. Many BSM theories predict the existence of LLP with all kinds of spin or quantum numbers. At the LHC, these are often difficult to detect or study, for instance because of pile-up and hadronic background noise.[23] A prime example for such new particles comes from supersymmetric scenarios with compressed

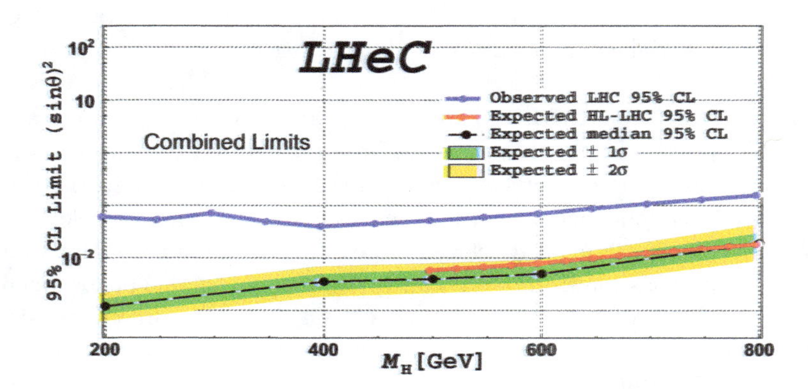


Fig. 3. Expected exclusion limits on the scalar mixing angle $\theta$ for a heavy scalar search at the LHeC[18] considering 1 ab$^{-1}$. The blue and red line denotes the LHC limit[19] and the forecast of the HL-LHC sensitivity,[20] respectively.

mass spectra, where the next-to-lightest supersymmetric particle (NLSP) is long lived because of phase space limitations. At the LHC, these are impossible to detect, due to the low amount of visible energy from the NLSP decay. At the LHeC these signatures, e.g. single pions with transverse momenta below a few GeV, can be tested.[13]

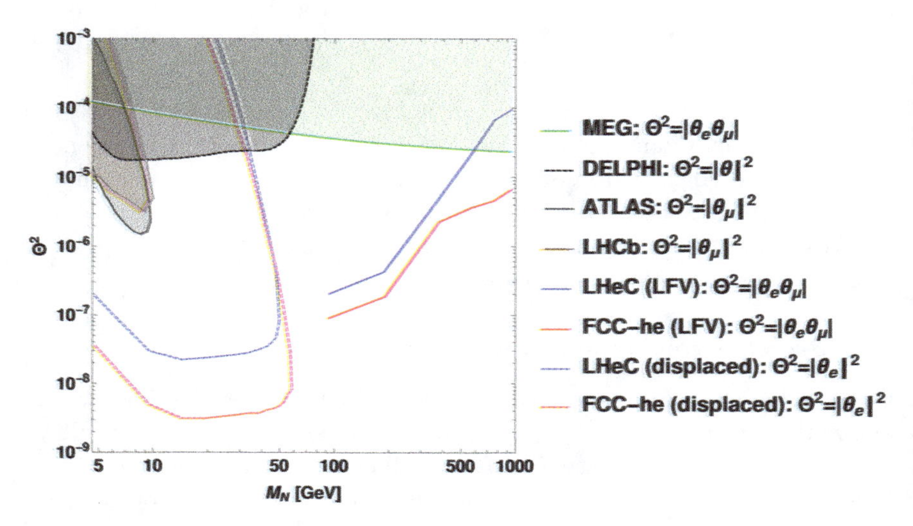


Fig. 4. Overview on prospected sterile neutrino searches. For a detailed discussion see Refs. 24 and 6.

A well-motivated class of BSM models predicting LLP are right-handed neutrinos, which address the big open question of the origin of neutrino masses. An overview of collider searches for right-handed neutrinos[24] shows a comparison of current limits and future LHeC and FCC-eh searches in Fig. 4. In these models the particles are LLP for masses below $M_W$, which makes displaced vertices the most promising way to detect and study them. Extensions with fermion triplets address the neutrino oscillations, and for very small Yukawa couplings their lifetimes can be sizeable and lead to interesting LLP signatures.

LLP also arise naturally in many theories that extend the SM with additional gauge sectors, in particular with the most common abelian $U(1)$ one. Such gauge extensions can be connected to a dark sector and imply a dark charge and a dark gauge boson. More importantly, the dark gauge boson can mix with the neutral gauge bosons of the SM and inherit interactions with charged SM particles that are scaled with the small mixing parameter $\epsilon$. A new heavy gauge boson, often called a dark photon, emerges which has lifetimes that are macroscopic for masses below the GeV scale. At the LHeC, these dark photons are produced via radiation from the electron and be searched for via an appearing vertex of two charged SM particles.[25]

### 2.4. *Leptoquarks*

Leptoquarks (LQ) were introduced in the Pati and Salam SU(4) model, where the lepton number was considered to be the fourth colour.[27] They can be scalar or vector particles with Yukawa couplings to leptons and quarks. These interactions allow, in principle, for the violation of lepton flavour at the tree level, which is why LQs provide feasible solutions to the so-called flavour anomalies[28] in the decays of heavy mesons.

One example of such scalar leptoquarks is a color triplet and electroweak doublet of hypercharge 7/6, denoted $R_2$, which dominantly couples to third generation fermions and can address the $R_{D^{(*)}}$ anomaly with the combination of coupling and mass parameters, see Fig. 5. Limits and future projections of LHC searches for $R_2$ are also shown in Fig. 5. This demonstrates that $R_2$ with $\mathcal{O}(1\text{ TeV})$ mass is not excluded at the LHC and could be studied at the LHeC, where it is produced via its small Yukawa couplings to first generation fermions and can be tested via, for example, the $b\tau$ final states.[26]

At the LHC, LQ are pair produced via the strong interaction. The current limits assume LQ with 100% branching ratios into SM fermions, and

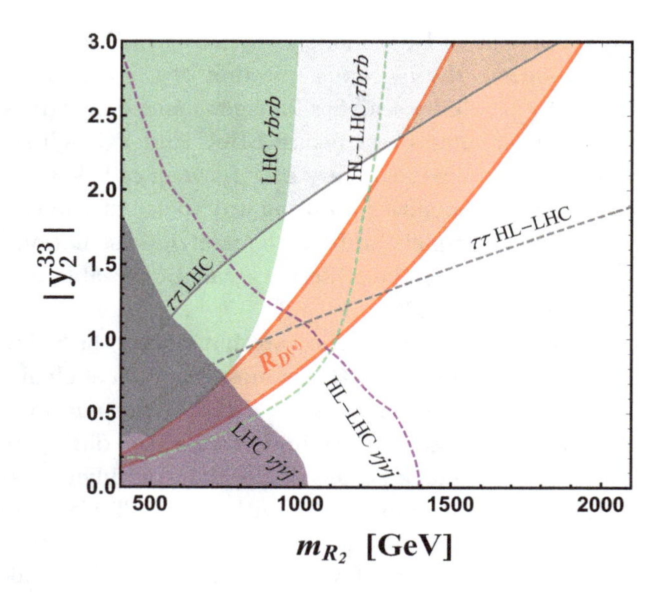


Fig. 5. Existing and projected LHC constraints at the 95% confidence level on the $y_2^{33}$ Yukawa coupling and mass of the $R_2$ leptoquark. Two other non-zero Yukawa couplings are $y_1^{23} = 1$ and $y_2^{11} \ll 1$. The red area denotes parameter combinations where $R_{D^{(*)}}$ can be explained. From Ref. 26.

exclude masses up to around 1 TeV where the limits are stronger (weaker) for the first (third) generation. Scenarios exist where LQ couple strongly to right-handed neutrinos, which in turn decay via a boosted topology that cannot easily be studied at the LHC. Such signatures can be addressed via displaced fat jets at the LHeC with great efficiency.[29]

## 2.5. *Further interesting subjects*

Many more enticing BSM theories have been studied over the years in the context of future *ep* colliders.[6] For example, couplings of gauge bosons and fermions may receive contributions from BSM at the loop level and deviate from the SM prediction. At the LHeC, one could study anomalous gauge couplings in interactions of $Wtb$, $t\bar{t}\gamma$ and $t\bar{t}Z$. Similarly, flavour changing neutral currents are strongly constrained from many experiments, but in interactions that involve top quarks the limits are not very strong, motivating studies of interactions like $tu\gamma$, $tuZ$ and $tHu$ via single top production. Also, anomalies in triple gauge boson vertices, $W^+W^-V$, $V = \gamma, Z$ can be tested at LHeC in great detail.

A unique opportunity comes with tests of charged lepton flavour violation in the tau sector at the LHeC. The BSM model under consideration could be a heavy $Z'$, scalar, or a generic contact interaction, and the tau lepton being correlated with the initial state electron makes this test more powerful than any other existing or proposed future experiment.[30,31]

Axion-like particles, motivated by the original idea of the QCD axion and dark matter candidates, are pseudoscalar particles assumed to be relatively light with QCD couplings. Interactions with other SM fields and mixing with the pion are possible, such that they can be produced via vector boson fusion processes, and improving on LHC results especially for masses below 100 GeV.[32]

It is not clear where new physics may appear. It ought to exist because the Standard Model, while being a phenomenologically successful gauge theory, it does not provide answers to a variety of fundamental questions as mentioned in the opening of this chapter and known to the field. It is very likely that the era of physics beyond the SM, wherever it commences, needs to be explored with different kinds of luminous colliders. Initial studies here sketched have illustrated the remarkable potential of high energy future *ep* colliders, of the LHeC, the HE-LHeC and later the FCC-eh, to find and understand new physics beyond their striking role for QCD, substructure, electroweak and Higgs physics. A next generation of colliders, much like LEP, HERA and the TeVatron, has crucial tasks ahead for a few decades hence. In the nearer future, the HL-LHC combined with its *ep* complement, the LHeC, provides a most attractive and affordable base for particle physics to be advanced, with surprises and possible discoveries that may change the route this science has hitherto taken.

## Acknowledgment

The work of J.B. has been supported by the FEDER/Junta de Andalucía project grant P18-FRJ-3735.

## References

1. G. Aad et al., Observation of a new particle in the search for the Standard Model Higgs boson with the ATLAS detector at the LHC, *Phys. Lett. B.* **716**, 1–29 (2012). doi: 10.1016/j.physletb.2012.08.020.
2. S. Chatrchyan et al., Observation of a New Boson at a Mass of 125 GeV

with the CMS Experiment at the LHC, *Phys. Lett. B.* **716**, 30–61 (2012). doi: 10.1016/j.physletb.2012.08.021.

3. F. Englert and R. Brout, Broken Symmetry and the Mass of Gauge Vector Mesons, *Phys. Rev. Lett.* **13**, 321–323 (1964). doi: 10.1103/PhysRevLett.13.321.
4. P. W. Higgs, Broken Symmetries and the Masses of Gauge Bosons, *Phys. Rev. Lett.* **13**, 508–509 (1964). doi: 10.1103/PhysRevLett.13.508.
5. G. S. Guralnik, C. R. Hagen, and T. W. B. Kibble, Global Conservation Laws and Massless Particles, *Phys. Rev. Lett.* **13**, 585–587 (1964). doi: 10.1103/PhysRevLett.13.585.
6. P. Agostini et al., The Large Hadron-Electron Collider at the HL-LHC, *J. Phys. G.* **48** (11), 110501 (2021). doi: 10.1088/1361-6471/abf3ba.
7. O. Fischer et al., Unveiling hidden physics at the lhc, *Eur. Phys. J. C.* **82** (8), 665 (2022). doi: 10.1140/epjc/s10052-022-10541-4.
8. A. Abada et al., FCC Physics Opportunities: Future Circular Collider Conceptual Design Report Volume 1, *Eur. Phys. J. C.* **79** (6), 474 (2019). doi: 10.1140/epjc/s10052-019-6904-3.
9. A. David, A. Denner, M. Duehrssen, M. Grazzini, C. Grojean, G. Passarino, M. Schumacher, M. Spira, G. Weiglein, and M. Zanetti, LHC HXSWG interim recommendations to explore the coupling structure of a Higgs-like particle (9, 2012).
10. M. Cepeda et al., Report from Working Group 2: Higgs Physics at the HL-LHC and HE-LHC, *CERN Yellow Rep. Monogr.* **7**, 221–584 (2019). doi: 10.23731/CYRM-2019-007.221.
11. J. de Blas et al., Higgs Boson Studies at Future Particle Colliders, *JHEP.* **01**, 139 (2020). doi: 10.1007/JHEP01(2020)139.
12. G. Aad et al., Search for invisible Higgs-boson decays in events with vector-boson fusion signatures using 139 $\text{fb}^{-1}$ of proton-proton data recorded by the ATLAS experiment (2, 2022).
13. D. Curtin, K. Deshpande, O. Fischer, and J. Zurita, New Physics Opportunities for Long-Lived Particles at Electron-Proton Colliders, *JHEP.* **07**, 024 (2018). doi: 10.1007/JHEP07(2018)024.
14. K. Cheung, O. Fischer, Z. S. Wang, and J. Zurita, Exotic Higgs decays into displaced jets at the LHeC, *JHEP.* **02**, 161 (2021). doi: 10.1007/JHEP02(2021)161.
15. M. Kumar, X. Ruan, R. Islam, A. S. Cornell, M. Klein, U. Klein, and B. Mellado, Probing anomalous couplings using di-Higgs production in electron–proton collisions, *Phys. Lett. B.* **764**, 247–253 (2017). doi: 10.1016/j.physletb.2016.11.039.
16. A. Jueid, J. Kim, S. Lee, and J. Song, Studies of nonresonant Higgs pair production at electron-proton colliders, *Phys. Lett. B.* **819**, 136417 (2021). doi: 10.1016/j.physletb.2021.136417.
17. H. Hesari, H. Khanpour, and M. Mohammadi Najafabadi, Study of Higgs Effective Couplings at Electron-Proton Colliders, *Phys. Rev. D.* **97** (9), 095041 (2018). doi: 10.1103/PhysRevD.97.095041.
18. L. Delle Rose, O. Fischer, and A. Hammad, Prospects for Heavy Scalar

Searches at the LHeC, *Int. J. Mod. Phys. A.* **34** (23), 1950127 (2019). doi: 10.1142/S0217751X19501276.
19. A. M. Sirunyan et al., Search for a new scalar resonance decaying to a pair of Z bosons in proton-proton collisions at $\sqrt{s} = 13$ TeV, *JHEP.* **06**, 127 (2018). doi: 10.1007/JHEP06(2018)127. [Erratum: JHEP 03, 128 (2019)].
20. CMS, Search for a new scalar resonance decaying to a pair of Z bosons at the High-Luminosity LHC, CMS-PAS-FTR-18-040 (2019).
21. S. Buddenbrock, A. S. Cornell, Y. Fang, A. Fadol Mohammed, M. Kumar, B. Mellado, and K. G. Tomiwa, The emergence of multi-lepton anomalies at the LHC and their compatibility with new physics at the EW scale, *JHEP.* **10**, 157 (2019). doi: 10.1007/JHEP10(2019)157.
22. A. Crivellin, Y. Fang, O. Fischer, A. Kumar, M. Kumar, E. Malwa, B. Mellado, N. Rapheeha, X. Ruan, and Q. Sha, Accumulating Evidence for the Associate Production of a Neutral Scalar with Mass around 151 GeV (9, 2021).
23. J. Alimena et al., Searching for long-lived particles beyond the Standard Model at the Large Hadron Collider, *J. Phys. G.* **47** (9), 090501 (2020). doi: 10.1088/1361-6471/ab4574.
24. S. Antusch, E. Cazzato, and O. Fischer, Sterile neutrino searches at future $e^-e^+$, $pp$, and $e^-p$ colliders, *Int. J. Mod. Phys. A.* **32** (14), 1750078 (2017). doi: 10.1142/S0217751X17500786.
25. M. D'Onofrio, O. Fischer, and Z. S. Wang, Searching for Dark Photons at the LHeC and FCC-he, *Phys. Rev. D.* **101** (1), 015020 (2020). doi: 10.1103/PhysRevD.101.015020.
26. G. Azuelos, O. Fischer, and S. Jana, Testing the $R_{D^{(*)}}$ anomaly at the LHeC, *Eur. Phys. J. C.* **81** (12), 1123 (2021). doi: 10.1140/epjc/s10052-021-09936-6.
27. J. C. Pati and A. Salam, Lepton Number as the Fourth Color, *Phys. Rev. D.* **10**, 275–289 (1974). doi: 10.1103/PhysRevD.10.275. [Erratum: Phys. Rev. D 11, 703–703 (1975)].
28. Y. S. Amhis et al., Averages of b-hadron, c-hadron, and $\tau$-lepton properties as of 2018, *Eur. Phys. J. C.* **81** (3), 226 (2021). doi: 10.1140/epjc/s10052-020-8156-7.
29. G. Cottin, O. Fischer, S. Mandal, M. Mitra, and R. Padhan, Displaced Neutrino Jets at the LHeC (4, 2021).
30. S. Antusch, A. Hammad, and A. Rashed, Probing $Z'$ mediated charged lepton flavor violation with taus at the LHeC, *Phys. Lett. B.* **810**, 135796 (2020). doi: 10.1016/j.physletb.2020.135796.
31. S. Antusch, A. Hammad, and A. Rashed, Searching for charged lepton flavor violation at $ep$ colliders, *JHEP.* **03**, 230 (2021). doi: 10.1007/JHEP03(2021)230.
32. C.-X. Yue, M.-Z. Liu, and Y.-C. Guo, Searching for axionlike particles at future $ep$ colliders, *Phys. Rev. D.* **100** (1), 015020 (2019). doi: 10.1103/PhysRevD.100.015020.

© 2024 The Author(s)
https://doi.org/10.1142/9789811280184_0025

# Chapter 25

# A New Experiment for the LHC

Peter Kostka
*University of Liverpool, Oxford Street, Liverpool L69 7ZE, UK*

Alessandro Polini
*INFN Bologna, via Irnerio 46, 40126 Bologna, Italy*

Yuji Yamazaki
*Graduate School of Science, Kobe University, 657-8501 Rokko-dai, Nada, Kobe, Japan*

An experiment for *ep* and *eA* collisions at the LHC is presented. Electrons are accelerated to about 50 GeV through a dedicated race-track Energy-Recovery Linac (ERL) and brought into collisions with one of the existing LHC hadron beams. In this chapter, we discuss the generic requirements a detector for such an $eh(= ep$ or $eA)$ experiment would need to fulfill and illustrate the present baseline design. Considerations and optimizations for the use with *hh* collisions are also presented.

## 1. Introduction

*ep* collisions at TeV scale centre-of-mass energies, as discussed in Chapters 23 and 24, provide a unique opportunity to measure precisely the partonic nucleon structure, study the Higgs and extend the search phase space for BSM physics. Such an experiment, the LHeC, can be realized by adding a 50 GeV $e^-$ beam to the existing LHC 7-TeV proton beam and delivering collisions at a centre-of-mass energy of $\sqrt{s} = 1.18$ TeV. In a similar way, with the LHC ion beam, the LHeC would allow for investigations into the partonic structure of nuclei, from small to large atomic numbers, understand the change in nucleon structure in its bound states or see the

This is an open access article published by World Scientific Publishing Company. It is distributed under the terms of the Creative Commons Attribution 4.0 (CC BY) License.

collective phenomena in nuclei as quark matter through $eA$ deep-inelastic scattering collisions at a centre-of-mass energy of $\sqrt{s_{eN}} = 0.74$ TeV, equivalent to collisions with 2.76 TeV per nucleon.

An accelerator scheme utilising an ERL technology for the electron beam, as described in Chapter 22, has benefits not only to the energy efficiency but also to the beam quality, namely stronger focusing for higher luminosity, while keeping the beam emittance within the limit for the energy recovery process. This enables the LHeC to serve as a high-luminosity machine up to $10^{34}$ cm$^{-2}$s$^{-1}$ for $ep$ collisions, making the machine more suitable for Higgs and BSM physics.

The same ERL complex would also be the preferred scheme for colliding the electron beam with the hadron beam of the FCC (FCC-$eh$), which would translate into an $ep$ centre-of-mass energy of 2.2 to 3.5 TeV with protons at 20 to 50 TeV. The foreseen FCC energies for the ion beam are 7.88 to 19.7 TeV, corresponding to $\sqrt{s}$ of 1.4 and 2.2 TeV, respectively.

A common feature among all these $ep$ or $eA$ scenarios is the large energy imbalance between the energy of the LHC or FCC beam and the colliding electrons, which would suggest an asymmetric detector design optimised for high energies in the "forward" direction, i.e. the direction of the outgoing hadron beam.

In this chapter, the requirements on the detector for such an experiment are summarised along with a discussion on the available technologies and the chosen baseline proposal.[1]

The LHeC experiment is planned to be realised at the LHC experimental hall IP2, after the current experiment for $hh$ collisions, the ALICE experiment, is finished at the long shutdown 4 (LS4) in the early 2030's. Recently, new interest for upgrading the ALICE experiment beyond LS4 has manifested, with emphasis on extending the overage of the tracking to $|\eta| < 4$ and PID capability through precise ToF measurement. As the wide tracking coverage is also one of the main requirements for the $eh$ detector, there is a natural interest in combining both detector proposals for $eh$ and for $hh$. This article briefly reviews the benefits of a combined design along with the requirements for such a detector.

## 2. Consideration for an *eh*-collider detector

### 2.1. *LHeC experimental environment*

The high-luminosity operation planned for *ep* collisions at LHeC gives non-negligible constraints on radiation field to be considered for its detector design. The collision rate per bunch crossing, however, is order of 0.1, meaning that the flow of hadronic particles is $O(10^{-3})$ with respect to that for the *pp* collisions on the same ring. This allows us to use moderately radiation-hard detector technologies, e.g., the ones with high performance but could not be used for HL-LHC detectors, or those developed for the ILC.

This also indicates that the events will be basically pile-up free and no special care is needed for degradation in reconstructing the event properties from the high pile-up, most notably the missing $E_{\mathrm{T}}$ reconstruction, a large part of which is measured and reconstructed by the calorimetry and it is difficult to remove particles from non-interest collisions at the LHC. This imposes relatively small demand on the trigger and data-acquisition system, which we therefore do not discuss further here.

On the other hand, severe constraints come from the interaction region where the electron and hadron beams are brought to collide. The design choice for the LHeC interaction point (IP) is to steer the electron beam and bring it into head-on collisions, i.e. with zero-degree crossing angle, by immersing the entire interaction region in a 0.2 T dipole field through a magnet system placed outside the 3 T solenoidal magnet for the central tracker.[1] The field is needed to not only achieve collisions and the optimal luminosity, but also to separate the beam and avoid

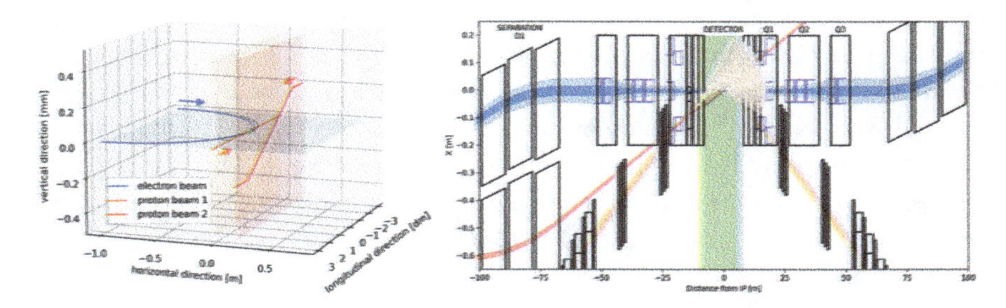


Fig. 1. Left: 3D sketch of the interaction region with the three beams (the electron, the interacting proton and the spectator proton beams). Right: $xz$ view of the interaction region. The central detector and the electron beam with its synchrotron radiation is highlighted.

parasitic collisions with the neighbouring bunches — which are spaced 25 ns apart at the LHC (Fig. 1(left)). The electron beam consequently radiates synchrotron light while passing the $\pm 10$ m dipole field area around the interaction point (Fig. 1(right)). A thin elliptic-shape beam pipe is required to accommodate this synchrotron radiation fan (see also Sec. 3.1).

### 2.2. *Detector requirement for precision measurements on DIS*

Neutral-current (NC) DIS events are characterised by a presence of a scattered electron and a recoiled hadronic system. The momentum transfer between the electron and incoming nucleon is expressed by $Q^2 = -q^2 = -(k - k')^2$, where $k$, $k'$ and $q$ are the four-momentum of the incoming and scattered electron and the exchanged state ($\gamma/Z^0$), respectively. The Bjorken variable $x = Q^2/(p{\cdot}q)$, where $p$ is the four-momentum of the incoming nucleon, represents the longitudinal momentum fraction of the scattered parton in the quark-parton model. The inelasticity variable $y = (p{\cdot}q)/(p{\cdot}k)$ gives the longitudinal fraction of the momentum transfer. The transverse momentum of the electron is balanced with that of the hadronic system from the scattered parton, $p_{T,h}$, which in most cases is observed as a jet (or as multi-jet final state when hard QCD radiation occurs).

In principle the kinematic variables can be fully determined by measuring the momentum of the scattered electron $k'$, or $E'_e$ and $\theta_e$, representing the energy and angle of the scattered electron, respectively. This naturally calls for good energy and position measurement for the scattered electron through both the central tracker and the EM calorimeter. However, as is well known from the HERA experiments, the reconstruction resolution of the DIS kinematic variables $x, y$ and $Q^2$ depends strongly on the value of the variables themselves. For example, at small $y$, limit $y \to 0$, or for large-$x$ events since $x = Q^2/sy$, the value of $y$ would be better reconstructed through the longitudinal momentum of the scattered parton, which is expected to be large due to the large momentum fraction in the initial state. The jet to be measured is boosted in the forward direction, in particular for low-$Q^2$ events, since $p^2_{T,h} \simeq (1 - y)Q^2$, or for high-$x$ events with its jet energy close to the incoming nucleon beam momentum.

For these reasons, good resolution in hadronic energy measurements to multi-TeV range is required. Also, effort should be made to extend the forward angular coverage as much as possible: at the LHeC, in the forward direction, up to 1° is aimed for. The forward coverage is important even for events with large angle jets, since QCD radiation between the scattered parton and the forward-going nucleon remnant may bring a significant part of the energy close to the beam pipe hole. This is particularly important when measuring the charged-current (CC) events, where only the hadronic final state is available for kinematic reconstruction. Precise determination of the total missing $p_T$ of the event is also necessary to distinguish the CC events from the photo-produced high-$E_T$ events, $\gamma p \to$ multijets, where some missing energy may arise due to e.g. mis-measurement of heavy quark jets.

To extend the sensitivity towards low-$x$, NC events with very small-angle scattering should be measured. The coverage for the scattered electron should, again, be as close to the electron beam direction: 179° is aimed for, to reach down to $Q^2 = 1$ GeV$^2$. In addition, measuring high-$y$ events is necessary for determining $F_L$, which is extracted from the $y$ dependence in the cross sections at given $(x, Q^2)$ points by measuring the cross sections in different centre-of-mass energies. This needs very high-$y$ events, or events with small energy deposit for the scattered electrons. Such events should be distinguishable from photoproduction processes, where a quasi-real photon from the electron collides with nucleons or nuclei. What is harmful there is the "resolved" photon process, where only a part of the photon collides with a nucleon and the rest of the system is hadronised and deposits energies in backward small-angle region. Fine granularity is necessary to distinguish between the backward-going mesons and low-energy scattered electrons there, since the cross section of resolved photoproduction is much larger than that from NC DIS.

Identification of the jet flavour, in particular bottom and charm quark jets, should be realised at the LHeC for flavour decomposition of the parton densities in NC and CC events. Good transverse impact parameter resolution at $O(10\mu\text{m})$ beyond $|\eta| < 3$ is required.

### 2.3. *Detector requirement for EW, Higgs, top and BSM physics*

Thanks to the high luminosity operation of LHeC by the ERL scheme has opened up a new possibility — to study the events with EW boson fusion, such as vector boson fusion (VBF) or scattering to study trilinear and quartic couplings. Higgs bosons are also produced primarily through the VBF processes, $WW \to H$ in CC events or $ZZ \to H$ in NC events. The characteristics of these processes are: a heavy final state from the EW vertex and additional forward jet produced through the emission of a vector boson. This requires a parton with $x$ at or beyond $O(10^{-2})$; the system is boosted towards the proton beam and the final states are peaked near $|\eta| \simeq 2$. The decay leptons from weak and Higgs bosons and heavy quarks from Higgs decay or diboson final state beyond $\eta = 3$ should be measured with purity. In addition, the forward jet close to $\eta = 5$ should be measured for high acceptance. Such capability for electrons, muons, jets and heavy flavour tagging should also serve for top quark studies and various BSM searches.

## 3. A detector for LHeC

Figure 2 shows a drawing of the LHeC baseline detector, originally presented in the CDR update[1] and then improved.[2] The detector consists of, from inside to outside, the central silicon tracker; the barrel electromagnetic Liquid Argon (LAr) calorimeter, surrounded by the solenoid and the dipole magnets sharing the cryostat with the LAr calorimeter; the steel-scintillator barrel hadron calorimeter; and the muon detector embedded inside the steel return yoke. Silicon-based EM and hadron sandwich calorimeters are located between the tracker and the muon detector covering the end-cap

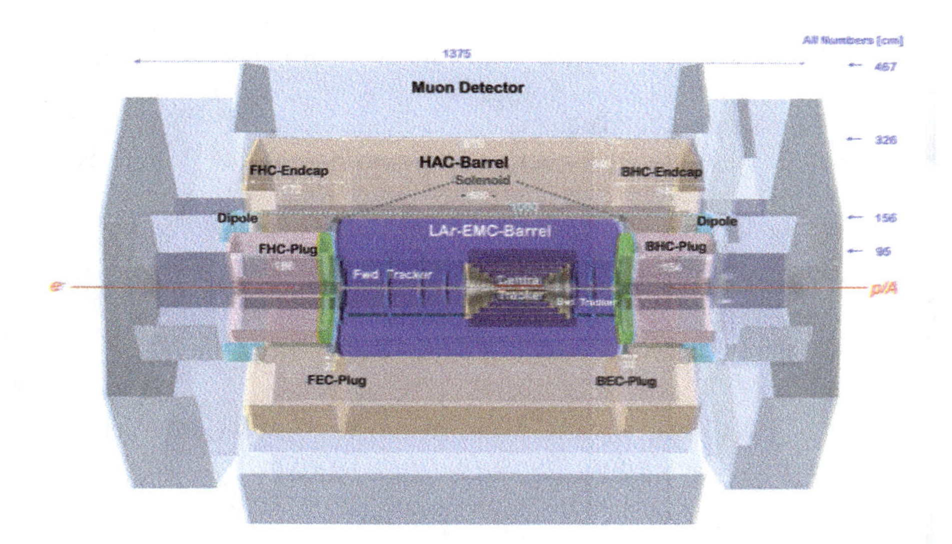


Fig. 2. Side view of the updated baseline LHeC detector concept, providing an overview of the main detector components, their locations and dimensions.[2]

regions. Not shown are the forward detectors for particles scattered or produced at very small angles. A brief explanation on these detectors is given below.

### 3.1. *The all-silicon central tracker*

Considering the radiation environment and the requirement on vertexing, the entire tracker should use silicon detectors. A resolution on the impact parameter of 10 $\mu m$ is achieved by using 10 layers of barrel sensors, covering $|\eta| < 1.4$. The first two layers are of pixel size of about $25 \times 50$ $\mu$m, followed by four $100 \times 400$ $\mu$m "Macropixel" layers and another four layers with 100 $\mu$m-pitch strips. In the end-caps the detector is asymmetric: up to seven silicon wheels are placed in the forward direction while five layers are adequate for the backward direction. The entire detector covers $-4.6 < \eta < 5.2$ with at least two hits, or six hits for $|\eta| < 3.6$.

The radiation level in the most forward angular region is large but manageable. The expected neutron flux in the most forward region is of the order of $10^{14}$ MeV $n_{eq}/\text{cm}^2$. This allows use of technologies developed for the HL-LHC but not the latest development — less radiation hard, therefore less cost intensive. The most crucial requirement is on the material budget, in particular in the forward direction, in order to minimise the multiple-scattering for low-momentum tracks, as required from flavour tagging performance. For that, CMOS-based technologies with integrated sensing element and readout electronics, would be one of the most attractive choice. For example, total sensor thickness of 50 $\mu$m is already achieved by Depleted Monolithic Active Pixel Sensors (DMAPS) processed with HV-CMOS technology. This sensor has 6 ns time resolution and is resistant up to $2 \times 10^{15}$ MeV $n_{eq}/\text{cm}^2$. Since the sensors are so thin, they can be bent by rollers to have curved layers.[6] This allows the innermost layer of the central tracker to follow the circular-elliptical-shaped LHeC beam-pipe (Fig. 3), which is necessary to accommodate the synchrotron radiation fan.

### 3.2. *Calorimetry*

Energies to be measured by the electromagnetic part of the calorimeter are relatively high at the LHeC. For example, the typical energy for scattered electrons is above 10 GeV for 50 GeV beam energy and the EM calorimeter does not need ultimate energy resolution: the balance between the resolution and fine granularity is more important. In the baseline detector, fine-segmented LAr (similar to the ATLAS design)[3,4] is chosen for longevity,

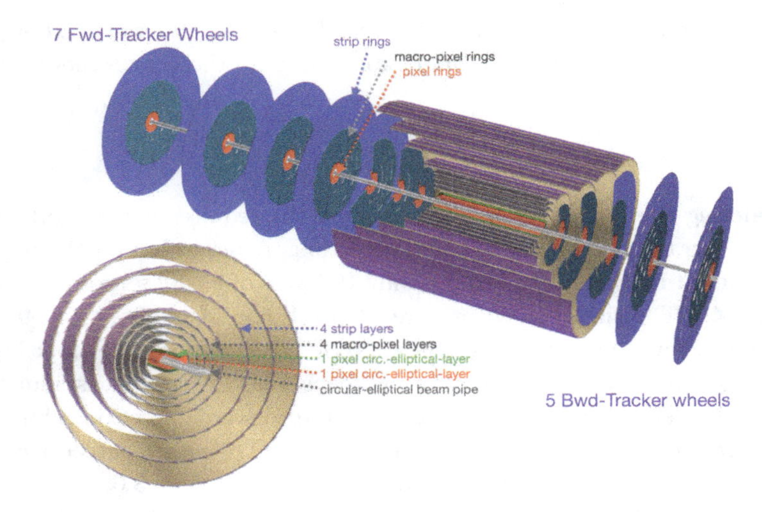


Fig. 3. Structure of the silicon tracker for the LHeC detector and an extended view of central barrel tracker with 2 innermost circular-elliptical silicon pixel layers following the shape of beam pipe.

fine segmentation and energy resolution. A sampling calorimeter with plastic scintillators as active layers is the preferred option as barrel hadronic calorimeter — mainly for the good $e/h$ ratio and hence good resolution for hadronic showers.

In the forward endcap region, a much finer segmentation is needed to resolve close-by particles, as well as for hadronic shower resolution — if we like to perform local shower weighting to compensate for the $e/h$ ratio. In addition, the calorimetry has to cope with a radiation of the order of $10^{15}$ MeV $n_{eq}/\mathrm{cm}^2$. The baseline choice here is to use silicon sensors since the energy leak should be minimised for better resolution along with tungsten for absorber layers, for the showers and the dimensions to be compact. Strip sensors are used for the EM part while the hadronic part is made of pads. For the backward direction, the requirement on energy leak is less stringent and the proposal is to use Pb absorber in the EM section while Cu is used for the hadronic part.

### 3.3. *Muon System*

Muons are produced from Higgs and weak vector boson decays, vector meson decays, $\tau$ decays and through direct dimuon production. Also, BSM physics such as leptoquarks and long-lived heavy particles may produce a muon or muon-like particle. They range from about 1 GeV to a few hundred GeV in transverse momentum.

Minimum requirement on the muon detector for TeV *ep* collisions is to identify and trigger i.e. tagging muons at high purity. Stations embedded in the return yoke would serve this purpose; the momentum measurement will mainly be provided by the central tracker. In the baseline proposal, an example from the inner station of ATLAS Phase-2 upgrade[5] is given. It consists of eight layers of 15 mm drift tubes providing precise tracking in $\eta$ direction and three layers of thin-gap RPC for timing and the tracking in the second coordinate.

### 3.4. *Forward detectors*

In photoproduction events ($Q^2 \approx 0$), the electron is deflected by a very small angle while it often loses significant fraction of energy ($E'_e/E_e \simeq y$). Such low-$Q^2$ electrons are also produced through a QED process, $eh \to e\gamma h$, where both the electron and photon in the final state are emitted in very small angles. The backward-going electrons in both cases are deflected towards inside the ring to be tagged by a detector there. The backward forward photon should also be measured by placing a calorimeter at $-180°$. These detectors are used for determining the luminosity through the counting rate of this process.

Similarly, a scattered proton or a proton inside a nucleus may lose some fraction of longitudinal momentum $\xi \simeq E'_p/E_p$ while the transverse deflection is typically small: $p_{T,p} \sim \Lambda_{\rm QCD}$. A Forward Proton Spectrometer should be placed, as has been done for the existing LHC experiments. Neutrons are similarly produced very forwardly, together with photons from the decay of $\pi^0$'s. They should be measured by a zero-degree calorimeter.

It is worth noting that the IP design is quite different from that for *hh* collision points and there may be a chance to place these forward detectors in more locations and/or with less constraint in space.

## 4. Extension for *hh* collisions

As described in Chapter 22 and elsewhere,[2] the latest machine optics for the $ep/eA$ interaction region is designed such that the standard LHC operation modes (proton-proton, ion-ion) can have the exact same interaction point. This opens up the option of a multipurpose detector capable of exploiting all available LHeC and LHC beam options. The baseline LHeC detector described above includes a tracking detector with requirements already very similar to those for the ALICE experiment after LS4 as it comes with a very good rapidity coverage, fine-pitch pixels to resolve close-by particles and small material budget for good momentum resolution for tracks down to $p \simeq 0.1$ GeV. By adding or replacing a few layers with more time sensitive sensors or using, for instance, most advanced 65 nm technology for sensor production, the detector would also be equipped with good PID capability. Option of lowering the solenoidal B-field should also be considered to avoid curl tracks and extend the tracking detector acceptance to lower $p_T$.

The LHeC baseline detector covers the same rapidity range as the tracking detector by a calorimeter system with high granularity and good energy resolution. In addition, high-purity muon tagging and momentum measurements are possible up to $|\eta| < 3$, possibly beyond. The combination of the tracking and the hermetic outer detectors allows for new possibilities in studying ion collisions at IP2, such as heavy meson tagging through e.g. $J/\psi \to \mu^+\mu^-$, jet physics related to jet quenching, particle flow towards very forward region etc., in combination with central tracking detector optimised for low-momentum particles.

The shape and detailed layout of the detector components are to be further optimised along with the IR design for concurrent *eh* and *hh* running. One obvious thing is that the baseline LHeC detector is asymmetric. It, however, already covers $-4.6 < \eta < 5.2$, as well the backward calorimeter is already deep enough for measuring low-energy particles up to $\sim 100$ GeV. But even a symmetric detector from point of view of geometry as well as functionality has been thought of.[2]

## References

1. P. Agostini *et al.* [LHeC and FCC-he Study Group], J. Phys. G **48** (2021) no. 11, 110501 doi:10.1088/1361-6471/abf3ba [arXiv:2007.14491 [hep-ex]].
2. K. D. J. André *et al.*, Eur. Phys. J. C **82** (2022) no. 1, 40 doi:10.1140/epjc/s10052-021-09967-z [arXiv:2201.02436 [hep-ex]].
3. [ATLAS], CERN-LHCC-96-41.

4. D. J. Mahon [ATLAS], JINST **15** (2020) no. 06, C06045 doi:10.1088/1748-0221/15/06/C06045
5. CERN-LHCC-2017-017, http://cds.cern.ch/record/2285580.
6. G. Rinella *et al.*, [arXiv:2105:13000v2 [physics.inst-det]] and ALICE 3 workshop October 2021 - https://indico.cern.ch/event/1063724/timetable/, talks: M. Mager, M. van Leeuwen, J. Klein

© 2024 The Author(s)
https://doi.org/10.1142/9789811280184_0026

# Chapter 26

# High Energy LHC Machine Options in the LHC Tunnel

Luca Bottura and Frank Zimmermann

*CERN, Esplanade des Particules 1, 1217 Meyrin, Switzerland*

The LHC infrastructure, i.e. the tunnel itself as well as the services associated with power, cooling, ventilation, network and access (among others), represent a considerable asset, and may be considered for hosting and supporting future versions of a "LHC" beyond its present configuration and the HL-LHC upgrade. In this chapter, we provide an overview of possible machine parameters and energy reach of a future higher-energy hadron collider in the LHC tunnel. We sketch four options with arc dipole magnetic fields in the range of 12 T to 24 T, each of which represents a well-defined discrete step in future accelerator magnet technology. We discuss the corresponding main machine and magnet parameters, and describe readiness, challenges and opportunities.

*Keywords*: High Energy LHC; $Nb_3Sn$ accelerator dipoles; HTS accelerator dipoles

## 1. Introduction

The LHC has a considerable value, which is customarily associated with the existing and running accelerator and experiments. Also, the site infrastructure in itself, extending from civil engineering to powering, cooling, ventilation and other auxiliary systems, is a noticeable asset. Hence, it is natural that several past analyses and studies discussed the possibility of using the LHC site infrastructure for upgrades well beyond the lifetime of the LHC and the HL-LHC.

The 2002 Feasibility Study for an LHC Luminosity and Energy Upgrade[1] defined an "LHC Phase 2", which consisted of installing new superconducting dipoles in the LHC arcs to reach a beam energy around

This is an open access article published by World Scientific Publishing Company. It is distributed under the terms of the Creative Commons Attribution 4.0 (CC BY) License.

12.5 TeV. The Study Report pointed out that "the energy upgrade is much easier to exploit than a luminosity upgrade, as it requires minimal changes to the detectors. Dipole magnets with a nominal field of 15 T and a safety margin of about 2 T can be considered a reasonable target...".

This idea was followed up in 2010, when a High-Energy LHC (HE-LHC) based on 20 T hybrid magnets was studied by a dedicated working group. This activity culminated in the HE-LHC'10 EuCARD workshop.[2] The HE-LHC'10 workshop also, for the first time, proposed a future Very High Energy LHC (VHE-LHC), a new ring with a larger $\sim$ 80 km circumference, which later became the Future Circular Collider hadron-hadron option (FCC-hh).[3]

The FCC-hh study took the proposal of the VHE-LHC further, developing a full design for a hadron collider with a tunnel length in the range of 80 to 100 km. An annex activity to FCC-hh was to study an energy upgrade in the LHC tunnel made possible by the magnet technology to be developed for the FCC-hh. The HE-LHC, based on the 16 T magnets of the FCC-hh case, including dispersion suppressor (DS), Interaction Regions (IRs), and collimations, is described in detail in the FCC design report volume 4.[4] This option was shown to require a new superconducting SPS (sc SPS) as injector to reach acceptable injection field and aperture.

Finally, a recent study considered the possibility of a partial energy increase of the present LHC that could be obtained by using the HL-LHC 11 T dipoles to replace one third of the present Nb-Ti dipoles by higher field $Nb_3Sn$ magnets.[5] This would result in a modest increase in energy reach, i.e. a centre-of-mass (COM) energy of 16.2 TeV vs. 14 TeV nominal for the LHC, and was not retained as an interesting investment by the authors of the study. Still, it is instructive to consider what would be the result of using the magnet technology under development for the ongoing HL-LHC upgrade, just short of 12 T, as a full replacement for the present LHC.

Taking into account the historical proposals and studies, we have selected four scenarios that could represent well-defined discrete and distinct options for a future High-Energy (HE) hadron collider in the LHC tunnel, following the completion of the HL-LHC physics programme, namely:

- A modest energy upgrade based on technology close to deployment, i.e., $Nb_3Sn$ 12 T dipoles, leading to a COM energy of 20 TeV. We refer to this option as HE20.
- The highest energy that could be reached by Low Temperature Superconductor (LTS) accelerator magnet technology, i.e., the

$Nb_3Sn$ 16 T dipoles being developed for the FCC-hh. In this case the COM energy would be about 27 TeV, and we denote this option by HE27.
- Higher dipole fields as could be reached if High Temperature Superconductor (HTS) dipole magnets can be developed as outlined in the HE-LHC proposal, namely 20 T, leading to a COM energy of 34 TeV. This option is called HE34.
- Finally, we consider an ultimate High Energy LHC, namely taking HTS magnets producing a dipole field of 24 T, which possibly is the highest field range that can be reached with such technology. The reason for setting this bound is that forces, stored energy and cost will be about one order of magnitude larger than for the LHC magnets, requiring extraordinary advances in material science and engineering. This option, with name HE41, would yield a COM energy of 41 TeV.

In the following sections, we will elaborate on the collider parameters for the four options, and discuss the dipole magnet designs that would correspond to such parameters. For the relevant magnet technology, we base the discussion on ongoing developments, while referring to the designs proposed in the past, but we also elaborate on a perspective where we consider advances and activities in other fields of application of superconducting magnet technology.

## 2. Main Collider Parameters

The options outlined in the previous section are detailed in Table 1, which compiles the main machine and magnet parameters. The figures reported there are a combination of results from the references quoted earlier, and scaling applied to such options. Note that for the discussion we only report the required dipole field, expanding, in a later section, on the possible magnet configurations and challenges. It is clear that a complete analysis of any option would require devising quadrupoles and dispersion suppressor magnets, as well as adapted insertions. Indeed, simple scaling from LHC and HL-LHC does not necessarily produce consistent and feasible configurations, as the optics for the different energy options may differ considerably. This point is illustrated by the study of the HE-LHC based on FCC-hh 16 T dipoles.[4] A detailed discussion on how the optics is modified by using different dipole configurations and strength can also be found in Ref. 6.

Still, in spite of the simple approach taken here, our basic considerations suffice to provide a good view of the perspective and challenges of an HE-LHC.

Table 1. Key parameters of four HE-LHC options compared with HL-LHC and LHC, for operation with proton beams. All values, except for the injection energy itself, refer to the collision energy. The ring circumference is 26.7 km and the straight section length 528 m, as for the existing LHC tunnel.

| Parameter | Unit | HE20 | HE27 | HE34 | HE41 | (HL-)LHC |
|---|---|---|---|---|---|---|
| Centre-of-mass energy | TeV | 20 | 27 | 34 | 41 | 14 |
| Injection energy | TeV | 0.6 | 0.8 | 1.0 | 1.2 | 0.45 |
| Peak arc dipole field | T | 12 | 16 | 20 | 24 | 8.33 |
| Beam current | A | 1.12 | | | | (1.12) 0.58 |
| Bunch population | $10^{11}$ | 2.2 | | | | (2.2) 1.15 |
| Bunches / beam | | 2808 | | | | (2760) 2808 |
| Rf voltage | MV | 16 | | | | (16) 16 |
| Rf frequency | MHz | 400 | | | | (400) 400 |
| momentum compaction | $10^{-4}$ | 5.8 | | | | (3.22) 3.22 |
| RMS bunch length | mm | $\sim 90$ | | | | (90) 75.5 |
| Bucket half height | $10^{-3}$ | 0.24 | 0.21 | 0.19 | 0.18 | 0.36 |
| RMS momentum spread | $10^{-4}$ | 0.77 | 0.67 | 0.59 | 0.54 | 1.129 |
| Longit. emit. ($4\pi\sigma_z\sigma_E$) | eVs | 2.9 | 3.3 | 3.7 | 4.0 | 2.5 |
| Bunch spacing | ns | 25 | | | | 25 |
| Norm. tr. rms emittance | µm | 2.5 | | | | (2.5) 3.75 |
| IP beta function $\beta^*_{x,y}$ | m | 0.25 | 0.45 | 0.55 | 0.60 | (0.15) 0.55 |
| Initial IP beam size $\sigma^*_{x,y}$ | µm | 7.7 | 8.8 | 8.7 | 8.3 | (7.1 min.) 16.7 |
| Half crossing angle | µrad | 160 | 104 | 84 | 74 | (250) 142.5 |
| Initial luminosity / IP | $\mathrm{nb}^{-1}\mathrm{s}^{-1}$ | 200 | 160 | 160 | 180 | (50, levelled) 10 |
| Total cross section | mbarn | 119 | 126 | 131 | 135 | 111 |
| Inelastic cross section | mbarn | 87 | 91 | 94 | 97 | 85 |
| Initial events / crossing | | 570 | 450 | 480 | 540 | (135) 27 |
| RMS luminous region | mm | $\sim 64$ | | | | (64) 45 |
| Stored energy / beam | GJ | 1.0 | 1.3 | 1.7 | 2.0 | (0.7) 0.36 |
| Energy loss / p / turn | keV | 28 | 93 | 230 | 470 | 6.7 |
| SR power / beam | kW | 30 | 100 | 251 | 532 | (7.3) 3.6 |
| SR power / length | W/m/ap. | 1.4 | 4.6 | 11.5 | 24.5 | (0.33) 0.17 |
| Transv. emit. damp. time | h | 8.8 | 3.6 | 1.8 | 1.0 | 25.8 |
| No. of high-luminosity IPs | — | 2 | | | | (2) 2 |
| Initial proton burn-off time | h | 3.5 | 4.4 | 4.1 | 3.6 | (15) 40 |
| Allocated physics time / yr | days | 160 | 160 | 160 | 160 | 160 (160) |
| Average turnaround time | h | 5 | | | | 4 (5) |
| Optimum run time | h | 4.5 | 5.2 | 4.7 | 4.3 | (18–13) ~10 |
| Accelerator availability | — | 75% | | | | (80%) 78% |
| Ideal luminosity / day | $\mathrm{fb}^{-1}$ | 4.1 | 4.2 | 4.3 | 4.4 | (1.9) 0.4 |
| Luminosity per year | $\mathrm{fb}^{-1}$ | 490 | 500 | 520 | 530 | (240) 55 |

In Table 1, the centre-of-mass collision energy increases in proportion to the arc dipole field. Higher injection energy, attainable from a new superconducting (sc) SPS, is required for adequate dynamic and physical aperture.

In case of $Nb_3Sn$ magnets, the field quality at injection can be improved by a superconductor with reduced low-field magnetization,[4] which would provide a solution for low-field injection. However, compared with the present LHC, at higher energy a more voluminous beam screen is required to intercept the synchrotron-radiation (SR) photons and to extract the increasing SR heat load, while still ensuring a good vacuum quality together with a low machine impedance. Consequently, the actual physical half aperture available for the injected beam shrinks from about 2 cm for the LHC to 1.3 cm for the HE-LHC.[4] Following Ref. 4, in view of the more challenging physical and dynamic aperture constraints, we assume that the injection energy needs to increase roughly in proportion to the collision energy.

For all scenarios we consider the same beam current $I_b$, bunch population $N_b$, transverse normalized emittance $\varepsilon_n$, and rms bunch length $\sigma_z$ as for the HL-LHC. These beam parameters are available from the LHC injector complex after its recent upgrade (LIU).[7]

In Ref. 4, two optics were studied. For the scenarios HE20 to HE41, we assume the optics with 18 FODO cells per arc and a FODO cell length of 137.33 m instead of 23 FODO cells with a cell length of 106.9 m, as for the LHC and HL-LHC, one advantage being a 5% higher energy reach at the same dipole field. For this optics, the momentum compaction factor $\alpha_C$ is more than doubled compared with the LHC. The bucket height scales as $\sqrt{V_{\mathrm{RF}}/(E_b \alpha_C)}$. Keeping the rf voltage constant, equal to 16 MV, for a given optics, the bucket height scales as the inverse square root of beam energy. In Table 1, we scale the rms energy spread in proportion to the bucket height, which makes the rms relative energy spread decrease as the inverse square root of the beam energy, and the longitudinal emittance rise as the square root of the beam energy $E_b$. This scaling also ensures longitudinal Landau damping.[8,9]

It is natural to assume high-luminosity collisions in $n_{\mathrm{IP}} = 2$ primary collision points, and possible lower-luminosity secondary collisions at two other IPs, as for the LHC and HL-LHC. We take the total number of bunches $n_b$ to be 2808, as was in the original LHC design, slightly larger than the 2760 value of the HL-LHC.[10] Higher beam energies will, however, require a revision of the dump and injection kicker system, which may have an impact on the maximum number of bunches permitted.

The HE-LHC optics design of Ref. 4 for a centre-of-mass energy of 27 TeV achieved an interaction point (IP) beta function $\beta^*$ of 0.45 m, with acceptable dynamic aperture. We assume the same value for HE27, recognizing that HL-LHC aims for 0.15 m, and interpolate between these values, as well as extrapolating for higher energies. This results in a roughly constant rms interaction-point (IP) beam size between 8 and 9 $\mu$m for all energies.

To maintain a constant impact of long-range collisions on the beam lifetime, the crossing angle $\theta_c$ scales as $\theta_c \propto 1/\sqrt{\beta^* E_b}$, that is in proportion to the IP beam divergence. In Table 1 we are assuming that the triplet length and the total number of long-range collisions stays approximately the same as for the HL-LHC. We note that for the 16 T HE27 scenario this scaling leads to a full crossing angle of 208 mrad, as indicated, whereas for the longer HE-LHC triplet considered in Ref. 4, and for the same IP beta function $\beta^*_{x,y} = 0.45$ m, a much larger crossing angle of 330 mrad was chosen. As for the HL-LHC and the HE-LHC of Ref. 4, crab cavities will be needed to avoid luminosity loss due to the crossing angle.

The initial luminosity is given by

$$L_0 = \frac{f_{\rm rev} n_b N_b^2}{4\pi {\sigma^*_{x,y}}^2}\,, \tag{1}$$

where $f_{\rm rev}$ denotes the revolution frequency, and $\sigma^*_{x,y} = \sqrt{\beta^* \varepsilon_n/\gamma}$ the rms beam size at the IP, assuming round beams ($\sigma^*_y = \sigma^*_x$), and $\gamma = E_b/(m_p c^2)$ the relativistic Lorentz factor, with $m_p$ the proton mass and $c$ the speed of light.

The total and inelastic proton-proton cross sections, $\sigma_{\rm tot}$ and $\sigma_{\rm inel}$, are weakly dependent on the collision energy as indicated. This dependence is described by Eqs. (6) and (7) in Ref. 11, which are based on Refs. 12–17. The total cross section $\sigma_{\rm tot}$ increases from about 111 mbarn at 14 TeV (LHC) to 135 mbarn at 41 TeV centre-of-mass energy (HE41), the inelastic cross section $\sigma_{\rm inel}$ from 85 to 97 mbarn. The inelastic cross section roughly relates to the number of events per bunch crossing recorded in the detector (the so called event pile up), as

$$n_{\rm event} = \frac{\sigma_{\rm inel} L_0}{n_b f_{\rm rev}}\,. \tag{2}$$

The initial pile up is about 500 per bunch crossing for all four HE-LHC versions considered. Especially for HE34 and HE41 the pile up will increase during the physics store. With perfect crab crossing and for Gaussian bunch profiles, the rms extent of the luminous region is equal to the rms bunch length divided by $\sqrt{2}$.

The total cross section $\sigma_{\rm tot}$ determines the initial proton burn off time $\tau_{\rm bu}$ as

$$\frac{1}{\tau_{\rm bu}} = -\frac{\dot{N}_b}{N_b} = \frac{\sigma_{\rm tot} L_0 n_{\rm IP}}{N_b n_b} \,. \tag{3}$$

The energy stored in the beam scales exactly with beam energy, and at highest beam energy (HE41) approaches a value of 2 GJ, which is about 3 times higher than for the HL-LHC.

The proton energy loss per turn due to synchrotron radiation grows as the fourth power of beam energy, increasing from 6.7 keV at the LHC to 470 keV at HE41. At constant beam current and bending radius, the total synchrotron-radiaton power also increases as the fourth power of energy. While for the nominal LHC, the SR power of one beam is 3.6 kW, for the HE41 it becomes 500 kW per beam, or about 1 MW in total, and the synchrotron radiation per unit length reaches 25 W/m per aperture, which is almost the same as the 29 W/m per aperture of the FCC-hh design. This implies that the SR heat can still be removed from inside the arcs with the FCC-hh/HE-LHC beam screen design. At even higher energies, the maximum allowable synchrotron radiation heat load would limit the maximum beam current.

The radiation damping time scales as $\rho E_b^{-3}$, where $\rho$ denotes the dipole bending radius. The interplay of proton burn off and radiation damping determines the optimum physics run time (i.e. the moment the two beams are dumped for a new injection) as a function of the average turnaround time (the time between the dump and the start of the new physics fill).

If the proton burn-off time is shorter than the transverse emittance damping time, as is the case for HE20, the beam-beam tune shift decreases during the store, and the calculation of the optimum run time $t_{\rm r}$ is based on Eqs. (2.2)–(2.4) of Ref. 4. In the opposite the case, both the luminosity and the beam-beam tune shift increase with time in store, and for the latter we must assume a maximum acceptable value, which, once reached, is maintained by controlled emittance blow up through transverse noise excitation, as is proposed for the FCC-hh.[11] The time evolutions of the luminosity and the optimum run length $t_{\rm r}$ then follow from Eqs. (33)–(54) in Ref. 11. This situation is encountered for HE27, HE34 and HE41. For the purpose of illustration, we consider a maximum beam-beam tune shift of 0.025, which is close to the value of 0.03 assumed for the "phase 2" of the FCC-hh.[3,11] The ideal evolution of instantaneous and integrated luminosity

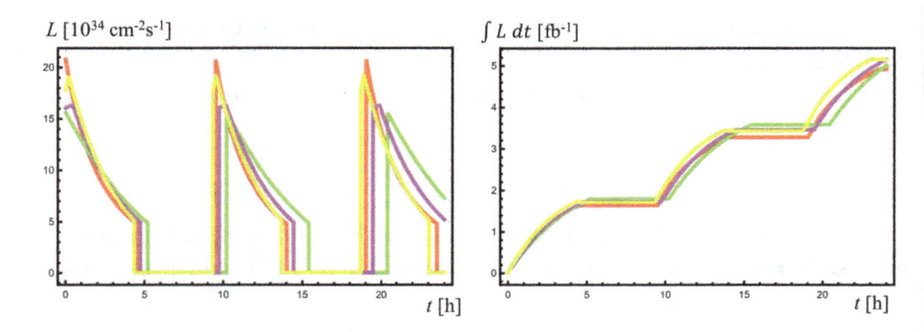


Fig. 1. Instantaneous (left) and integrated luminosity (right) as a function of time during 24 hours with 25 ns bunch spacing, for HE-LHC options HE20 (red), HE27 (green), HE34 (purple), and HE41 (yellow); considering a maximum total beam-beam tune shift of 0.025.

during 24 h is shown in Fig. 1, for all four HE-LHC versions. The increase of the instantaneous luminosity during the early store for HE34 and HE41 also is a measure of the increase of the event pile up from its initial value, which is of order 10%.

For the average turnaround time and for the number of physics days per year, we adopt the canonical values of Ref. 18 (5 hours and 160 days). The ideal integrated luminosity per day is then computed for the optimum run time $t_{\mathrm{r}}$ and the assumed average turnaround time. The luminosity delivered per year is finally obtained by multiplying the latter with the number of physics days scheduled and the postulated availability of 75%,[18] which is slightly lower than for the LHC and HL-LHC.

## 3. Dipole Magnets for a LHC Energy Upgrade

The technology driver of the magnet system of a collider are the main bending dipoles, determining the maximum energy that can be reached, and representing, by far, the most expensive item. In general, the dipole magnets also represent the main challenges in the whole magnet system well, including the arc and final focus quadrupoles. Our discussion revolves around dipole concepts suitable for the collider options outlined in Table 1. We are aware that this is only a partial picture of the whole magnet system, though appropriate for our scope.

The idea of developing new dipoles for an energy upgrade of the LHC has gone hand in hand with the considerations of new machine configurations described in the introduction. Several magnet concepts and designs for a high energy LHC have been considered and presented in the past thirty years, starting even before the LHC construction was completed.

One of such works first documented is Ref. 19, proposing, in 2005, a 24 T dipole magnet for an energy tripler solution, fitting the requirements for the HE41 option described above. At the 2010 HE-LHC workshop[20] a somewhat more modest field was considered; a dipole magnet generating 20 T.[21,22] This dipole magnet corresponds to the HE34 option reported in Table 1. More recently, as a part of the Future Circular Collider study, 16 T magnet designs were derived from the FCC-hh and adapted for installation in the LHC tunnel.[23] The adapted FCC dipoles are those that yield a collider with the characteristics of the HE27 option. Finally, the present HL-LHC construction work[24] is producing dipoles and quadrupoles with ultimate and peak fields in the range of 12 T,[25,26] i.e. the field required for the HE20 collider option of Table 1.

Below, we describe the designs from the quoted references, including some adjustments when necessary, paying special attention to available technology, or expected advances from on-going R&D. We will not go into details of the designs, but rather, we will concentrate on field goals, broad design choices and other prime characteristics such as electromechanical forces and stored energy density, including considerations on the planned and necessary developments to achieve such goals. We will then summarize the main results and discuss alternative approaches that could be of interest in the long term.

Table 2 presents a summary of the main parameters and characteristics of the designs referenced, including field, current, selected operating temperature, coil dimensions, cold mass and cryostat diameter, structural support concept, total forces (acting on a quarter of the dipole coil), stored energy, and energy density. The values of horizontal force and energy density are also plotted in Figs. 2 and 3, together with analogous values of collider and development dipole magnets. This is useful, putting the selected dipole designs in the perspective of past realizations and on-going developments. In fact, Figs. 2 and 3 are a good representation for the main challenges of high field dipoles, namely mechanical stresses and quench protection under increased electromagnetic loads and stored magnetic energy.

In Table 2, we also include a Technology Readiness Level (TRL) indicator, and a relative cost indicator. The TRL provides a standard measurement of the maturity of a technology, ranging from the lowest readiness level 1 (basic principle demonstration) to the highest readiness level 9 (proven technology in application). The cost indicator is given by referring

Table 2. Summary of main design parameters, characteristics, and performance indicators for the four dipole designs described here.

| Parameter | Unit | HE20 | HE27 | HE34 | HE41 |
|---|---|---|---|---|---|
| Dipole field | T | 12 | 16 | 20 | 24 |
| Operating current | kA | 16 | 22 | N/A | 33 |
| Operating temperature | K | 1.9 | 1.9 | 1.9 | 4.2 |
| Superconducting material | — | $Nb_3Sn$ | $Nb_3Sn$ | $Nb_3Sn$ REBCO | Nb3Sn Bi-2212 |
| Supporting structure concept | — | Collars | Bladder-and-key | | |
| Aperture | mm | 50 | 50 | 40 | 40 |
| Operating current density | $A/mm^2$ | 480 | 480 | 380 | 580 |
| Coil cross section (one aperture) | $mm^2$ | 6500 | 10000 | 18000 | 12500 |
| LTS cross section | $mm^2$ | 6500 | 10000 | 12500 | 6100 |
| HTS cross section | $mm^2$ | 0 | 0 | 5500 | 6400 |
| Cold mass outer diameter | mm | 570 | 8600 | 800 | 750 |
| Cryostat outer diameter | mm | 914 | 1200 | >1200 | 1200 |
| Horizontal force | kN/m | 2727 | 5470 | 10063 | 12149 |
| Vertical force | kN/m | 2270 | 4335 | 7287 | 9041 |
| Stored energy | kJ/m | 522 | 1171 | 2448 | 2578 |
| Stored energy density | $J/cm^3$ | 80 | 115 | 136 | 270 |
| TRL | — | 4 | 3 | 2 | 2 |
| Cost at present | a.u. | 1.2 | 1.6 | 5 | 13 |
| Cost perspective | a.u. | 0.5 | 0.7 | 1.5 | 5 |

to the cost of the $Nb_3Sn$ magnets for HL-LHC, taken equal to one. For this evaluation we have removed the cost of R&D, tooling and infrastructure, as well as fringe costs related to the small-scale HL-LHC production. The resulting cost per m of magnet is in the range of 400 kCHF/m, which is consistent with the projection of using 11 T dipoles for a partial energy upgrade of the existing LHC.[5] Two assumptions were made to evaluate the relative cost indicator. The first assumption is based on superconductor costs per unit weight as they currently are, where the cost of REBCO per unit mass is about four times that of $Nb_3Sn$, while the cost of Bi-2212 is about twelve times higher. The corresponding values have been indicated as "*present*" relative cost indicator. The second assumption is based on a cost reduction that could result from on-going developments and a perspective production scale up.

For $Nb_3Sn$ the targeted reduction is a factor of three,[27] in which case the cost per kg would be comparable to that of the ITER production. This would correspond to a superconductor cost equal to three times the cost of raw material, i.e. a factor P = 3, as discussed in Ref. 28 where P is defined as the ratio of conductor to raw material cost. The cost reduction assumed

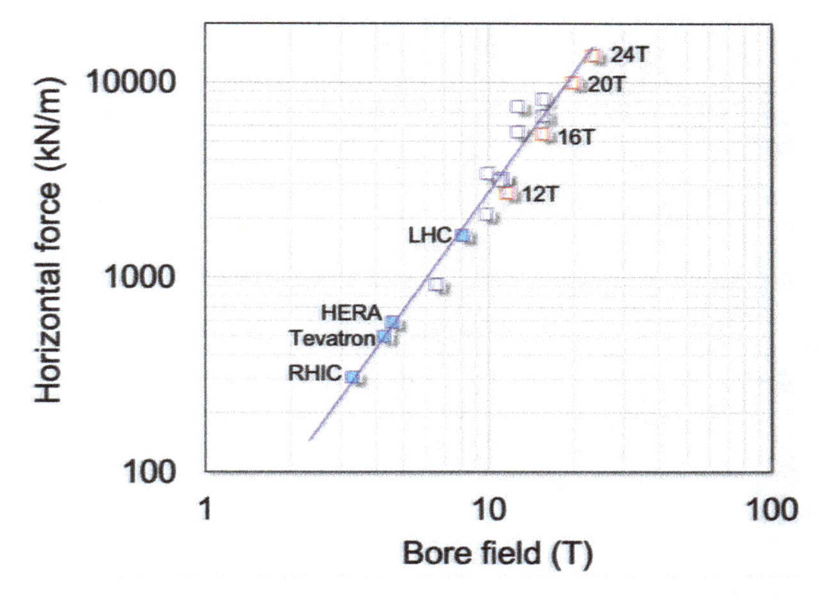


Fig. 2. Scaling of horizontal electromagnetic force per unit length (one coil quarter) vs. magnetic field. The scatter plot reports values for built colliders (blue, solid), magnet models and designs (blue, empty), and the four dipole designs described here (red empty).

for REBCO is a factor four, which would yield a cost per unit weight equal to the present $Nb_3Sn$ for HL-LHC. This would still correspond to a P factor of several hundreds, recognizing the higher process complexity of REBCO manufacturing. At the same time, such a high P factor points to a remarkable potential for cost reduction in case capacity is further scaled up, and the process simplified. For Bi-2212 the cost reduction assumed is a factor 2.5, and a resulting cost five times that of present $Nb_3Sn$ for HL-LHC. This is justified by the fact that the cost of Ag is one order of magnitude higher than that of Cu. As a result, in the case of Bi-2212, the projected cost reduction would correspond to a factor $P = 10$, i.e. comparable to that of present $Nb_3Sn$. This also shows, contrary to REBCO, that there is not much room for a further substantial reduction in this case. The resulting costs with these assumptions have been identified as "*perspective*".

### 3.1. *12 T dipole*

The first, more modest, energy upgrade version of Table 1, HE20, requires collider dipoles operating at 12 T. This field is beyond the reach of Nb-Ti,

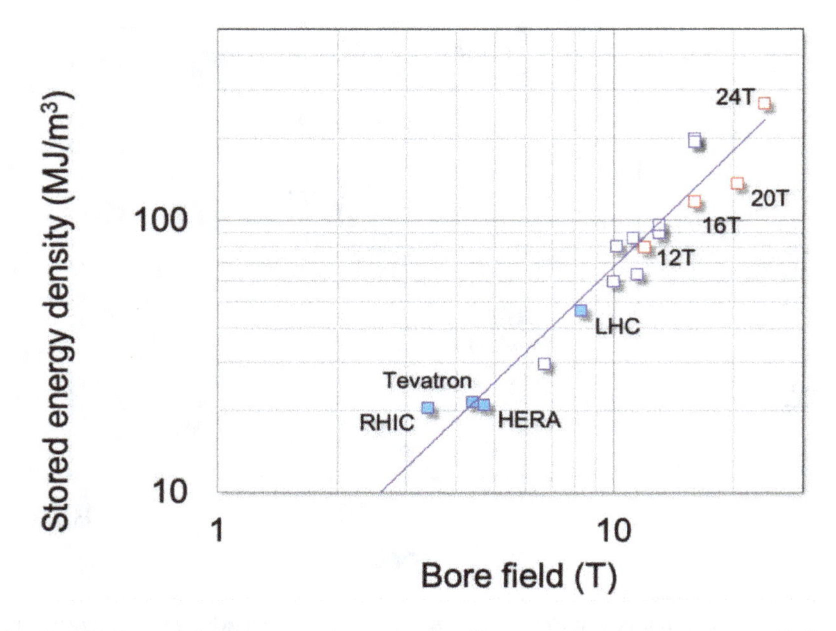


Fig. 3. Scaling of stored magnetic energy density per unit volume vs. magnetic field. The scatter plot reports values for built colliders (blue, solid), magnet models and designs (blue, empty), and the four dipole designs described here (red empty).

the well established accelerator magnet technology for all past and present colliders. A bore field of 12 T can be reached using $Nb_3Sn$ using existing wires and cables, as demonstrated by the results achieved with the 11 T short models and long magnets.[25,26] For this reason we take the 11 T dipole as a good basis for extrapolation.

The HL-LHC 11 T development, initially launched as a collaboration between CERN and FNAL,[29] is presently on-going within the scope of HL-LHC. A cross section of the 11 T magnet is shown in Fig. 4, showing the cos-theta coils, enclosed and supported by a collared coil structure, and the twin aperture assembly in a single iron yoke. The main design and manufacturing features of this magnet, and the results achieved by the development program, can be found in Ref. 26.

The 11 T dipole was designed for compatibility with the existing LHC magnets. This imposed a number of strong constraints on several aspects. The geometry, including length, outer diameter, and inter-beam distance, were fixed to fit with the envelope of a LHC main dipole. The operating current and corresponding magnetic field resulted from the need to be

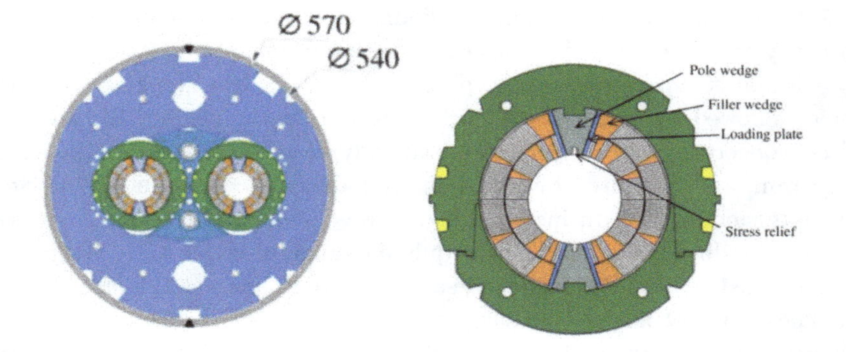


Fig. 4. Cross section of the HL-LHC 11T twin aperture dipole, showing the twin aperture cold mass (left), and the collared coil (right), reproduced from Ref. 26.

powered in series with the LHC main dipole circuit. For the same reason the 11 T dipoles were designed to operate in superfluid helium, at 1.9 K, as the rest of the LHC arc. The operating point resulting from the challenging design optimization was set at 80% of critical conditions along the magnet load line. This corresponds to a fraction of critical current of about 50%, and a temperature margin of at least 4.5 K. The 11 T dipoles are built as straight magnets of 5.31 m magnetic length. The coil aperture, 60 mm, larger than that of the LHC main dipoles, allows for a comfortable space for the beam sagitta. Finally, given the limited number of magnets to be installed (2 units consisting of 2 magnets each, compared to 1230 main dipoles in the rest of the LHC), the specification on the field quality could be relaxed. The persistent current contribution to the normal sextupole is of the order of 40 units, with a variation by more 10 units during the ramp, an order of magnitude larger than in the LHC. While this was found to be acceptable from the point of view of beam performance in the LHC, it will surely need to be reduced for a collider based on this magnet type, also because of AC loss, i.e. the energy dissipation associated with the change of magnetization.

Three out of four 11 T series magnets, units fully equipped for installation in the LHC, achieved the nominal current at the first thermal cycle, and sustained repeated simulated ramps, demonstrating that this field level is within reach with an adequate operating margin.[30] Further tests have shown, however, that performance retention through powering and thermal cycles is an issue. Degradation and some erratic quenches were observed in these full-size magnets, likely related to localized breakage of the brittle $Nb_3Sn$ superconducting filaments. R&D is actively pursued to identify causes and mitigate this effect, which is presently attributed to coil stress and strain concentration, possibly already at the stage of construction, followed by variations of such state during the magnet lifetime. In fact, concerns on stress peaks were already identified earlier in the 11 T program, and mitigated by intentionally reducing the amount of coil pre-compression applied during collaring. It was found that the coil pre-stress could be reduced below the value required to guarantee that the coil remains compressed against the collar pole at full current, and yet the magnet still reaches nominal field. This suggests that full pre-stress may not be necessary for the impregnated and stiff $Nb_3Sn$ coils, a major change in the design paradigm for collared, cos-theta coils.

In the perspective of a new accelerator realization, the constraints on coil and magnet characteristics can be relaxed to a certain degree. This would allow for a reduction of stress and stored energy density, both of which would be beneficial to resolving the concern of robustness of the present 11 T, and easing magnet protection from quench. This evolution is in fact one of the two main avenues pursued by the R&D Programme on High Field $Nb_3Sn$ accelerator Magnets (HFM),[31] that seeks to demonstrate that $Nb_3Sn$ technology is fit for deployment on large scale.

A modification of the 11 T design to produce a 12 T dipole is outlined in Table 2, where we report approximate coil dimensions, field and current, and structure selection. A suitable cable option could be a scaled-up version of the 11 T cable, going from a 0.7 mm strand to a 0.85 mm strand, and a 40 strands cable, identical to that used for the HL-LHC interaction region quadrupoles (QXF).[32] The coil width increases by approximately 20%, and we maintain the same operating margin, which has been shown to result in limited training to operating conditions, enough for operation. To achieve this margin, the cable current density is reduced by about 10% with respect to the 11 T dipole.

The operating temperature of this dipole is set to 1.9 K, as in the LHC. The main reason is to profit from the good heat transfer properties of stagnant superfluid helium acting as thermal vector for the cold mass. Unlike Nb-Ti, the additional operating margin of $Nb_3Sn$ at 1.9 K with respect to liquid helium conditions, 4.2 K, would not, by itself, justify the additional cooling effort for a 12 T dipole. This is also true because heat transfer from an impregnated coil, the standard for $Nb_3Sn$, is rather governed by heat conduction than by heat transfer to the helium bath. Alternatives to a pool bath of helium could hence be devised to reduce the helium inventory of a 12 T dipole.

For the modest field increase to 12 T we can still assume that collars provide a suitable structural support, under the assumption that pre-loading of the coil is limited to the maximum acceptable conductor transverse stress, with large margin (e.g. setting a maximum in the range of 100 MPa on the coil midplane). The cold mass diameter can be maintained at the standard of the LHC dipole, 570 mm, as was done for the 11 T dipole, at the expense of some field leakage and marginally degraded field quality at high field. These effects should be minor and manageable. The result is an outer cryostat diameter identical to that of the present LHC, which can be integrated into the existing tunnel infrastructure.

The increased conductor width partially compensates for the increase in forces and results in lower accumulated electromagnetic stress on the midplane, by about 10%. Lower pre-stress and electromagnetic stress allow increased robustness to geometric errors, assembly tolerances and local stress intensification factors. These choices all lead to easier manufacturing and, in the end, cheaper magnets. Note, however, that despite the increased structural margin, significant development is still necessary to simplify manufacturing steps and tooling, reduce manipulation, reducing conductor cost.

It is for this reason that we assign a TRL of 4 to this dipole option in Table 2 (*technology validated in lab*). As for the relative cost, in present conditions this 12 T dipole would cost 1.2 times the reference HL-LHC cost. Provided that the $Nb_3Sn$ R&D is successful in achieving its cost reduction targets,[27] we can expect that the 12 T dipole option described here to be 0.5 times the reference HL-LHC cost. A relatively high TRL and moderate relative cost can be maintained as the main result of this dipole design, to be balanced by the modest reach of a HE20 collider.

### 3.2. *16 T dipole*

The second energy upgrade version, HE27, requires collider dipoles operating at 16 T, also built with $Nb_3Sn$. This is the field level targeted by the Future Circular Collider, and the design of this option is described in detail in Ref. 4. For the FCC-hh a cos-theta baseline is assumed, built with four layers, with graded cables, and assembled in a bladder and key structure. In fact, as mentioned in Ref. 4, several alternatives are possible, also detailed in Ref. 33, and it is not yet clear which conductor, coil and structure configuration is the optimal choice for this field level. Indeed, reaching this operating field level in a model accelerator dipole magnet has not yet been demonstrated, although tests on conductors and racetrack coils, assembled in block coil configuration, have shown that it is possible to generate such fields with $Nb_3Sn$.[34,35]

Given the good performance of the block coil demonstrators, we take, for our discussion, the configuration shown in Fig. 5 as a suitable option of a 16 T dipole built with $Nb_3Sn$. Block coils are built as double pancakes with flared ends, stacked around the desired aperture. A description of the general features of block magnets and typical geometries can be found in Refs. 36–38. The main advantage of the block coil configuration is that it decouples the location of peak field, usually close to the inner coil perimeter, from the location of maximum stress, which in the case of block coils tends to be towards the outside perimeter, where the electro-magnetic force accumulates. Another interesting feature of block coils, important to the design and manufacturing of magnets in this field range, is that the coils can be made larger by simply winding more turns. The additional ampere-turns increase the operating margin, while a wider coil reduces the stress. This is not the case for a cos-theta coil, whose thickness is locked once the cable width is selected. The downside of block coils is that for maximum field efficiency the cable width must fit the given magnet aperture, which in our case results in large cables (ideally 25 mm cable width for a 50 mm aperture). This is however not a serious issue, given that large cables within this range have already been manufactured and tested within the scope of the ongoing efforts towards high field $Nb_3Sn$ magnets.[37]

The configuration shown in Fig. 5 is an evolution of the program that has led CERN to the successful demonstration of 16.5 T peak field in racetrack

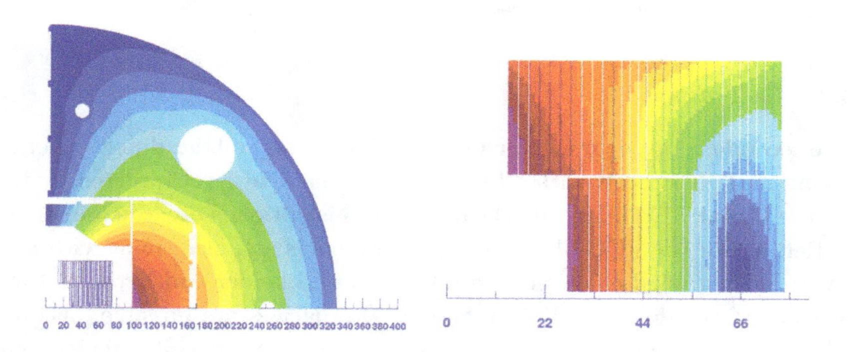


Fig. 5. Cross section of the 16 T block dipole designed as a demonstrator for FCC, a quarter of the whole magnet (left) and detail of the graded coil (right).[41]

coils quoted earlier (eRMC and RMM),[35] as well as the construction of the 14.6 T, 100 mm bore dipole for the FRESCA2 test facility, designed and built as a collaboration between CERN and CEA.[38] Conductors of characteristics required are being developed for the construction of the US Test Facility Dipole (TFD) at LBNL,[39] which should be used in the High Field Vertical Magnet Test Facility at FNAL to test future LTS and HTS cables.[40]

The operating temperature of this dipole is chosen at 1.9 K, as for the 12 T option. Besides the good heat transfer properties of a stagnant bath of superfluid helium permeating the cold mass already discussed, a 16 T field is at the upper bound within the reach of $Nb_3Sn$ (in dipole magnets). In this case it is hence mandatory to aim for the lowest practical operating temperature to increase the operating margin, and keep the coil cross section as compact as possible.

The design consists of two pancake coils with flared ends, graded to reduce the coil cross section. The structure best adapted for this magnet and field level is based on the bladder-and-key principle. This system provides a pre-load that increases as the magnet cools down, owing to the differential of thermal contraction between the Al-alloy outer magnet shell and the internal magnet structure built of iron and steel. This is opposite to the collaring used in present accelerator magnets that tends to lose the pre-load as the magnet cools down. Bladder-and-keys have a large range of tuneability and allow the desired stress state to be reached gradually, without the need to over-stressing the coil during magnet assembly. Alternative systems have been proposed, all based on the same concept of exploiting differential thermal contraction.

Comparing the main magnet characteristics in Table 2, the 16 T dipole has significantly larger coil than the 12 T dipole — to compensate for the reduced critical current. The large coil helps to maintain both mechanical stress and stored energy density at reasonable level. The larger structure compared to the 12 T dipole, and the need to return a substantially larger magnetic flux, result in a larger cold mass diameter, 800 mm. This in turn requires an increase of the cryostat dimension, reaching values in the range of 1200 mm which is considered to be the largest diameter that can be integrated in the present LHC tunnel.

At this point, we need to note that the development towards a dipole of this level of field is substantial. As hinted above, demonstrations have, so far, been successful, but much work is required to improve conductor performance (to reduce magnet cost), engineer the magnet ends (still a limitation in block coils), demonstrate effective conductor grading (required to make compact coils), and in general to make an accelerator magnet out of this promising concept (protection, field quality, alignment, heat transfer and cooling). For this reason a modest value of 3 is assigned to the TRL indicator (experimental proof of concept). Based on present conditions the relative cost indicator for this magnet is 1.6 times the HL-LHC reference. Assuming the $Nb_3Sn$ superconductor cost reduction of a factor 3, the cost indicator would decrease to 0.7 of the present HL-LHC reference. This is substantial, but would still result in a magnetic system more than twice as expensive as the LHC — a relatively high price tag for a factor two increase in beam energy.

### 3.3. *20 T dipole*

Dipole fields in excess of 16 T, as required for the HE33 collider option, are well beyond the projected reach of LTS materials, and need a switch to HTS. Ideas on how to build a 20 T dipole were presented at the 2010 HE-LHC workshop in Malta,[20,21] and reviewed a few years later, together with other high field magnet options.[22] The basic idea of the design developed there is to profit from the HTS ability to generate high fields, but limit the amount of HTS material, still significantly more expensive than LTS. LTS can be used to generate a large portion of the field, grading the coil with different materials rather than different cable dimensions. This is akin to what done in high field solenoid magnets, where a small high field insert made in HTS is installed in a large bore outer magnet made of LTS. No specific HTS material was selected for the conceptual magnet design in Ref. 21, though at the time it seemed that Bi-2212, in round isotropic wires, could be a suitable choice. Several geometries and grading were considered in the following work.[22] We show in Fig. 6 the simplest among the options considered, with only two material grades, HTS and $Nb_3Sn$. This configuration is not the most efficient among all those studied, but retains a level of simplicity that is interesting from the point of view of manufacturing and cost reduction. An interesting consideration in Ref. 22 is the fact that field quality for any very high field dipole is not an issue, given that the coil cross section is forcibly large and the coil tends to naturally generate a good dipole field.

The main parameters of this magnet design are reported in Table 2. We note the expected increase of forces and stored energy per unit length, a factor about 4 with respect to the 12 T dipole described earlier. The design

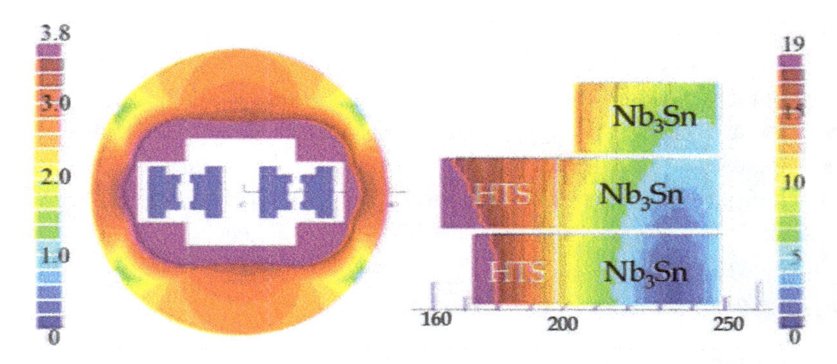


Fig. 6. Cross section of the 20 T block dipole designed in Ref. 22 based on two material grades only, twin aperture magnet (left) and one quarter of the coil (right).

assumptions in Ref. 22 is aimed at stress reduction and magnet protection margin, extrapolating from present engineering and applying conservative limits. The result is a modest current density, below 400 A/mm$^2$. This explains the relatively large coil cross section, and the value of the stored energy density in the coil being comparable to that of the 16 T $Nb_3Sn$ dipole (as well as mechanical stress, not reported here). The design in Ref. 22 projected an outer diameter of the iron yoke of 800 mm. This should be increased to include the structural features necessary for the cold mass. Given the large level of electromagnetic force and the block structure, a bladder and key support and loading concept would be suitable, but would need the addition of a stiff shell around the iron yoke. Typical shell thickness would be around 60 to 70 mm, and integration of a cold mass of 950 mmm diameter in a cryostat with outer diameter of 1200 mm may pose challenges.

The relatively large coil area, driven by the modes current density, results in a significant cost increase. Even in the optimistic scenario of successful cost reduction, the projected cost of the 20 T dipole would be 1.5 times that of the HL-LHC reference (REBCO was assumed for the HTS material).

As for technology readiness, this is presently at the level of a conceptual study, and the TRL assigned is hence low, a value of 2 (technology concept formulated). Indeed, no such dipole has ever been built. Although it is thought to be possible to boost the field of an LTS outsert with an HTS insert, by building them separately and assembling them once completed, the electromagnetic and mechanical interaction of the two magnets is by no means trivial. In addition to the development required for the LTS part, already listed in the section on the 16 T dipole, a suitable HTS insert technology should be validated beyond the initial results available at present.[42] Integration of the HTS insert in the LTS background implies that the hybrid dipole is designed as a whole, rather than two separate magnets. A good example of successful integration is provided by HTS/LTS UHF NMR solenoids[43] and solenoid magnets for high-field science.[44]

### 3.4. *24 T dipole*

The most ambitious design is the 24 T dipole of Ref. 19, a hybrid LTS/HTS built as an assembly of block coils, and shown schematically in Fig. 7. The outer low field grade coils, below 16 T, are made of $Nb_3Sn$, while the inner high field coils are made of Bi-2212. Besides the use of bladder and keys already described earlier, this 24 T dipole relies on stress management to deal with the spectacular increase of electromagnetic force. This is achieved by introducing structural supports within the blocks, whose purpose is to intercept part of the load and avoid accumulation. A second interesting feature is the use of a flux plate inserted between the lower and upper coils of a pole that provides means to compensate for persistent current effects at low field, i.e. injection conditions where the beam is most sensitive. Finally, in order to keep the outer diameter of the magnet small, small Nb-Ti windings are placed at the periphery of the iron yoke. These windings

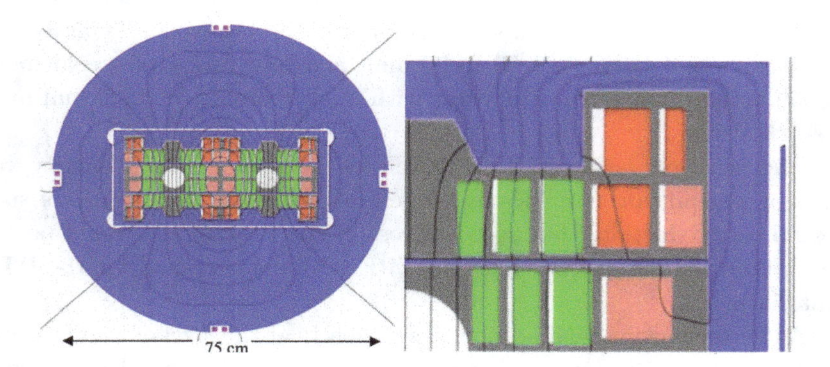


Fig. 7. Cross section of the 24 T LTS/HTS hybrid block dipole proposed in Ref. 19. Twin aperture assembly (left) showing Nb-Ti windings for flux cancellation at the outer diameter of the iron yoke, and one quarter coil cross section (right) showing the HTS (green) and LTS (red) grades, the stress management structure (grey) and the flux plate between lower and upper coils.

are powered to cancel the leaking flux and thus reduce stray field in the vicinity of the cold mass.

As we see from the summary values of Table 2, the coil cross section is kept very compact when compared to the 20 T dipole described above (note that we do not count the stress management structure in the coil cross section). This yields a relatively high current density, and stored energy density. While mechanical stress may be reduced and controlled thanks to stress management, the combination of high stored energy (large inductance), current density (fast heating rate in case of quench) and energy density (high hot-spot temperature) will pose a protection challenge. The cold mass diameter was designed in Ref. 19 using active magnetic shielding, and limited to an iron yoke diameter of 750 mm. Structural components are not considered in this diameter, but given the increase in outer dimension brought by active shielding, an integration into a 1200 mm cryostat seems possible.

Just as for the 20 T dipole, in this case the level of technology readiness is low, a value of 2 is assigned (technology concept formulated), and the issues to be resolved are essentially the same. Additional, in this case the projected cost is further driven up by the HTS material. Under our assumptions, and even with the projected cost reduction, a 24 T dipole built with $Nb_3Sn$ and Bi-2212 would cost around the order of 5 times the HL-LHC reference. This is mainly driven by the contribution of Bi-2212. At such absolute cost, ranging in the several tens of MCHF for a 15 m long dipole, a HE41 would not be an interesting option.

## 4. Discussion

Beyond the technical feasibility of the designs presented above, it is interesting to look at the relative costs indicators to guide towards the most interesting long-term developments. Restricting ourselves to the hypotheses and studies reported, at first view only a 12 T and 16 T $Nb_3Sn$ dipole magnet system would be at an affordable level, i.e. a unit cost below that of the HL-LHC $Nb_3Sn$ magnets. The LTS/HTS hybrid options are presently out of a reasonable cost range. Even assuming a rather optimistic reduction of the superconductor cost, dipole magnet systems in the range of 20 T to 24 T are still nearly two to five times as expensive as the current HL-LHC $Nb_3Sn$ magnet. The main reason is that the field generated by a dipole coil of given operating current density is proportional to the amount of conductor, hence to its mass and, in last instance, its cost. In addition, the decrease of critical current density with field, the increased electromagnetic forces (scaling like $B^2$), and the increased stored energy (also scaling like $B^2$) tend to reduce the maximum allowable operating current density, further increasing the required amount of conductor and cost.

Given these simple considerations, there is only one way to reduce the cost of high field magnets — increasing their current density. This is in fact the first main asset of HTS materials, whose engineering critical current densities nowadays exceed 2000 A/mm$^2$ in the range of 20 T and at 4.2 K.[45] Besides, HTS magnets do not train, profiting from the large enthalpy reserve deriving from their high critical temperature (compared to LTS). It is hence not necessary to assume large operating margins with respect to the critical surface, as was assumed for LTS. Research is also on-going on how to make HTS winding solid and self-protecting, using a combination of structural and electrical ingenuity such as non-insulated (NI), metal-insulated (MI) or, more in general, controlled-insulation (CI) coils.[46] Such windings have no insulating layers, they are soldered, forming a solid mechanical component. Small demonstrators have shown that it is possible to exploit this technique to generate large solenoid fields, from 20 T to record values in excess of 45 T,[47] with winding current densities just short of 1000 A/mm$^2$. CI windings do not respond to ramps like a classical accelerator magnet, exhibiting field delays, drifts, and large remnant fields. It is nonetheless unquestionable that tapping on such potential for an increase of operating current density, by a good factor of two with respect to the assumptions taken for the design of the dipoles described earlier, would significantly change the perspective. A 16 T dipole built with CI HTS would be cheaper than an HL-LHC $Nb_3Sn$ magnet, the 20 T dipole would have similar unit cost, and the 24 T dipole would be a factor of 1.3 more expensive, i.e. all in range of consideration for a HE-LHC.

The second main asset of HTS materials, especially driven by the recent developments for fusion application, is that they can operate at temperatures significantly higher than liquid helium. The range of 10 K to 20 K is of particular interest, because the loss in critical current density at high field at this temperature is limited (by a factor of two in the worst case of 20 K operation). In this range it is possible to devise cooling with gaseous helium, or other solutions such as *dry winding* with thermal links to a cold sink, e.g. a long pipe cooled by gas flow. The increased temperature would improve cryogenic efficiency, and reduce power consumption, by an estimated factor of two to four. At the same time the helium inventory would be reduced by a large amount, possibly up to an order of magnitude.

This would help mitigate the risks of helium availability and cost volatility, which are presently recognized as a definite concern for any future cryogenic installation, especially at the scale considered here.

Given these considerations on such large unexploited potential, it seems that development of HTS for the next step collider should be given high priority. This would offer magnet options to increase the LHC energy by two to three times, well beyond the reach of $Nb_3Sn$, but with a comparable projected cost, increased cryogenic efficiency, and reduced helium inventory.

## 5. Other Collider Challenges

In addition to the arc magnets, several other magnet systems will be required. Most important and most challenging are the sc final-focusing quadrupoles in the high-luminosity interaction regions, the sc separation dipoles, sc dipole magnets for the dispersion suppressors, and the (possibly warm) quadrupoles and dipoles in the collimation insertions.[4] The accelerator footprint must fit into the existing tunnel; see some pertinent discussions in Ref. 6. The collimation insertion itself poses several new challenges.[48]

An rf voltage of 16 MV is required per beam, the same value as for the present LHC. The rf power demand depends on the speed of acceleration. With a total stored beam energy of 4 GJ, a ramp duration of e.g. 30 minutes implies an rf power of at least 2 MW.

Crab cavities are needed to realize effective head-on collisions. Since the crossing angle decreases roughly as the inverse energy, the crab-cavity rf voltage, scaling as the product of crossing angle and beam energy, is similar to the HL-LHC's. However, crossing angle and required crab-cavity voltage also depend on the length of the final quadrupole triplet, which may grow with increasing beam energy. A novel Nb/Cu crab cavity consisting of a ridged waveguide resonator with wide-open apertures ("WOW")[49] could be an interesting option for the HE-LHC.

For the collider vacuum system, the rather voluminous beam screen developed for the FCC-hh,[50] which was successfully tested with FCC-hh like synchrotron radiation at a beamline in Karlsruhe,[51] provides excellent vacuum performance[52] and, by efficiently shielding the pumping slots, a low beam impedance.[53] In addition, suppression of electron-cloud build up may require either laser ablated surface engineering (LASE) treatment,[54] or amorphous carbon coating (a-C),[55] on parts of the inner beamscreen wall. An intriguing proposal is to coat either all or the remaining uncoated portions of the inner beamscreen chamber with a thin layer ($\sim$1 $\mu$m) of high-temperature superconductor, to reduce the resistive-wall impedance.[56,57]

Injection into the HE-LHC at beam energies from 0.6 to 1.2 TeV requires a new sc SPS in the existing SPS tunnel. A conceptual design for such a machine with a top energy of 1.3 TeV was developed.[58] The higher beam energies also imply upgrades and technology developments for the injection system and for the beam dump; see e.g. Ref. 59.

Numerous other accelerator systems, such as cryogenics, electric distribution, cooling and ventilation, infrastructures for the experiments, etc., deserve consideration. The HE-LHC Conceptual Design Report[4] provides a comprehensive overview and helpful starting points for further development, as do the existing LHC systems.

## 6. Conclusions

In this chapter, building on past studies, we have explored how future high-field dipole magnets of various fields between 12 and 24 T, if (or when) they become available, could be used to construct a High-Energy Large Hadron Collider (HE-LHC) in the existing LHC tunnel. For each field and energy level, we discussed the collider performance that can be achieved, and looked at the corresponding magnet designs, their estimated technology readiness, and relative cost projection.

The main result of the analysis is that none of the HE-LHC options considered seems to come out as a true *sweet spot* for a higher energy LHC. An *affordable* but still costly 12 T dipole, which could be produced with technology at hand, would only yield a very modest centre-of-mass energy of 20 TeV. This is barely 50% above the present LHC, and the effort and resources necessary are hard to justify, while lacking a solid discovery perspective within this energy range. Any other option, and in particular pushing towards the high end of the dipole field, e.g., 24 T thanks to the use of HTS, would extend the energy reach by a good factor. At the same time, the scaling based on standard accelerator magnet technology would assign a price tag too excessive to these magnets to be considered for an energy upgrade at all, let alone the technology development required to reach this field level.

However, looking into the magnet cost drivers, we can see a major opportunity in the use of HTS. Adopting new magnet technology, i.e. compact winding with high current density thanks to specific features of HTS, may break standard scaling. This direction is similar to on-going work in other domains of science and energy applications, and may produce the cost benefit required for a future application.

In fact, it is clear that the technology development outlined above would be beneficial not only for an energy upgrade of the LHC, but would also produce interesting alternatives to the baseline FCC-hh magnetic system, as well as the technology solutions sought for a muon collider. As we discussed briefly, compact HTS magnets could be significantly more energy efficient than an LTS magnet system, and reduce the long term risk of helium availability and cost, definite bonuses along the lines of sustainable science.

These considerations call for an increased effort towards HTS accelerator magnet R&D, seeking specifically conceptual designs and demonstration beyond present standards. Reaching this *technology hinge* may answer the question which, if any, of the various collider options, could, or should, be built and when.

Finally, for a higher energy hadron collider in the LHC tunnel, the beam parameters required at injection are already available today, and the beam dynamics at higher energy poses no particular challenges. A new feature compared to the present LHC is the much higher synchrotron radiation power. The resulting heat could be more efficiently removed if the arc magnets operate at a temperature higher than 1.9 K, which would be supported by HTS magnet technology. Another consequence of the enhanced synchrotron radiation is that the HE-LHC luminosity evolution during a physics fill will be determined by the combined effects of proton burn-off and significant radiation damping. Counteracting the latter, both longitudinal and transverse emittance blow up by controlled noise excitation are likely necessary to maintain longitudinal Landau damping and an acceptable beam-beam tune shift, respectively.

## References

1. O. S. Brüning, R. Cappi, R. Garoby, O. Gröbner, W. Herr, T. P. R. Linnecar, R. Ostojic, K. Potter, L. Rossi, F. Ruggiero, K. Schindl, G. R. Stevenson, L. Tavian, T. Taylor, E. Tsesmelis, E. Weisse, F. Zimmermann, LHC Luminosity and Energy Upgrade: A Feasibility Study, Tech. rep., CERN, Geneva (Dec 2002). URL `http://cds.cern.ch/record/601847`
2. E. Todesco, F. Zimmermann (Eds.), EuCARD-AccNet-EuroLumi Workshop: The High-Energy Large Hadron Collider: Villa Bighi, Malta, Republic of Malta 14–16 Oct 2010, CERN, CERN, Geneva, 2011, 29 lectures, 156 pages, published as CERN Yellow Report. `doi:10.5170/CERN-2011-003`. URL `http://cds.cern.ch/record/1344820`

3. M. Benedikt, M. Capeans Garrido, F. Cerutti, B. Goddard, J. Gutleber, J.M. Jimenez, M. Mangano, V. Mertens, J.A. Osborne, T. Otto, K. Poole, W. Riegler, D. Schulte, L.J. Tavian, D. Tommasini, F. Zimmermann, FCC-hh: The Hadron Collider, Eur. Phys. J. Spec. Top. 228 (2019). URL https://doi.org/10.1140/epjst/e2019-900087-0
4. F. Zimmermann, M. Benedikt, M. Capeans Garrido, F. Cerutti, B. Goddard, J. Gutleber, J. M. Jimenez, M. Mangano, V. Mertens, J. A. Osborne, T. Otto, J. Poole, W. Riegler, L. J. Tavian, D. Tommasini, HE-LHC: The High-Energy Large Hadron Collider: Future Circular Collider Conceptual Design Report Volume 4. Future Circular Collider, The European Physical Journal Special Topics 228 (2019) 1109. doi:10.1140/epjst/e2019-900088-6. URL http://cds.cern.ch/record/2651305
5. Brüning, Oliver et al., LHC Full Energy Exploitation Study: Upgrade for Operation Beyond Ultimate Energy of 7.5 TeV, Tech. rep., CERN, Geneva (Sep 2020). URL https://cds.cern.ch/record/2729796
6. J. Keintzel et al., Lattice and optics options for possible energy upgrades of the Large Hadron Collider, Phys. Rev. Acc. Beams 23 (2020). URL https://journals.aps.org/prab/abstract/10.1103/PhysRevAccelBeams.23.101602
7. H. Damerau, A. Funken, R. Garoby, S. Gilardoni, B. Goddard, K. Hanke, A. Lombardi, D. Manglunki, M. Meddahi, B. Mikulec, G. Rumolo, E. Shaposhnikova, M. Vretenar, J. Coupard, LHC Injectors Upgrade, Technical Design Report, 2014. doi:10.17181/CERN.7NHR.6HGC. URL http://cds.cern.ch/record/1976692
8. E. N. Shaposhnikova, T. Argyropoulos, T. Bohl, C. M. Bhat, P. Baudrenghien, A. C. Butterworth, T. Mastoridis, J. Esteban Muller, G. Papotti, J. Tuckmantel, W. Venturini Delsolaro, U. Wehrle, Loss of Landau Damping in the LHC (2011) 3 p. URL http://cds.cern.ch/record/1378465
9. P. Baudrenghien, A. Butterworth, M. Jaussi, T. Mastoridis, G. Papotti, E. Shaposhnikova, J. Tuckmantel, Longitudinal emittance blow-up in the LHC (2011) 4 p. URL http://cds.cern.ch/record/1378464
10. G. Iadarola, HL-LHC filling schemes: possible optimization, 140th HiLumi WP2 Meeting, 29 January 2019.
11. M. Benedikt, D. Schulte, F. Zimmermann, Optimizing integrated luminosity of future hadron colliders, Phys. Rev. ST Accel. Beams 18 (2015) 101002. doi:10.1103/PhysRevSTAB.18.101002. URL https://link.aps.org/doi/10.1103/PhysRevSTAB.18.101002
12. C. O. Domínguez Sánchez de la Blanca, Electron cloud studies for the LHC and future proton colliders, PhD thesis, EPFL Lausanne (Jan 2014). URL http://cds.cern.ch/record/1645669
13. J. R. Cudell, V. V. Ezhela, P. Gauron, K. Kang, Y. V. Kuyanov, S. B. Lugovsky, E. Martynov, B. Nicolescu, E. A. Razuvaev, N. P. Tkachenko, Benchmarks for the forward observables at rhic, the tevatron-run ii, and the lhc, Phys. Rev. Lett. 89 (2002) 201801. doi:10.1103/PhysRevLett.89.201801. URL https://link.aps.org/doi/10.1103/PhysRevLett.89.201801

14. Latino, Giuseppe, Summary of physics results from the totem experiment, EPJ Web of Conferences 49 (2013) 02005. `doi:10.1051/epjconf/20134902005`. URL `https://doi.org/10.1051/epjconf/20134902005`
15. M. Menon, P. Silva, An updated analysis on the rise of the hadronic total cross-section at the lhc energy region, International Journal of Modern Physics A 28 (20) (2013) 1350099. `arXiv:https://doi.org/10.1142/S0217751X13500991`, `doi:10.1142/S0217751X13500991`.
URL `https://doi.org/10.1142/S0217751X13500991`
16. G. Antchev et al., Luminosity-independent measurements of total, elastic and inelastic cross-sections at $\sqrt{=}7$ TeV, EPL (Europhysics Letters) 101 (2) (2013) 21004. `doi:10.1209/0295-5075/101/21004`.
URL `https://doi.org/10.1209/0295-5075/101/21004`
17. G. Antchev et al., Luminosity-independent measurement of the proton-proton total cross section at $\sqrt{s} = 8$ TeV, Phys. Rev. Lett. 111 (2013) 012001. `doi:10.1103/PhysRevLett.111.012001`.
URL `https://link.aps.org/doi/10.1103/PhysRevLett.111.012001`
18. F. Bordry, M. Benedikt, O. Bruning, J. Jowett, L. Rossi, D. Schulte, S. Stapnes, F. Zimmermann, Machine Parameters and Projected Luminosity Performance of Proposed Future Colliders at CERN, Tech. rep., CERN, Geneva, * Temporary entry * (Oct 2018). `arXiv:1810.13022`.
URL `http://cds.cern.ch/record/2645151`
19. P. M. McIntyre, A. Sattarov, On the Feasibility of a Tripler Upgrade for LHC (2005). URL `http://cds.cern.ch/record/926779`
20. R. Assmann, R. Bailey, O. Brüning, O. Dominguez Sanchez, G. de Rijk, J. M. Jimenez, S. Myers, L. Rossi, L. Tavian, E. Todesco, F. Zimmermann, First Thoughts on a Higher-Energy LHC, Tech. rep., CERN, Geneva (Aug 2010). URL `http://cds.cern.ch/record/1284326`
21. L. Rossi, E. Todesco, Conceptual design of 20 T dipoles for high-energy LHC, arXiv 1108.1619 (Aug 2011). `doi:10.5170/CERN-2011-003.13`.
URL `http://cds.cern.ch/record/1373969`
22. E. Todesco, L. Bottura, G. De Rijk, L. Rossi, Dipoles for High-Energy LHC, IEEE Trans. Appl. Supercond. 24 (2014) 4004306. 6 p. `doi:10.1109/TASC.2013.2286002`. URL `http://cds.cern.ch/record/1662724`
23. D. Tommasini et al., Status of the 16 T Dipole Development Program for a Future Hadron Collider, IEEE Transactions on Applied Superconductivity 28 (2018) 4001305.
24. O. Aberle et al., High-Luminosity Large Hadron Collider (HL-LHC): Technical design report, CERN Yellow Reports: Monographs, CERN, Geneva, 2020. `doi:10.23731/CYRM-2020-0010`. URL `http://cds.cern.ch/record/2749422`
25. Alexander V. Zlobin, $Nb_3Sn$ 11 T Dipole for the High Luminosity LHC (FNAL), in *$Nb_3Sn$ Accelerator Magnets: Designs, Technologies and Performance* (eds. D. Schoerling and A.V. Zlobin), Springer International Publishing (2019) 193–222, `doi:{10.1007/978-3-030-16118-7\_8}`.
26. Bernado Bordini et al., $Nb_3Sn$ 11 T Dipole for the High Luminosity LHC (CERN), in *$Nb_3Sn$ Accelerator Magnets: Designs, Technologies and Perfor-*

*mance* (eds. D. Schoerling and A.V. Zlobin), Springer International Publishing (2019) 223–258, doi:{10.1007/978-3-030-16118-7_9}.

27. A. Ballarino, L. Bottura, Targets for R&D on $Nb_3Sn$ conductor for High Energy Physics, IEEE Trans. Appl. Supercond. 25 (2015) 6000906. 6 p. doi: 10.1109/TASC.2015.2390149. URL http://cds.cern.ch/record/1987573
28. L. D. Cooley, A. K. Ghosh, R. M. Scanlan, Costs of high-field superconducting strands for particle accelerator magnets, Supercond. Sci. Technol. 18 (2005) R51–65. URL http://cds.cern.ch/record/909109
29. M. Karppinen et al., Design of 11 T twin-aperture $Nb_3Sn$ dipole: Demonstrator magnet for LHC upgrades, IEEE Trans. Appl. Sup. 22 (3) (2012) 4901504. URL https://doi.org/10.1109/TASC.2011.2177625
30. A. Devred, Status of the 11 T dipole and CERN magnet programs beyond HiLumi, presented at the 9$^{th}$ HL-LHC Collaboration Meeting, Fermilab, 14–16 Oct. 2019 (October 2019). URL https://indico.cern.ch/event/806637/contributions/3487461/
31. C. Adolphsen et al., European Strategy for Particle Physics - Accelerator R&D Roadmap. European Strategy for Particle Physics – Accelerator R&D Roadmap, Vol. 1 of CERN Yellow Reports: Monographs, 2022. doi:10.23731/CYRM-2022-001. URL http://cds.cern.ch/record/2800190
32. P. Ferracin et al., Magnet design of the 150 mm aperture low-$\beta$ quadrupoles for the High Luminosity LHC, IEEE Trans. Appl. Sup. 24 (3) (2013) 4002306. URL https://doi.org/10.1109/TASC.2013.2284970
33. D. Schoerling, A. Z. (eds.), $Nb_3Sn$ accelerator magnets, Particle Acceleration and Detection Series, Springer, Cham, 2019. URL https://doi.org/10.1007/978-3-030-16118-7
34. J. C. P. et al., 16 T $Nb_3Sn$ racetrack model coil test result, IEEE Trans. Appl. Supercond. 26 (4) (2015) 4004906. URL https://doi.org/10.1109/TASC.2016.2530684
35. C. News, A demonstrator magnet produces a record magnet field (March, 2020). URL https://home.cern/news/news/accelerators/demonstrator-magnet-produces-record-magnet-field
36. P. McIntyre, A. Sattarov, $Nb_3Sn$ accelerator magnets, Particle Acceleration and Detection Series, Springer, Cham, 2019, Ch. Block-Type $Nb_3Sn$ Dipole R&D at Texas A&M University. URL https://doi.org/10.1007/978-3-030-16118-7_12
37. G. Sabbi, $Nb_3Sn$ Accelerator Magnets. Particle Acceleration and Detection, Springer, 2019, Ch. The HD Block-Coil Dipole Program at LBNL, pp. 285–310.
38. E. Rochepault, P. Ferracin, $Nb_3Sn$ accelerator magnets, Particle Acceleration and Detection Series, Springer, Cham, 2019, Ch. CEA–CERN block-type dipole magnet for cable testing: FRESCA2. URL https://doi.org/10.1007/978-3-030-16118-7_12
39. I. P. et al., Cable design and development for the high-temperature superconductor cable test facility magnet, IEEE Trans. Appl. Supercond. 31 (7) (2021) 4804505. URL https://doi.org/10.1109/TASC.2021.3094410
40. J. L. R. Fernández, D. Arbelaez, P. Ferracin, R. Hafalia, R. Lee, P. Mallon,

S. Prestemon, G. Sabbi, T. Tristan, G. Vallone, Engineering design of a large aperture 15 t cable test facility dipole magnet, IEEE Transactions on Applied Superconductivity 32 (6) (2022) 1–5. doi:10.1109/TASC.2022.3158642.

41. S. Izquierdo-Bermudez, private communication (2017).
42. L. Rossi, C. Senatore, HTS accelerator magnet and conductor development in Europe, Instruments 5 (2021) 8. URL https://doi.org/10.3390/instruments5010008
43. Bruker Presse release, World's first 1.2 GHz high-resolution protein NMR data. URL http://cern.ch/go/Kn9F
44. H. Weijers et al., Progress in the development of a superconducting 32 t magnet with rebco high field coils, IEEE Transactions on Applied Superconductivity 24 (3) (2014) 4301805.
45. A. Molodyk et al., Development and large volume production of extremely high current density $YBa_2Cu_3O_7$ superconducting wires for fusion, Scientific Reports 11 (5) (2021) 2084.
URL https://doi.org/10.1038/s41598-021-81559-z
46. S. Hahn et al., Current Status of and Challenges for No-Insulation HTS Winding Technique, TEION KOGAKU (J. Soc. Jpn Cryo. Super) 53 (1) (2018) 2–9.
47. S. Hahn et al., 45.5-tesla direct-current magnetic field generated with a high-temperature superconducting magnet, Nature 570 (2019) 496–499.
48. M. Varasteh, R. Bruce, F. Cerutti, M. Crouch, F. Zimmermann, Impact of betatron collimation losses in the high-energy large hadron collider, Phys. Rev. Accel. Beams 24 (2021) 041601. doi:10.1103/PhysRevAccelBeams.24.041601.
URL https://link.aps.org/doi/10.1103/PhysRevAccelBeams.24.041601
49. K. Papke, A. Amorim Carvalho, C. Zanoni, A. Grudiev, Design studies of a compact superconducting RF crab cavity for future colliders using Nb/Cu technology (Jul 2019). URL http://cds.cern.ch/record/2683894
50. I. Bellafont, M. Morrone, L. Mether, J. Fernández, R. Kersevan, C. Garion, V. Baglin, P. Chiggiato, F. Pérez, Design of the future circular hadron collider beam vacuum chamber, Phys. Rev. Accel. Beams 23 (2020) 033201. doi:10.1103/PhysRevAccelBeams.23.033201.
URL https://link.aps.org/doi/10.1103/PhysRevAccelBeams.23.033201
51. L. A. Gonzalez, M. Gil-Costa, V. Baglin, P. Chiggiato, C. Garion, R. Kersevan, S. Casalbuoni, E. Huttel, I. Bellafont, F. Perez, Commissioning of a beam screen test bench experiment with a future circular hadron collider type synchrotron radiation beam, Phys. Rev. Accel. Beams 22 (2019) 083201. doi:10.1103/PhysRevAccelBeams.22.083201.
URL https://link.aps.org/doi/10.1103/PhysRevAccelBeams.22.083201
52. L. A. González, V. Baglin, P. Chiggiato, C. Garion, R. Kersevan, S. Casalbuoni, A. Grau, D. S. de Jauregui, I. Bellafont, F. Pérez, Photo-stimulated desorption performance of the future circular hadron collider beam screen, Phys. Rev. Accel. Beams 24 (2021) 113201. doi:10.1103/PhysRevAccelBeams.24.113201.
URL https://link.aps.org/doi/10.1103/PhysRevAccelBeams.24.113201

53. S. Arsenyev, D. Schulte, Broadband Impedance of Pumping Holes and Interconnects in the FCC-hh Beamscreen, J. Phys.: Conf. Ser. 1067 (2018) MOPMF030. 5 p. doi:10.18429/JACoW-IPAC2018-MOPMF030. URL http://cds.cern.ch/record/2647705
54. R. Valizadeh, O. B. Malyshev, S. Wang, S. A. Zolotovskaya, W. Allan Gillespie, A. Abdolvand, Low secondary electron yield engineered surface for electron cloud mitigation, Applied Physics Letters 105 (23) (2014) 231605. arXiv:https://doi.org/10.1063/1.4902993, doi:10.1063/1.4902993. URL https://doi.org/10.1063/1.4902993
55. C. Yin Vallgren, G. Arduini, J. Bauche, S. Calatroni, P. Chiggiato, K. Cornelis, P. C. Pinto, B. Henrist, E. Métral, H. Neupert, G. Rumolo, E. Shaposhnikova, M. Taborelli, Amorphous carbon coatings for the mitigation of electron cloud in the cern super proton synchrotron, Phys. Rev. ST Accel. Beams 14 (2011) 071001. doi:10.1103/PhysRevSTAB.14.071001. URL https://link.aps.org/doi/10.1103/PhysRevSTAB.14.071001
56. S. Calatroni, E. Bellingeri, C. Ferdeghini, M. Putti, R. Vaglio, T. Baumgartner, M. Eisterer, Thallium-based high-temperature superconductors for beam impedance mitigation in the future circular collider, Superconductor Science and Technology 30 (7) (2017) 075002. doi:10.1088/1361-6668/aa6bd0. URL https://doi.org/10.1088/1361-6668/aa6bd0
57. S. Calatroni, R. Vaglio, High-Temperature Superconductor Coatings for Beam Impedance Reduction in Particle Colliders: Nonlinear Effects, IEEE Trans. Appl. Supercond. 31 (2021) 3500208. 8 p. doi:10.1109/tasc.2021.3053299. URL http://cds.cern.ch/record/2776595
58. F. Burkart, W. Bartmann, M. Benedikt, B. Goddard, A. Milanese, J. Schmidt, Conceptual Design Considerations for a 1.3 TeV Superconducting SPS (scSPS) (2017) WEPVA033. 4 p doi:10.18429/JACoW-IPAC2017-WEPVA033. URL http://cds.cern.ch/record/2289465
59. W. Bartmann et al., Injection and Dump Systems for a 13.5 TeV Hadron Synchrotron HE-LHC, J. Phys. Conf. Ser. 1067 (5) (2018) 052021. doi:10.18429/JACoW-IPAC2018-TUPAF060.

© 2024 The Author(s)
https://doi.org/10.1142/9789811280184_0027

# Chapter 27

# Physics at Higher Energy at the Large Hadron Collider

Monica D'Onofrio
*Department of Physics, University of Liverpool, Liverpool, UK*

## 1. Introduction

The High Luminosity LHC programme will have reached its completion by around 2040, resulting in outstanding measurements of fundamental SM parameters and studies of the heavy-flavour sector, as well as extensive searches for new physics at the energy frontier. Even if not a discovery, the HL-LHC will have provided a unique understanding of the Higgs boson properties and their relation to the EWSB mechanism, the potential observation of the SM Higgs-pair production process, and the exploration of a variety of BSM scenarios, including new resonances, additional Higgs bosons, candidates for dark matter, feebly-interacting new particles arising in a hidden sector, and lepton flavour violation. The direct reach in mass and coupling for most new particles predicted by beyond SM theories will have increased by at least 20–50% thanks to the large *pp* datasets collected by ATLAS and CMS, and the LHCb programme will have enabled precision searches for BSM physics through loop processes at unprecedented level. High-density QCD studies with ion and proton beams will have allowed characterisation of quark-gluon plasma properties. Finally, a suite of non-collider experiments,[1,2] if realised, will have provided exciting opportunities to complement the LHC programme in searches for feebly-interacting particles and measurements of high-energy neutrinos produced at the LHC.

The possibility of increasing the centre-of-mass energy of the LHC machine, turning the current accelerator complex into a High Energy (HE) machine, is certainly an appealing option for a future hadron collider project

This is an open access article published by World Scientific Publishing Company. It is distributed under the terms of the Creative Commons Attribution 4.0 (CC BY) License.

at CERN after the end of the High Luminosity physics programme. A HE-LHC immediately following the HL phase would allow collection of large *pp* datasets at a centre-of-mass energy at least two times that of the HL-LHC for both *pp* ($\sqrt{s}$ = 27–30 TeV) and heavy-ion collisions (*PbPb*: $\sqrt{s_{NN}}$ = 10.6 TeV; *pPb*: $\sqrt{s_{pN}}$ = 17 TeV). This would lead to a substantial extension of the HL-LHC reach in direct searches for new physics, approximately doubling the reach in mass of potential new particles. Through energy recovery technologies, collisions of one of the proton or ion beam from the HE-LHC with an intense electron beam ($E_e$ = 60 GeV) could allow concurrent electron-proton (HE-LHeC[3]) and proton-proton operations, extending the reach of deep-inelastic scattering to unprecedented high centre-of-mass energies (*ep*: $\sqrt{s}$ = 1.7 TeV, *ePb*: $\sqrt{s}$ = 1.1 TeV). Finally, the infrastructure, yet to be build, hosting forward-physics, non-collider experiments could be exploited further.

In this chapter, the physics reach of a possible upgrade of the LHC to high energies is presented, with emphasis given to its discovery potential and its perspective value for the future of particle physics. The studies mostly refer to those performed in preparation for the European Strategy for Particle Physics update (ESPPU) process[4,5] and during the Snowmass and Particle Physics Project Prioritization Panel (P5) process.[6]

## 2. The High Energy LHC Physics Potential: an overview

One of the primary objectives of a HE-LHC programme would be to establish the structure of the symmetry-breaking Higgs potential through measurements of the Higgs self-coupling, and improve the precision of the HL-LHC measurements on Higgs properties, EW and flavour sectors. Concurrent *ep* operations would complement the results of *pp* experiments, offering unique access to rare Higgs decay modes in a relatively clean environment, and providing precision measurements of fundamental EWK SM parameters and PDFs. The latter is particularly relevant for classification of the strong interactions dynamics, especially at high parton densities (low-$x$). In the context of new physics models, if tentative signs of discovery arise from the LHC experiments by the end of the HL programme, the HE-LHC could corroborate the signs of discovery, and allow exploration of their properties in greater detail. Much better sensitivity to dark matter and hidden sectors could be achieved depending on the models. In case new phenomena are revealed indirectly, i.e. through anomalies or deviations from SM predictions, or in other experiments (e.g. DM, neutrino or LFV experiments), the HE-LHC experiments could study and possibly identify their underlying origin — even more effectively by considering *pp* and *ep* collisions together. Finally, heavy-ion increased centre-of-mass collisions and *ePb* collisions would offer an excellent endeavor for the studies of denser and hotter strongly-interacting systems.

The physics potential of a HE-LHC facility is depicted through examples[7] of expected results for measurements of the Higgs boson properties,[8] searches for new physics,[9] including scenarios considered to quantify the HE-LHC's ability to characterise potential new physics, precision measurements[10,11] in the EW and flavour sectors, and heavy-ion physics measurements.[12] Rather than a direct comparison with the obviously more powerful and ambitious FCC-hh project, the expected benefits that a HE facility will bring after the HL-LHC has finished operation are emphasised.

No detailed design for a HE-LHC detector is currently available. The experimental environment in terms of radiation flux, track densities and event pile-up is expected to be more challenging than the one at the HL-LHC, but less demanding than the one at the FCC-hh. Hence, while upgrades of the detector designs and novel technologies may be needed — especially for the innermost tracking systems, it is not inconceivable that currently available or in-development approaches can be adopted, *i.e.* for outer trackers, calorimeter systems and muon detectors. At this stage, physics studies assume the typical performances of the ATLAS and CMS HL-LHC detectors for *pp* collisions, modelled by a simplified simulation software, and neglecting the impact of the expected higher pile-up. For an electron-proton(ion) experiment, the detector dimensions and acceptances scale with the logarithm of the proton energy, such that the same technologies and very similar resolution assumptions can be made for a HL and a HE *ep* detector. It should also be noted that in some cases, the physics studies are simply of a phenomenological nature, or extrapolated from (HL-)LHC (prospective) current results. The integrated luminosity benchmark is set at 15 $ab^{-1}$ for *pp* and 1 $ab^{-1}$ for *ep*, to be compared with 3–4 $ab^{-1}$ at HL-LHC and 0.001 $ab^{-1}$ delivered at the first *ep* collider HERA. This is consistent with the accelerator projections and with the possibility to combine the results of two experiments for *pp* and one experiment for *ep*, respectively.

## 3. Higgs properties and EW symmetry breaking

The study of the Higgs boson properties and their connection to EW symmetry breaking will remain one of the most important targets of particle physics well beyond the HL-LHC. At the HE-LHC, the reach of the Higgs physics programme will be substantially extended; the number of collected Higgs events will increase by a factor between 10 and 25 (depending on the production process), with the largest increases occurring for the production of a Higgs boson in association to top-quark pairs, and for Higgs boson pair production.

### 3.1. *Higgs boson couplings to bosons and fermions*

Measurements of Higgs boson couplings to photons, gluons, $W$, $Z$, taus, and $b$-quarks at HE-LHC will reach a percent-level precision by virtue of the increase in Higgs boson production rates. Higher yields will lead to an overall reduction in the statistical uncertainties of Higgs boson properties measurements by factors of 3 to 5, with the biggest improvements occurring for the $t\bar{t}H$ channel where the HL-LHC is statistically limited. Couplings to top quarks through the study of $t\bar{t}H$ processes could be measured with a 3% precision, which could be further reduced to 1–2% measuring ratios of couplings (e.g. $t\bar{t}H$ to $t\bar{t}Z$ ratio). The overall precision in Higgs boson couplings to bosons and fermions at HE-LHC will be limited by the theoretical uncertainty on the signal predictions, hence significant improvements in the precision of theoretical calculations will be required. In this respect, the concurrent operations of a HE-LHeC, for example, would lead to substantially improved PDFs and $\alpha_S$ measurements that can, in turn, provide a significant boost to the achievable precision of Higgs boson properties. Standalone measurements of the Higgs boson coupling to $b$-quarks at the HE-LHeC would also allow the $H \to b\bar{b}$ signal strength to be constrained to less than 1%.

The available HE-LHC statistics will also considerably improve the sensitivity to elusive or rare decays (i.e. Higgs boson decay into second generation quark or lepton pairs, or into a $Z$-boson and a photon), and to hypothetical invisible decays (i.e. Higgs boson decay into a pair of dark matter candidates). For rare processes, such as the decay into a muon pair, a precision of approximately 2% on the coupling could be achievable, whilst for Higgs boson decaying into a charm-quark pair, inclusive searches similar to the ones carried out at the LHC could offer good sensitivity since the signal to SM background ratio is improved at HE-LHC. The HE-LHeC would measure the Higgs to charm quarks decay to 4%, giving the possibility of setting sensible constraints on the Higgs interactions with charm quarks.[3]

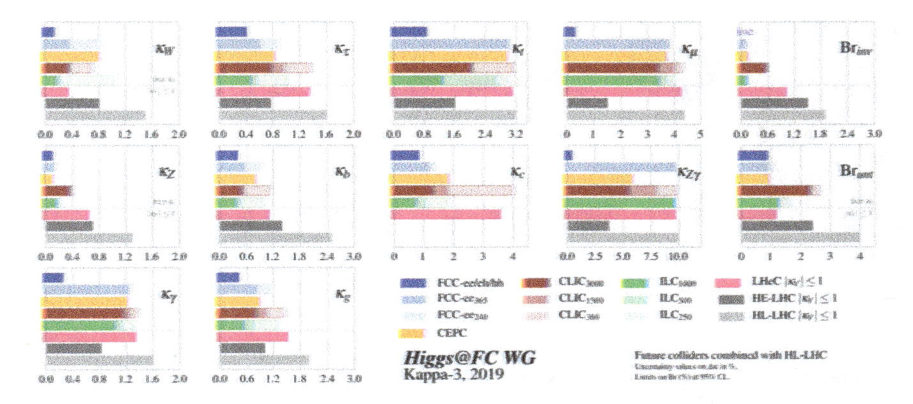


Fig. 1. Expected relative precision of the $\kappa$ parameters and 95% CL upper limits on the branching ratios to bosons, fermions, and invisible and untagged particles for various colliders. The HE-LHC prospect results are depicted in dark gray. All future colliders are combined with HL-LHC. From Ref. 13.

The expected precision for each coupling, including that to invisible particles, can be expressed via the $\kappa$ framework introduced in Chapter 15 about the Higgs Chapter. Figure 1 shows the relative precision of the $\kappa$ parameters and 95% CL upper limits on the branching ratios to bosons, fermions, and invisible and untagged particles for various colliders. HE-LHC prospects, combined with HL-LHC results, are especially competitive for couplings involving top-quarks or rare decays. The figure also depicts prospects from the *ep* facility proposed for HL (LHeC) which are estimated to reach half of the HE-LHC uncertainty, as presented in Ref. 3.

The significantly larger dataset and the increase in centre-of-mass energy at the HE-LHC would allow measurement of differential Higgs boson production cross sections, which are sensitive probes for physics beyond the SM. Thus far, projections and results obtained depend on the assumed detector layout due to challenging pile-up conditions. However, it is unambiguous that sufficient statistics could be collected to perform searches for new physics through deviations from SM predictions of high-transverse momenta Higgs boson production, especially in the gluon-gluon fusion loop — the dominant Higgs boson production mechanism at *pp*.

### 3.2. *Higgs boson self-coupling and rare decays*

The statistics available for Higgs boson pair production (HH) process at the HE-LHC will be about 20 times higher than at the HL-LHC, allowing

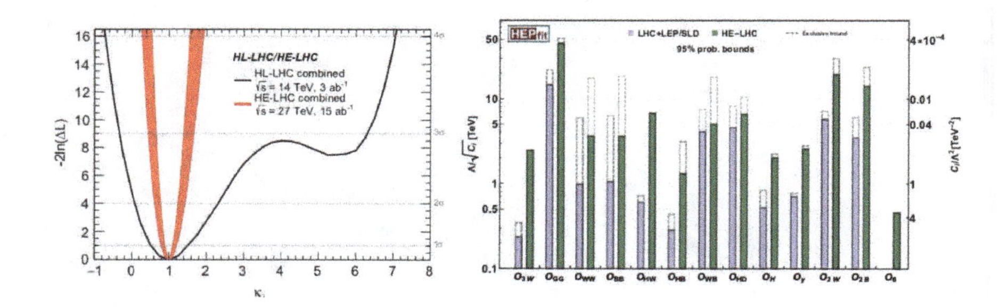


Fig. 2. Left: Projected precision for the measurement of the Higgs trilinear coupling through the measurement of Higgs pair production via gluon-gluon fusion at the HE-LHC. The HL-LHC is shown for comparison. Right: Summary of constraints on the EFT operators considered. The shaded bounds arise from a global fit of all operators, those assuming the existence of a single operator are labeled as "exclusive". From Ref. 5.

constraint and measurement of Higgs boson trilinear coupling, $\lambda_{HHH}$, with considerable precision. Sensitivity studies have been carried out on the basis of the methods and channels considered at the HL-LHC, in particular considering one Higgs boson decaying into a pair of $b$ quarks and the other into a pair of photons or of tau leptons ($b\bar{b}\tau^+\tau^-$ and $b\bar{b}\gamma\gamma$ channels, respectively). The results are reported in Fig. 2(left). At the HE-LHC the HH signal would be observed unambiguously, leading to a precision in $\lambda_{HHH}$ of 10–20% just by combining these two channels, assuming no BSM contributions. The second minimum of the likelihood distribution would be unambiguously excluded at the HE-LHC. Additional sensitivity to the Higgs self-coupling is expected when other channels are also considered, such as if both Higgs bosons decay into a $b$-quark pair, or if one decays into a pair of $W$-bosons, or considering rarer decays that could become relevant at the HE-LHC and when adding the sensitivity of $ep$-based studies. Therefore, the HE-LHC could be the most realistic option to access the Higgs potential without considering any future colliders.

Several BSM theories predict deviations from the SM Higgs couplings, which would be sizable at the HE-LHC. The Higgs boson could decay into invisible particles (DM candidates) or into particles decaying promptly or long-lived and belonging to a wider hidden sector, such as light scalars, dark photons or axion-like particles. Direct searches for these exotic rare Higgs decays would result in a substantially higher reach than at the HL-LHC, and branching ratios one or two order of magnitude smaller could be probed. Precision measurements of the Higgs properties can also help to constrain new physics models. Within the effective field theory (EFT) framework, where the SM Lagrangian is supplemented with higher dimension operators, BSM effects can be systematically parametrised and their deviations from SM processes, estimated. Figure 2(right) shows the summary of constraints on the EFT operators comparing LHC and HE-LHC reaches. Constraints on the operators can translate into sensitivity to new physics models: for example, at the HE-LHC a Higgs compositness scale below 2 TeV would be excluded, corresponding to a new physics mass scale of 25 TeV for an underlying strongly coupled theory.

## 4. Searches for new physics

The HE-LHC is expected to double the mass reach for the discovery of new particles, when compared to the HL-LHC. In several interesting new physics scenarios this is sufficient to cover a large fraction of the relevant parameter space. This covers SUSY models based on the principle of naturalness, extended Higgs sector models and DM models predicting the presence of mediator particles. Direct searches for heavy new particles can be complemented by precision studies of SM observables, and deviations from predictions can provide evidence of new physics. The copious amount of HE-LHC datasets would enable these studies such that, in the possible absence of direct observation of new physics, the EFT formalism could provide a framework to study BSM physics that is realized at a scale $\Lambda$ much larger than the collider $\sqrt{s}$. In the following, a few examples of direct searches for new physics possible at HE-LHC are given, focusing on new heavy resonances, supersymmetric particles and dark matter candidates. In the case of heavy resonances, a brief description of how possible future discoveries at the LHC could be further scrutinised at the HE-LHC is also given.

### 4.1. *New resonances*

Several BSM theories, ranging from new models of EWSB to extensions of the SM gauge group, predict multi-TeV resonances to exist. Typical scenarios include singly-produced resonances with integer spin, or pair-produced heavy resonances. Since direct access requires the centre-of-mass energy of the collider to be large enough to produce them, searches of this kind greatly benefit from the increased energy of the HE-LHC. A qualitative estimate of the improved sensitivity with respect to the expected HL-LHC reach can be obtained by extrapolating the partonic luminosities that are

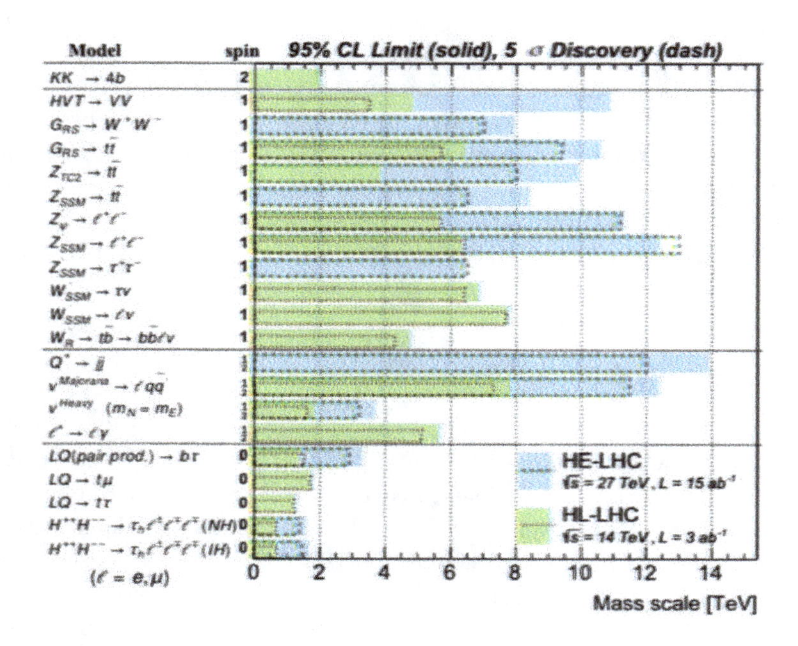


Fig. 3. Summary of the expected mass reach for 5 $\sigma$ discovery and 95% C.L. exclusion at the HE-LHC (blue) as a result of feasibility studies carried out on a variety of models predicting new resonances. Results for HL-LHC are also depicted in green. From Ref. 9.

relevant for the production of various final states. In general, for a system mass that can be probed in searches for new particles at HE-LHC and given an established reach at HL-LHC, an approximate factor of two increase in mass reach is expected.

Dedicated studies for specific new physics models verify (if not exceed in reach) qualitative results obtained via extrapolation. Figure 3(right) shows a summary of the expected mass reach for $5\sigma$ discovery and 95% CL exclusion at the HE-LHC as a result of feasibility studies carried out on a variety of models predicting new resonances and documented extensively in Ref. 9. The HL-LHC projections are also shown. One of the most widely used benchmark scenarios predicts a new high-mass vector (spin-1) boson, the $Z'$, whose couplings to SM quarks and leptons are assumed either as in the SM (SSM) or modified ($\psi$, TC2). Di-leptonic and di-top-quark final state events are searched for resonances and results are interpreted depending on the couplings, with the best discovery reach (13 TeV) achieved for a SSM $Z'$ in the dilepton ($e^+e^-$, $\mu^+\mu^-$) channel. In the di-tau channel, where hadronic $\tau$ reconstruction is more challenging, the discovery reach is about 6 TeV. In the di-top channel, a $Z'$ decaying into a top-quark pair could be discovered up to 6 TeV and excluded up to 8 TeV. The same search interpreted in terms of Randall–Sundrum Kaluza–Klein heavy particles lead to a sensitivity up to 10.7 TeV. This is a 4 TeV extension with respect to HL-LHC and discovery of masses up to around 10 TeV would be possible. The HE-LHC would approximately double the HL-LHC mass reach for dijet resonances. For instance, the reach for an excited quark $Q^*$ decaying to two jets will be improved to 14 TeV, with discovery potential up to 12 TeV. Expected sensitivity on the production and decay of spin-0 and -2 particles decaying into several different SM final states are also studied. Models considered include, among others, resonant double-Higgs production and heavy scalar singlets that could mix with the Higgs boson. Leptoquark (LQ) models, that can give rise to lepton universality violating decays of heavy mesons at tree level, are also studied for couplings to $b$-quark and $\tau$ or $\mu$ leptons. Masses up to about 3 TeV can be reached at the HE-LHC for $\tau$ final states, while reinterpretations of $Z'$ searches in LQ models in $b\mu$ final state events indicate that masses up to about 4 TeV can be excluded, depending on the coupling's strength (2 TeV at HL-LHC).

The $Z'$ has been used as reference to evaluate the capability at HE-LHC to determine the nature and properties of a new hypothetical di-leptonic resonance if discovered at the HL-LHC in the $e^+e^-$ or $\mu^+\mu^-$ channels. This is because, as opposed to supporting evidence or claiming a discovery, the complete identification of the properties of a new particle requires large datasets and possibly higher centre-of-mass energy. Results show that high statistics HE-LHC samples are sufficient to study angular and rapidity distributions and discriminate different models of $Z'$ by exploiting forward-backward asymmetry and other observables.

### 4.2. *Supersymmetry*

Despite the excellent sensitivity of searches in the strong and electro-weak sectors, supersymmetry might remain elusive at the HL-LHC. As one of the most plausible beyond the SM scenarios, providing the only known dynamical solution to the Higgs naturalness problem that can be extrapolated up to very high energies, a potential DM candidate, and the possible reconciliation of gravity and other forces, supersymmetry will certainly be at the core of the HE-LHC programme. Similarly to many scenarios of new physics, SUSY presents a spectrum of multiple states distributed over a broad mass range such that, in case a deviation from SM predictions is found at HL-LHC, the doubling of the LHC energy will be crucial to complement a discovery.

The increase in centre-of-mass energy leads to a large increase in the production cross section of heavy coloured states, such that a 3.5 TeV gluino has a nearly 400-fold increase in production cross section. For supersymmetric spectra without compression, the HE-LHC has 95% CL sensitivity to gluinos up to masses of 6 TeV and a discovery potential up to 5.5 TeV. If the coloured states are close in mass to the lightest supersymmetric particle (LSP), the amount of missing transverse momentum is decreased. Prospect studies in such a compressed scenario show that if, for example, the gluino-LSP mass splitting is assumed to be 10 GeV, gluino masses can be excluded up to 2.6 TeV exploiting the so-called monojet searches. Exclusion and discovery reaches for top squarks are up to 3.5 TeV and 3 TeV, respectively. If the top-squark and the LSP masses are close, final states include very off-shell $W$ and $b$-jets, and masses up to about 1 TeV could be excluded, extending the HL-LHC reach by about a factor of two. Results of this kind would certainly shed light on "natural" supersymmetric models, and potentially discover or exclude them conclusively.

The SUSY electroweak sector presents a considerable challenge for any proton-proton collider, due to the chargino and neutralino (collectively called ewkinos) cross sections depending on the mixing parameters and typically much smaller than those of coloured SUSY particles. If the LSP is a pure higgsino or wino, a very small neutralino-chargino mass splitting is expected (340 MeV or 160 MeV, respectively) and the chargino has a correspondingly long lifetime ($c\tau \simeq 5$ or 1 cm, respectively). The value of the missing transverse momentum is small unless the produced electroweakinos recoil against an ISR jet. A search for monojet signature would give a sensitivity for exclusion (discovery) of winos up to about 600 GeV (300 GeV) and of higgsinos up to about 400 GeV (150 GeV). Taking advantage of the long lifetime of the charginos, searches for disappearing charged tracks can also be performed. Considering a detector similar to the ones available for HL-LHC, winos below 1.8 TeV (1.5 TeV) can be excluded (discovered), while the equivalent masses for higgsinos are 500 (450) GeV (see Fig. 4). Sensitivity to disappearing track signatures relevant for SUSY models at intermediate or shorter lifetime can be complemented by HE-LHeC searches, thanks to the absence of pile-up and the low levels of backgrounds. Feasibility studies[3] have been carried out for LHC- and FCC-like proton beam

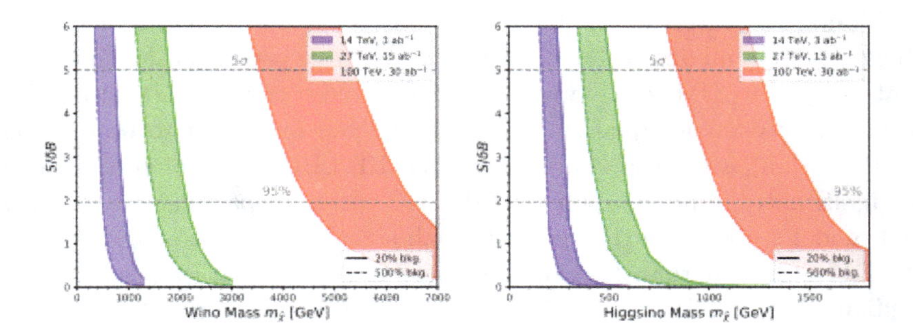


Fig. 4. The reach of HE-LHC in the search for a wino (left) or higgsino (right) DM WIMP candidate, using a disappearing charged track signature.[9] The bands limited by the solid and dashed lines show the range obtained by modifying the central value of the background estimate by a factor of five. The results are compared to the reach of HL-LHC and FCC-hh.

energies. While these results do not reach the mass values required for electroweakinos to be a thermal relic and account for all of DM, the mass range accessible to HE-LHC greatly extends the HL-LHC potential and can be complementary to the indirect detection probes, e.g. using gamma rays from dwarf-spheroidal galaxies.

### 4.3. *Dark matter and dark sector searches*

Generic weakly interactive dark matter candidates are predicted by several DM models beyond compressed SUSY scenarios. All of those could be targeted at the HE-LHC utilising monojet signatures,[4] as this channel is a useful probe for DM production through the exchange of a neutral mediator that couples to the SM, or for a dark sector that contains heavy coloured particles nearly degenerate with the DM and decaying to DM and SM coloured particles. The reach to these kinds of models is strongly dependent on the choice of couplings. Should an excess be observed, the identification of the spin and colour representation of the mediator sector will require high-order precision for the predictions of the SM backgrounds. Analyses of double-ratios of cross sections at varying transverse momenta of the jet could be utilized to partially cancel uncertainties, benefiting from the large HE-LHC datasets. Similarly, searches for monophoton and vector-boson-fusion production can be used, with the latter potentially being more dependent on the robustness of the selection against pile-up, the optimisation of the analysis, and the capability to reduce theoretical uncertainties.

Models characterized by the presence of an extended Higgs sector, with Higgs doublets mixing with an additional scalar or pseudo-scalar mediator that couples to DM, can be searched for in more complex signatures involving heavy-flavour quarks. Prospect searches for associate production of DM with a pair of top quarks show that HE-LHC is expected to be able to significantly improve upon the HL-LHC reach, in particular for spin-0 DM model realisations that predict small signal cross sections. Assuming the DM is lighter than half the mediator mass, a scalar or pseudoscalar mediator can be ruled out at 95% CL up to 900 GeV, a factor of two higher in mass compared to the HL-LHC bounds.

Finally, there is a vast landscape of theoretical models, motivated by DM, for which particles responsible for the still unexplained phenomena are below the EW scale and interact very weakly with SM particles. These particles would belong to a new hidden (or dark) sector where thermal DM is coupled to mediators through portal operators and often have long lifetimes. Prospect searches at future hadron colliders,[4] including the HE-LHC and HE-LHeC, have been carried out for dark photons, dark higgses, axion-like particles and heavy neutral lepton models, exploiting unconventional signatures such as reconstructed leptons, jets, tracks or vertices displaced from the primary interaction point. While it is difficult to firmly conclude on the sensitivity for each specific model, it is expected that a higher centre-of-mass energy and the exploitation of new technologies for the detectors would significantly extend upon the HL reach.

## 5. QCD and EW processes at the highest energies

The increase in energy and integrated luminosity at the HE-LHC will allow us to probe QCD at the highest values of $Q^2$ and search for potential deviations induced by new physics at energy scales well beyond the reach of the HL-LHC. The impact of precise PDFs measured at the HE-LHeC, especially at high $x$, would radically increase the potential of these searches.

In terms of strong interactions, the HE-LHC kinematic reach would extend up to 10 TeV in the jet transverse momentum, $p_T$ (see Fig. 5(left)), and up to about 20 TeV in the di-jet invariant mass, $m_{jj}$. The inclusive production of isolated-photons and jets will reach up to 5 TeV in the photon transverse energy and jet $p_T$, respectively, and up to 12 TeV in the photon-jet invariant mass.

Among the EWK processes that can be studied at unprecedented energies is the scattering of two massive vector bosons $V = W, Z$ (vector boson

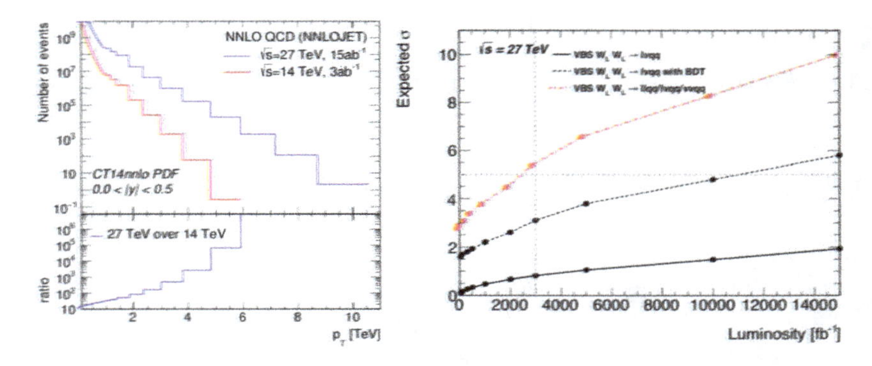


Fig. 5. Left: Predicted number of inclusive jet events as a function of the jet $p_T$. Right: Observed significance as a function of the luminosity and expected uncertainty for the EW $W_L W_L$ signal assuming a 10% fraction predicted from MC. One line shows the results obtained by fitting a single variable, the total invariant mass of the system and the other one shows the expected significance using the BDT. The third line shows the expected significance assuming the combination of all three semi-leptonic channels with the same sensitivity. From Ref. 10.

scattering, VBS), which provide a key opportunity to probe the nature of EWSB mechanism as well as BSM physics. At the HE-LHC, it will be possible to probe longitudinal $VV$ scattering and verify whether the Higgs boson preserves unitarity at all energies. Figure 5(right) shows the significance as a function of the luminosity and expected uncertainty for the EW $W_L W_L$ signal assuming a 10% fraction predicted from MC. The different lines show the results using different analysis approaches and combining three channels exploiting semi-leptonic events. In the latter case, the HE-LHC will reach the 5 $\sigma$ sensitivity with 3 ab$^{-1}$. The various VBS processes also provide excellent probes for the structure of gauge boson interactions, in particular for the quartic gauge couplings. Deviations from SM predictions can be parameterised by an effective Lagrangian, and operators of energy dimension-8, which do not give rise to anomalous trilinear couplings, can used for a parameterisation of anomalous quartic gauge couplings (aQGC). Using observables sensitive to high $VV$ invariant-mass regions, 95% CL bounds on aQGC can be obtained: up to a factor of 10 reduction with respect to the expected HL-LHC bounds is found on all relevant coefficients of dimension-8 operators for HE-LHC.[10]

## 6. Top-quark sector and heavy flavour physics

The HE-LHC would significantly strengthen the role of high-$p_T$ measurements in flavour physics. In the top-quark sector, the typical increase in event yields for inclusive $t\bar{t}$ pair production will be about a factor 15–20, and about a factor 500 if a selection for top-quark with $p_T > 2$ TeV is applied. The large data sets of top quarks will also improve the sensitivity to top-quark FCNC decays by one order of magnitude, relative to the HL-LHC. The production of four top-quark events will see an increase by about a factor 40, leading to an expected 1% statistical uncertainty in the cross section measurement. This process is particularly interesting as it provides direct ways to constrain the top quark Yukawa coupling as well as SM Effective Field Theory parameters sensitive to the quartic couplings between top quarks. The $t\bar{t}t\bar{t}$ signature also arises in models for dark matter involving a two-Higgs-doublet extended sector together with an additional pseudoscalar mediator to DM, and the HE-LHC could provide, if not a discovery, stringent constraints on these models.

It is expected that HE-LHC would be equipped with an upgraded LHCb-like detector fully dedicated to flavour physics studies. An integrated luminosity of about or exceeding 3 ab$^{-1}$ could be collected during the HE LHC phase, hence guaranteeing a significant increase with respect to the HL-LHC, associated to the doubling of the $b$-quark production cross section as a result of the higher centre-of-mass energy. Processes such as rare kaon decays could also be exploited to find new sources of CP violation in the charm sector. Ratios of $B_d$ and $B_s$ process rates such as $\mathrm{BR}(B_d \to \mu^+\mu-)/\mathrm{BR}(B_s \to \mu^+\mu-)$ would be measured with precision at the percent level and would be a probe for new physics at very high energy (O(10–100) TeV). Finally, the CKM phase $\gamma$ could be determined to $< 0.1^o$ using a variety of rare hadronic $B$ decays, and the phase $\beta$ could be extracted from several tree-level decays. Together, these measurements would result in uniquely stringent tests for CKM unitarity.

## 7. Heavy Ion physics

Heavy-ion collisions[12] would be possible at the HE-LHC with the same injected beams as the HL-LHC. Based on extrapolations of LHC performances, the increase in integrated luminosity per run with respect to the LHC is expected to be about a factor 2: about 6 nb$^{-1}$ could be collected in a typical one-month run. A larger increase in terms of nucleon–nucleon luminosity could be achieved with collisions of nuclei lighter than Pb, such as Xe, which would retain production of Quark-Gluon Plasma (QGP) with a volume and energy density similar to those of *PbPb* at LHC energies.

An increased centre-of-mass energy in heavy-ion collisions lead to the creation of initially denser (a factor about 1.4 from LHC to HE-LHC) and hotter strongly-interacting systems that expand for a longer duration and over a larger volume, thereby developing stronger collective signals. The HE-LHC collision energies would be closer to the range of temperatures ($T \simeq 1$ GeV) where charm quarks start to contribute as active thermal degrees of freedom in the QGP equation of state in addition to $u$, $d$, $s$, quarks, thus playing a novel role in QCD equilibration processes. High-energy partons produced in heavy-ion collisions also undergo jet quenching, a marked reduction of the energy of the emerging jets. Jet quenching measurements provide quantitative information on the transport properties of hot and dense QCD matter, hence they are a fundamental target of the heavy-ion programme at the highest possible energy.

The HE-LHC would provide much larger abundance of hard-scattering processes than the LHC, as well as novel probes such as boosted top quarks and, potentially, the Higgs boson. Figure 6(left) shows NNLO cross sections and event yields for various hard-scattering processes as a function of $\sqrt{s_{NN}}$. Photon–photon collisions can also arise: a higher centre-of-mass energy would allow the HE-LHC to reach higher diphoton masses and be sensitive to BSM physics through new heavy charged particles contributing to the virtual loop such as, e.g., from SUSY particles.

The higher centre-of-mass energy of HE-LHC gives access, in the initial state of heavy-ion collisions, to a wide, previously uncharted, kinematic range at low $x$ and $Q^2$, where parton densities become very large and may reach the non-linear QCD regime known as "parton saturation". Through the potentially available electron beam from the ERL, high energy, high luminosity deep inelastic electron-ion scattering physics could also be realised concurrently, giving new insight to the partonic substructure and dynamics inside nuclei. Figure 6(right) shows the kinematic regions in the $x - Q^2$ plane that could be explored by an electron beam colliding against a HL-LHC proton beam or a 20 TeV and 50 TeV beam. The HE line, not on this plot, would cover an area only slightly smaller than the 20 TeV case. With a kinematic reach at the TeV scale, the electron–nucleus option at the HE-LHC could provide conclusive evidence for the existence of a new non-linear regime of QCD, possibly uncovering the chromodynamic origin of the QGP. It would be clearly complementary with the *pPb* case, leading

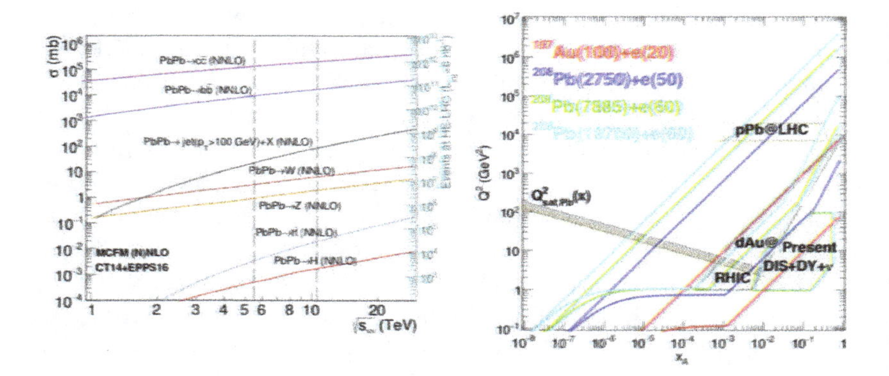


Fig. 6. Left: NNLO cross sections and event yield for various hard-scattering processes as a function of $\sqrt{s_{NN}}$, from Ref. 12. Right: Kinematic regions in the $x - Q^2$ plane explored by different data sets compared to the ones achievable at the EIC (red), the LHeC (ERL against the HL-LHC beams, dark blue) and two FCC-eh versions (with *Pb* beams corresponding to proton energies $E_p = 20$ TeV - green and $E_p = 50$ TeV - light blue). Details on assumptions in Ref. 3.

to a precise knowledge on the partonic structure of nucleons and nuclei and on the small-$x$ dynamics.

## 8. Concluding remarks and summary

The possibility of utilising the LHC tunnel and accelerator complex for a higher energy proton-proton collider immediately after the completion of the HL-LHC programme has been historically one of the first (and most obvious) ideas for a future large-scale project at CERN. The physics potential of such HE-LHC has been briefly illustrated in this chapter, assuming a doubled proton-beam energy with respect to the LHC, and 20 years of operations delivering 10–15 ab$^{-1}$ integrated luminosity to two general purpose detectors. The possibility of having a dedicated flavour physics experiment and a heavy-ion collisions programme are also factored in. Furthermore, the importance of a facility that allows concurrent electron-proton/ion collisions through the addition of a 60 GeV electron beam provided by an energy recovery linac has been illustrated.

It is clear that searches for new physics and precision measurements of Higgs properties, SM parameters and SM processes would greatly benefit from the increased partonic energy of a 27–41 TeV centre-of-mass energy machine. The 20-fold increase in statistics available for the Higgs boson pair production process with respect to the HL-LHC would allow constraint and measurement of the Higgs boson trilinear coupling with O(10%) precision. Heavy new particles up to masses around 10 TeV could be discovered i.e. if resulting in di-lepton or di-jet resonances, and several BSM models could be explored and uncovered.

The main technical challenges are posed by the need to replace the existing LHC dipole magnets with higher magnetic field dipole magnets, as described subsequently. Consequently, the HE-LHC would need civil engineering and technical infrastructure work, and substantial upgrades of the detectors to overcome the higher level of event pile-up, radiation flux, track densities, trigger and readout rates. Still, should the construction of a 100-km Future Circular Collider not be possible in the foreseeable future, the upgrade of the currently available and unique LHC facility to the highest energy possible can indeed be considered as one of the best routes to guarantee the future of collider particle physics.

## References


1. J. Beacham et al., Physics Beyond Colliders at CERN: Beyond the Standard Model Working Group Report, *J. Phys. G.* **47** (1), 010501 (2020). doi: 10.1088/1361-6471/ab4cd2.
2. J. L. Feng et al., The Forward Physics Facility at the High-Luminosity LHC, *J. Phys. G.* **50** (3), 030501 (2023). doi: 10.1088/1361-6471/ac865e.
3. P. Agostini et al., The Large Hadron–Electron Collider at the HL-LHC, *J. Phys. G.* **48** (11), 110501 (2021). doi: 10.1088/1361-6471/abf3ba.
4. E. P. P. S. Update. European particle physics strategy update 2018–2020. URL `https://europeanstrategyupdate.web.cern.ch` (2018).
5. A. Abada et al., HE-LHC: The High-Energy Large Hadron Collider: Future Circular Collider Conceptual Design Report Volume 4, *Eur. Phys. J. ST.* **228** (5), 1109–1382 (2019). doi: 10.1140/epjst/e2019-900088-6.
6. M. Narain et al., The Future of US Particle Physics - The Snowmass 2021 Energy Frontier Report (11, 2022).
7. M. M. et al. The physics potential of he-lhc. URL `https://twiki.cern.ch/twiki/pub/LHCPhysics/HLHELHCWorkshop/HEreport.pdf` (2019).
8. M. Cepeda et al., Report from Working Group 2: Higgs Physics at the HL-LHC and HE-LHC, *CERN Yellow Rep. Monogr.* **7**, 221–584 (2019). doi: 10.23731/CYRM-2019-007.221.
9. X. Cid Vidal et al., Report from Working Group 3: Beyond the Standard Model physics at the HL-LHC and HE-LHC, *CERN Yellow Rep. Monogr.* **7**, 585–865 (2019). doi: 10.23731/CYRM-2019-007.585.
10. P. Azzi et al., Report from Working Group 1: Standard Model Physics at the HL-LHC and HE-LHC, *CERN Yellow Rep. Monogr.* **7**, 1–220 (2019). doi: 10.23731/CYRM-2019-007.1.

11. A. Cerri et al., Report from Working Group 4: Opportunities in Flavour Physics at the HL-LHC and HE-LHC, *CERN Yellow Rep. Monogr.* **7**, 867–1158 (2019). doi: 10.23731/CYRM-2019-007.867.
12. Z. Citron et al., Report from Working Group 5: Future physics opportunities for high-density QCD at the LHC with heavy-ion and proton beams, *CERN Yellow Rep. Monogr.* **7**, 1159–1410 (2019). doi: 10.23731/CYRM-2019-007.1159.
13. J. de Blas et al., Higgs Boson Studies at Future Particle Colliders, *JHEP.* **01**, 139 (2020). doi: 10.1007/JHEP01(2020)139.

© 2024 The Author(s)
https://doi.org/10.1142/9789811280184_0028

# Chapter 28

# HE-LHC operational challenges

Frederick Bordry and Markus Zerlauth
*CERN,*
*Esplanade des Particules 1, 1211 Meyrin, Switzerland*

## 1. Introduction

A further increase in the operating energy of today's LHC will inevitably imply new operational challenges and exacerbate those already known from the development and operation of the current LHC up to its nominal energy of 7 TeV. As discussed in detail in Chapter 26, various options for the choice of the main bending field for a High Energy LHC are being developed by currently ongoing magnet R&D as a function of the desired target energy. While the choice of the detailed lattice might be further optimized for each energy option,[1] the fixed geometry of the existing LHC tunnel will dictate a more or less similar number of main dipole and quadrupole magnets to be installed along the 27 km circumference. In addition, the demanding requirements for beam steering and the quality of the magnetic fields will require a large number of distributed corrector magnets to be installed.

Combining magnets of the same type and powering them in series is an elegant way to optimize the number of power converters and auxiliary protection equipment required for such large scale facilities, as well as having a beneficial impact on machine availability. However, this increases the stored magnetic energy and voltage to ground for the operation of these magnet circuits considerably, both of which already present major challenges and design constraints for the magnets and powering components of today's LHC.

It is for this reason that, early on during the design process of the cold

This is an open access article published by World Scientific Publishing Company. It is distributed under the terms of the Creative Commons Attribution 4.0 (CC BY) License.

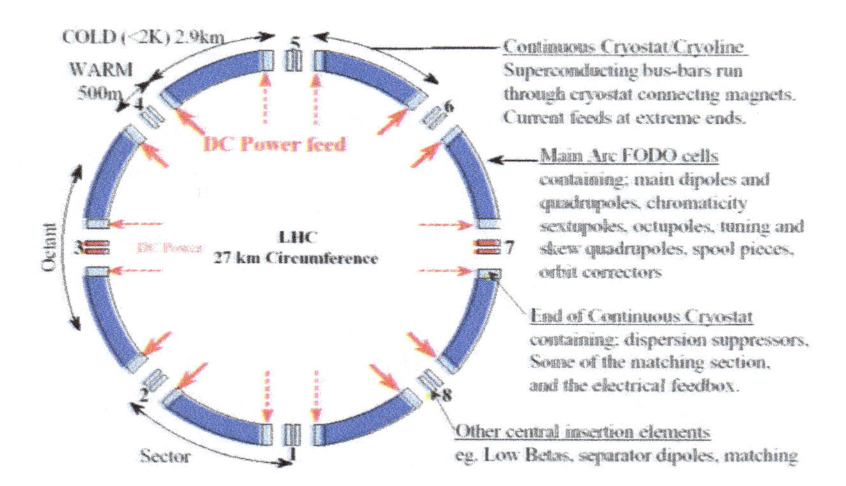


Fig. 1. Powering of the eight arc cryostats. Main circuits and most of the corrector circuits are powered from even LHC insertions, while some additional correctors are being powered from odd insertions.

powering system of the LHC, the decision was made to divide the machine into 8 independent and symmetrical sectors (see Fig. 1), with the main dipole and two main quadrupole magnet families being powered in series, composing three independent main circuits in each of them. This sectorisation comes in some sense naturally, as the superconducting magnets are interleaved in the case of the LHC with conventional, normal conducting magnets around the interaction regions due to the higher radiation levels. The superconducting magnets are housed within more than 40 different cryostats, including the eight 3 km long arc crystats. Even with this split, the inductance of each dipole circuit in one octant is 15 H with a stored energy at nominal current of more than 1 GJ.

Segmentation of the powering of such extended electrical systems also allows for easier installation, testing, commissioning and operation. This has proven to be a major asset, providing flexibility for parallel activities not only during the initial installation and commissioning period of the LHC but also for subsequent campaigns that need to be regularly executed after end-of year technical stops and the long maintenance shutdowns following the 3-year long operational periods. The following operational advantages with a sectorized machine have been confirmed during the initial two operational runs of the LHC machine, and are expected to remain equally valid for an HE-LHC upgrade, if not further enhanced due to the cryogenic separation of the Long Straight Sections from the arc as already realized for HL-LHC:

- Early commissioning experience from first powering subsectors, allowing the development of automatic tools that considerable enhanced efficiency for initial hardware commissioning period after installation
- Parallel commissioning and magnet training in several machine sectors (provided cryogenic capacity is available). Considering the unexpectedly large number of re-training quenches after thermal cycles, this has proven a vital asset for LHC re-commissioning after long shutdowns.
- Increased flexibility during preventive maintenance periods as well as for corrective actions during regular operation. Only concerned sectors will be switched off, while normal activities can resume elsewhere.

Initial concerns about the required tracking precision of the main dipole and quadrupole currents between the 8 independent sectors have been shown to be well within reach of today's state-of-the-art power converter controls electronics.[2,3] Similarly, the increase of the required high current powering equipment has not been found to have a significant detrimental effect on overall machine availability.[4]

## 2. Architecture and powering of magnet circuits in the HE-LHC

In the following, we will consider the second energy upgrade option described in Chapter 26, HE27, as an example to illustrate possible solutions to the challenges of stored energy and extraction voltage when using 16 T main dipole magnets.[5] Considering the use of this magnet technology as well for the HE-LHC, the stored energy in the superconducting magnet system will increase to about 41 GJ. In order to safely handle such energies, the magnets must be powered in several independent powering sectors inside a given LHC octant. In today's LHC, the energy stored in the 154 dipole magnets of one of the eight sectors is in the order of 1.1 GJ. In order to maintain the powering and magnet protection systems similar to the ones of the LHC (including the cold by-pass diode ratings), the baseline of the HE-LHC powering would be subdivided into 32 independent dipole circuits as detailed in Table 1. This will also allow the voltage and net power

Table 1. HE-LHC versus LHC dipole circuit parameters.

| | LHC | HE-LHC |
|---|---|---|
| Number of circuits $N_{cir}$ | 8 | 32 |
| Nominal current | 11.9 kA | 11.4 kA |
| Magnets in series | 154 | 41 |
| Energy | 1.1 GJ | 1.6 GJ |
| Apparent inductance | 15 H | 24 H |
| Ramp up time | 20 min | 20 min |
| Inductive boost voltage required from PC | 150 V | 230 V |
| Max PC net power during ramping | 1.8 MW | 2.6 MW |

requirements for the power converters to be maintained in a comparable order of magnitude, while obviously adding additional complexity due to the required synchronization of 32 power supplies instead of the previous 8.

The subdivision into 32 independent circuits will allow voltages to ground during energy extraction to be maintained within acceptable limits and avoid excessive insulation requirements during design and operation of the magnets and the associated circuit powering components (in particular during a fast discharge of the energy after e.g. magnet quenches). Increasing the insulation voltage to ground beyond the present 3 kV for the LHC dipoles at cold would require developments in magnet technology, and would also make the interconnection and bus bar insulation significantly more difficult, a non-trivial task given the restricted space offered by the present LHC tunnel.

If a circuit powering scheme as depicted in (Fig. 2) is adopted, a single energy extraction system per dipole circuit could be envisaged. The extraction time could be further reduced compared to a resistor-based system by performing the extraction at constant maximum voltage, at the same time

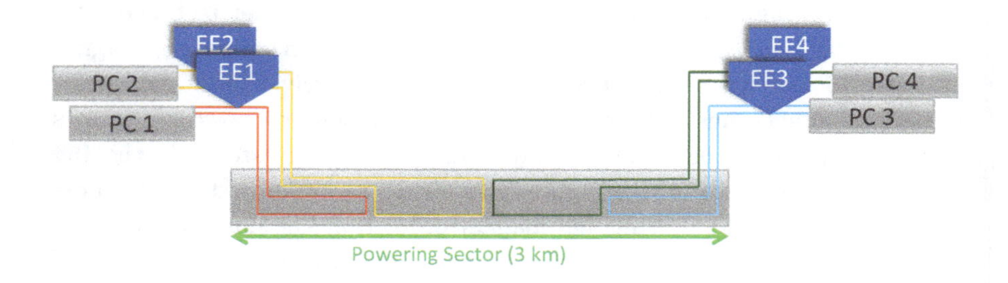


Fig. 2. Schematic of the dipole circuit architecture for HE-LHC.

will have to be carefully reconsidered as shown in (Table 2). To avoid long and costly superconducting links, additional cold busbars would have to be foreseen inside the cold masses of the main dipole and quadrupole magnet to feed the magnets in the middle of the long arc cryostat. The insertion of a second energy extraction system in the middle of the arc cryostat could be possible, but will require the creation of an additional small alcove in this mid-arc region, or, alternatively, the connection of the mid-point to the underground galleries in the arc extremities via an additional cold busbar system. Protection of quadrupole and corrector magnets for a HE-LHC is well within the capabilities of the current protection hardware already deployed in the LHC.

Table 2. Fast power abort of HE-LHC dipole circuit compared with LHC.

| | Half of EE voltage | Discharge time constant | MIITs | Busbar copper cross-section |
|---|---|---|---|---|
| LHC | 0.45 kV | 100 s | $7 \cdot 10^3$ MA$^2$s | 270 mm$^2$ |
| HE-LHC | 1.3 kV | 106 s | $7 \cdot 10^3$ MA$^2$s | 235 mm$^2$ |

## 3. Beam operation challenges

The increase of the center of mass energy for HE-LHC, assuming again HE27 as an example, will imply a factor of 4 higher stored energy in the particle beams than todays LHC, for which the damage limit can be derived from detailed energy deposition studies.[6] At top energy of 14 TeV and assuming similar normalized emittance to today's LHC, a localized loss of only $1.6e10$ protons (equivalent to about 7% of a nominal HE-LHC bunch) would already damage accelerator equipment. As for the LHC, the machine protection system (MPS) for HE-LHC should therefore be designed to prevent any uncontrolled release of energy stored in the magnet system and the particle beams. In view of the reduced quench margin in the superconducting magnets, the protection system must also be able to prevent, or at least minimize, beam induced quenches of the superconducting magnets. This includes resilience against beam losses from interactions of the main beam with dust particles that are inevitable present in the vacuum chamber [refer Chapter 6, Section 2.5], which will become an ever more challenging issue with increasing beam energy and intensity.

Additionally, the effect of flux-jumps that are an inherent feature of the proposed Nb3Sn based magnet technology, will have to be thoroughly studied in terms of quench protection as well as their potential effect on beam orbit, emittance growth and ultimately machine performance.[7,8]

The main principles of the LHC design are nevertheless still valid for HE-LHC, namely to define the aperture limitation in the ring and transfer lines by collimator's, to detect abnormal equipment and beam conditions with fast and reliable instrumentation, to provide passive protection for specific fast failures by beam absorbers and collimator's and to provide — wherever possible — diverse redundancy for the detection of the most critical failures. If the injection energy for HE-LHC remains 450 GeV, a modest upgrade of the current injection protection system should be adequate. However, due to the increased energy swing of the machine, the aperture at 450 GeV appears challenging for collimation. Future studies would be required for the correction of the non-linear errors at lower injection energies to reach the target dynamic aperture.[9,10] An increase of the injection energy (to 900 GeV or 1.3 TeV) implies the need to review the robustness of the concerned absorbers and collimators or to limit the number of bunches per injection.

To survive an asynchronous dump at nominal energy, different techniques are being explored, such as decreasing the kicker rise-time, modifying the optics around the extraction region and upgrading the robustness of the impacted absorbers. Similar studies, as already conducted for HL-LHC, are required for the HE-LHC to determine the acceptable level of halo population for collimators and absorbers to survive these most critical failure cases. The installation of dedicated fast beam loss monitors with nano-second resolution close to the injection and extraction absorbers would allow understanding and possibly reducing and mitigating some of the ultra-fast losses.

## References

1. J. Keintzel, R. Tomás, R. Bruce, M. Giovannozzi, T. Risselada, and F. Zimmermann, Lattice and optics options for possible energy upgrades of the large hadron collider, *Physical Review Accelerators and Beams.* **23** (10) (2020).
2. H. Thiesen, M. Cerqueira Bastos, G. Hudson, Q. King, V. Montabonnet, D. Nisbet, and S. Page, High precision current control for the LHC main power converters, *Proceedings of IPAC'10, Kyoto, Japan* (May, 2010).
3. F. Bordry and H. Thiesen, RST digital algorithm for controlling the LHC magnet current, *Electrical Power Techonology in European Physics Research EP2, Grenoble, France* (Oct, 1998). URL `https://fcc-cdr.web.cern.ch/#HELHC`.

4. A. Apollonio, M. Brugger, L. Rossi, R. Schmidt, B. Todd, D. Wollmann, and M. Zerlauth, Roadmap towards high accelerator availability for the CERN HL-LHC era, *Proceedings of IPAC15, Richmond, US* (2015). URL `https://cds.cern.ch/record/2141848/files/tupty053.pdf`.
5. F. Z. et al., Future circular collider study, volume 4: The high energy lhc (he-lhc) conceptual design report, *CERN accelerator reports, published in Eur. Phys. J. ST.* (Dec, 2018). URL `https://fcc-cdr.web.cern.ch/#HELHC`.
6. Y. Nie, R. Schmidt, V. Chetvertkova, G. Rosell-Tarrago, F. Burkart, and D. Wollmann, Numerical simulations of energy deposition caused by 50 MeV–50 TeV proton beams in copper and graphite targets, *Physical Review Accelerators and Beams*. **20** (8) (2017).
7. J. C. de Portugal, R. Tomás, L. Fiscarelli, D. Gamba, and M. Martino, Impact of flux jumps on future circular colliders, *Physical Review Accelerators and Beams*. **23** (1) (2020). URL `https://journals.aps.org/prab/abstract/10.1103/PhysRevAccelBeams.23.011001`.
8. R. Tomás, J. Keintzel, and S. Papadopoulou, Emittance growth from luminosity burn-off in future hadron colliders, *Phys. Rev. Accel. Beams*. **23**, 031002 (Mar, 2020). doi: 10.1103/PhysRevAccelBeams.23.031002. URL `https://link.aps.org/doi/10.1103/PhysRevAccelBeams.23.031002`.
9. M. Hofer, M. Crouch, J. Keintzel, T. Risselada, R. Tomás, and F. Zimmermann, Integrated he-lhc optics and its performance, *Proceedings of IPAC2018, Vancouver, Canada* (2018). URL `http://accelconf.web.cern.ch/ipac2018/html/sessi0n.htm`.
10. R. Tomás, M. Benedikt, M. Hofer, J. Keintzel, and F. Zimmermann, Towards future circular colliders, *Proceedings of Sixth Annual Conference on Large Hadron Collider Physics, Bologna, Italy* (2018). URL `https://pos.sissa.it/321/268/pdf`.

© 2024 The Author(s)
https://doi.org/10.1142/9789811280184_0029

# Chapter 29

## Vacuum challenges at the beam energy frontier

V. Baglin, P. Chiggiato and R. Kersevan
*Vacuum Coatings and Surfaces Group, Technology Department*
*CERN, Geneva, Switzerland*

### Introduction

Designing the vacuum system for a machine like the HE-LHC is a very challenging task, since it demands the vacuum designer to cope with synchrotron radiation (SR) critical energy and power at levels much higher than those of the LHC, and a linear photon flux (photons/s/m) 80% higher than that estimated for the FCC-hh,[1] see Table 1.

Table 1. Synchrotron radiation (SR) characteristics in the arcs of LHC, HE-LHC and FCC-hh.

| Parameter | LHC | HE-LHC | FCC-hh |
|---|---|---|---|
| Linear SR power (W/m) | 0.25 | 5.5 | 35 |
| Linear photon flux ($10^{16}$ photons/m/s) | 5 | 27 | 15 |
| Critical photon energy (eV) | 44 | 320 | 4300 |

A number of critical features of the design of the vacuum system for HE-LHC are discussed below.

### Beamscreen design

As the cold-bore diameter and the length of the dipoles are identical for HE-LHC and FCC-hh, the cross-section of the beam-screen (BS) studied so far are the same for the both accelerators. This way, the design

This is an open access article published by World Scientific Publishing Company. It is distributed under the terms of the Creative Commons Attribution 4.0 (CC BY) License.

optimisation and the experimental validation performed for the FCC-hh during the first dedicated study program funded by Horizon 2020[2,3] can be applied to the HE-LHC. It should be noted that contrary to the dipoles of FCC-hh which have a straight yoke and cold-bore, the HE-LHC must have curved yoke and cold bore like the LHC, since the beam orbit sagitta in the middle of the dipole would otherwise be 8.6 mm for a straight magnet, causing the beam halo to touch the BS, especially at injection energy when the beam cross-section is bigger. For comparison, the beam sagitta in the middle of an FCC-hh dipole is only 2.3 mm. This necessitated the creation of new 3D models for the Monte Carlo ray-tracing simulations of the SR and molecular density distribution as compared to FCC-hh.[2,4] The modelled lattice version is the "23 × 90", with 3 dipoles and 1 quadrupole/corrector package in each half cell, ~ 54 m-long.[1] The resulting molecular density profiles are qualitatively similar to those calculated for FCC-hh, with a ~ 2 m long density spike corresponding to the position of a short SR absorber placed at the very end of the dipole which masks the dipole-dipole or dipole-quadrupole connection area, where the RF fingers inside the bellows

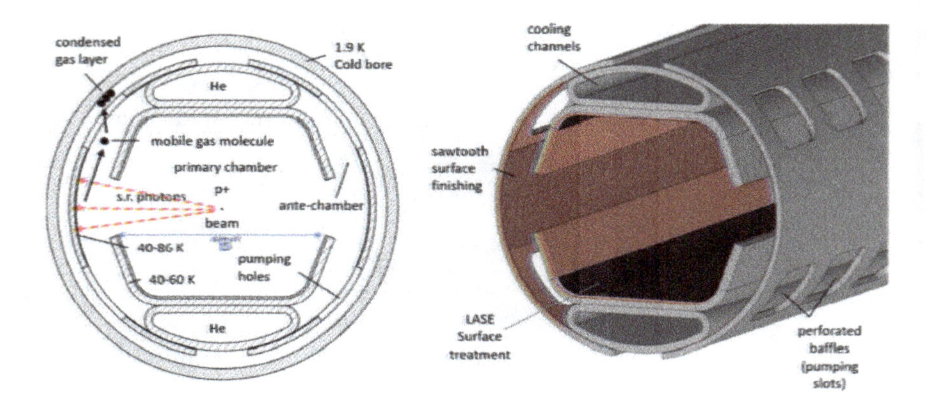


Fig. 1. Design of the HE-LHC BS. The main functions and parts are indicated, together with nominal temperature range of the different parts, presently under review. Width "B" is 27.55 mm. The pumping slots are much larger than those of the LHC, giving an effective linear pumping speed for $H_2$ of ~ 900 l/s/m vs ~ 480 l/s/m for the LHC. The vacuum behaviour of such a BS has been tested at KARA (formerly ANKA) light source, see Refs. 2–4. Only the primary SR photon fan is indicated, but low energy reflected photons can and will be scattered around the BS, finally irradiating all surfaces and generating photon stimulated desorption (PSD) of gas molecules everywhere on all surfaces of the BS. The aim is to minimize those generated on the inner surfaces of the BS in direct view of the beam.

and the beam-position monitor electrodes are placed, see Fig. 4.40 in Ref. 4. The Monte Carlo ray-tracing simulations for the gas density profiles show average values below the maximum allowed value of $1 \cdot 10^{15}$ $H_2$-equivalent molecules/m$^3$, corresponding to nuclear beam-gas scattering lifetime higher than 100 hours as required.

### Electron-cloud

In addition to the higher PSD generated by the SR fan, the HE-LHC is also challenging because of its high beam current, 1.12 A compared to the 0.5 A of the FCC-hh, making the electron-cloud (EC) countermeasures even more important. As shown in Refs. 1,3,4 experimental tests and calculations have been carried out extensively to make sure that the EC can be kept under control. The EC is currently one of the major problems with increasing the beam intensity in the LHC as it is a source of beam instabilities and an issue for cryogenic heat load.[5] Various surface treatments and thin-film deposition techniques have been proposed and validated experimentally, such as laser ablation and amorphous carbon coating of the BS internal surfaces.[3]

### Impedance

The resistive-wall impedance in such a narrow BS geometry is also a very important issue.[1] Several collaborations and tests have been set up recently to determine whether high-temperature superconductor (HTS) inserts could be fixed onto the 6 flat inner sides of the BS geometry.[6] A robust program of study using numerical simulations is under way,[6] as well as experimental validation using a light source beamline.[3]

### Ion-stimulated desorption

Another effect to take care of is ion-stimulated desorption (ISD), which is known to depend on several quantities, such as the gas composition, the beta functions, bunch spacing and separation, and applied pumping speed and beampipe conductance. The large linear pumping speed given by the pumping slots helps, keeping the molecular density low. In addition, tests at 80 K carried out at KARA light source[3] have shown that $\sim 90\%$ of the gas generated via PSD is hydrogen, with only the remaining 10% being

CO and $CO_2$. This should help keep the ISD effect under control, as per calculations carried out for FCC-hh.[4]

## Summary

The design of the HE-LHC vacuum system relies on the large body of literature generated for the vacuum system of FCC-hh. There are some features of the BS which are specific to the HE-LHC which will need to be tested and validated in the future, such as the use of HTS for minimizing the resistive-wall impedance. Additionally, the effect of an increased BS temperature and its effects on the cryogenic heat load, technology, and operating costs will need to be ascertained, with possible implications on energy saving and related operational costs, which are becoming more and more important.

## References

1. A. Abada et al., HE-LHC: The High-Energy Large Hadron Collider – Future Circular Collider Conceptual Design Report Volume 4, Eur. Phys. J. Special Topics 228, 1109–1382 (2019).
2. I. Bellafont, Photon Ray-Tracing and Gas Density Profile in the FCC-hh, Proc. FCC Week 2017, Berlin, 2017, https://indico.cern.ch/event/556692/contributions/2487660/.
3. L. A. Gonzalez Gomez, Test of FCC-hh beam screens at the ANKA beamline, Proc. FCC Week 2017, Berlin, 2017, https://indico.cern.ch/event/556692/contributions/2487658/.
4. I. Bellafont, Study of the beam induced vacuum effects in the cryogenic beam vacuum chamber of the Future Circular Hadron Collider, PhD thesis Univ. Politecnica de Catalunya, Nov 2020, https://cds.cern.ch/record/2791779/files/CERN-THESIS-2020-396.pdf.
5. G. Skripka, G. Iadarola, L. Mether et al., Non-monotonic dependence of heat loads induced by electron cloud on bunch population at the LHC, Eur. Phys. J. Plus **137**, 849 (2022). https://doi.org/10.1140/epjp/s13360-022-02929-8.
6. T. Puig et al., Coated conductor technology for the beamscreen chamber of future high energy circular colliders, 2019 Supercond. Sci. Technol. 32 094006, DOI 10.1088/1361-6668/ab2e66.

© 2024 The Author(s)
https://doi.org/10.1142/9789811280184_0030

# Chapter 30

# The LHC as FCC injector

Michael Benedikt and Brennan Goddard
*CERN*

The re-use of the modified LHC is presently the reference baseline for the High Energy Booster (HEB) injector[1] into the FCC-hh hadron collider,[2] although a number of other promising options exist. As part of the FCC study, the transformation of the LHC into the last acceleration stage into the new collider has been investigated in some detail, including the key aspects of new insertion designs and faster ramping. Performance aspects including energy reach, flexibility and filling time of the collider have been considered, and the transfer lines linking the HEB to the FCC have been defined. This chapter describes the required performance, the required changes which would be needed to the existing LHC machine and discusses the remaining challenges for LHC operation as FCC-hh injector. The study was based on the FCC-hh machine layout defined in the Conceptual Design Report in 2018:[3] the design continues to evolve, which could have an impact on the changes needed in specific LHC straight sections, but the main considerations on the feasibility of reuse of LHC remain valid.

## 1. Requirements for FCC-hh injector

The FCC-hh concept is for an accelerator of 91 km circumference which will collide protons and ions at about 50 TeV per beam. The HEB should be able to fill roughly 80% of the collider with 3.3 TeV protons in about ~30 minutes, several times per day. This corresponds to 10400 bunches spaced by 25 ns with a bunch intensity of $1 \times 10^{11}$ protons and a normalized emittance of 2.2 $\mu$m. A list of the most relevant beam parameters for the FCC-hh injector complex is given in Table 1. A 5 ns option has also been considered, which would mean some changes to the upstream machines in

This is an open access article published by World Scientific Publishing Company. It is distributed under the terms of the Creative Commons Attribution 4.0 (CC BY) License.

Table 1. Beam parameters for the FCC injector complex.

| Parameter | Baseline | Ultimate |
|---|---|---|
| Injection energy | 3.3 TeV | 3.3 TeV |
| Number of bunches | 10,400 | 52,000 |
| Bunch spacing | 25 ns | 5 ns |
| Bunch intensity | $1 \times 10^{11}$ p | $0.2 \times 10^{11}$ p |
| Normalized emittance | 2.2 $\mu$m | 0.44 $\mu$m |
| Turn around time | 5 h | 4 h |
| Max. FCC filling time | 30 min | 30 min |
| LHC duty cycle for FCC filling | 0.10 | 0.125 |

the present LHC injector chain, or a new set of pre-injectors to match any changing physics needs.

The baseline FCC-hh injection energy is 3.3 TeV. Higher energy is favourable in terms of impedance, beam stability, aperture and energy (field) swing, but a lower energy is favourable for transfer to FCC and simplicity of the injector complex, lower capital and operating cost for the HEB, as well as opening more options for its realisation.

The FCC-hh injection energy also determines the number of bunches which can be transferred safely to the FCC,[4] because of damage limits of the injection protection absorbers. This limit scales non-linearly with beam energy, as the energy deposition in the absorber also depends on the secondary shower development. At the baseline energy of 3.3 TeV, only ~80 bunches can safely impact the absorbers, so a staggered transfer is necessary, affecting the kicker design parameters and filling scheme. Around 100 of these multiple extractions are needed to fill the FCC collider.

Other important requirements are that the HEB should be reliable with highest possible availability, and also that it should be considerably easier and cheaper to operate than FCC itself.

The duty cycle for FCC filling is (for the ultimate beam) only about 12%, which means that the LHC could be available for the remainder of the time for its own dedicated physics program — either at 3.3 TeV or at full energy, if the new insertion designs with crossings remain compatible. A discussion of possible alternatives is beyond the scope of this chapter, see e.g. Ref. 5 for more details.

## 2. Reuse of existing LHC as 3.3 TeV HEB

The study baseline of reusing LHC as HEB is conceptually the most straightforward way to inject at 3.3 TeV into FCC-hh. In this scenario using LHC as HEB injector for FCC would still rely on the whole existing injection chain, including the SPS. In the version studied, to make space for the extractions towards FCC two beam crossings in experimental Interaction Points (IPs) have to be removed. Depending on which crossing points will be removed, the total circumference of Beam 1 and Beam 2 in the LHC might no longer be identical. This is not an issue for the HEB but might impact the potential remaining LHC physics program beyond the FCC injection. Suppressing the crossing in IR2 and IR8 keeps the circumference of the two LHC beams identical. The locations and layouts of the existing RF, collimation and beam dump systems are maintained; however, keeping the orientation of the beam dump while removing two crossings means that injection will have to be shifted from the outer rings to the inner rings. The physics experiments and low beta insertions will have to be decommissioned. The changes in LHC layout are depicted in Fig. 1. In view of other possible uses of the LHC, the study aimed at keeping the energy reach of the LHC to 7 TeV, e.g. in the design of the beam crossings, while designing extraction and transfer lines to 3.3 TeV is needed for the FCC-hh.

The other important requirement is to speed up the LHC ramp, which will have to be improved by roughly a factor 5, to 50 A/s, to keep the overall FCC filling time in the ~1 hour regime. This requires new main power

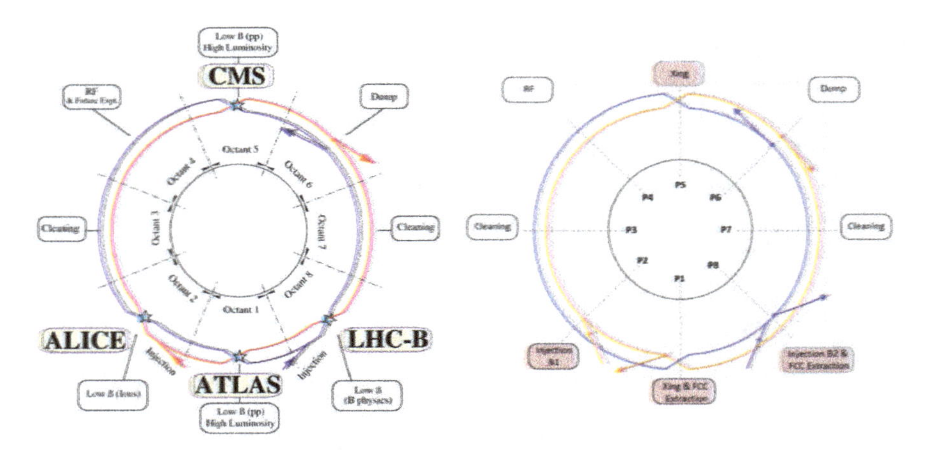


Fig. 1. Existing (left) LHC layout, illustrating possible changes for use as HEB (right).

converters as well as changes to other systems, such as quench protection. At this stage, it is not clear if such a modification of the power converters will be implemented with the same accuracy that is required for storage ring operation. These aspects all need to be considered in the overall optimisation. The existing LHC RF system, with a voltage of 16 MV at 400 MHz, is able to accelerate an LHC beam to 3.3 TeV in 1 minute, but the limiting factor is the ramp of the main dipoles.[6] After the modifications these will be able to ramp to 3.3 TeV in about 3 minutes, so the RF will not need any changes. As far as the RF system is concerned, the FCC-hh requirements are still less demanding then the nominal LHC and HL-LHC operation.

The early part of the ramp will also need to be changed from the baseline operational LHC Parabolic-Exponential-Parabolic-Linear (PELP), which would otherwise take over half of the total ramp duration. In one of the few FCC tests in the LHC machine, a simpler Parabolic-Parabolic-Linear-Parabolic (PPLP) ramp was developed and tested successfully in 2017,[7] and shown to save 1.5 minutes per ramp with no adverse effect on the beam. This has actually been adopted as the operational LHC ramp since 2018, and has saved a total of about 10 hours per year of operation (a direct benefit of the FCC studies for the present LHC physics program).

### 2.1. *Insertion modifications*

For compatibility with the FCC-hh version studied, the main layout features per LHC straight section are summarised below:

- IR1: new beam 1 extraction system plus beam crossing plus decommissioning of ATLAS;
- IR2: injection to inside ring plus decommissioning of ALICE;
- IR3: no changes to momentum collimation;
- IR4: no changes to RF system;
- IR5: decommissioning of CMS, plus beam crossing;
- IR6: no changes to beam dump;
- IR7: no changes to betatron collimation;
- IR8: injection to inside ring plus new beam 2 extraction system plus decommissioning of LHCb.

#### 2.1.1. *IR1*

The new extraction system in IR1 can be very similar to the current beam dump extraction system in LSS6. However, a modification is needed by opening up the space between the Q4 and Q5 downstream the septum to accommodate the needed beam crossing. In the easier case of leaving all distances between the Q5 upstream the septum to the Q4 downstream the septum unchanged, a space of 75 meters is available between the downstream Q4 and Q5 which is not enough to facilitate a beam crossing at 7 TeV.

There are several possibilities for realizing the beam crossing in this region, the details of which are discussed in [3]. Assuming four of the 11 T dipoles developed within the HL-LHC framework can be used in this crossing, 98 meters are needed between the downstream quadrupoles (a simple copy of the IR6 dump extraction would allow crossing only up to 4.5 TeV). Shifting the downstream Q4 and the septum further upstream reduces the maximal extraction energy but increases the maximal energy at which the crossing can operate. By changing the drift after the kicker to 123 m and the one after the septum to 121 m, enough space is created between Q4 and Q5 to facilitate the 7 TeV crossing.

This would be a realistic layout for 3.3 TeV extraction, although the needed kicker switch technology in terms of dI/dt exceeds present technology by a factor 2.6 which means technological advances in this area would be a key R&D topic. Higher energy transfer would need further improvement.

### 2.1.2. *IR2*

The injection in IR2 needs to be changed from the outer to the inner ring. In order to do so, while still maintaining the optics at the injection elements, all injection elements need to be shifted by 16 meters. This shift of the septum, Q5, protection devices and kicker along with optics matching at the septum entrance and preserved Q5 strength are essential to this proposed layout. Q4 and other quadrupoles can be changed, as long as the optics constraint of 90 degree phase advance between kicker and injection protection device (TDI) is respected. The TDI may be moved to facilitate this. However, if Q4 and the TDI location are changed, or if extra quadrupoles are added between Q4 and the TDI, new studies of the injection system failure cases are needed to ensure proper machine protection.

The only other elements present in this region are the added quadrupoles, used to introduce a FODO-like structure that keeps the optics functions close to those for the arc. Hence there are no spatial restrictions, which allows us to replicate the current LHC injection system without introducing new constraints to the maximal LHC energy.

Note that even though the added quadrupoles have been located in approximately symmetric and at equal distances, their placement does not influence the injection system. Some flexibility in placement of these quadrupoles for optics considerations, since they will not have any effect on the injection system.

#### 2.1.3. *IR8*

In IR8, in addition to the injection system that needs to be changed from the outer to the inner ring, an extraction system may need to be installed. The injection system is thus to be moved by 16 meters, as in IR2. The FCC layout chosen will determine if the extraction system should be on the inner ring or the outer ring. In the following, the extraction on the outer ring is illustrated.

Similar to the situation in IR2, optics matching at the septum entrance and preserved Q5 strength are important for these layouts, while other quadrupole strengths can be changed, as long as the optics constraint of 90 degree vertical phase advance between kicker and TDI is respected.

While extraction on the outer ring is easier than the case where injection and extraction are on the same ring, the layout with the injection system does not have enough space for the present dump extraction system. This is because the quadrupoles used around the injection kicker on the inner ring now determine the distance in which the extraction system needs to reach a large enough clearance. Another problem encountered is the required 90 degree phase advance between the kickers and their respective protection elements. Due to this requirement it becomes necessary to move the protection device further away.

This layout is a realistic one for 3.3 TeV extraction, requiring a factor 2.1 advance in kicker technology. If we would again assume that the kicker limit changes by a factor 5, then if needed we could reach an extraction energy of 7.0 TeV by moving the septum back by 13 meters and adding three more septum modules.

### 2.2. *Other considerations*

The design of the beam transfer lines from the LHC to FCC-hh[8] is dependent on the location of the FCC tunnel. The version of the FCC tunnel described for this study passes directly under the LHC tunnel, which could allow for normal-conducting transfer line magnets depending on the detailed FCC-hh layout and orientation.

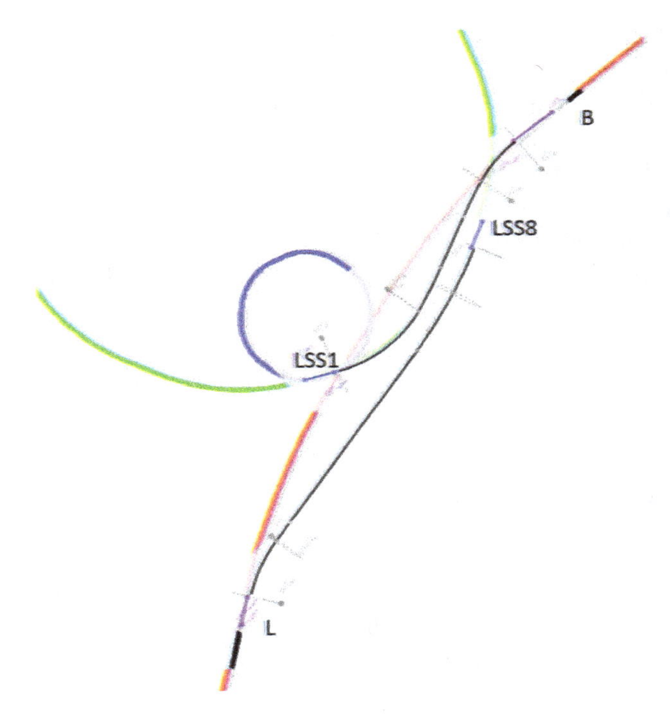


Fig. 2. Transfer lines from 3.3 TeV LHC HEB (red) to FCC, from LSS1 and LSS8. With this layout version the transfer lines could be fully normal-conducting.

Decommissioning of the LHC experiments is likely to be one of the cost-drivers for the conversion of LHC to FCC HEB. The activation levels in IR1 and IR5 at the end of LHC physics operation are expected to be at around 1 mSv/h level, with a large amount of material to decommission. Work on dose estimation, handling and decommissioning work is required, using the available detailed dose rate maps to compute job doses once work scenarios are available.

With the most recent FCC implementation roadmap, the construction and operation of FCC-ee before FCC-hh would mean that LHC as FCC-hh injector would not be required for about 20 years after the end of HL-LHC operation. Although it would leave more time for cool-down of radio-active equipment, this long delay poses important feasibility issues in terms of keeping equipment and expertise available. It therefore reduces the attractiveness of directly reusing LHC as FCC- hh injector.

The possibility of operating LHC as a Fixed-Target facility between filling the FCC has also been studied;[5] here the problems are the integration of a slow-extraction system into the superconducting, small aperture LHC machine, the issues associated with designing the extraction hardware at such a high rigidity, and also the small annual number of protons on target that could be realised due to the long LHC ramp time.

One important remaining concern for the reuse of LHC will be the high power consumption for the LHC cryogenic system, in addition to the operating resource cost and also the age and related availability of the LHC when FCC-hh comes on line.

### 2.3. *3.3 TeV Superconducting (4 T) HEB 27 km in LHC tunnel*

If 3.3 TeV injection energy is mandatory, another interesting option would be to replace LHC in the same tunnel with a fast-ramping, relatively low-cost, superferric or superconducting machine, which would need a field of 4 T to reach 3.3 TeV. The machine could be twin-aperture, although this would increase the cost substantially, both for the magnets and also for powering, instrumentation and other ancillary systems. Alternatively it could use polarity reversal to minimise the length of transfer line to FCC, depending on the detailed cost trade-off. Such a machine would follow the existing LHC geometry and layout, re-using the injection and dump transfer lines, but differently configured in terms of injection and extraction systems.

Although the capital cost might be higher than modifying the present LHC, this could be an attractive option in terms of operating cost, consumption and maintainability, compared to reusing the existing LHC magnet system.

## References

1. B. Goddard, W. Bartmann, M. Benedikt, W. Herr, M. Lamont, P. Lebrun, M. Meddahi, A. Milanese, M. Solfaroli Camillocci, and L. Stoel, Possible Reuse of the LHC as a 3.3 TeV High Energy Booster for Hadron Injection into the FCC-hh. p. THPF094. 4 p (2015). URL `https://cds.cern.ch/record/2141905`.
2. M. Benedikt, B. Goddard, D. Schulte, F. Zimmermann, and M. J. Syphers, FCC-hh Hadron Collider - Parameter Scenarios and Staging Options. p. TUPTY062. 4 p (2015). URL `http://cds.cern.ch/record/2141745`.
3. M. Benedikt, M. Capeans Garrido, F. Cerutti, B. Goddard, J. Gutleber, J. M. Jimenez, M. Mangano, V. Mertens, J. A. Osborne, T. Otto, J. Poole, W. Riegler, D. Schulte, L. J. Tavian, D. Tommasini, and F. Zimmermann,

FCC-hh: The Hadron Collider: Future Circular Collider Conceptual Design Report Volume 3. Future Circular Collider. Technical report, CERN, Geneva (Dec, 2018). URL `https://cds.cern.ch/record/2651300`.

4. B. Dalena, R. Alemany-Fernandez, A. Chancé, B. Holzer, J. Payet, and D. Schulte, First Considerations on Beam Optics and Lattice Design for the Future Hadron-Hadron Collider FCC-hh. p. WEBB2. 3 p (2015). URL `http://cds.cern.ch/record/2141857`.
5. B. Goddard, G. Isidori, F. Teubert, M. Bai, A. Ball, B. Batell, T. Bowcock, G. Cavoto, A. Ceccucci, M. Chrzaszcz, A. Golutvin, W. Herr, J. Jowett, M. Moulson, T. Nakada, J. Rojo, and Y. Semertzidis, Physics Opportunities with the FCC-hh Injectors. Physics Opportunities with the FCC-hh Injectors, *CERN Yellow Report.* pp. 693–705. 13 p (Jun, 2017). doi: 10.23731/CYRM-2017-003.693. URL `https://cds.cern.ch/record/2271775`. 13 pages, Chapter 5 in Physics at the FCC-hh, a 100 TeV pp collider.
6. A. Milanese, B. Goddard, and M. Solfaroli Camillocci, Faster ramp of LHC for use as an FCC High Energy hadron Booster. Technical report, CERN, Geneva (Oct, 2015). URL `http://cds.cern.ch/record/2057723`.
7. A. Milanese, B. Goddard, and M. Solfaroli Camillocci, Faster Magnet Ramps for Using the LHC as FCC Injector, *ICFA Beam Dyn. Newsl.* **72**, 113–121. 9 p (2017). URL `https://cds.cern.ch/record/2315724`.
8. W. Bartmann, M. Barnes, M. Fraser, B. Goddard, W. Herr, J. Holma, V. Kain, T. Kramer, M. Meddahi, A. Milanese, R. Ostojic, L. Stoel, J. Uythoven, and F. Velotti, Beam Transfer to the FCC-hh Collider from a 3.3 TeV Booster in the LHC Tunnel. p. THPF089. 4 p (2015). URL `http://cds.cern.ch/record/2141901`.

© 2024 The Editor(s)
https://doi.org/10.1142/9789811280184_bmatter

# About the Editors

OLIVER BRÜNING is a senior scientist at CERN, who was born in 1964 in Hamburg, Germany. He specializes in accelerator physics and has been working on several flagship accelerator projects since 1991. His involvement in accelerator projects ranges from non-linear beam dynamics studies and commissioning of the HERA electron-proton collider at DESY, to the LEP-II upgrade and the LHC design and commissioning at CERN. He was one of the initial LHC Commissioning coordinators from 2008 until 2013 and has led the CERN Accelerator Beam Physics group from 2005 until 2015. Since 2010 he has been the deputy project leader of the HL-LHC upgrade and has been leading the project since 2021.

He is a member of the EPS-AG and has chaired the Accelerator Group from 2008 until 2011. He was also the Scientific Program Chair of the EPAC 2008 conference and chaired the 2011 IPAC conference.

He has coordinated the accelerator design of the LHeC and FCC-eh studies since 2008 and since 2012, he has been promoting the development of Energy Recovery Linacs as a vital ingredient for the LHeC and FCC-eh studies. In this role, he has been participating in several international advisory boards on ERL related studies.

This is an open access article published by World Scientific Publishing Company. It is distributed under the terms of the Creative Commons Attribution 4.0 (CC BY) License.

MAX KLEIN has worked on theory, detectors, data analysis and coordination of particle physics experiments at several accelerators at CERN (Geneva, Switzerland) and at DESY (Hamburg, Germany). As a postdoc, he spent several years at JINR (Dubna, Soviet Union).

Max was initially employed by the Institute for High Energy Physics at Zeuthen near Berlin, which is now a part of DESY. He has been a Professor at the University of Liverpool since 2006, Emeritus since 2022. For many years, Max has been coordinating the development of the Large Hadron electron Collider and the FCC-eh projects, options for luminous, energy frontier electron-hadron scattering experiments at CERN.

Max served as Spokesperson of the H1 Collaboration at the electron-proton collider HERA. Recently, he chaired the Collaboration Board of the ATLAS experiment at CERN as well as an accelerator expert panel laying out a European Roadmap for the Research and Development on Energy Recovery Linacs (ERL). He currently is the Spokesperson of the high power ERL facility PERLE, which is being built at Irene Joliot Curie Laboratory, Orsay (France).

His contributions to Particle Physics has earned him the Max-Born-Prize of the Institute of Physics and the German Physics Society.

LUCIO ROSSI is full professor of Experimental Physics at the Physics department of the University of Milano. He was at CERN from 2001 to 2020, where he led the LHC superconducting magnets construction and then he proposed and led the High Luminosity LHC project. After retirement from CERN, at University and INFN-Milano, he has been pursuing new advanced technologies for cancer therapy with heavy ion particles (a superconducting gantry) and for FCC-hh and the Muon Collider projects for the post-HiLumi LHC era. He is also co-leader of IRIS, a large program in Italy of the Next Generation Europe, with the goal of developing new research infrastructures in Milano, Genova, Frascati, Napoli, Salerno and Lecce, and of design and building two new lag demonstrator; a 1 GW Superconducting line (with MgB2) for green energy and an energy saving 10 T magnet with HTS conductor.

Lucio has authored more than 200 publications in international journals and reviews and he is active in public outreach on science and large projects, developing themes like the relation between science and technology, "certainty and truth."

He has recently published an autobiographical book: Lucio Rossi, La conoscenza è un'avventura, edizioni Bietti, Milano, 2022.
http://www.bietti.it/negozio/la-conoscenza-e-unavventura/

PAOLO SPAGNOLO has been recently appointed as director of INFN-Pisa, Italy. He is a member of CMS and is in the Italian review board of the LHCb Experiment and the LHC Computing. He is also the author and co-author of about 1350 papers with a h-index=184. He is also a co-founder of the CERN spin-off Planetwatch.

Paolo has gathered many years of expertise in Particle Physics, data analysis and detector development, from his time working at CERN, where he was Research Fellow and Scientific Associate within the ALEPH and the CMS collaborations.

After the Higgs discovery, to address the issue of the Physics beyond the Standard Model, he started to work on searching for light Dark Matter beyond the Colliders and proposed STAX: a new experimental concept to search for Axions in laboratory with microwaves high power source, through the Primakov effect.